Extrusion Cooking for Food and Feed

Extrusion Cooking for Food and Feed

Biochemistry, Technology, and Applications

Edited by

Amit Baran Das

Department of Food Engineering and Technology,
Ghani Khan Choudhury Institute of Engineering & Technology,
Malda, West Bengal; Department of Food Engineering
and Technology, Tezpur University, Tezpur, Sonitpur, Assam, India

Poonam Singha

Department of Food Process Engineering,
National Institute of Technology Rourkela,
Rourkela, Odisha, India

Kasiviswanathan Muthukumarappan

Agricultural and Biosystems Engineering,
South Dakota State University,
Brookings, SD, United State

ELSEVIER

ACADEMIC PRESS

An imprint of Elsevier

Academic Press is an imprint of Elsevier
125 London Wall, London EC2Y 5AS, United Kingdom
50 Hampshire Street, 5th Floor, Cambridge, MA 02139, United States

ISBN: 978-0-443-29987-2

For information on all Academic Press publications visit our website at https://www.elsevier.com/books-and-journals

Publisher: Mara Conner
Acquisitions Editor: Vaishali Singh
Editorial Project Manager: Sadhna Rajan
Production Project Manager: Gomathi Sugumar
Cover Designer: Mark Rogers

Typeset by Integra Software Services

Contents

Contributors

Nasser Al-Habsi Department of Food Science and Nutrition, College of Agricultural & Marine Sciences, Sultan Qaboos University, Muscat, Oman

Maha Al-Khalili Department of Food Science and Nutrition, College of Agricultural & Marine Sciences, Sultan Qaboos University, Muscat, Oman

Carlos Alberto Avila-Orta Department of Advanced Materials, Research Center in Applied Chemistry, Saltillo, Mexico

Srijita Banerjee Department of Food Process Engineering, National Institute of Technology Rourkela, Rourkela, Odisha, India

Bhesh Bhandari School of Agriculture and Food Sustainability, The University of Queensland, Brisbane, QLD, Australia

Suvendu Bhattacharya Food Engineering Department, CSIR-Central Food Technological Research Institute, Mysore, India

Shubham Chandrakar National Institute of Food Technology, Entrepreneurship and Management (NIFTEM-T), Thanjavur, India

Carlos Andrés Covarrubias-Gordillo Faculty of Chemical Sciences, Autonomous University of Coahuila, Saltillo, Mexico

Victor Javier Cruz-Delgado Department of Plastics Transformation Processes, Research Center in Applied Chemistry, Saltillo, Mexico

Amit Baran Das Department of Food Engineering and Technology, Ghani Khan Choudhury Institute of Engineering & Technology, Malda, West Bengal; Department of Food Engineering and Technology, Tezpur University, Tezpur, Assam, India

Abhishek Dasore Department of Biological and Agricultural Engineering, Universiti Putra Malaysia, Serdang, Malaysia; School of Computer Science and Artificial Intelligence, SR University, Warangal, Telangana, India

Ed de Souza Wenger Mfg, Inc., Sabetha, KS, United States

Heidi Andrea Fonseca-Florido Department of Advanced Materials, SECIHTI, Research Center in Applied Chemistry, Saltillo, Mexico

Saphal Ghimire Department of Food, Agriculture, and Bioresources, Asian Institute of Technology, Thailand

Kishore Kumar Gopalakrishnan Department of Food Process Engineering, National Institute of Technology Rourkela, Rourkela, Odisha, India

Pratik Madhukar Gorde Department of Food Process Engineering, National Institute of Technology Rourkela, Rourkela, Odisha, India

Victoria Graciano-de la Cruz Department of Chemical Engineering and Biochemistry, Mexico Technology Institute/Durango Technology Institute., Durango, Mexico

Norhashila Hashim Department of Biological and Agricultural Engineering, Universiti Putra Malaysia, Serdang, Malaysia

Sushil Koirala School of Agriculture and Food Sustainability, The University of Queensland, Brisbane, QLD, Australia

Ankansha Kumari Laboratory of Applied Food Chemistry, Microbiology, and Process Engineering, Centre for Food Engineering and Technology, Department of Chemical Engineering, Birla Institute of Technology Mesra, Ranchi, Jharkhand, India

Sagar Loni Department of Food Process Engineering, National Institute of Technology Rourkela, Rourkela, Odisha, India

R. Mahendran National Institute of Food Technology, Entrepreneurship and Management (NIFTEM-T), Thanjavur, India

Akash Kusum Maiti Department of Food Process Engineering, National Institute of Technology Rourkela, Rourkela, Odisha, India

Kasivishwanathan Muthukumarappan Agricultural and Biosystems Engineering, South Dakota State University, Brookings, SD, United States

Pankaj B. Pathare Department of Soils, Water & Agricultural Engineering, College of Agricultural & Marine Sciences, Sultan Qaboos University, Muscat, Oman

Srutee Rout Department of Agricultural and Food Engineering, Indian Institute of Technology Kharagpur, Kharagpur, India

Anupam Roy Laboratory of Applied Food Chemistry, Microbiology, and Process Engineering, Centre for Food Engineering and Technology, Department of Chemical Engineering, Birla Institute of Technology Mesra, Ranchi, Jharkhand; Centre for Rural Development and Technology, Indian Institute of Technology Delhi, New Delhi, Delhi, India

Shubhajit Sarkhel Laboratory of Applied Food Chemistry, Microbiology, and Process Engineering, Centre for Food Engineering and Technology, Department of Chemical Engineering, Birla Institute of Technology Mesra, Ranchi, Jharkhand, India

Sushil Kumar Singh Department of Food Process Engineering, National Institute of Technology Rourkela, Rourkela, Odisha, India

Poonam Singha Department of Food Process Engineering, National Institute of Technology Rourkela, Rourkela, Odisha, India

Florentino Soriano-Corral Department of Plastics Transformation Processes, Research Center in Applied Chemistry, Saltillo, Mexico

Prem Prakash Srivastav Department of Agricultural and Food Engineering, Indian Institute of Technology Kharagpur, Kharagpur, India

Soubhagya Tripathy Department of Agricultural and Food Engineering, Indian Institute of Technology Kharagpur, Kharagpur, India

Vigneshwaran V. Department of Biotechnology, Alagappa College of Technology, Anna University, Chennai, Tamil Nadu, India

Vaibav Laboratory of Applied Food Chemistry, Microbiology, and Process Engineering, Centre for Food Engineering and Technology, Department of Chemical Engineering, Birla Institute of Technology Mesra, Ranchi, Jharkhand, India

Beijia Wan College of Biosystems Engineering and Food Science, Zhejiang University, Hangzhou, China

Zhengzong Wu Qilu University of Technology, Hangzhou, China

Enbo Xu College of Biosystems Engineering and Food Science, Zhejiang University, Hangzhou, China

Siyu Yao College of Biosystems Engineering and Food Science, Zhejiang University, Hangzhou, China

Foreword

Extrusion technology stands today as one of the most transformative and versatile processes in the food and feed industries. Its capacity to convert diverse raw materials into nutritious, safe, and value-added products has made it indispensable, from breakfast cereals and snacks to pet foods and livestock feeds. In an era of dynamic consumer demands, sustainability challenges, and innovations in food science, a comprehensive understanding of extrusion cooking is not just beneficial; it is essential.

Extrusion Cooking for Food and Feed by **Dr. Amit Baran Das**, **Dr. Poonam Singha**, and **Dr. Muthukumarappan Kasiviswanathan** is a remarkable contribution to this critical field. This work brings clarity to a complex subject, blending foundational principles with the latest scientific insights and practical applications. The authors have woven together multidisciplinary perspectives, which encompass food engineering, chemistry, nutrition, and process design, into a coherent and engaging narrative that will serve a wide audience of students, researchers, and industry professionals.

What sets this book apart is its balance of depth and accessibility. The fundamental science behind extrusion includes: material behavior under heat and shear, rheological transformations, and the influence of formulation variables. This is presented with precision, yet without sacrificing readability. At the same time, the authors thoughtfully link these concepts to real-world challenges and opportunities, including product development, quality control, sustainability considerations, and innovations in feed applications. This integrated approach enables readers to not only understand how extrusion works but also how it can be harnessed to create better, more sustainable products.

The global food and feed landscape is rapidly evolving, driven by changing dietary patterns, resource constraints, and technological breakthroughs. In this context, mastering extrusion cooking is synonymous with mastering a foundation for the topic of modern food processing. This book equips its readers with the knowledge they need to excel, whether in academia, research laboratories, or industrial settings.

I congratulate the authors on producing a text of such breadth and utility. Their expertise and thoughtful organization have created a resource that will undoubtedly become a benchmark in the field. It is with great enthusiasm that I recommend *Extrusion Cooking for Food and Feed* to all who seek a deeper understanding of this vital technology.

Dr. Roger Ruan, Professor
Director of the Bioproducts and Biosystems Science,
Engineering and Management Graduate Program
Department of Bioproducts and Biosystems Engineering,
University of Minnesota, Eckles Ave, St. Paul, MN,
United States of America

Preface

Extrusion cooking has become one of the most important and versatile technologies in the modern food and feed industries. Over the past few decades, extrusion has evolved from a simple mechanical forming process into a sophisticated high-temperature short-time processing technology capable of producing a wide range of products, including snack foods, breakfast cereals, texturized vegetable proteins, pet foods, and aquaculture feeds. The ability of extrusion to integrate several unit operations—mixing, kneading, cooking, shearing, shaping, and partial drying—into a single continuous process makes it highly efficient, flexible, and economically viable for large-scale production.

With the rapid expansion of processed food and feed sectors worldwide, extrusion technology has gained increasing attention from researchers, academicians, and industry professionals. Continuous innovations in extruder design, ingredient functionality, product development, and process optimization have significantly expanded the scope of extrusion cooking. At the same time, there is a growing need for a comprehensive resource that clearly explains the fundamental principles, biochemical transformations, engineering aspects, and industrial applications of extrusion technology.

This book, *Extrusion Cooking for Food and Feed*, has been developed to address this need by presenting a systematic and comprehensive overview of extrusion science and technology. The book covers the fundamental principles of extrusion, types and components of extruders, process parameters, and the physicochemical changes that occur in food and feed materials during extrusion. It also highlights the effects of extrusion on major food constituents, including starch, proteins, lipids, vitamins, and phytochemicals. In addition, the book discusses various industrial applications, including the development of value-added food products, texturized proteins, and pet food.

The content of this book is intended for undergraduate and postgraduate students in food engineering, food technology, and agricultural engineering, as well as researchers and professionals working in the food and feed processing industries. By combining theoretical foundations with practical applications, the book aims to provide readers with a clear understanding of extrusion technology and its growing role in modern food and feed production.

The authors sincerely acknowledge the support and encouragement received from colleagues, students, and industry professionals during the preparation of this book. We hope that this work will serve as a useful reference and inspire further research and innovation in extrusion processing.

Amit Baran Das
Poonam Singha
Kasiviswanathan Muthukumarappan

Introduction to extrusion cooking

1

Amit Baran Das[a,b]
[a]Department of Food Engineering and Technology, Ghani Khan Choudhury Institute of Engineering & Technology, Malda, West Bengal, India
[b]Department of Food Engineering and Technology, Tezpur University, Tezpur, Assam, India

1.1 Introduction

Extrusion technologies play an important role in the food industry as efficient manufacturing processes (Maskan and Altan, 2012). Their primary purpose was to convey and shape fluid forms of processed raw materials, such as doughs and pastes. Extrusion cooking technologies are utilized for cereal and protein processing in the food industry and, closely related, in the pet food and feed sectors (Guy, 2001). The processing units have evolved from simple conveying devices to highly sophisticated systems over the last decade. Today, their processing functions may include conveying, mixing, shearing, separation, heating or cooling, shaping, coextrusion, venting volatiles and moisture, flavor generation, encapsulation, and sterilization (Alam et al., 2016). They can be used for processing at relatively low temperatures, as with pasta and half-product pellet doughs, or at very high temperatures, as with flatbreads and extruded snacks (Ganjyal, 2020). The pressures used in extruders to control shaping, to keep water in a superheated liquid state, and to increase shearing forces in certain screw types may vary from around 15 to over 200 atmospheres. The history of the extruder, its working principles, types of extruders, and various applications have been extensively studied (Schmid et al., 2022).

The most important feature of an extrusion process is its continuous nature (Alvarez, 2020). It operates in dynamic steady-state equilibrium, where the input variables are balanced with the outputs. Therefore to obtain the required characteristics in an extrudate, the multivariate inputs must be set at the correct levels to achieve the dependent physical conditions and chemical process changes within the barrel of the machine (Muthukumarappan and Swamy, 2020). These dependent system variables determine the extrudate variables, which are reflected in the product variables (Alam et al., 2016). Once the relationships between the independent variables and the dependent variables within the processor are established for an individual product type, they must be maintained close to their optimum levels, within a small processing window, to ensure that the extrudate variables are also kept at the required levels.

In the development of extrusion processes, there have been improvements in extrusion equipment and the ancillary processing units, which form a complete

Extrusion Cooking for Food and Feed. https://doi.org/10.1016/B978-0-443-29987-2.00001-4

processing line for an individual product type (Bouvier and Campanella, 2014; Moscicki, 2011). However, in recent years, the development of a better understanding of the sequence of processes occurring within extruders has been equally important (Arora et al., 2020). For instance, determining the impact of specific machine parameters on commercial extruders and quantifying how raw materials affect dependent processing variables. Raw materials have been shown to play an active role in determining the magnitude of variables such as pressure, temperature, and motor load, as well as providing the structure-forming materials, which are developed in the extruder to form the extrudate (Guy, 2001; Yu, 2011). This new understanding within the industry has led to better use of existing extrusion machines and modifications to improve their function. The development of extruders has moved forward from a purely empirical approach, which led to the development of products on the early single-screw machines from 1940 onward (Bender and Böger, 2009). Extrusion technologists are now more likely to use mathematical modeling for different applications of extruders (Adekola, 2016).

In the food system, extrusion cooking is an important and challenging aspect. Webster's dictionary describes cooking as, among other things, the heating process of food for consumption (Webster, 1855). This suggests that (1) cooking, or applying heat to food, is regarded as food processing, giving rise to the term "processed food" and (2) it considers the thermal energy needed to raise the food to a specific temperature so that a chemical reaction can occur, according to Gray and Chinnaswamy (1995).

According to Maskan and Altan (2012), the most controversial and important aspect of cereal processing, for example, is the conversion of the batch cooking process into a continuous process. More specifically, it is the conversion of batch cooking to extrusion cooking without a change in texture or quality, as defined (Cheftel et al., 1992). In order to closely examine the cooking process, we must review the elements of an equation developed to specify and define the term "cooked." In reviewing the total cooking process, five components play a major role: moisture content, granule size of the particulate making the dough, residence time, total energy input, and the type of energy input (Pauly et al., 2013). Hydration of the product plays a major role in the gelatinization of starch granules, in protein texturization and denaturation, and in the elasticity of such components in the presence of shear (Scott and Awika, 2023). Moisture also serves as a plasticizing agent, which allows the starch granules to change phases from the crystalline stage, which can be easily fractured, to the amorphous stage during extrusion. Residence time is a key factor in providing the conduction of heat energy and also the hydration of particulates, because a very narrow residence time distribution provides a homogeneous balance so that all particulates have access to the hydration, conduction, and conversion energies simultaneously (Adekola & Engineering, 2016).

Extrusion cooking has gained popularity over the last two decades for a number of reasons (Alam et al., 2016):

- Versatility: Changing the ingredients, extruder working conditions, and dies allowsor the production of a large variety of products, many of which are difficult to make using any other method.

- Cost: Compared to other heating and forming methods, extrusion is more productive and has lower processing costs.
- Productivity: Extruders can operate continuously with high throughput.
- Product quality: By applying high temperatures for a short period of time, extrusion cooking preserves a variety of food ingredients that are sensitive to heat.
- Environmentally friendly: As a low-moisture process, extrusion cooking does not produce significant process effluents, reducing water treatment costs and levels of environmental pollution.

1.2 History of extruder

The early 20th century witnessed the invention of food extrusion, which was influenced by metalworking and plastics extrusion technology (Table 1.1). Extrusion was initially used in food manufacturing in the 1930s when cold extrusion was used to make breakfast cereals and dry pasta. These early extruders formed food items without requiring a lot of heating.

A significant advancement was made in the 1950s and 1960s when the high-temperature, short-time (HTST) procedure, known as extrusion cooking, was developed. This invention made it possible to produce more cooked, expanded foods, including pet food, ready-to-eat cereals, and puffed snacks. In contrast to single-screw designs, twin-screw extruder systems, which were invented by companies like

Table 1.1 History of extruder.

Era	Development	Description	Impact	Reference
1870s	Early extruders	First food-specific extruder inspired by the Archimedean screw, used for sausage production	Introduced a new method for shaping and forming food, limited application	Riaz (2000)
1930s–1950s	Rise of single-screw extrusion	Single-screw extruders adopted for: dry pasta production, breakfast cereal production. Development of the first commercially extruded pet food (Purina Dog Chow in 1957)	Revolutionized production of dry, shelf-stable food items, increased efficiency and consistency, expanded variety of processed food options, created a new market for extruded pet food	Tannenbaum (1942)

Continued

Table 1.1 History of extruder—cont'd

Era	Development	Description	Impact	Reference
1960s	Expansion and development	Development of low-cost, single-screw extruders by companies like Insta-Pro for on-farm grain and bean processing	Made extrusion technology more accessible for smaller producers, increased on-farm processing capabilities	Alam et al. (2016)
1960s–1999	Introduction of twin-screw extruders	Development of twin-screw extruders offering greater Versatility for handling diverse ingredients, control over cooking conditions, pressure, and shear forces, creation of new food products: Baby food, snacks, texturized vegetable proteins (meat analogs), candies (e.g., jelly beans)	Opened doors for innovation in food texture, nutrition, and functionality, enabled production of complex food structures and textures, expanded the range of extruded food applications, increased the nutritional value of certain extruded products pen spark	Tayeb et al., (1989)
2000 onward	Modern extruder	Ultra-high-speed extruder Hybrid extruder	Introduced new approaches for extrudate and new food products	(Ek and Ganjyal, 2020)

Wenger Manufacturing and Clextral, provided better control over mixing, shearing, and thermal treatment.

Food extrusion expanded quickly during the 1970s and 1980s, entering industries such as infant food, instant noodles, textured vegetable protein, and aquatic feeds. Extruders are essential for mass manufacturing because of their adaptability, energy efficiency, and capacity to process a variety of raw materials.

Food extrusion has played a key role in the development of sustainable and functional meals in recent decades, such as fortified foods, gluten-free snacks, and high-protein meat substitutes. In order to properly manage temperature, moisture, and product structure, modern extruders are outfitted with sophisticated sensors and automation.

1.3 Function of food extruder

Food extruders serve several functions in food processing, facilitating the creation of various products (Riaz, 2000). Here is a breakdown of their functions:

- **Mixing:** Food extruders thoroughly mix various ingredients. This ensures that the ingredients are distributed evenly throughout the finished product, which is essential for quality and consistency.
- **Cooking/heating:** Extruders can heat or cook the contents as they move through the equipment. This cooking process helps to gelatinize starches, denature proteins, and eliminate harmful microorganisms, ensuring food safety and improving digestibility.
- **Extrusion**: The main purpose of a food extruder is to create particular shapes and textures by extruding cooked and prepared dough or paste through a molded die. This method is responsible for the unique shapes of items such as pasta, snacks, breakfast cereals, and pet treats.
- **Texturizing**: By adjusting parameters such as temperature, pressure, and moisture content throughout the extrusion process, food extruders can produce a broad variety of textures in the finished product. This enables producers to develop foods with different levels of crunchiness, chewiness, or crispiness.
- **Forming**: Food extruders shape the extruded dough or paste into specific shapes, sizes, and lengths. This process is particularly useful for creating products like pasta, noodles, breadsticks, and certain types of snacks.
- **Enrichment/addition**: Nutrients, flavors, colors, and functional ingredients can be added during the extrusion process. This allows manufacturers to fortify products with vitamins, minerals, fiber, or other nutritional elements and enhance their taste and appearance.
- **Puffing**: Some food extruders have the capability to puff products by rapidly depressurizing them after extrusion. This creates airy, puffed snacks like cheese puffs or expanded cereals, which are popular among consumers.
- **Cutting/slicing**: Food extruders often include mechanisms for cutting or slicing the extruded product into the desired lengths or shapes before packaging. This ensures that the final products are uniform in size and appearance.
- **Degassing:** Degassing using an extruder is a common technique used in food industries to remove volatile substances such as moisture, solvents, and trapped air from the material being processed. This is critical for ensuring the quality and consistency of the final product.
- **Agglomeration**: Extruders are a powerful tool for agglomeration, which is the process of forming small particles into larger, more manageable ones. Agglomerated materials are easier to handle, transport, and store due to their increased density and reduced dust. Animal feed pellets, coffee, instant beverages, and breakfast cereals are often produced using extrusion.
- **Expansion**: Extruders can be surprisingly useful for achieving expansion in various materials. The expansion process significantly increases the product's volume, which can be beneficial for things like puffed snacks or creating lower-density materials. This is a prime example: puffed cereals, breakfast bites, and certain snack foods are all created using extrusion for expansion.

1.4 Advantages of extruder

Food extruders offer several advantages in food processing, making them a popular choice for manufacturers. Some of the key advantages include:

- **Versatility**: Food extruders can process a wide range of ingredients, including grains, legumes, vegetables, meats, and dairy products. This versatility allows manufacturers to produce a diverse array of food products, from snacks and cereals to pasta and pet foods.
- **Efficiency**: Extrusion processes are often continuous, allowing for high production volumes with minimal downtime. This efficiency helps manufacturers meet the demand for their products while keeping production costs relatively low.
- **Consistency**: Food extruders provide precise control over processing parameters such as temperature, pressure, and moisture content. This consistency ensures uniformity in product quality, taste, texture, and appearance, which is essential for consumer satisfaction and brand loyalty.
- **Nutritional enhancement**: Food extrusion can improve the nutritional profile of food products by increasing their bioavailability, digestibility, and nutrient content. Manufacturers can incorporate vitamins, minerals, fiber, and other functional ingredients into extruded products to enhance their health benefits.
- **Texture modification**: Extrusion allows manufacturers to create a wide range of textures in food products, from crunchy snacks to chewy bars. By adjusting processing parameters, such as temperature and pressure, manufacturers can achieve desired textures that appeal to consumers.
- **Extended shelf life**: The cooking and extrusion process can help extend the shelf life of certain food products by reducing microbial contamination and enzyme activity. This improves product stability and reduces the need for preservatives, enhancing food safety and quality.
- **Reduced waste**: Food extruders can utilize a variety of raw materials, including by-products and off-spec ingredients, which might otherwise go to waste. This helps manufacturers minimize waste and optimize resource utilization, making food production more sustainable.
- **Customization**: Food extruders offer flexibility in product design, allowing manufacturers to tailor products to meet specific consumer preferences and market trends. This customization capability enables innovation and differentiation in the marketplace.
- **Continuous reactor:** Instead of simply melting and shaping a material, the extruder becomes the reaction vessel. Reactants are fed into the extruder along with the base material.
- **Process scale-up:** Using an extruder for process scale-up involves transitioning from smaller-scale production to larger-scale production while maintaining or improving product quality and efficiency. Extruders offer excellent control over process parameters such as temperature, pressure, and mixing, which ensures consistent product quality across batches. This is crucial when scaling up production to meet larger demand without compromising quality.

1.5 Extruder classification

A variety of extruders have been developed to manufacture different types of food products (Fig. 1.1). On the basis of the demand for food processing, the extruder is categorized. According to Harper and Clark (1979), the extruder can be classified into the following types.

1.5.1 Pasta extruders

The deep-flighted screw, smooth barrel, and low screw speed of the pasta extruder make it ideal for working the moistened semolina flour and pressing it through a die with little or no cooking (Abecassis et al., 1994; Le Roux et al., 1995). In the pasta

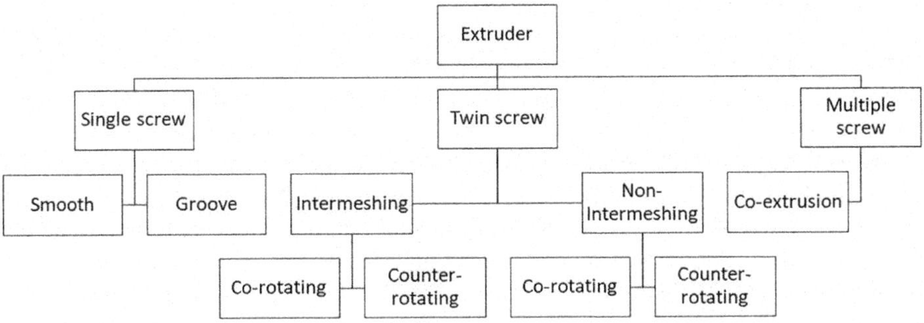

Fig. 1.1 Types of extruders.

extruder, little input energy is dissipated because of the low shear rate of the product and the smooth barrel used. In addition, similar extruders are used in the manufacture of hot dogs, pastry doughs, cookies, and certain types of candy products.

1.5.2 High-pressure forming extruders

Many preshaped food products are made by extruding pregelatinized cereal doughs through a die to form un-puffed pellets, which are subsequently partially dried and puffed in a fryer, puffing gun, or toaster. To produce high pressures, these extruders normally have grooved barrels to prevent slip at the wall and achieve greater compression, which results in their ability to produce higher pressures at the die (Hung and Hung, 2000). Excessive temperature production in the screw can lead to unwanted puffing at the die and unsatisfactory performance.

1.5.3 Low-shear cooking extruders

The soft-moist pet foods are prepared on extruders that have moderate shear, high compression, and grooved barrels to enhance mixing. Heat can be applied to the barrel or screw to heat the product, since little viscous dissipation to heat occurs due to the relatively low viscosity of the material being extruded (Riaz, 2000).

1.5.4 Collet extruders

Rapid viscous dissipation of mechanical energy input to the short screw (Length/Diameter~3:1) occurs in the collet extruder because of the high shear in a screw with shallow flights operating in a grooved barrel to prevent slipping at the walls. Collet extruders normally extrude relatively dry feed materials and heat them rapidly to temperatures exceeding 175°C, where the starch is gelatinized and partially dextrinized (Chang et al., 2015). When this material exits the die, the rapid change in pressure and normal forces from the non-Newtonian dough causes substantial expansion of the product and subsequent loss of moisture, forming a crisp, expanded curl or collet. The most common ingredient for collet extruders is defatted corn grits.

1.5.5 High-shear cooking extruders

These extruders were designed to produce a large variety of precooked, gelatinized, or heat-treated food products with their tactile components restructured. Initially, plastic extruders having long barrels (L/D=15 to 20:1), high compression ratios, and the ability to heat or cool the product externally through the barrel were used (Feng et al., 2022). Such extruders provide a significant range of operating capabilities, being able to accept a wide range of initial product moistures and product ingredients, with an ability to control the desired processing conditions such as temperature and puffing. Product applications include pet foods, cereals, and snacks. High-shear cooking extruders have been classified as HTST devices. In most applications, feed ingredients are preheated with steam or hot water and then processed through the high-shear cooking extruder to further work the product and increase its temperature rapidly. Nearly instantaneous cooling occurs once the product leaves the die, which is coupled with the short residence time in the screw.

A thermodynamic classification of single-screw extruders was also developed by Rossen and Miller, 1973), which is given in the following.

1.5.6 Autogenous extruders

The entire heat input to the extruder results from the viscous dissipation of mechanical energy inputs, and little or no heat is added or removed from the barrel (SAHAGU & Harper, 1979). Collet extruders and some high-shear cooking extruders are examples of autogenous extruders. Since temperatures are controlled by feed composition and screw configuration, autogenous extruders tend to have less flexibility and are more difficult to control (Frolov et al., 2019).

1.5.7 Isothermal extruders

Constant temperatures are maintained throughout the length of the barrel. Forming extruders typically fall into this category. To maintain isothermal conditions, heat is normally removed through jackets surrounding the barrel (Broyer and Tadmor, 1972). Since dough conditions remain relatively constant in isothermal extruders, they lend themselves to easier mathematical description.

1.5.8 Polytropic extruders

In reality, all extruders are polytropic, although some nearly operate as autogenous or isothermal extruders. Cooking extruders that have jacketed barrels, where heat is alternatively added or extracted, operate in the polytropic regime (Harper and Clark 1979).

1.5.9 Heat source

Extruders are typically not classified based on heating methods themselves, but rather by how they achieve the desired temperature for the material being extruded. The following are the two main ways to achieve this:

Electrical heating: This is the most common method, where external heaters are placed around the barrel of the extruder to raise the temperature. The heaters can be band heaters, ceramic heaters, or other types (Hsiao, 2017). The specific type of heater used will depend on the material being extruded and the required temperature profile.

Adiabatic heating: In this method, the heat generated by the shearing of the material itself is used to raise the temperature. This is a more energy-efficient method, but it is limited to materials that generate enough heat through shear. Adiabatic extruders are often used for food processing applications (McKelvey, 1954).

1.5.10 Interrupted-flight extruder

An interrupted-flight extruder is a variation of the traditional single-screw extruder design. In an interrupted-flight extruder, the screw is characterized by interrupted or discontinuous flights along its length, as opposed to a continuous helical flight seen in standard single-screw extruders (Williams, 2000). These interruptions create gaps or spaces between the flights, which can vary in length and depth. interrupted-flight extruders find applications in various industries where specific material processing requirements need to be met. They are commonly used in the plastics industry for compounding, mixing, and processing of materials with additives or fillers (C. Rauwendaal, 1991).

1.5.11 Single-segmented screw extruder

Single-segmented screw extruders are a type of extrusion machinery used in manufacturing processes, particularly in industries like plastics, food processing, and pharmaceuticals (Barreto-Curiel et al., 2018; Riaz, 2000). They are an adaptation of traditional single-screw extruders, designed with segmented screws rather than a continuous, smooth screw. Instead of having a continuous helical flight along the entire length of the screw, segmented screw extruders have distinct sections or segments along the screw. Each segment may have a different shape, pitch, or configuration tailored to specific processing requirements (Sotelo-Rodríguez et al., 2018). Single-segmented screw extruders find applications in a wide range of industries and processes, including plastic compounding, recycling, food extrusion, pharmaceutical granulation, and chemical processing.

1.5.12 Twin-screw extruder

Twin-screw extruders are a type of machine used in the food industry to create various food products. They are essentially two screws working together to process the food. There are two main ways to categorize twin-screw extruders: by screw rotation direction and by screw geometry.

1.5.12.1 By screw rotation direction

- **Corotating twin-screw extruders:** Both screws rotate in the same direction. This type is ideal for applications that require high mixing and kneading action, such as dough mixing, compounding, and blending (Lawal and Kalyon, 1995) (Fig. 1.2).
- **Counterrotating twin-screw extruders:** The screws rotate in opposite directions. This creates a self-wiping action that is ideal for applications requiring high shear and dispersion, such as devolatilization (removing gas bubbles) and the addition of flavors or colors (Wilczyński et al., 2012).

1.5.12.2 By screw geometry

- **Parallel twin-screw extruders:** The screws have a constant diameter along their length. This type is versatile and can be used for a wide range of applications (Ohara et al., 2020).
- **Conical twin-screw extruders:** The screws have a decreasing diameter toward the discharge end. This design creates a self-densifying effect that is useful for applications requiring high pressure, such as cooking and shaping (Rauwendaal, 1981).

In addition to these two main categories, there are also subcategories of twin-screw extruders, such as:

- **Fully intermeshing:** The screws intermesh completely along their length (Fig. 1.3) (Andersen et al., 2012).
- **Partially intermeshing:** The screws only intermesh in certain sections (Sastrohartono et al., 1990).
- **Nonintermeshing:** The screws do not intermesh at all (Mudalamane et al., 2003).

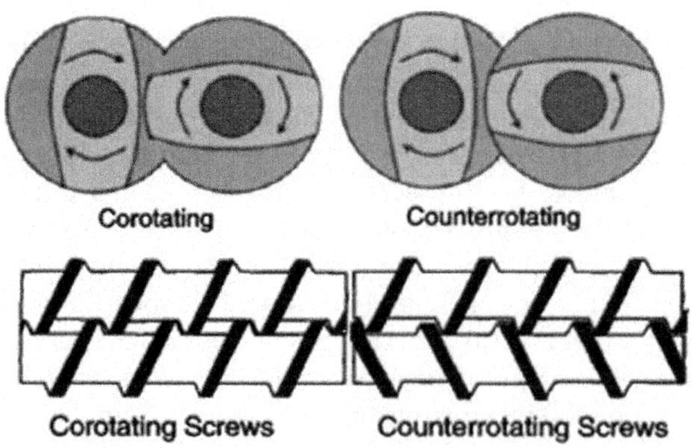

Fig. 1.2 Rotation direction of a twin-screw extruder, showing both corotating and counterrotating screws.

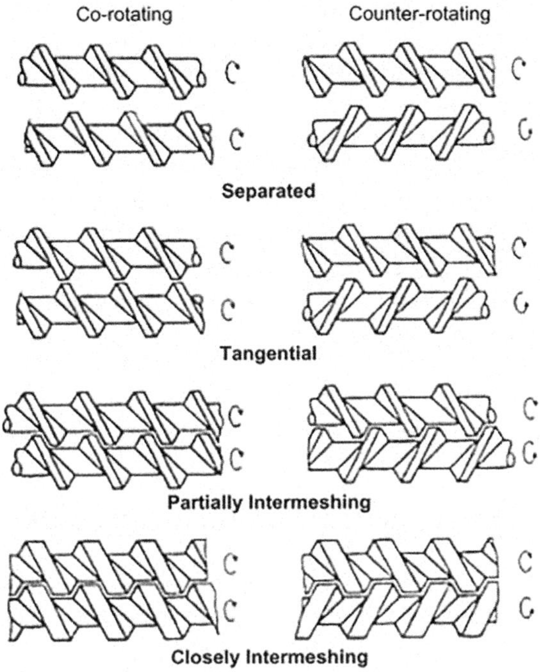

Fig. 1.3 Rotation behavior of the twin screw. Intermeshing behavior of the twin screw.

The specific type of twin-screw extruder that is used for a particular application will depend on the desired outcome. For example, a corotating parallel twin-screw extruder might be used to mix dough for bread, while a counterrotating conical twin-screw extruder might be used to cook and shape pasta.

1.6 New-generation extruder

A new-generation food extruder builds on the traditional extruder design but incorporates advancements to address the growing demands of the food industry. The following are the two types of new-generation extruders.

High-speed extruders: Developed in 2003, high-speed extruders are now a proven solution in the production of thermoforming sheets from polymers. This type of extruder can also be used in the production of feeding pyrolysis reactors for recycling postconsumer waste (Christiano et al., 2012; Farahanchi et al., 2016). These extruders are known for their energy efficiency, short start-up times, and ease of use.

Modular extruders: Another example of a “new-generation extruder” is a modular extruder. These extruders are designed with interchangeable components, allowing for greater flexibility and customization. This can be beneficial for manufacturers who need to produce a variety of products with different requirements (Eitzlmayr et al., 2014).

Hot-melt extruder: Hot-melt extrusion is one of the most widely applied processing technologies in the food industry (Azad et al., 2022). Hot-melt extrusion has provided an opportunity for the use of materials to mask the bitter taste of active substances. Hot-melt extrusion involves the compaction and conversion of blends from a powder or granular mix into a product of uniform shape (Maniruzzaman et al., 2012). During this process, biopolymers are melted and formed into products of different shapes and sizes by forcing biopolymeric components and active substances, including any additives or plasticizers, through an orifice or die under controlled temperature, pressure, feeding rate, and screw speed.

Extrusion-based 3D printer: A 3D food printer is a digitally controlled piece of equipment in which food materials are deposited in multiple layers, one placed above the other. The extrusion-based printers are commercially used for 3D food printing to create various innovative designs for personalized food products (Hussain et al., 2022).

1.7 Benefits of new-generation food extruders

New-generation food extruders offer several advantages over traditional models, making them a valuable tool for modern food manufacturers. Here is a breakdown of the key benefits:

- **Reduced energy consumption:** New designs optimize components and processes to minimize energy use during food production. This translates to lower operational costs and a smaller environmental footprint.
- **Sustainable ingredient handling:** These extruders are adept at handling delicate ingredients like fruits, vegetables, and ancient grains with minimal damage. This preserves nutrients and reduces waste during processing.
- **Minimized food waste:** Advanced control systems ensure consistent product quality and improve overall yield, leading to less food waste.
- **Multifunctionality:** Some new extruders integrate tasks such as metering, mixing, and shaping into a single unit, streamlining the production process and reducing equipment needs.
- **Advanced texturization:** Innovative screw designs and processing techniques allow for the creation of unique textures and shapes for food products. This caters to consumers' preferences for variety and novel eating experiences.
- **Precise ingredient incorporation:** Improved control over mixing and residence time within the extruder enables the consistent incorporation of functional ingredients like vitamins, minerals, or fibers. This allows for the development of fortified and enriched foods.
- **High-moisture extrusion:** This technique allows the processing of food with higher moisture content, leading to the development of healthier and fresher-tasting products.
- **Coextrusion:** New extruders can combine different food materials into a single product, creating visually appealing and texturally interesting food items, such as layered snacks or combination fillings.
- **Microencapsulation:** These extruders can be used to encapsulate sensitive ingredients, like probiotics or flavors, within a protective coating. This improves their shelf life and stability while maintaining functionality.

1.8 Advantages of single-screw extruder

Single-screw extruders offer several advantages, making them widely used in various manufacturing processes, particularly in the plastics industry. Here are some key advantages:

- **Simplicity**: Single-screw extruders have a relatively simple design compared to other types of extruders, such as twin-screw or multiscrew extruders. This simplicity makes them easier to operate, maintain, and troubleshoot.
- **Cost-effectiveness**: Due to their simpler design and construction, single-screw extruders tend to be more cost-effective, in terms of both initial investment and ongoing maintenance, compared to more complex extrusion systems.
- **Versatility**: Single-screw extruders can process a wide range of materials, including thermoplastics, thermosets, elastomers, and food products. They are used in various industries for applications such as compounding, extrusion coating, film and sheet production, pipe and profile extrusion, and pelletizing.
- **High output rates**: Single-screw extruders can achieve high output rates, making them suitable for large-scale production. By adjusting screw speed, temperature, and other process parameters, manufacturers can optimize throughput while maintaining product quality.
- **Ease of scale-up**: Scaling up production with single-screw extruders is relatively straightforward. Manufacturers can increase output by simply using larger extruder machines without significant changes to the process setup or parameters.
- **Energy efficiency**: Single-screw extruders typically consume less energy compared to other extrusion technologies, especially in applications where high shear rates or specialized mixing capabilities are not required.
- **Consistent performance**: With proper setup and maintenance, single-screw extruders can deliver consistent and uniform product quality over long production runs. This consistency is essential for meeting quality standards and customer specifications.
- **Flexibility in design**: Single-screw extruders can be designed with various screw geometries, including standard, barrier, and grooved designs, to accommodate different processing requirements and material properties.

Twin-screw extruders offer several advantages over single-screw extruders, particularly in certain applications and industries:

- **Mixing capability**: Twin-screw extruders are excellent for mixing multiple materials together uniformly. They can achieve better dispersion and distribution of additives, fillers, and reinforcements compared to single-screw extruders. This makes them highly suitable for compounding polymers with additives or blending different polymers.
- **Higher throughput**: Twin-screw extruders generally have higher throughput rates than single-screw extruders of comparable size. This is advantageous for large-scale production, where high volumes of material need to be processed efficiently.
- **Flexibility in process control**: Twin-screw extruders offer more flexibility in controlling process parameters such as temperature, shear rate, and residence time. This control is crucial for optimizing the processing conditions for specific materials and achieving desired product properties.
- **Better homogeneity**: Due to their efficient mixing capabilities, twin-screw extruders can achieve better homogeneity in the final product. This is important for applications where uniformity of composition and properties across the material is critical.

- **Handling of viscous materials**: Twin-screw extruders can handle more viscous materials effectively compared to single-screw extruders. This capability is beneficial in processing materials that require higher shear rates or have high viscosity.
- **Lower energy consumption**: In some cases, twin-screw extruders can be more energy-efficient than single-screw extruders, especially when considering the amount of material processed per unit of energy consumed.
- **Processing of temperature-sensitive materials**: Twin-screw extruders typically offer better control over temperature profiles during processing, which is advantageous for materials sensitive to thermal degradation or requiring precise temperature management.
- **Wider range of applications**: While single-screw extruders are versatile and widely used, twin-screw extruders are particularly favored in industries such as polymer compounding, food processing, pharmaceuticals, and chemical processing due to their ability to handle complex formulations and achieve specific product characteristics.

1.9 Conclusion

Extrusion cooking is a flexible and effective food processing method that is frequently used to make textured protein products, morning cereals, snacks, and pet food. Among its many benefits are its high production, versatility with regard to a range of raw ingredients, and capacity to produce food products with a wide range of textures, shapes, and flavors. The process is energy-efficient and minimizes waste, making it an environmentally friendly option. Extrusion cooking enables the modification of nutritional properties by enhancing digestibility, reducing antinutritional factors, and incorporating functional ingredients such as vitamins and minerals. However, to preserve product quality and avoid certain disadvantages, like nutrient loss or an unwanted texture, meticulous control of process parameters—such as temperature, pressure, and moisture content—is essential.

References

Abecassis, J., Abbou, R., Chaurand, M., Morel, M.H., Vernoux, P., 1994. Pressure in the extruder and on pasta quality. Cereal Chem. 71 (3), 247–253.

Adekola, 2016. Engineering review food extrusion technology and its applications. J. Food Sci. Eng. 6 (3), 149–168.

Alam, M.S., Kaur, J., Khaira, H., Gupta, K., 2016. Extrusion and extruded products: changes in quality attributes as affected by extrusion process parameters: a review. Crit. Rev. Food Sci. Nutr. 56 (3), 445–473.

Alvarez, J.C.M., 2020. Engineering aspects of extrusion: Extrusion processing as a multiple-input and multiple-output system. In: Extrusion cooking. Woodhead Publishing, pp. 29–71.

Andersen, P., Kanarski, R., Wagner, J.R., 2012. Troubleshooting the co-rotating fully intermeshing twin-screw compounding system. In: Handbook of Troubleshooting Plastics Processes: A Practical Guide, pp. 57–67.

Arora, B., Yoon, A., Sriram, M., Singha, P., Rizvi, S.S.H., 2020. Reactive extrusion: a review of the physicochemical changes in food systems. Innov. Food Sci. Emerg. Technol. 14668564. 64, 102429. doi:10.1016/j.ifset.2020.102429.

Azad, M.O.K., Adnan, M., Sung, I.J., Lim, J.D., Baek, J., Lim, Y.S., Park, C.H., 2022. Development of value-added functional food by fusion of colored potato and buckwheat flour through hot-melt extrusion. J. Food Process. Preserv. 0145-8892. 46 (5). doi:10.1111/jfpp.15312.

Barreto-Curiel, F., Ramirez-Puebla, S.T., Ringø, E., Escobar-Zepeda, A., Godoy-Lozano, E., Vazquez-Duhalt, R., Sanchez-Flores, A., Teresa Viana, M., 2018. Effects of extruded aquafeed on growth performance and gut microbiome of juvenile Totoaba macdonaldi. Anim. Feed Sci. Technol. 03778401. 245, 91–103. doi:10.1016/j.anifeedsci.2018.09.002.

Bender, W., Böger, H.H., 2009. A Short History of the Extruder in Ceramics. Springer Science and Business Media LLC, pp. 85–128. doi:10.1007/978-3-540-27102-4_5.

Bouvier, J., Campanella, O.H., 2014. Extrusion Processing Technology: Food and Non-Food Biomaterials. Wiley. 9781444338119. doi:10.1002/9781118541685.

Broyer, E., Tadmor, Z., 1972. Solids conveying in screw extruders part I: A modified isothermal model. Polym. Eng. Sci. 12 (1), 12–24.

Chang, C., Yang, C., Samanros, A., Lin, J., 2015. Collet and cooking extrusion change the soluble and insoluble β-glucan contents of barley. J. Cereal Sci. 66, 18–23.

Cheftel, J.C., Kitagawa, M., Quéguiner, C.J.F.R.I., 1992. New protein texturization processes by extrusion cooking at high moisture levels. Food Rev. Int. 8 (2), 235–275.

Christiano, J.P., Davis Standard, L., Pawcatuck, C., 2012. Examination of the performance of a high speed single screw extruder for several different extrusion applications. In: Proceedings of the Annual Technical Conference—ANTEC, Conference Proceedings, Orlando, FL, USA (Vol. 2).

Eitzlmayr, A., Koscher, G., Reynolds, G., Huang, Z., Booth, J., Shering, P., Khinast, J., 2014. Mechanistic modeling of modular co-rotating twin-screw extruders. Int. J. Pharm. 03785173. 474 (1-2), 157–176. doi:10.1016/j.ijpharm.2014.08.005.

Ek, P., Ganjyal, G.M., 2020. Basics of extrusion processing. In: Extrusion cooking. Woodhead Publishing, pp. 1–28.

Farahanchi, A., Malloy, R., Sobkowicz, M.J., 2016. Effects of ultrahigh speed twin screw extrusion on the thermal and mechanical degradation of polystyrene. Polym. Eng. Sci. 56 (7), 743–751.

Feng, R., van den Berg, F.W., Lillevang, S.K., Ahrne, L., 2022. High shear cooking extrusion to create fibrous mozzarella cheese from renneted and cultured curd. Food Res. Int. 157, 111192.

Frolov, D.I., Kurochkin, A.A., Garkina, P.K., Zimnyakov, V.M., Kukharev, O.N., 2019. Improving the efficiency of extrudate dehydration in a vacuum chamber of a modernized extruder. Volga Region Farmland 2, 87–94.

Ganjyal, G.M., 2020. Extrusion Cooking: Cereal Grains Processing, Elsevier, Elsevier.

Gray, D.R., Chinnaswamy, R., 1995. Role of Extrusion in Food Processing. Elsevier BV, pp. 241–268. doi:10.1016/b978-044481500-2/50015-3.

Guy, R., 2001. Extrusion Cooking: Technologies and Applications, vol. 61. Woodhead Publishing.

Harper, J.M., 1979. Food extrusion. Crit. Rev. Food Sci. Nutr. 00990248. 11 (2), 155–215. doi:10.1080/10408397909527262.

Harper, J.M., Clark, J.P., 1979. Food extrusion. Crit. Rev. Food Sci. Nutr. 11 (2), 155–215.

Hsiao, K.-L., 2017. Combined electrical MHD heat transfer thermal extrusion system using Maxwell fluid with radiative and viscous dissipation effects. Appl. Therm. Eng. 13594311. 112, 1281–1288. doi:10.1016/j.applthermaleng.2016.08.208.

Hung, J.-C., Hung, C., 2000. The design and development of a hydrostatic extrusion apparatus. J. Mater. Process. Technol. 09240136. 104 (3), 226–235. doi:10.1016/s0924-0136(00)00593-8.

Hussain, S., Malakar, S., Arora, V.K., 2022. Extrusion-based 3D food printing: technological approaches, material characteristics, printing stability, and post-processing. Food Eng. Rev. 14 (1), 100–119.

Lawal, A., Kalyon, D.M., 1995. Mechanisms of mixing in single and co-rotating twin screw extruders. Polym. Eng. Sci. 35 (17), 1325–1338.

Le Roux, D., Vergnes, B., Chaurand, M., Abecassis, J., 1995. A thermomechanical approach to pasta extrusion. J. Food Eng. 26 (3), 351–368.

Maniruzzaman, M., Boateng, J.S., Snowden, M.J., Douroumis, D., 2012. A review of hot-melt extrusion: process technology to pharmaceutical products. Int. Sch. Res. Not. 2012 (1), 436763.

Maskan, M., Altan, A., 2012. Advances in Food Extrusion Technology, CRC press Taylor and Francis group, CRC press Taylor and Francis group.

McKelvey, J.M., 1954. Theory of adiabatic extruder operation. Ind. Eng. Chem. 46 (4), 660–664.

Moscicki, L., 2011. Extrusion-cooking Techniques: Applications, Theory and Sustainability, John Wiley & Sons, John Wiley & Sons.

Mudalamane, R., Bigio, D.I., Tomayko, D.C., Meissel, M., 2003. Behavior of fully filled regions in a non-intermeshing twin-screw extruder. Polym. Eng. Sci. 43 (8), 1466–1476.

Muthukumarappan, K., Swamy, G.J., 2020. Bioconversions in Extrusion Cooking. Elsevier BV, pp. 445–478. doi:10.1016/b978-0-12-815360-4.00014-6.

Ohara, M., Sasai, Y., Umemoto, S., Obata, Y., Sugiyama, T., Tanifuji, S.-I., Kihara, S.-I., Taki, K., 2020. Experimental and Numerical Simulation Study of Devolatilization in a Self-Wiping Corotating Parallel Twin-Screw Extruder. Polymers. 2073-4360. 12 (11), 2728. doi:10.3390/polym12112728.

Pauly, A., Pareyt, B., Fierens, E., Delcour, J.A., 2013. Wheat (Triticum aestivum L. and T. turgidum L. ssp. durum) kernel hardness: II. Implications for end-product quality and role of puroindolines therein. Compr. Rev. Food Sci. Food Saf. 12 (4), 427–438.

Rauwendaal, C., Mixing in single screw extruders. Mixing in polymer processing. (1991),

Rauwendaal, C.J., 1981. Analysis and experimental evaluation of twin screw extruders. Polym. Eng. Sci. 21 (16), 1092–1100.

Riaz, M.N., 2000. Introduction to Extruders and Their Principles Extruders in Food Applications. CRC Press, United States, 1–23. Available from: https://www.taylorfrancis.com/books/e/9781482278859.

Rossen, J.L., Miller, R.C., 1973. J. Food Technol. 27 (8), 46.

Sastrohartono, T., Esseghir, M., Kwon, T.H., Sernas, V., 1990. Numerical and experimental studies of the flow in the nip region of a partially intermeshing co-rotating twin-screw extruder. Polym. Eng. Sci. 30 (21), 1382–1398.

Schmid, E.M., Farahnaky, A., Adhikari, B., Torley, P.J., 2022. High moisture extrusion cooking of meat analogs: A review of mechanisms of protein texturization. Compr. Rev. Food Sci. Food Saf. 21 (6), 4573–4609.

Scott, G., Awika, J.M., 2023. Effect of protein–starch interactions on starch retrogradation and implications for food product quality. Compr. Rev. Food Sci. Food Saf. 22 (3), 2081–2111.

Sotelo-Rodríguez, J.I., Rombenso, A.N., Barreto-Curiel, F., Mata-Sotres, J.A., Viana, M.T., 2018. Insights on feed characteristics of tuna moist extruded aquafeeds using glycerol. Anim. Nutr. Feed Technol. 0974181X. 18 (1), 37–53. Available from http://www.indianjournals.com/ijor.aspx?target=ijor:anft&volume=18&issue=1&article=004&type=pdf. doi:10.5958/0974-181X.2018.00004.5.

Tannenbaum, A., 1942. The Genesis and Growth of Tumors: III. Effects of a High-Fat Diet. Cancer Res. 15387445. 2 (7), 468–475.

Tayeb, J., Vergnes, B., Valle, G.D., 1989. A basic model for a twin-screw extruder. J. Food Sci. 54 (4), 1047–1056.

Webster, T., 1855. An Encyclopæaedia of Domestic Economy: Comprising Such Subjects as are Most Immediately Connected with Housekeeping As, the Construction of Domestic Edifices, with the Modes of Warming, Ventilating, and Lighting Them; a Description of the Various Articles of Furniture; a General Account of the Animal and Vegetable Substances Used as Food, and the Methods of Preserving and Preparing Them by Cooking; Making Breads Materials Employed in Dress and the Toilet; Business of the Laundry; Description of the Various Wheel Carriages, Vol. 2. Preservation of Health, Domestic Medicines, &c . Harper & Brothers.

Wilczyński, K., Lewandowski, A., Wilczyński, K.J., 2012. Experimental study of melting of LDPE/PS polyblend in an intermeshing counter-rotating twin screw extruder. Polym. Eng. Sci. 52 (2), 449–458.

Williams, M.A., 2000. Interrupted-flight Expanders-extruders Extruders in Food Applications. CRC Press, United States, 63–79. Available from https://www.taylorfrancis.com/books/e/9781482278859.

Yu, L., 2011. Extrusion Processing of Protein Rich Food Formulations. McGill University, Canada.

Design and modeling of single-screw extruder

2

Amit Baran Das[a,b]
[a]Department of Food Engineering and Technology, Ghani Khan Choudhury Institute of Engineering & Technology, Malda, West Bengal, India
[b]Department of Food Engineering and Technology, Tezpur University, Tezpur, Assam, India

2.1 Introduction

A food extruder comprises a flighted Archimedes screw that rotates within a tightly fitted cylindrical barrel. Preground raw ingredients are blended prior to introduction at the feed end of the extrusion screw (Harper, 1979). These ingredients are frequently moistened and partially cooked in a preconditioning chamber, which can be either an atmospheric cooker or a pressure cooker. By pushing the food product forward, the flights on the screw combine the ingredients, resulting in a viscous mass that resembles dough (Riaz, 2000a). Heat is added to the food dough as it passes through the screw by one or more of three mechanisms: (1) viscous dissipation of mechanical energy being added to the shaft of the screw, (2) heat transfer from steam or electrical heaters surrounding the barrel, and (3) direct injection of steam, which is mixed with the dough in the screw (Harper, 1990). A restriction at the barrel's discharge, caused by one or more orifices or shaped apertures called a die, causes the pressure inside the barrel to grow as the hot, moist food material passes through the extruder. Typically, discharge pressures in extruder range from 30 to 60 atm. At these elevated pressures, boiling or flashing of moisture does not occur within the confines of the barrel because of that the pressure exceeds the vapor pressure of water at the extrusion temperature. When the food comes out of the die, the pressure is released, causing the product to expand as moisture flashes off. The food ingredients cool adiabatically as a result of the product losing moisture; it reaches a temperature of around 80°C in a few seconds, hardens, and sets, frequently maintaining its enlarged shape (Harper, 1979). The food product is exposed to high shear rates as it is delivered and flows by the screw's motion because the flights on a food extruder's screw are fully loaded. Long molecules in the food ingredients are more likely to align in these high shear rate regions, which can lead to cross-linking or restructuring and give the extruded meal its distinctive texture. Similar effects may also be produced by the shear environment in the extruder's die assembly.

Extrusion Cooking for Food and Feed. https://doi.org/10.1016/B978-0-443-29987-2.00002-6

2.1.1 Types of single-screw extruders

Single-screw extruders are used extensively in the processing of polymers and food, and they are available in a wide range of types, each with a unique purpose. The main types include the following.

- **Plasticating single-screw extruders**

These extruders are designed to melt and convey polymers. They start with solid polymer pellets, which are fed into the extruder barrel, where they are heated and melted. The screw conveys the molten polymer through the barrel to the die for shaping (Covas et al., 1999).

- **Melt-conveying single-screw extruders**

The main purpose of these extruders is to move molten polymers. Instead of melting the polymer from a solid state, they focus more on moving the molten material to the intended spot along the manufacturing line than plasticating extruders do (Marschik et al., 2022).

- **Mixing single-screw extruders**

Even though single-screw extruders are less capable of mixing than twin-screw extruders, they can still be used for basic mixing work. These extruders are commonly used to incorporate colorants or additives into polymer melts (Mohr et al., 1957).

- **Compounding single-screw extruders**

The purpose of these extruders is to combine or blend various polymers or additives. They might have unique screw patterns installed to improve mixing and guarantee a uniform material mixture (Wang and Huang, 1996).

- **Extrusion cooking single-screw extruders**

These extruders are used in the food sector to shape and cook food items. As the food ingredients are pushed through the extruder barrel and die, they are cooked by high temperatures and pressures (Yeh and Jaw, 1999).

2.1.2 Cooking mechanism

Food is cooked in an extruder by combining pressure, heat energy, and mechanical energy to turn raw ingredients into a cooked, formed product. High-viscosity dough-like fluids or particle-moist solids are the materials treated in cooker extruders (Moscicki, 2011). As the screw rotates, the flights drag the material toward the exit. The flow channel is delimited by two solid surfaces, namely, the screw and the barrel. Friction with the moving material occurs on both surfaces. Ideally, the friction at the barrel surface should be the strongest of the two for internal shear to occur. Where the friction is weak at the barrel surface, e.g., as a result of lubrication, and strong at the screw surface, the material would stick to the feed section.

Feed section: This segment primarily serves as a screw conveyor, moving the material from the feed entry to the sections that follow. In this portion, the mass is hardly compressed or modified (Steiner and Haight, 2018).

Transition section: The material is heated and compressed in this section. In this section, the material is compressed by screws as it passes through the barrel's reducing gaps. High pressure keeps the mixture in contact with the heated barrel, which ensures homogeneous cooking and compaction of the material (Riaz, 2000b).

Metering section: Most of the extrusion process's goals—such as melting, texturization, kneading, and chemical reactions—are accomplished in this stage by shearing and mixing. A large portion of the power needed to turn the screw shaft is lost as heat through friction inside the material. A fraction of the heat that is transferred to the product is, therefore, produced onsite. Additional heat is supplied by the direct injection of live steam and transmitted from the barrel surface, which has been heated externally. The majority of the energy input in single-screw extruders consists of heat produced internally. As a consequence, cooking extruder heating is fairly quick (Lekakou and Brandao, 1996). Because of the high pressure in the extruder, the moist material can be heated to temperatures well above 100°C (sometimes up to 180–200°C). When the pressure is suddenly released at the exit from the die, some of the water in the product is flash-evaporated and, as a result, the product is puffed. The degree of puffing can be controlled by releasing some of the pressure and cooling the mass in the last section of the extruder, before the die (Berk, 2016). As a result of compression, a pressure gradient is built in the opposite direction to the average movement of the mass. Therefore the flow along the continuous flow channel contains two components: drag flow from the feed end to the die end, caused by the mechanical thrust of the screw flights, and pressure-driven back-flow in the opposite direction, caused by the pressure difference between the two extremities of the extruder (Mount, 2011). The difference between the flow rates associated with the two velocity fields is known as the net flow rate. These two opposing flow patterns are mostly responsible for mixing in the single-screw extruder. The resistance of the die and additional constraints imposed on the flow channel determine the degree of backflow and, consequently, the extruder's "pumping efficiency."

2.2 Extruder components

Because extruders are special pieces of equipment used in food processing, it is essential to pay close attention to their nomenclature. By defining the components and features that the extruder uses, a deeper comprehension of its operation can be attained, and the foundation for future discussions of food extrusion can be established. Much of the nomenclature and terminology used in food extrusion borrows heavily from plasticating extrusion processes and technology.

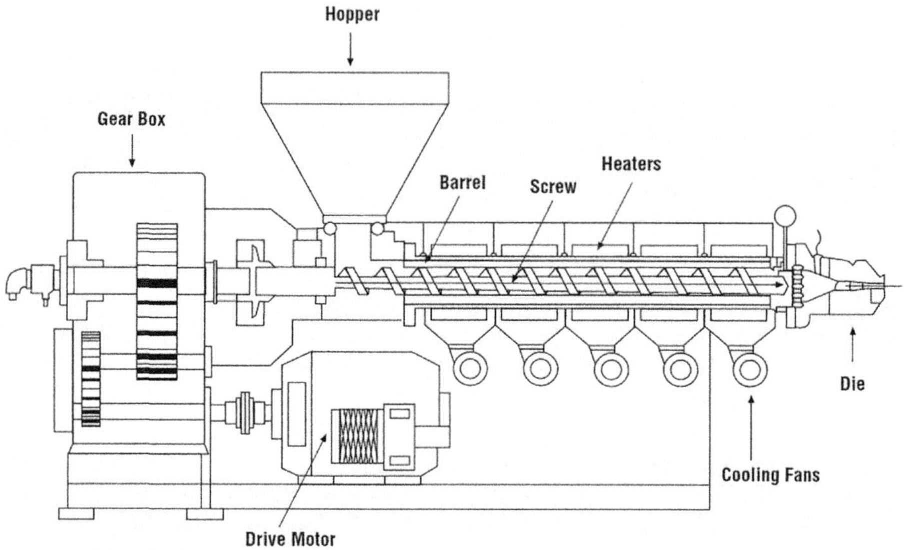

Fig. 2.1 Various components of a single-screw extruder.

2.2.1 Extrusion drive

Fig. 2.1 shows a typical food extruder and the major components of the stand, drive, and transmission, as defined in the following sections.

2.2.1.1 Support or stand

A steel or cast-iron frame, which can be fastened to the floor, supports the extruder. The frame holds the main extruder parts, such as the feeders, screw, barrel, and drive, and positions the working members of the extruder for easy operation.

2.2.1.2 Drive motor

Food extruders are often powered by electric motors. Depending on the extruder's capacity, the motor's size can range up to 300 kW (400 horsepower).

2.2.1.3 Speed variation

The extruder's screw speed variation is an important control parameter. Continuous variations can be produced mechanically, electrically, or magnetically. The need for constant speed control, without the wear and tear of mechanical parts that eventually result in imprecise control, has led to the increasing use of silicon-controlled rectifier DC motors.

2.2.1.4 Transmission

Food extruders typically have a screw speed of <500 rpm. A transmission is used to decrease speed while increasing the drive motor's torque proportionately (Giberti, Cinquemani and Legnani, 2010). Belts or chains are sometimes used as reducers, but gear reducers with positive lubrication are more frequently employed for the high loads encountered in extrusion.

2.2.1.5 Thrust bearing

A bearing is required to support and center the extrusion screw and absorb the thrust exerted by the screw (Kramer and Steward, 1991). Since the screw forces food forward against a back pressure, a rearward thrust is produced on the screw, which must be absorbed by the thrust bearing. The magnitude of the thrust is approximated by the maximum extrusion pressure multiplied by the cross-sectional area of the barrel. The thrust bearing must be able to sustain the load produced under normal extrusion conditions, providing an expected life of 20,000–50,000 h.

2.2.2 Feed assembly

A crucial step in an extrusion process is feeding, blending, and preconditioning, which involves heating and/or moisturizing components. For an extruder to operate consistently, food and other ingredients must be fed in a consistent and uniform manner.

2.2.2.1 Hoppers or bins

Above feeders, food components are stored in steel bins. Bin vibrators are frequently used to facilitate the ingredients' passage into the feeders and avoid bridging. Food items are moved into the hopper or bin's discharge throat by these vibrators, which can be oscillators, air jets, or internally mounted screw conveyors (Nowak, 2018). Hoppers, equipped with devices at their discharges to ensure a continuous flow of ingredients, are said to have "live-bottoms."

2.2.2.2 Feeder

It is a device for providing a uniform delivery of food ingredients, which are often sticky, non-free-flowing substances. A number of feeding devices are used, including the following:

Vibratory feeders: A vibrating pan that has either variable frequency or stroke and is used to control the feed rate from a hopper (Han et al., 2012).

Variable-speed auger: Screw conveyors equipped with a variable-speed auger can be used to volumetrically meter relatively free-flowing food ingredients.

Weigh belts: Food ingredients are measured gravimetrically using a moving belt that has a weighing instrument under its central part. There are two kinds of weigh belts: one with a variable-speed belt and a fixed gate and the other with a variable feed gate opening to the belt. To achieve a consistent feed rate (White and Wolf, 1995), signals from the weighing apparatus modify the belt speed or the gate opening.

2.2.3 Slurry tanks

Slurry tanks are used to store a variety of liquid substances or slurries that are added to the extrudate. For slurry tanks to maintain materials warm, evenly distributed, and free-flowing, agitators and heating jackets are frequently needed.

2.2.4 Liquid feeders

The uniform metering of liquid ingredients is accomplished in a variety of ways (White and Wolf, 1995).

Water wheels: A metering device can be made out of a revolving wheel with several cups fastened to its edge. The reservoir into which the cups are dipped needs to maintain a steady level of liquid. The discharge rate is directly correlated with the wheel's rotational speed.

Positive displacement metering pumps: Liquids are metered at a constant rate by a positive displacement metering pump, whose length of stroke or speed of rotation can be varied.

Variable orifice: A needle valve is used to create a variable orifice opening to control liquid flow (Habbal, 2021). A diaphragm actuator can be used to automatically position the needle valve. Its position is determined by a feedback controller that responds to a signal proportional to the flow rate, as determined by a primary sensor. In these systems, a regulator keeps the liquid stream pressure behind the control valve constant.

Variable head: The pressure, or more commonly, the height of the liquid entering a fixed orifice, can be used to control liquid flow rate.

Liquid flow control systems are classified as either feedforward or feedback systems. Feedforward systems operate on the basis that a setting will give a fixed and constant feed-rate response, while feedback systems adjust the position of the primary control element on the basis of a measurement of the actual flow, which is fed back to the primary control element.

2.2.5 Batch feed system

The proper ratio of individual feed ingredients is maintained by weighing them into a large weigh hopper on a batch basis. This batch of ingredients is mixed and fed as a preblend to the extruder.

2.2.6 Continuous feed systems

Individual ingredient streams are continuously combined in the proper proportion or ratio. Individual feeders on each stream control the rate of addition of the various ingredient types to a continuous blender before they enter the extruder.

2.2.7 Preconditioner

It is a closed vessel where the feed ingredients are mixed with water, steam, and/or other liquid ingredients for the purpose of increasing their moisture and/or temperature. In some cases, the preconditioner can serve as a continuous blender for several individual feed streams. The preconditioner can be designed and operated as either a pressure or an atmospheric vessel. When it functions under pressure, a rotary valve or a similarly functioning device must be used to maintain the pressure differential between the preconditioner and the surroundings. A variety of mixing devices are used within the preconditioner.

Ribbons: Helical ribbon flights are attached to the central shaft of the preconditioner. The pitch of the ribbon flights serves to convey the feedforward within the preconditioner toward the extruder inlet.

Paddles: Small, flat-bladed paddles can extend from the central preconditioner shaft(s). The angle of the blades make with the drive shaft is often adjustable to alter the conveying action within the preconditioner.

2.2.8 Rotary or star valve

It is used as a continuous feed port that maintains a seal between the internal pressure of the preconditioner and the external atmospheric pressure. The rotary valve consists of a driven rotating wheel with deep pockets that fit tightly within a valve body designed with opposing feed and discharge ports. Feed material falls into a pocket under the feed port and, by the rotation of the wheel, is transported to the discharge port, where it falls out. The vanes on the rotating wheel serve as the physical barrier between the high- and low-pressure sides of the device.

2.2.9 Feed transition

It is a device that accepts the feed ingredients directly from the preconditioner. In the case of free-flowing ingredients, this may be as simple as an open hopper. In instances where the ingredients are sticky, a forced feeder consisting of a feed screw or a hopper with paddles may be used. To maintain a pressure difference between the preconditioner and the feed port on the extrusion screw, the feed transition may incorporate a rotary valve or another pressure-differentiating device.

2.3 Extruder screw

The screw is the central portion of a food extruder. It accepts the feed ingredients at the feed port, conveys, works, and forces them through the die restriction at the discharge. The extrusion screw is divided into three sections (Fig. 2.2), whose names correspond to the function each section plays.

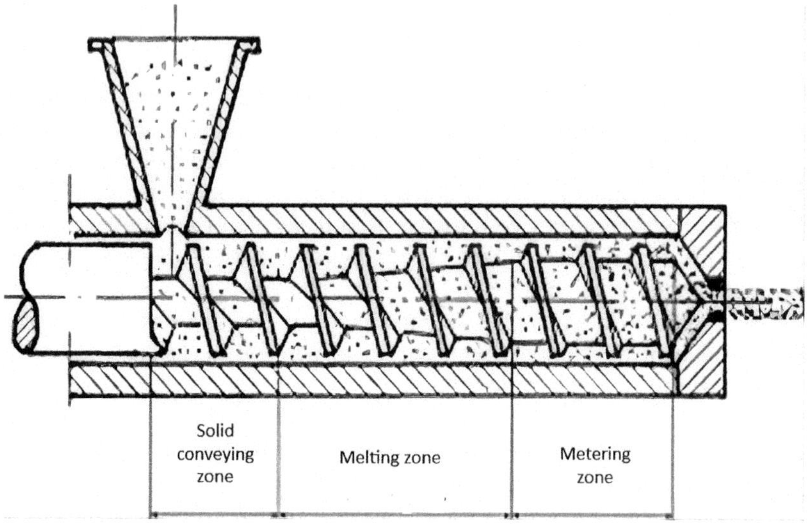

Fig. 2.2 Various zones in a single-screw extruder.

2.3.1 Feed section

It is the portion of the screw that accepts the food materials at the feed port or throat. Usually, the feed section is characterized by deep flights so that the product can easily fall into the flights. The function of the feed section is to ensure sufficient material is moved or conveyed down the screw and the screw is completely filled. When a screw is only partially filled with feed materials, the condition is termed "starved feeding." The feed section typically constitutes 10–25% of the total length of the screw.

2.3.2 Compression section

It is the portion of the screw between the feed section and the metering section. Sometimes, the transition section is referred to as the compression section. Compression is achieved in several ways, as shown in Fig. 2.2. The most common method is the gradual decrease in the flight depth in the direction of discharge. Another method involves a decrease in the pitch in the transition section. The food ingredients are typically heated and worked into a continuous dough mass during passage through the transition section. The character of the feed materials changes from a granular or particulate state to an amorphous or plasticized dough. It is common for the transition section to be the longest portion of the screw, comprising approximately half its length.

2.3.3 Metering section

The portion of the screw nearest the discharge of the extruder is normally characterized by having very shallow flights. The shallow flights increase the shear rate in the channel to the maximum level within the screw. The viscous dissipation of mechanical energy is typically large in the metering section, causing the temperature to increase rapidly. The high shear rate in the screw also enhances internal mixing, producing temperature uniformity in the extrudate. In some cases, pins or cut flights are employed to increase mixing and mechanical energy dissipation. The critical screw dimensions are shown in the sectional drawing of an extrusion screw in Fig. 2.2. The dimensions, terminology, and their interrelationships are described in the following.

Flight: The helical metal rib wrapped around the screw, whose action on the food mechanically advances it toward the discharge of the screw. In most instances, the flight is continuous, but when it is broken, it is called a cut flight.

Root: The continuous central shaft of the screw, which is usually cylindrical or conical in shape.

Land: The surface at the radial extremity of the flight, which constitutes the outer periphery of the screw. In some cases, the screw land is specially hardened by flame induction or by depositing a hard-facing metal surface.

Leading flight edge: The pushing face of the flight, which faces the extruder discharge. Wear on the flight is first observed on the leading flight edge.

Trailing flight edge: The face of the flight that faces toward the discharge of the extruder.

Flight shape: Flights are usually rectangular or trapezoidal in shape. The trapezoidal shape results in a channel that is wider at the surface of the screw than at the root, thus reducing stagnation of the product in the corners between the flight and screwroot.

Screw shank: The rear protruding portion of the screw to which the extruder drive is attached. Normally, the shank passes through a seal in the support stand before the thrust bearing, which prevents leakage of food materials or grease. The shank is fitted with a keyway that mates with a drive coupling. Soft keys or shear pins can also serve as overload protection devices.

Screw diameter, *D*: Nominally, the internal diameter of the barrel in which the screw rotates is denoted as D. The actual screw diameter, accounting for the clearance, δ, is

$$D_s = D - 2\delta \tag{2.1}$$

Flight height, *H*: Nominally, the dimension between the barrel surface and the root of the screw is defined as H. The actual flight height of the screw is

$$H_s = H - \delta \tag{2.2}$$

Root diameter, *D*: Diameter of the root of the screw.

$$D_r = D - 2H = D_s - 2H_s \tag{2.3}$$

Diametric screw clearance, 2δ: The difference between the diameter of the screw and the bore of the barrel.

$$2\delta = D - D_s \tag{2.4}$$

Radial screw clearance, δ: Half the diametric clearance.

Lead, l: The axial distance from the leading edge of a flight at its outside diameter to the leading edge of the same flight in front of it.

Helix angle, θ: The angle that the flight makes with a plane normal to the axis of the screw.

$$\theta = \tan^{-1}\frac{1}{\pi D_s} = \tan^{-1}\frac{1}{\pi D} \tag{2.5}$$

Channel: The helical opening that extends from the feed to the discharge end of the screw. The channel is formed on two sides by screw flights, on the bottom by the screwroot, and enclosed at the top by the inner barrel surface.

Axial channel width, $\boldsymbol{B}$: The axial distance from the leading edge of one flight to the trailing edge of the same flight one complete turn away at the diameter of the screw.

Channel width, W: The channel width measured as above, but perpendicular to the flight.

$$W = B.\cos\theta \tag{2.6}$$

Axial flight width, $\boldsymbol{b}$: The axial width of a flight measured at the diameter of the screw.

$$b = l - B \tag{2.7}$$

Flight width, $\boldsymbol{e}$: The flight width measured perpendicular to the face of the flight.

$$e = b.\cos\theta \tag{2.8}$$

Reference axes, $\boldsymbol{x}$**,** $\boldsymbol{y}$**,** $\boldsymbol{z}$: The following directions are adopted by convention: x, cross-channel direction, z, down the channel direction along the screw helix, and y, direction perpendicular to the root of the screw. A is in the direction of the axis of the screw or along its length.

Channel length, $\boldsymbol{Z}$: The length of the screw channel in the Z direction. The length of the channel in one complete turn of the helix is given by:

$$Z = l.\sin\theta \tag{2.9}$$

Peripheral speed or tip velocity, V: The speed of the screw tip.

$$V = \pi D_s N \approx \pi D N \tag{2.10}$$

Number of flight turns $\boldsymbol{p}$: Total number of single flights in an axial direction.

Single-flighted screw: A screw having a single helical flight. In this case, the parameter $p=1$.

Multiple-flighted screw: A screw having more than one helical flight in parallel, such as a double-flighted screw, where $p=2$. Viewed from the discharge tip or nose of the screw, two flights can be seen terminating.

Axial area of screw channel: The cross-sectional area of the channel, measured in a plane through the axis of the screw.

Developed volume of screw channel: The volume developed by the axial area of the screw channel during one revolution of the screw axis.

Compression ratio (CR): The factor obtained by dividing the developed volume of the screw channel at the feed opening by the developed volume of the last full flight prior to discharge. Sometimes, CR is incorrectly given as the ratio of H in the feed section to H in the metering section, which is actually the channel depth ratio. Typical ranges in CR are from 1 to 5:1.

Height-to-diameter ratio, H/D: The ratio of the flight height to screw diameter, normally calculated for the metering zone.

Hollow screw: Sometimes, extrusion screws have a hollow internal core to accept either a heating or cooling medium, thus providing extra heat transfer surface area.

Nose: The discharge tip of the extruder screw, which can be pointed or blunt depending on the design of the die head assembly.

2.4 Extruder barrel

The extruder barrel is the cylindrical member that fits tightly around the rotating extruder screw. The nomenclature and features of the barrel are discussed in the following.

2.4.1 Bore, D

It is the inside diameter of the barrel and the nominal size of the extruder. Internal diameters range from 5 cm to 25.4 cm.

2.4.2 Length-to-diameter ratio, L/D

The distance from the rear edge of the feed opening to the discharge end of the barrel bore is divided by the bore diameter to express a ratio where the diameter is reduced to 1. Food extruders typically have L/Ds ranging from 1:1 to 20:1.

2.4.3 Segmented barrel

A barrel is made up of segments. The barrel is fabricated by clamping or bolting together several segments. Such an arrangement makes it relatively easy to alter the interior configuration of the barrel and to replace the discharge section, which wears out the most rapidly. Segmented barrels have the disadvantage that they tend to leak between the segments if not carefully assembled.

2.4.4 Feed throat or opening

It is a hole through the feed end of the barrel for the introduction of feed material, usually at least one diameter in length.

Vertical: The feed throat has vertical sides that are tangent to the sides of the bore.

Sloped: The side of the throat toward which the screw rotates is inclined. Such a design improves the manner in which feed materials are accepted into the screw.

Undercut: The area open to feed is cut under the screw and improves the feeding of soft or rubbery-like materials.

Side: An opening that feeds material into the side of the screw.

2.4.5 Barrel liner

It is a removable sleeve within the barrel. Usually, the liner is of specially hardened or cast materials that resist wear.

2.4.6 Hardened material of construction

Barrels are constructed of special hard alloys, such as Xaloy 306 and stainless steel 431 (heat-treatable). Nitriding is also used as a method to produce a very hard surface material that is wear-resistant.

2.4.7 Grooves or splines

The interior surface of the barrels of food extruders often has small grooves to prevent slippage of the food material at the walls. The presence of grooves increases the ability of the extruder to pump food materials against high back pressures. The grooves are normally shallow and can take on a variety of configurations:

Straight: Grooves that run axially down the length of the barrel.

Spiral: Helical grooves inside the bore, which normally have a hand opposite to that of the screw.

2.4.8 Vent

An opening at an intermediate point in the extruder barrel permits the removal of air, steam, or other volatile materials from the material being processed. Vented extruders tend to act like two extruders placed in series.

2.4.9 Jacket

Hollow cavities around the outside of the barrel, in which a heat transfer medium can be circulated, such as water, oil, or steam. To enhance heat transfer, fins are often cast into the outer surface of the barrel. The fins have the ability to increase the outside heat transfer surface of the barrel and the velocity of the circulating heat transfer medium.

2.4.10 Heaters

Under conditions where only heating is required, a variety of heating elements can be placed directly around the barrel to accomplish this task. Electrical resistance band heaters are common, but tubular resistance heaters or induction coils have also been used.

2.5 Extruder discharge

Once the food material leaves the end of the extrusion screw, it enters the discharge section of the extruder, which normally holds the extruder die, cutters, and take-away devices. Each of these important elements is discussed in the following.

2.5.1 Die head assembly

It is the holder for the extrusion die, this discharge valve sometimes serves as the support for the cutter.

2.5.2 Breaker plate

This perforated plate, located at the end of the extruder, can serve as a mechanical seal between the barrel lining and die and also provides a more even pressure distribution behind the die plate. The thickness of the breaker plate is normally 20% of the barrel diameter and contains a number of countersunk holes ranging from 3 to 6 mm in diameter. In some cases, screens called a screen pack are placed on the upstream side of the breaker plate, which acts as their support. The function of the screen pack is to remove large or uncooked chunks of food, which might otherwise clog the small openings in the die.

2.5.3 Gate assembly

In certain instances, an assembly can be attached directly to the discharge of the extruder, which contains an adjustable discharge valve. Gate assemblies are designed for pressures greater than 400 atm.

2.5.4 Dead volume

This refers to the volume of the die head assembly. If the dead volume is large, uneven temperatures can occur within the head, causing nonuniform flow through the holes of the die.

2.5.5 Die

Extrusion dies are small openings that shape the food material as it flows out of the extruder. The shape of the dies varies, the simplest being a hole. Annular openings and slits are also common. Often, the opening on the feed side of the die is streamlined to improve the uniformity of the extrudate.

2.5.6 Die plate

This heavy plate contains numerous holes that can receive individual die inserts containing the actual die opening. Should one hole become damaged, an individual die insert can be replaced rather than the entire die. Therefore the use of the plate/insert approach is quite popular.

2.5.7 Die insert

Individual die openings exist as inserts that slip into holes in a die plate. The insert is positioned in the die plate from the rear, so its collar positions and holds the insert in place. Special Teflon coatings are used to reduce dough drag in the actual die opening.

2.5.8 Quick-change dies or breaker plates

A new die plate or breaker plate can be quickly slipped into place by means of a hydraulic piston when the original die plate ceases to operate properly or becomes clogged. The quick-change operation can be likened to changing a slide in a projector. Using the quick-change device, little production time is lost.

2.6 Die design

The die area is the section of the extruder that occurs after the food material leaves the extruder screw. It normally consists of transition, distribution, and die plate sections. There are basic formulas of fluid dynamics that can be used to model the flow through the die.

Although these equations give a good basic approximation, they do make several assumptions. These are:

(a) The flow is laminar (Re<2000)
(b) The density is constant (incompressible)
(c) The flow is independent of time (steady state)
(d) The flow is Newtonian
(e) End effects are negligible
(f) There is no slip at the wall
(g) There is no phase separation

In the absence of slip, the pressure drop in nonelastic viscometric flow through a circular hole is related to the flow rate, radius, and wall shear stress by

$$P_d = \frac{2L}{R} T_w \tag{2.11}$$

where T_w is the shear stress at the wall, R is the radius of the die hole, and L is the land length.

The volumetric flow rate from the die can be calculated from the pressure drop across a die as a function of output viscosity and die resistance:

$$Q = K_D \frac{P_D}{\mu_D} \tag{2.12}$$

where Q is the volumetric flow rate, P_D is the total pressure drop, and μ_D is the viscosity of the dough at the die.

If the viscosity is constant and independent of the shear rate, then for a die of circular cross-section:

$$K_D = \frac{TR^4}{8L_D} \tag{2.13}$$

For a rectangular slit cross section:

$$K_D = \frac{2WC^3}{3LD} \tag{2.14}$$

For a circular annulus:

$$K_D = \frac{\pi(R_o + R_i)(R_o - R_i)^3}{12L_D} \tag{2.15}$$

where R is the radius of the circular cross-section, L_o is the land length of the die opening, W is the long dimension of the rectangular slit, C is the narrow dimension of the rectangular slit, R_o is the outer radius of the annulus, and R_i is the inner radius of the annulus.

2.6.1 Cutters

A common cutter used with extruder operations is called a face cutter, meaning that the cutting knives revolve in a plane parallel to the face of the die. The relative speed of rotation of the cutter knives and the linear speed of the extrudate result in the control of the length of the extruded pieces. Depending on the location of the cutter drive motor and the length of the arms holding the cutting knives or blades, two different types of cutters exist.

Fly cutters: Motors are located outside the central axis of the die. Relatively long cutter arms rotate at very high tip velocities.

Central cutter: Cutter blades rotate about the central axis of the die assembly, are relatively short, and operate at a lower tip velocity than those on fly cutters.

2.6.2 Take-away

Once the extruded pieces are cut, they normally fall into some type of take-away system. This can be pneumatic or mechanical when small pieces are involved. Care must be exercised in the design of the take-away system to avoid clumping of the sticky, hot-extruded pieces. Extruded sheet goods are removed from the extruder by a series of driven rolls, which serve as the take-away system.

2.7 Flow

2.7.1 Isothermal Newtonian flow

2.7.1.1 Net flow

Net flow in an extruder is the overall effective flow rate of the material being processed. It can be calculated by combining drag flow and pressure flow and subtracting any leakage flow. The formula is:

$$Q_{Net} = Q_{Drag} + Q_{Pressure} - Q_{Leakage} \tag{2.16}$$

where Q_{Net} is the net flow rate, Q_{Drag} is the drag flow rate, $Q_{Pressure}$ is the pressure flow rate, and $Q_{Leakage}$ is the leakage flow rate.

2.7.1.2 Leak flow

Leakage flow occurs due to pressure differences that cause material to escape through clearances in the extruder. It can be calculated using the formula:

$$Q_{Leakage} = k_i - \Delta p \tag{2.17}$$

where k_i is the leakage coefficient, which is dependent on the clearance and material properties, and Δp is the pressure difference driving the leakage.

2.7.1.3 Drag flow

Drag flow is the movement of material due to the rotation of the screw within the extruder barrel. It is influenced by screw speed, screw design, and material viscosity. The formula for drag flow is:

$$Q_{Drag} = k_d.N \tag{2.18}$$

where k_d is a constant dependent on the screw geometry and material properties and N is the screw speed.

2.7.1.4 Pressure flow

Pressure flow is generated by the pressure build-up within the extruder as the material is pushed through the die. The formula for pressure flow is:

$$Q_{Pressure} = k_p.\Delta p \tag{2.19}$$

where k_p is a pressure flow coefficient dependent on the die geometry and material properties and Δp is the pressure difference across the die.

2.7.1.5 Power consumption

The power consumption of a single-screw extruder depends on the die. When a single-screw extruder works without a die, the specific energy consumption (E) of the extruder is

$$E = \frac{2\pi N T_m}{Q_m \rho_m} \tag{2.20}$$

where N (rpm) is the screw speed; T_m is the torque required to rotate the screw; Q_m (m^3/s) is the flow rate of extrudate from the extruder without a die; and ρ_m is the density of the extrudate.

When the die is fitted with a barrel extruder, the specific energy consumption (E_d) of the extruder is

$$E_d = \frac{2\pi N T_d}{Q_d \rho_d} \tag{2.21}$$

where T_d is the torque required to rotate the screw; Q_d (m^3/s) is the flow rate of extrudate from the extruder without a die; and ρ_d is the density of the extrudate.

2.7.1.6 Extrudate viscosity

It is assumed that the extrudate in a single-screw extruder shows Newtonian viscosity. Thus the viscosity of the extrudate is

$$\mu = \frac{\pi d^4 P}{128 l Q_d} \tag{2.22}$$

where d (m) and l (m) are, respectively, the diameter and length of the die fitted to the barrel and P (Pa) is the gauge pressure at the upstream end of the die.

2.8 Conclusion

The design of a single-screw extruder is discussed in this chapter. The design and modeling of a single-screw extruder play a crucial role in optimizing the extrusion process for various applications, including polymer processing, food production, and metal extrusion. The study and analysis of key design parameters, such as screw geometry, barrel design, material flow, and temperature distribution, ensure improved efficiency and performance. The development of an efficient single-screw extruder depends on a balance between theoretical design principles and practical experimentation. Future advancements in material science and computational techniques will further improve extruder performance, leading to more energy-efficient and high-quality extrusion processes.

References

Berk, Z., 2016. Food extrusion. In: Engineering Foods for Bioactives Stability and Delivery. Springer, New York, pp. 309-339.

Covas, J.A., Cunha, A.G., Oliveira, P., 1999. An optimization approach to practical problems in plasticating single screw extrusion. Polym. Eng. Sci. 0032-3888. 39 (3), 443–456. doi:10.1002/pen.11434.

Giberti, H., Cinquemani, S., Legnani, G., 2010. Effects of transmission mechanical characteristics on the choice of a motor-reducer. Mechatronics. 09574158. 20 (5), 604–610. doi:10.1016/j.mechatronics.2010.06.006.

Habbal, O., Continuously Variable Orifice Extrusion for Build Time Reduction in Fused Element Fabrication Additive Manufacturing. (2021).

Han, L., Li, J., Zhang, L.X., 2012. Design and Study on the Automatic Extrusion Device for Curved Chassis of Vibratory Feeder. Appl. Mech. Mater. 1662-7482. 198-199, 1298–1301. doi:10.4028/www.scientific.net/amm.198-199.1298.

Harper, J.M., 1979. Food extrusion. CRC Crit. Rev. Food Sci. Nutr. 00990248. 11 (2), 155–215. doi:10.1080/10408397909527262.

Harper, J.M., 1990. Extrusion of Foods, Vol. 1. CRC Press.

Kramer, W.A., Steward, E.L., 1991. Extrusion Processes. Springer Sci. Bus. Media LLC, 79–132. doi:10.1007/978-1-4615-7604-4_4.

Lekakou, C., Brandao, J., 1996. Extrusion of polypropylene. Part II: process analysis of the metering zone. Polym. Eng. Sci. 0032-3888. 36 (1), 56–64. doi:10.1002/pen.10384.

Marschik, C., Roland, W., Osswald, T.A., 2022. Melt conveying in single-screw extruders: modeling and simulation. Polymers 14 (5), 875.

Mohr, W.D., Saxton, R.L., Jepson, C.H., 1957. Theory of Mixing in the Single-Screw Extruder. Ind. Eng. Chem. 00197866. 49 (11), 1857–1862. doi:10.1021/ie50575a031.

Moscicki, L., 2011. Extrusion-cooking Techniques: Applications, Theory and Sustainability, John Wiley & Sons.

Mount, E.M., 2011. Extrusion Processes. In: Applied Plastics Engineering Handbook. Elsevier Inc., United States, pp. 227–266. Available from http://www.sciencedirect.com/science/book/9781437735147. doi:10.1016/B978-1-4377-3514-7.10015-7.

Nowak, S., 2018. Feeding Technology and Material Handling for Pharmaceutical Extrusion Pharmaceutical Extrusion Technology. CRC Press, pp. 105–131.

Riaz, M.N., 2000a. Extruders in Food Applications. CRC Press, United States, pp. 1–228. Available from https://www.taylorfrancis.com/books/e/9781482278859.

Riaz, M.N., 2000b. Introduction to extruders and their principles. In: Extruders in Food Applications. CRC Press, United States, pp. 1–23. Available from https://www.taylorfrancis.com/books/e/9781482278859.

Steiner, R., Haight, B., 2018. Extruder Design Pharmaceutical Extrusion Technology. CRC Press, pp. 37–52.

Wang, Y., Huang, J.-S., 1996. Single screw extrusion compounding of particulate filled thermoplastics: state of dispersion and its influence on impact properties. J. Appl. Polym. Sci. 0021-8995. 60 (11), 1779–1791. doi:10.1002/(sici)1097-4628(19960613)60:11<1779::aid-app1>3.0.co;2-i.

White, D.H., Wolf, D., 1995. Direct Biomass Liquefaction by an Extruder-Feeder System. Chem. Eng. Commun. 15635201. 135 (1), 1–19. doi:10.1080/00986449508936335.

Yeh, A., Jaw, Y., 1999. Effects of Feed Rate and Screw Speed on Operating Characteristics and Extrudate Properties During Single-Screw Extrusion Cooking of Rice Flour. Cereal Chem. 0009-0352. 76 (2), 236–242. doi:10.1094/cchem.1999.76.2.236.

Design and modeling of twin-screw extruder

3

Sagar Loni, Pratik Madhukar Gorde, and Sushil Kumar Singh
Department of Food Process Engineering, National Institute of Technology Rourkela, Rourkela, Odisha, India

3.1 Introduction

Extrusion is a thermochemical process where raw ingredients are shaped by feeding them through screws at specified high temperatures and pressures to produce various food products. The extruder is classified into two types based on the number of screws used: single-screw extruders and twin-screw extruders. Both work on quite similar principles, but the twin-screw extruder differs from single-screw extruders in its complex construction structure, better mixing ability, material movement in the barrel, and advanced level of applications including food processing, pharmaceuticals, plastics, and chemical compounding. It may not be available in underdeveloped and developing countries due to its high capital investment, advanced construction features, the need for trained personnel to operate it, and its high maintenance cost. These are the main reasons for the poor growth of the twin-screw technology field (Sobowale et al., 2017).

Twin-screw extruders are strong, adaptable devices that provide unmatched versatility and efficiency in the processing of materials, making them essential instruments in contemporary manufacturing. Their ability to handle complex material compositions makes them indispensable in sectors where product quality and process precision are paramount. Continuous technological advancements have broadened the application range and operational potential of twin-screw extruders. They are used across different sectors, such as food, chemicals, plastics, and pharmaceuticals. They are made up of two corotating or counterrotating screws that are barrel-mounted. This arrangement facilitates effective material mixing, compounding, and processing.

The twin-screw extruder is flexible for altering the arrangement of screws in different configurations, which influences the pressure parameters as per the requirements needed; therefore the twin-screw extruder is multipurpose equipment. The development of the twin-screw extruder is crucial because it produces a combined effect of operations such as mixing, melting, and reacting between ingredients within a single pass. This leads to major improvements in the efficiency of the manufacturing process.

Extrusion Cooking for Food and Feed. https://doi.org/10.1016/B978-0-443-29987-2.00003-8

The feeding unit, the extruder's screw speed limit, and torque limit define the throughput of twin-screw extruders, which typically run on starved feed. As a result, the screw length will only be partially comprised of the melt pumping section's actual length. The screw design, screw speed, throughput, viscosity of the material, and total flow resistance will all have an impact. Along the melt section, the material's temperature rises quickly, changing the material's viscosity. This makes it impossible to calculate the necessary pressure and, consequently, the length of the melt pumping section. Thus a method of trial and error was used in earlier days. Modern scientific technologies are now available to design twin-screw extruder parts in accordance with specifications. In a short period of time, real-time problems can be readily solved by utilizing modeling and simulation techniques (Yacu, 1985).

3.2 Fundamentals of extrusion

3.2.1 Principle

The twin-screw extruder revolves around the integrated action of two intermeshing screws within a barrel, generating combined effects of mechanical shearing, pressure, and heat. These conditions promote efficient mixing, melting, and plasticization of raw materials within the barrel, yielding products of the desired form. In addition, screw elements such as kneading blocks (KBs) and mixing zones are used to enhance the mixing efficiency, ensuring uniform distribution of additives, fillers, and various ingredients.

At the onset, raw materials like pellets, powders, doughs, or granules are fed into the extruder's hopper and undergo a combination of various processes, which result in thermochemical changes in the raw materials through compression and compaction actions, and the processed products come out from the other end by passing through a die, which influences the physical properties of the product.

Food processing equipment, such as variously designed twin-screw extruders, is used to create a large variety of final goods in the shortest period while utilizing affordable raw materials and consuming less energy. Extrusion produces the very least amount of waste, and the method is highly flexible in terms of choosing ingredients and producing a wide variety of products with different shapes and textures. Extruders function mechanically in a dynamic steady state, wherein the outputs and inputs must be balanced to produce goods of the appropriate caliber. The various extruder variables, such as barrel temperature, screw speed, die size, and input material properties, such as feed moisture level, composition, and structure, must be set at the desired levels to give the machine's barrel the dependent physical conditions and chemical changes needed to produce the desired results in extruded foods. The complex biological materials used to produce processed foods include moisture, carbohydrates, protein, lipids, salts, water, vitamins, fibers, and other elements that can change chemically and physically when exposed to high pressure, temperature, and shear in the extruder barrel. If these variables are not correctly calculated and set at the necessary levels, it will eventually affect the quality of the finished products.

Therefore how well the various process variables are established and altered during the cooking process will determine the final product of the extrusion cooking process. Due to the response variables' dependence on input variables like raw materials and machine parameters, or the quality of the food, quantitative relationships in the form of mathematical models could be built for the optimization of processes and product quality (Danbaba et al., 2018).

Reverse screw threads act as mixing components, whereas forward screw threads encourage material flow forward. The screw thread structure alters in the melting and cooking phase to raise temperature and pressure. The temperature of the material is raised until it melts entirely through external heating of the barrel and shear between the screws and barrel. To obtain the required texture and qualities, the proteins are denatured, and the starch is thoroughly gelatinized. The molten material is evenly distributed to the die in the forming portion. The pressure is swiftly released as the material comes out of the die, which causes free water to evaporate quickly. The material solidifies into the final extruded shape by expanding and cooling rapidly (Danbaba et al., 2018).

3.2.2 Types of twin-screw extrusion

Classification of the twin-screw extruder is based on the relative rotation of the screws inside the barrel. Fig. 3.1 illustrates the classification of a twin-screw extruder.

3.2.2.1 Co-rotating

Co-rotating twin-screw extruders, which are illustrated in Fig. 3.2, are the most widely used ones in which both the screws present in the barrel rotate in the same direction, either clockwise or anticlockwise, and they also offer flexibility in arranging the different screw elements for specific designs. The speed of rotation is high, which results in maximum output. The following are the two types of screw arrangements that exist

(a) Intermeshing: In these flights, one screw is engaged with other screws. There are numerous attractive benefits associated with the co-rotating intermeshing twin-screws. These

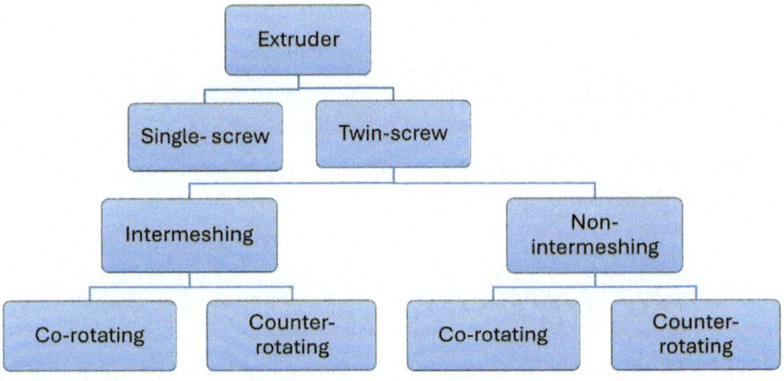

Fig. 3.1 Classification of the extruder.

include a short residence time, remarkable mixing performance, and self-wiping qualities, all of which immediately result in higher yield and better product quality. Moreover, the process's solventfree characteristic is especially helpful in regular manufacturing since it lessens the significant expenses related to the use, recovery, separation, and disposal of solvents. It also avoids chemical adulteration. Furthermore, the modular screw design's widespread use increases operational flexibility, making it a flexible and adaptable solution for a range of industrial requirements. This co-rotating intermeshing type's main disadvantage is that pressure develops rather slowly. The major challenge with intermeshing co-rotating twin-screw extruders is creating a screw configuration that is suitable for the real requirements of the process, which presents a significant challenge in the manufacturing process. Achieving the right screw configuration demands a deep understanding of the specific process requirements, as well as extensive experience in experimental and empirical work. This also includes careful consideration of factors such as material type, viscosity, moisture content, temperature, and other process variables to ensure optimal performance. The intricate nature of this task underscores the need for thorough research and testing to arrive at the most effective screw configuration (Eitzlmayr et al., 2014).

(b) Nonintermeshing: This type of arrangement does not have selfwiping capabilities. As a result, it often has lower mixing efficiency, and material can flow around the screws without being exposed to the significant shearing action and mixing pressures required for homogeneity, rendering it ineffective. Therefore this configuration is not used in practice (Eitzlmayr et al., 2014).

3.2.2.2 Counterrotating

The counterrotating twin-screw extruders, which are illustrated in Fig. 3.3, are specifically designed for producing high pressure, in which both screws rotate slowly in opposite directions to each other, resulting in the buildup of very high pressure; therefore they are recommended where high pressure is needed. The screw arrangements are the same as those of co-rotating extruders. The fully intermeshing design generates extremely high pressure, potentially causing air entrapment, product quality degradation, and increased wear and tear due to the counterrotation of the screws. Consequently, this leads to reduced operating speed and minimal output. In the intermeshing twin-screw extruder, backflow of material is minimal, allowing it to convey materials efficiently compared to nonintermeshing extruders (Dhaval et al., 2022).

Fig. 3.2 Co-rotating twin-screw assembly with two parallel intermeshing screws featuring continuous helical flights along their length.

Fig. 3.3 Counter-rotating twin-screw assembly with two parallel screws featuring opposing helical flight orientations along their length.

3.2.3 Components of twin-screw extrusion (elements/long screws)

The co-rotating twin-screw extruder, which is illustrated in Fig. 3.4, is the most widely used extruder machine in producing high-quality extrudates:

- Barrel
- Screw elements
- Conveying zone
- Mixing zone
- Kneading zone
- Feeder
- Die
- C Chamber

During the extrusion process, raw ingredients undergoes feeding, melting, mixing, venting, homogenization, and discharge processes, as shown in Fig. 3.5, to obtain the required end product.

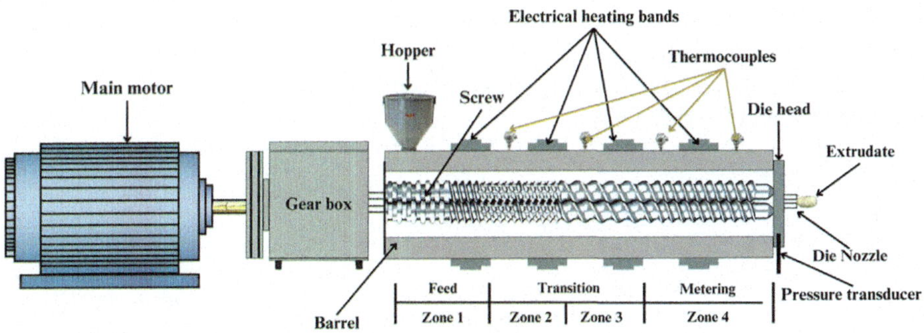

Fig. 3.4 Schematic diagram of co-rotating twin-screw extruder.

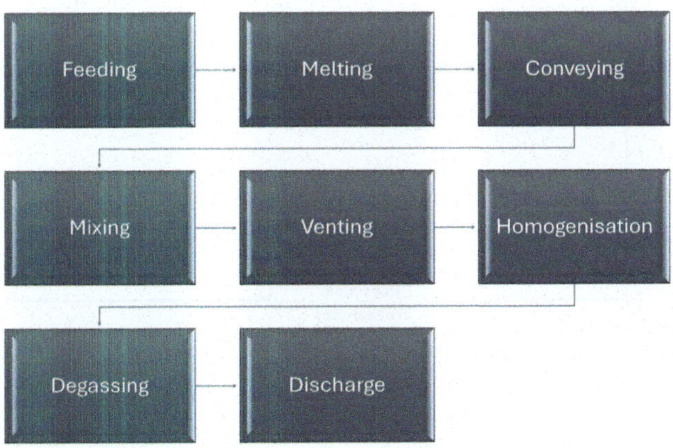

Fig. 3.5 Extrusion process.

3.3 Design and control considerations

3.3.1 Feeding control systems

Conveying screw: It is located at the entry point and exit point of the barrel, which helps in the forward movement of material. The pitch screw has a high pitch size and is flighted in the forward direction. This type of design is favorable for proper material feeding, reduces pressure drop, and restricts the exiting of material from the opening for a fairly short residence time. The recommended pitch size is 14 mm (Senanayake and Clarke, 1999).

3.3.2 Pressure control systems

The pressure control system in a twin-screw extruder is an important component because it facilitates monitoring the pressure inside the barrel, and this pressure helps in establishing the relationship between pressure and other controllable parameters like temperature, die size, moisture content, and speed; however, it does not directly affect the quality of the result. A four-arm strain gauge is essential for measuring the pressure force in a cantilever beam. The pressure control system in a die, as shown in Fig. 3.6, eliminates overheating and prevents damage to the extruder motor (Senanayake and Clarke, 1999).

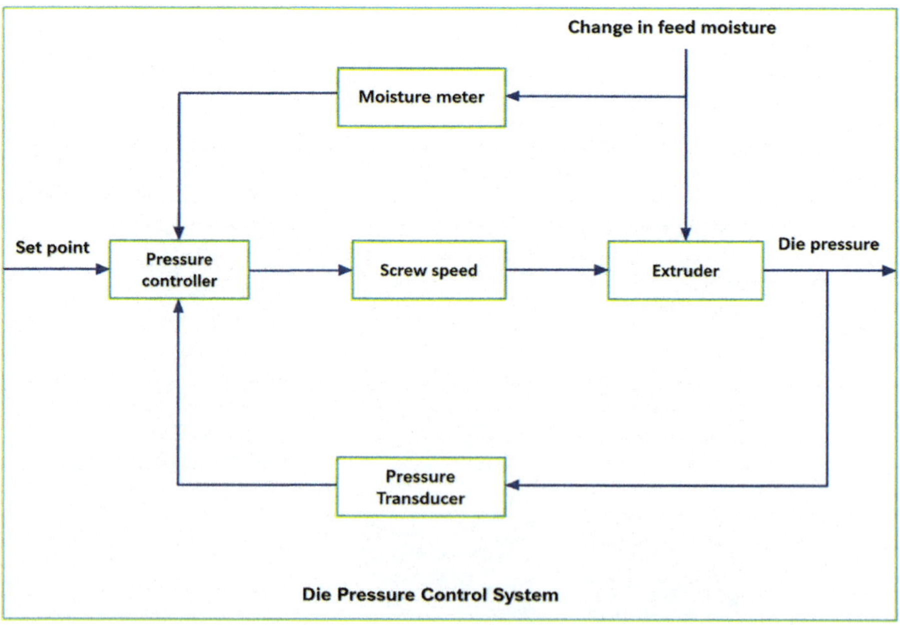

Fig. 3.6 Die pressure control system.

3.3.3 Heating and cooling of the barrels control system design

The heating of raw ingredients is a crucial process parameter in extrusion. A pair of cartridge heaters with 800 W and 1200 W capacity are used, placed in the grooves chiseled out of the top and bottom of the barrel toward the die end. Temperature sensors are installed along with a thermocouple, which measures the temperature closest to the die plate temperature (Senanayake and Clarke, 1999).

Cooling and sizing of extrudates is very important to produce the required extrudates with exact specifications, which is achieved by using different types of cooling and sizing hardware. All extrudates need to be cooled and sized by some metallic element, except for blown film and strand profiles.

3.3.4 Screw speed control systems

The twin-screw extruder utilizes an electric motor of 2.2 kW with a belt drive mechanism between the motor and gearbox. An electric speed controller is used to control the speed over a specific range. In gearboxes designed to withstand high axial loads, a lubricated phosphor bronze thrust bearing is used along the shaft to support the material (Senanayake and Clarke, 1999).

3.3.5 Barrel geometry

The barrel is made of high-strength stainless steel grades such as 304 and 316, which are commonly used to withstand high pressure, shear, and compressive forces. A corrosion-resistant liner consists of food-grade steel or ceramic coatings that prevent contamination. Heat is generated in the barrel either by an external electronic heater or by internal electronic cartridges, with the temperature range between 101°C and 149°C. The length is measured in terms of the ratio of the overall length to the outer diameter of the screw. Various specifications are available and are selected based on the required handling capacity of specific raw materials being used. The entire length of the barrel is divided into different zones based on specific temperatures necessary for the extrusion process. Barrel sections are 4D long for twin-screw extruders. The standard extruder length-to-diameter *(L/D)* ratios available for commercial use are 24, 48, and 60. The barrel design should be such that it minimizes the residence time.

3.3.6 Screw

The geometry of the screw is completely defined by five parameters. They are the distance between the two screw shafts, the external radius, the flight width at the top and bottom of the channel, and the pitch of the screw. The ratio of outer to inner diameter and channel depth are significant screw parameters, which control the volume available for material and the torque transferred to the material during the process. As the inner diameter increases, channel depth decreases, which results in maximum torque production. Two-, three-, and four-flight conveying screws are available. The screw speed is in the range of 208–393 rpm. The throughput capacity of a twin-screw extruder is mostly dependent on the screw design. Higher pitch and more flights on a screw can enhance shear and mixing forces, which can increase throughput capacity. This is because the increased number of flights and pitches enhances the mixing and compounding of the material, allowing for more efficient processing and higher production rates. The throughput capacity of a twin-screw extruder can also be increased by using multiple screw elements including conveying screws, kneading blocks, and mixing elements. This is because greater production rates and improved material processing are possible due to the numerous elements and sections' increased options for material mixing and compounding. As a result, a twin-screw extruder's throughput capacity is largely dependent on the screw design (Cheng and Friis, 2010).

3.3.6.1 Conveying screw

It is located at the entry point and exit point of the barrel, which helps in the forward movement of material. The pitch screw has a high pitch size and is flighted in the forward direction. This type of design is favorable for proper material feeding, pressure drop, and restricting the exiting of material from the opening for a fairly short residence time. The recommended pitch size is 14 mm (Sobowale et al., 2017). It also has an additional mechanism of self-wiping, which is required for easy transportation.

3.3.6.2 Mixing element

It is an integral part of a screw; these are the staggered discs at a specific angle that are used to proceed with the melting, mixing, and homogenization. By increasing the offset angle, the mixing properties are also enhanced. If the increasing length of the mixing zone increases residence time, it may degrade the quality of the end product. To minimize the formation of large clumps and break them into powder form properly, a conveying element is fixed after the kneading zone. Therefore screw configuration plays a major role in maintaining the quality of the product (Dhaval et al., 2022).

The full-flight screw and the KB, which are illustrated, are the conventional mixing elements used during the extrusion process. Mixing plays a critical role in achieving the desired quality of products. Twin-screw extruders are equipped with a variety of screw elements. Over the past decades, technological advancements have led to the development of specialized elements that aim to optimize mixing performance while operating at high speeds to maximize output (Agunbiade et al., 2022). Mixing elements in Fig. 3.7, such as conveying elements (a) and KBs (b), are most frequently used to enhance the mixing efficiency and the conveying capacity of raw materials.

One of the key screw elements used for mixing is the screw mixing element, which has become a staple in extrusion processes. This element features a standard screw

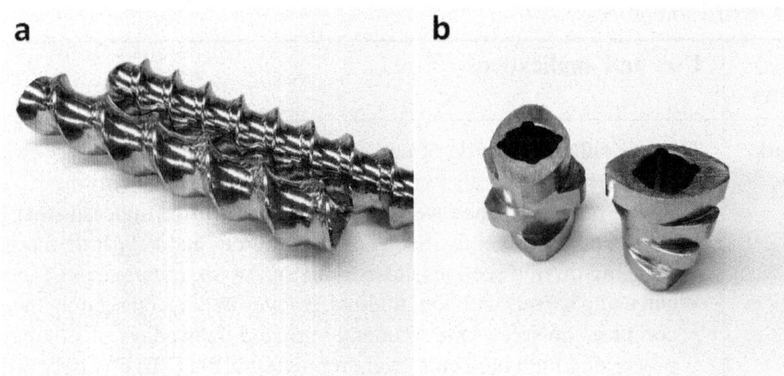

Fig. 3.7 Mixing elements.

profile with strategically placed slot cuts across the flight tip. These slot cuts are designed to facilitate improved mixing by increasing the leakage flow within the extruder. This heightened flow dynamics enhances the residence time distribution of materials, which ultimately contributes to the overall quality and consistency of the extruded products. Through a series of experimental studies, the performance of the screw mixing element was thoroughly evaluated. The findings revealed that the combination of a steep pressure gradient and significant back mixing resulting from the screw mixing element led to an increase in residence time distribution. This indicates the potential for improved mixing and reaction kinetics within the system when utilizing this element (Ishikawa et al., 2002).

The extrusion cooking process has many uses and cutting-edge technology, and the extrusion cooking method is becoming more and more popular. The co-rotating twin-screw extruder is the most often used because of its exceptional performance. In a co-rotating twin-screw extruder, screw configuration is basically divided into four categories, with individual categories having different roles in the extrusion process. They are discussed in Table 3.1 (Potente et al., 1994).

3.3.7 Die

It is located toward the end part of the barrel and directly attached to the screw's end through which the product is coming out. Different types of die configurations exist based on our requirements to get the desired shape, which not only influence the product's shape but also control the physical properties like surface texture, expansion, and density of the end product. When the product emerges from the die, it undergoes rapid expansion due to a sudden pressure drop. Consequently, the product takes on a shape that differs from the dies. The product's shape is primarily influenced by the length ratio to the die orifice's cross-sectional area. The removal of dies is compulsory after every run of the extruder to clean the die properly before proceeding to another run. The die plate is machined in a double cone shape for a smooth flow of material; it helps prevent the clogging of the material (Senanayake and Clarke, 1999).

Table 3.1 Extrusion process.

Screw categories	Uses and applications
Low work screw	There is either very little or no compression applied by these screws; therefore these are also known as zero-compression screws. They are preferred when working with heat-sensitive materials that need to be compounded gently. Such screws provide relatively little mixing, but by using mixing sections close to the screw tip, mixing can be increased. When using barrels with longitudinal grooves in the feed section, these zero-compression screws are frequently utilized. Nowadays, it's thought that processing high molecular weight poly ethylene (PE) and poly propylene (PP) at high throughput rates requires these grooved barrel assemblies.
Standard screw	Standard screws are employed for compounding and melting operations because they are made with a standard compression ratio. The so-called standard screws' designs differ among machine builders.
High-shear screw	Screw and barrel assemblies often operate in extremely harsh conditions, which can lead to serious wear issues. Temperature variations in the screw channel will increase, and the screw's pumping effectiveness will decline as the clearance between the flight and the barrel increases. This screw is typically made of high-quality tensile steel, like AISI 4140, that has been oil-quenched, tempered, or hardened by induction heating; this steel resists torque and distortion. High shear rate-generating components are used in these screws to enhance dispersion behaviors. This is accomplished, in the case of co-rotating twin screws, by using kneading blocks and backward-conveying elements. High shear rates are produced by kneading blocks, especially when partial filling is used.
Vented screw	These are mainly used in the polymer extrusion process when leftover monomers or other reaction-related by-products need to be removed from the melt flow. Vented screws can be utilized to remove gases from the polymer melt.

3.3.8 C Chamber

It is an accessory used in counterrotating twin-screw extruders. It resembles the alphabet "C," which acts as an isolated channel segment to hold the two screws together by interlocking between the intermeshing due to the presence of material, resulting in the development of pressure inside the barrel, which is essential for extrusion cooking. The "C"-shaped chamber's shear rate and throughput are determined by the distance between the two screws. Higher shear rates from smaller gaps can enhance homogenization and mixing, but can also weaken materials that are susceptible to heat. Larger gaps result in lower shear rates, which are gentler on the material but might not mix or melt the materials held in this portion as fully (Ravikumar et al., 2021).

3.3.9 Formulas or equations for designing the components of twin-screw extruder (Sobowale et al., 2017)

- Barrel length $L_b = \frac{L}{D} \times D$
  L is the screw length
  D is the screw diameter

- The height of beam $Y_{\max} = \frac{P_0 b^a}{192EI}$
  $Y_{\max}$ is the height of the beam
  P_0 is the initial circle pitch diameter
  b is the thickness of the beam
  E is the flight width
  I is the face length of the beam

- The overall power consumption (kW) $P_t = P_s + V_d \Delta P$
  P_s is the amount of power utilized during the dissipation of viscous material related to feed shear dynamics (kW)
  V_d is the speed diameter
  ΔP is the change in the pressure (Nmm^{-2})

- The screw power rotations (rpm) $N_p = \frac{P_s}{\rho L D^4 N^s}$
  P_s is the amount of power utilized during the dissipation of viscous material related to feed shear dynamics (kW)
  ρ is the density of the extrudate (kg)
  N is the screw speed (rpm)
  D is the screw diameter
  L is the screw length

- The speed ratio (rpm) $N_r = \frac{D_2}{D_1}$
  D_1 is the driving pulley diameter
  D_2 is the driven pulley diameter

- $N_1 D_1 = N_2 D_2$
  N_1 is the driving pulley speed (rpm)
  D_1 is the driving pulley diameter
  N_2 is the driven pulley speed (rpm)
  D_2 is the driven pulley diameter

- The barrel length $B_l = \pi D_2 N_2$
  $\pi = 3.142$
  D_2 is the driven pulley diameter
  N_2 is the driven pulley speed (rpm)

- The width of flight (%)
  D is the screw diameter

- The radial flight clearance (%) $\in = \frac{10D}{100}$
  D is the extruder barrel's the inside diameter

- The radial flight clearance (%) $\delta_f = 0.1 - \frac{0.3D_b}{100}$
  D_b is the extruder barrel's the inside diameter

- The helix angle at the root of the screw (degrees (°)) $\theta_s = \theta_b + 2.8^{\circ}$
  θ_b is the helix angle at the root of the bolt (degrees (°))
- The channel width at the root of the screw $W_s = W_b - 0.003$
  W_b is the channel width at the root of the bolt
- $W_p = M_p$
  W_p is the pulley weight (N)
  M_p is the mass of the pulley (kg)
- The volume of the hopper (m^3) $V = \pi \Delta r^2 h$
  Δr is the shaft radius difference
  h is the shaft height
- The shaft diameter $D_b = \sqrt[a]{\frac{16T}{\pi r}}$
  T is the permissible shear stress of the shaft (Nm)
  r is the shaft radius
- The extruder barrel's the inside diameter $D_b = \sqrt{\frac{m}{2.3NG\,H_m}}$
  m is the mass flow rate (kg/h)
  N is the screw speed (rpm)
  H_m is the channel depth of metering
  G is the dimensionless parameter used to characterize the geometric relationship between the screw and the barrel (dimensionless quantity)
- The dynamic bearing capacity of thrust bearing (kN) $C_{req} = f_d F_{ax} \left[\frac{60N\,L_f}{10^6}\right]^{\frac{3}{10}}$
  f_d is the factor for sense of rotation (dimensionless quantity)
  L_f is the life duration the bearing
  N is the screw speed (rpm)

3.4 Modeling and simulation

3.4.1 An overview of modeling and simulation techniques

The modeling and simulation technique is the most trending tool nowadays to solve experimental problems and develop mathematical equations. Creating novel extruded food items and enhancing the consistency and quality of existing food products, automatic control over food extrusion is essential. To effectively implement automatic control, one must understand the dynamics of the system and the relationship between controllable variables and product quality.

A model is a practical application of a statistical equation that uses the relationship between independent and dependent variables unique to a process or product to be mathematically represented. Models have the ability to predict actual situations and offer real economic benefits. They possess the inherent capacity for parametric transferability and reusability. As a result, food process modeling saves significant time and operational costs by enabling the investigation of product properties under both extreme and intermediate process conditions without requiring the use of real material quantities. There has been an increase in the modeling of process phenomena in food systems with the invention of extremely efficient extruders for food processing. Due to the multidimensional nature of food extrusion processes, modeling them has proven to be a challenging issue. As a result, most studies focus on comprehending and forecasting the changes that occur in the extruder and their potential effects on product quality in particular food formulations. This is because it makes it simple to create a mathematical relationship between input and output variables and their attributes through a step-by-step process with a few experimental runs. Central composite rotatable design in the response surface method modeling approach is the most commonly utilized technique in food systems. Since it is completely observation-based, it is more accommodating to operations involving several components, phases, and mechanics. It has been effectively used in the development, enhancement, and optimization of biological processes in which controllable factors such as temperature, screw speed, moisture content, and feed composition affect response variables either separately or in combination (Danbaba et al., 2018).

Some of the factors that make an extruder difficult to control include the absence of precise mathematical models that explain food extrusion, the system's time-varying and nonlinear nature during the extrusion process, and the scarcity of trustworthy online sensors that can identify state variables like viscosity and moisture content. Recently, there has been a focus on using adaptive control and online system identification techniques, such modeling and simulation, and machine learning, to address these issues (Moreira et al., 1990). The field of modeling and simulation involves creating a computational representation of the physical and chemical processes of extruders. This helps predict their characteristics under various operating conditions and optimize parameters for high-quality products and increased efficiency. The following are the two types of models:

- **The black-box model**: It uses statistical analysis to develop equations based on parameters like geometry, ingredient types, and machine set points.
- **Comprehensive model**: This considers internal conditions and thermochemical reactions inside the extruder, taking into account the pressure, shear rate, and temperature as functions of residence time to develop the generalized model (Tayeb et al., 1989).

3.4.2 Material modeling and its importance in twin-screw extrusion

The theoretical approach to modeling the twin-screw extruder is divided into four sections: a conveying section, a melt pumping section, a shearing section, and a die. The models are then set in series for the computation model (Tayeb et al., 1989). The dimensional analysis method is often used to reduce physical problems to their simplest form for the engineering process. This theory includes the Rayleigh method and the Buckingham Pi method. Physical variables like screw speed, barrel, moisture content, and temperature are used to develop the physically relevant equation and are expressed in terms of independent fundamental quantities. It can be challenging to describe the generated empirical model in an accessible format. Using dimensional analysis with the Buckingham Pi theory can provide a concise way to present the data, which will help in developing a theoretical model to solve the problem (Cheng and Friis, 2010).

3.4.3 Process modeling and its importance in twin-screw extrusion

In the process modeling of the twin-screw extruder process, it is essential to adopt a comprehensive approach. This involves placing significant emphasis on understanding the intricacies of transporting raw solid material, the process of material melting, and the complex behavior of molten-state material during flow. However, it is evident that there is a lack of clarity surrounding the phenomenon of material transportation and melting. Therefore it becomes imperative to develop a model that can effectively elucidate and explain the melting phenomena, as it serves as the cornerstone for achieving accurate global process modeling of the twin-screw extruder. The intermeshing co-rotating extruder, which is identical to the single-screw extruder in most cases, is the primary model used to study solid transport in twin-screw extrusion. There are two transport processes in this model, which are as follows:

- Transportation in the screw channel
- Transportation in the intermeshing area

The existing solid transportation model for co-rotating twin-screw extruders provides valuable insights into the prediction of screw filling based on geometry and process operating parameters. However, it is important to note that there is currently a scarcity of research on solid transport in counterrotating machines (Lewandowski and Wilczyński, 2022).

3.5 Material behavior in twin-screw extrusion

The material properties taken into account during the development of the model are rheological properties, flow behavior, heat transfer, mass transfer, density, viscosity, thermal conductivity, and heat capacity. The rheology of the material, the screw's design and configuration, and the processing parameters are some of the variables that affect the complicated behavior of materials in a twin-screw extruder. The material goes through several conveying, melting, mixing, and devolatilization processes as it passes through the extruder. The screws in the feed portion move the solid material into the barrel from the hopper. After that, as the material passes through the melting zone, the barrel and screw temperatures rise, and the material melts. The molten material is subsequently put through a vigorous shearing and mixing process in the extruder's kneading and dispersive stages. This aids in homogenizing the melt, dispersing additives, and breaking down agglomerates. The particular screw elements, such as KBs and reverse elements, that are utilized in these sections are critical in establishing the degree of mixing and shear that is applied to the material. The melt is forced through the die at a regulated pressure in the extruder's last segment. The screw design, the viscosity of the substance, and the die geometry all affect the pressure profile in the extruder. For the melt flow at the die exit to be steady and constant, proper pressure control is necessary. The temperature and viscosity of the material vary dramatically during the extrusion process. It is essential to comprehend and manage these modifications to maximize the extrusion process and ensure product quality.

3.5.1 Rheological properties of materials

Rheology is the study of matter flow and deformation, which is directly related to material viscosity. Viscosity is an important characteristic that determines the ideal operating parameters to manufacture high-quality goods. The viscosity of the material in the extruder is determined by its composition, moisture content, and particle type. During the process, materials having high moisture content cause gelatinization of starch and produce a viscous mass due to high temperature, which increases the viscosity. However, due to intense shearing action, the breakdown of macromolecules into smaller units results in reduced viscosity (Fellows, 2022).

3.5.2 Flow behaviors in the extruder

Flow behavior is completely raw material-specific; materials including starch, protein, and polymer melts produce various kinds of flow even though the same operating conditions are maintained in the same extruder. This is because of differences in the quantity and quality of starch, protein, moisture, and other additives that produce different viscosities and flow characteristics. A common ingredient in many materials that are processed in twin-screw extruders is starch and protein. The kind and caliber of starch can have a big effect on how the flow behaves. Higher amylose starches, for instance, typically result in more viscous pastes, which may impact the processing-related flow rate and pressure needed. Flow behavior and rheological properties are interrelated. Proteins can also change a material's viscosity and rheological characteristics, which can have an impact on flow behavior. Increased viscosity from a higher protein concentration can make the material more challenging to process (Sathyanarayana et al., 2013).

A research gap exists in the modeling of melt flow in a twin-screw extruder. Drag flow, which is also observed in single-screw extruders, is the mechanism of melt flow in co-rotating twin-screw extruders. Prior to developing a non-Newtonian flow model using the flow analysis network and 3D finite element method simulations, the co-rotating twin-screw extruder first develops a Newtonian flow model in the fully filled elements model. This model is then used to evaluate the velocity and temperature distributions, as well as the pressure and flow rate relationships in twin-screw extrusion. However, these simulations require significant computing resources and time, making them unsuitable for global process modeling involving hundreds of computational iterations. To overcome this challenge, screw pumping characteristics were developed to more effectively characterize flow behavior. These properties, which are defined by dimensionless flow rate and pressure gradient, can be illustrated using regression models. Reasonable computing speed and accuracy can be achieved by incorporating these features into iterative processes. As a result, knowledge of screw pumping characteristics is now essential to understanding how twin-screw extruders flow (Lewandowski and Wilczyński, 2022).

3.5.3 Heat and mass transfer phenomena

The design of extrusion and scale-up processes depends heavily on mass and heat transfer mechanisms. Because the contents inside the barrel flow in nonNewtonian and nonisothermal ways, it is difficult to model heat and mass transfer in food extruders. The primary challenge in creating a model for heat transfer lies in accounting for all variables involved in the extrusion process. However, due to computational complexities, it is practically impossible to consider every single variable. As a result, it becomes necessary to exclude insignificant variables based on certain considerations. This means deriving a solution that is used under specific operating conditions by making some assumptions. However, it is not considered a generalized model. Even though some models explain the heat transfer mechanism better under certain considerations, they are limited only to the metering section, which also assumes the material should be in the molten state. Due to the greater complexities involved in the feeding and compression sections, no work has been done on this mechanism to date (van Zuilichem et al., 1990).

Heat transfer coefficients: For the energy balance development, three types of heat transfer coefficients (HTCs) are required, which are as follows:

HTC between screws and melt (α_b)

HTC between melt and barrel (α_b)

HTC between environment and barrel (α_e)

The barrel absorbs the heat dissipated when the screw section is fully loaded with raw material. Therefore the heat transfer to the barrel needs to balance the viscous dissipation. It is directly proportional to the square of the screw speed and the melt viscosity (Eitzlmayr et al., 2014).

$$(\eta \times n^2) \propto \{\alpha_b \times A_b m \times (T_m - T_b)\}$$

where,

η is the melt viscosity

n is the screw speed

α_b is the HTC between melt and barrel

A_{bm} is the barrel surface area

T_m is the melt temperature

T_b is the barrel temperature

3.6 Challenges and future trends

3.6.1 Current challenges in twin-screw extruder design

Temperature, mechanical stress, and residence time are the primary parameters influencing final product quality. Prolonged exposure to high temperatures may damage the nutritional quality of bioactive compounds such as proteins, starches, and sugars. Conversely, insufficient residence time may prevent heat-resistant compounds from reaching their melting point. Since these compounds must attain a molten state to undergo the necessary chemical transformations, inadequate processing conditions can compromise product quality and increase material waste. Some smooth-textured materials lose their complete texture profile due to mechanical stress during the extrusion process; along with this, sensitive vitamins and minerals may deteriorate. It is difficult to accurately simulate the rheological behaviors of the materials being processed, particularly in food extrusion, complex polymer melts, and nonNewtonian fluids. The impacts of shear, temperature, and other processing parameters must be included for the models to produce accurate predictions. For consistent product quality, it's critical to accurately forecast the flow patterns and pressure profiles inside the extruder barrel and die. Experts are working to create sophisticated computational fluid dynamics (CFD) models that can represent these intricate flow events. These are some of the challenges faced by twin-screw extruders. The primary objective for the modeling and design process is to optimize design parameters such as residence time, temperature, mechanical stress, and the reactive extrusion process in such a way as to develop high-quality end products. The reactive extrusion process presents significant challenges due to the complexities involved in managing chemical reactions and maintaining specific conditions during extrusion. Understanding all aspects of this process and elucidating the interactions between numerous variables is particularly challenging. While numerical modeling offers an efficient solution to address these difficulties, implementing it in twin-screw extrusion remains a notable challenge (Vergnes and Berzin, 2006).

Developing a twin-screw extruder for plasma fueling applications faces challenges due to the complexities of creating multiple design iterations that meet specific requirements. The primary goal is to minimize the volume and viscous dissipation rate to achieve optimal output. Modeling is complicated due to the intermeshing of the screws and the nonNewtonian behavior of the solid material, resulting in nonlinear flow characteristics (Ravikumar et al., 2021). The flow in partially filled screws and the change from granular to molten state have not yet been subjected to well-established, comprehensive simulation techniques (Eitzlmayr et al., 2014).

3.6.2 Emerging trends and technologies

Continuous innovation is central to advancing twin-screw extrusion technology, and accepting the need to upgrade and modify equipment is essential for the productive development of food items, ensuring the preservation of tastes, nutrients, and functional qualities. Twin-screw extruder performance and efficiency have increased due to advancements in screw design and the materials used for the construction of parts of the extruder. The design of the screw mixing element, the screw driving torque, and the screw channel depth have all evolved, greatly increasing the throughput capacity of today's twin-screw extruders compared to earlier designs. The utilization of vacuum feed technology and high-torque twin-screw extruders can further aid in improving throughput efficiency. Here, we discuss some of the most trending technologies of twin-screw extruders, which are as follows:

1. Hot melt extrusion is an advanced processing technology that involves the application of heat and pressure to mix and melt a formulation. This process improves the stability and digestibility of food products by preserving their bioavailability. In addition, hot melt extrusion allows for precise control over the inactivation of enzymes, microorganisms, and antinutritional properties, thereby enhancing the overall quality and safety of the product.
2. Supercritical carbon dioxide is used in the extrusion process in the twin-screw extruder to act as a leavening agent, which shortens production times and yields consistent final products (Prabha et al., 2021).
3. A twin-screw extruder that rotates in synchronization with the 3D printer. Extrusion-based 3D printing, another technological first, involves the layer-by-layer extrusion of foodgrade or edible printing material from a heated or unheated nozzle. The extruded product is heated above its melting temperature for the easy adhesion of the printing material (Agunbiade et al., 2022).

3.7 Future directions for research and development in this technology

Extruded food products are becoming more and more popular in the market, in terms of both consumption and commercialization. Twin-screw extruders are in high demand; thus it is critical to overcome the drawbacks to optimize the variables, produce more, and create higher-quality products. Modeling and designing twin-screw extruders with the help of advanced methods like artificial intelligence, deep learning, and machine learning algorithms might be advantageous.

One ground-breaking technique that can propel the expected improvements in the twin-screw extrusion process is modeling. More precise and effective models can be developed through methods like process optimization and enhanced computational models. Artificial intelligence and machine learning can be used to forecast results based on input parameters and optimize extruder performance. Using innovative techniques for in-process material assessment in the extruder is another goal to obtain a thorough understanding of the material's behavior under various processing conditions. This entails integrating high-precision sensors with modern data-gathering technologies to dynamically control and continually monitor the process. Future work should focus on maximizing screw design and configuration to work with various materials and accomplish specific product attributes. To maximize longevity and improve performance under various operating conditions, the study will also investigate new screw geometries and materials. To reduce the impact on the environment, we will focus on developing designs and processes that use less energy. Furthermore, we are focused on developing methods for producing laboratory results on an industrial scale without sacrificing effectiveness or quality. We are exploring how twin-screw extruders might be used in a variety of industries, including bioplastics, food processing, and pharmaceuticals. To address these intricate twin-screw extrusion challenges, mechanical engineers, material scientists, chemists, and data scientists are collaborating. This includes the design and optimization of complete extrusion processes through the application of systems engineering techniques. Creating incredibly effective extruders that can handle recycled materials is essential to supporting efforts toward a circular economy. In addition, to comprehend the properties of sustainable and biodegradable feedstocks for twin-screw extruders, a thorough investigation into these feedstocks is necessary. Furthermore, we aim to create comprehensive training courses for engineers and operators to improve their knowledge of twin-screw extrusion technology. To support hands-on learning and experimentation for the manufacturing of extruder products, we also need to focus on developing advanced simulation tools and augmented reality environments.

3.8 Conclusion

Twin-screw extruders' modular screw design gives them a great deal of versatility. Simple conveying to sophisticated reactive extrusion can be accomplished with these machines by carefully selecting and organizing the screw parts. The superior mixing capabilities of the intermeshing co- or counterrotating screws make them appropriate for procedures requiring a constant blending of ingredients. This is particularly useful for the food, pharmaceutical, and plastics sectors. Extruder precision in regulating temperature and shear guarantees that materials are processed under ideal circumstances, minimizing deterioration and optimizing the intended transformations. The barrel's sophisticated heating and cooling mechanisms aid in this control effortlessly. The utilization of CFD and additional simulation instruments is highly valuable in precisely forecasting the behavior of the material inside the extruder. Improved design and operational strategies result from an extensive comprehension of the complex interactions between screw design, material properties, and processing conditions made possible by these models. Experiments must be conducted to validate simulation models, which in turn ensures the accuracy and reliability of the models. Through iterative testing and refining procedures, twin-screw extruder performance may be precisely tuned to meet specific application needs. Due to their versatility, twin-screw extruders are widely used in a variety of commercial sector industries. These are essential to the food sector because they enable the production of a vast range of goods, from wholesome pet feeds to delicious snacks. These multipurpose devices are also essential for compounding and processing materials in the plastics industry. They also make it possible to continuously manufacture and treat solid dispersions in the pharmaceutical business. Twin-screw extruder design and operation require an extensive knowledge of process dynamics, material behavior, and mechanical design concepts. Outstanding performance and efficiency can be attained in a variety of industrial applications by utilizing advanced modeling and simulation approaches along with experimental validation.

References

Agunbiade, A.O., Song, L., Agunbiade, O.J., Ofoedu, C.E., Chacha, J.S., Duguma, H.T., Hossaini, S.M., Rasaq, W.A., Shorstkii, I., Osuji, C.M., Owuamanam, C.I., Okpala, C.O.R., Korzeniowska, M., Guine, R.P.F., 2022. Potentials of 3D extrusion-based printing in resolving food processing challenges: a perspective review. J. Food Process Eng. 17454530. 45 (4). Available from: http://onlinelibrary.wiley.com/journal/10.1111/(ISSN)1745-4530. doi:10.1111/jfpe.13996.

Cheng, H., Friis, A., 2010. Modelling extrudate expansion in a twin-screw food extrusion cooking process through dimensional analysis methodology. Food Bioprod. Process. 09603085. 88 (2-3), 188–194. doi:10.1016/j.fbp.2010.01.001.

Danbaba, N., Nkama, I., Badau, M., Idakwo, P., 2018. Statistical Modeling and Optimization of Processing Conditions of Twin-Screw Extruded Rice-Legume Instant Breakfast Gruel. Arid Zone J. Eng. 14 (4), 693–712.

Dhaval, M., Sharma, S., Dudhat, K., Chavda, J., 2022. Twin-Screw Extruder in Pharmaceutical Industry: history, Working Principle, Applications, and Marketed Products: an In-depth Review. J. Pharm. Innov. 19398042. 17 (2), 294–318. Available from: https://link.springer.com/journal/12247. doi:10.1007/s12247-020-09520-7.

Eitzlmayr, A., Koscher, G., Reynolds, G., Huang, Z., Booth, J., Shering, P., Khinast, J., 2014. Mechanistic modeling of modular co-rotating twin-screw extruders. Int. J. Pharm. 03785173. 474 (1-2), 157–176. doi:10.1016/j.ijpharm.2014.08.005.

Fellows, P.J., 2022. Food Processing Technology: principles and Practice Food Processing Technology: principles and Practice. Elsevier 1–779. Available from: https://www.sciencedirect.com/book/9780323857376. doi:10.1016/C2019-0-04416-0.

Ishikawa, T., Amano, T., Kihara, S.I., Funatsu, K., 2002. Flow patterns and mixing mechanisms in the screw mixing element of a co-rotating twin-screw extruder. Polym. Eng. Sci. 00323888. 42 (5), 925–939. Available from: http://onlinelibrary.wiley.com/journal/10.1002/(ISSN)1548-2634. doi:10.1002/pen.11002.

Lewandowski, A., Wilczyński, K., 2022. Modeling of Twin Screw Extrusion of Polymeric Materials. Polymers. 2073-4360. 14 (2), 274. doi:10.3390/polym14020274.

Moreira, R.G., Srivastava, A.K., Gerrish, J.B., 1990. Feedforward control model for a twin-screw food extruder. Food Control. 09567135. 1 (3), 179–184. doi:10.1016/0956-7135(90)90009-2.

Potente, H., Ansahl, J., Klarholz, B., 1994. Design of Tightly Intermeshing Co-Rotating Twin Screw Extruders. Int. Polym. Process. 2195-8602. 9 (1), 11–25. doi:10.3139/217.940011.

Prabha, K., Ghosh, P., S, A., Joseph, R.M., Krishnan, R., Rana, S.S., Pradhan, R.C., 2021. Recent development, challenges, and prospects of extrusion technology. Future Foods. 26668335. 3. Available from: www.journals.elsevier.com/future-foods. doi:10.1016/j.fufo.2021.100019.

Ravikumar, P.S., Arumugam, S.K., Gangradey, R., Mukherjee, S., Srinivasan, K., Sadasivan, S., Gupta, V., Aggarwal, M.C., 2021. Development of Cryogenic Extrusion Techniques and Modelling of a Twin Screw Extruder: a Review. J. Fusion Energy. 15729591. 40 (1).

Available from: http://www.kluweronline.com/issn/0164-0313/. doi:10.1007/s10894-021-00297-2.

Sathyanarayana, S., Olowojoba, G., Weiss, P., Caglar, B., Pataki, B., Mikonsaari, I., Hübner, C., Henning, F., 2013. Compounding of MWCNTs with PS in a twin-screw extruder with varying process parameters: morphology, interfacial behavior, thermal stability, rheology, and volume resistivity. Macromol. Mater. Eng. 14392054. 298 (1), 89–105. doi:10.1002/mame.201200018.

Senanayake, S.A.M.A.N.S., Clarke, B., 1999. A simplified twin screw co-rotating food extruder: design, fabrication and testing. J. Food Eng. 02608774. 40 (1-2), 129–137. doi:10.1016/s0260-8774(99)00049-7.

Sobowale, S.S., Adebo, O.A., Adebiyi, J.A., 2017. Development of a twin screw extruder. Agric. Eng. Int. CIGR J. 16821130. 19 (4), 181–186. Available from: http://www.cigrjournal.org/index.php/Ejounral/article/viewFile/4254/2623.

Tayeb, J., Vergnes, B., Valle, G.D., 1989. A Basic Model for a Twin-Screw Extruder. J. Food Sci. 17503841. 54 (4), 1047–1056. doi:10.1111/j.1365-2621.1989.tb07941.x.

van Zuilichem, D., van der Laan, E., Kuiper, E., 1990. The development of a heat transfer model for twin-screw extruders. J. Food Eng. 02608774. 11 (3), 187–207. doi:10.1016/0260-8774(90)90027-6.

Vergnes, B., Berzin, F., 2006. Modeling of reactive systems in twin-screw extrusion: challenges and applications. Comptes Rendus. Chimie. 1878-1543. 9 (11-12), 1409–1418. doi:10.1016/j.crci.2006.07.006.

Yacu, W.A., 1985. Modeling a twin screw co-rotating extruder 1. J. Food Process Eng. 0145-8876. 8 (1), 1–21. doi:10.1111/j.1745-4530.1985.tb00095.x.

Reactive extrusion processing

4

Kishore Kumar Gopalakrishnan, Poonam Singha, and Sushil Kumar Singh
Department of Food Process Engineering, National Institute of Technology Rourkela, Rourkela, Odisha, India

4.1 Introduction

Extrusion is a versatile manufacturing process utilized to produce various types of food products, such as snacks, ready-to-eat cereals, confectionery items, texturized protein products, and pet foods. It involves continuous kneading, mixing, and shaping of ingredients through a specialized die to create products with consistent textures and shapes (Zhao et al., 2011). The extrusion process occurs in three types of extruders: corotating, counterrotating, and twin-screw extruders. Extrusion is widely utilized as a thermomechanical process due to its adaptability, flexibility, rapidity of manufacturing, affordability, and energy efficiency. This method allows for the continuous production of a wide range of food products and the effective utilization and value addition of by-products from industries, offering numerous advantages in terms of scalability and customization to meet diverse consumer demands (Dey et al., 2021).

Reactive extrusion was developed in the 1980s and is generally used for the production and modification of polymers. It is a widely used technique that combines the conventionally distinct chemical processes of polymer synthesis and modification with extrusion processes, such as melting, mixing, structuring, devolatilization, and shaping, into a single process executed within the extruder (Maliger and Halley, 2014). The process alters the extrudate's characteristics by combining mass transfer, heat, and chemical interactions (Gutiérrez and Alvarez, 2017). Continuous reactant feeding via a heated shear mixer is made possible by an extruder. This mixer creates high pressure and shear, which facilitate reactions that are difficult to accomplish under standard air mixing conditions (Tremaine et al., 2014). The enzymes used in the reactive extrusion process efficiently act as biocatalysts on substrates with high levels of structural strength, polymerization, and crystallinity. By combining thermal and mechanical energy, this process enhances enzymatic reactions and transforms high-molecular-weight products into low-molecular-weight polymers, which have significant industrial applications (Xu et al., 2020). The extruder offers significant operational flexibility, accommodating a wide range of processing conditions, including temperature (70–500°C) and pressure (0–500 bar). It features multiple injection ports and allows for control over residence time and mixing intensity (Xie et al., 2006).

Extrusion Cooking for Food and Feed. https://doi.org/10.1016/B978-0-443-29987-2.00004-X

Functional food-based ingredients are now being prepared using various novel technologies, such as bioreactors. Functionalization of food ingredients is achieved through extrusion technology, where the extruder can act as a continuous bioreactor. The food material flows through the metal barrel of the extruder. To facilitate the transportation and shearing of the food stuff, the metal barrel is equipped with a single or double screw. As the material flows through the barrel, the shearing action and pressure generate friction, causing high-temperature development and resulting in the cooking of the material. During the process, various physical transformations and chemical reactions take place in the feed, through which the functionalization is achieved (Singh and Muthukumarappan, 2017). The functional properties are imparted, altered, and modified (Pavani et al., 2024). These reactions are directed by reactive extrusion by adding reactants to the feed. These reactants may be catalysts, modifiers, or other agents that induce or enhance chemical reactions during the extrusion process (Arora et al., 2020). Different kinds of extruded products produced through reactive extrusion and the chemical reactants added in the extrusion process are listed in Table 4.1. Extruder parameters such as barrel length, residence time, and screw speed are critical factors that influence the chemical reactions occurring in the extruder. It is necessary to fully comprehend the extruder settings and reaction kinetics to control these physical changes and chemical reactions. In this chapter, the

Table 4.1 Extruded products developed through the reactive extrusion process.

Extrudate product	Reactants	References
Thermoplastic starch/Poly lactic acid blends	Maleic anhydride (MA) with a dicumyl peroxide (DCP) initiator	(Wang et al., 2007)
Graft copolymers of starch and polyacrylamide	Ammonium persulfate	(Willett and Finkenstadt, 2003)
Native and oxidized corn (*Zea mays*) starch/ polystyrene	Zinc octanoate	(Gutiérrez and Alvarez, 2017)
Chemical modification of starch	Citric acid	(Ye et al., 2019)
Esterified sorghum starch	Acetic acid	(García-Gurrola et al., 2019)
Starchy noodles (spaghetti-type pastas)	Sodium propionate	(Hernandez-Hernandez et al., 2022)
Novel maleated thermoplastic starch (MTPS)	Maleic anhydride (MA)	(Raquez et al., 2008)
Microcrystalline cellulose extracted from soybean hulls	NaOH and H_2SO_4	(Merci et al., 2015)
Starch acetates with a high degree of substitution (DS)	Acetic anhydride using NaOH as the catalyst	(Xu et al., 2004)

Extrudate product	Reactants	References
Polystyrene-grafted starch, homopolymer of polystyrene, and starch	Benzoyl peroxide and $K_2S_2O_8$	(De Graaf and Janssen, 2000)
Soy protein modified with a functional monomer	Maleic anhydride, glycidyl methacrylate, and styrene	(Liu et al., 2008)
Polylactic acid (PLA)/ cellulose nanocrystal (CNC)-based cast films	Dicumyl peroxide (DCP) and compatibilizers such as carbon nanofibers (CNFs), nanosilica, alumina, montmorillonite (MMT), and cellulose	(Dhar et al., 2017)

principles, kinetics involved, and the physicochemical transformations that take place in food material, such as gelatinization, polymerization, retrogradation, crystallization, esterification, and hydrolysis in the reactive extrusion process, are discussed, and critical parameters affecting the extrusion process, along with the changes that take place in starch, protein, lipids, dietary fibers, and bioactive components, are also described.

4.2 Principle, kinetics, and rheology

Reactive extrusion is a process of enhancing the chemical reactions of polymers or other materials by utilizing the conditions of the extruder. The shearing action provides intensive mixing, and the high temperature facilitates faster chemical reactions (Moad, 2011). The extruder can maintain and control precise reaction temperatures, which accelerates reaction kinetics, as different types of products require specific temperatures. In reactive extrusion, uniform mixing is achieved, which facilitates better interaction between the reactants, ensuring a consistent and thorough reaction throughout the material. In a typical reactive extrusion process, reactants are added along with the feed through the hopper. The reactants are then heated to initiate the chemical reactions (Beyer, Hopmann and Cohnen, 2017). In the screw extruder, the material passes through three zones: the primary feed zone, the melt zone, and the pump zone. As the material passes through the extruder, it experiences laminar flow inside the screw channels, pressure-induced flow (back pressure created by the die fixed at the end), leakage flow (between the barrel wall and the edge of the screw flight), and drag flow, which results from the reduced slip of material at the barrel wall (Xie et al., 2006).

In the extruder, due to the pressure, temperature conditions, and mechanical actions, the feed material undergoes structural and chemical changes, such as gelatinization, protein denaturation, lipid oxidation, degradation of phytochemicals and vitamins, and an increase in dietary fiber solubility. These reactions are governed by reaction time, reaction order, and rate constant. Kinetic studies are crucial for controlling extrusion operations and predicting product quality. The temperature of the material, its moisture content, and the shear stress applied to it are some of the variables that affect the rate constant during food extrusion (Singh and Muthukumarappan, 2017). The extrusion process can be optimized to maximize desired reactions and minimize undesirable ones by understanding how the rate constant relates to these variables (Zhao et al., 2011).

The compound concentration follows a general rate law:

$$-\frac{dC}{dt}=kC^{n} \tag{4.1}$$

where k denotes the rate constant, t represents the reaction time, C stands for the concentration, and n indicates the reaction order.

This equation provides information on the rate of chemical reaction in relation to reactant concentration. This formula helps in understanding how concentration of reactants changes over time. Reaction rate ($\frac{dC}{dt}$) depends on reaction rate constant (k) and reaction order (n).

The temperature has an influence on the reaction rate; the relationship is given by the Arrhenius equation:

$$k=k_0\ exp\left(\frac{-E}{RT}\right) \tag{4.2}$$

where T denotes the temperature, R denotes the universal gas constant, E denotes the activation energy, and k_0 denotes the preexponential factor. In extrusion, the temperature of the material is often high and varies along the extruder; this equation helps us understand the rate of chemical reactions under different temperature conditions.

The following formula is used to determine the extruder's residence time ($\bar{t}$):

$$\bar{t}=\frac{L}{u} \tag{4.3}$$

where L denotes the axial length of the section and u stands for the axial velocity of the material.

The axial velocity is calculated by

$$u=\pi DN\frac{\tan\,\theta\ \ \tan\,\phi}{\tan\,\theta+\ \ \tan\,\phi} \tag{4.4}$$

where N represents the screw's rotational speed, D its diameter, ϕ its pitch angle, and θ its traction angle (Zhao et al., 2011).

Lipid oxidation in pet foods, snacks, and dry foods follows zero-order kinetics, and the reaction rate is unaffected by the concentration of reactants (Lin et al., 1998):

$$C = C_0 \pm kt \tag{4.5}$$

where C_0 represents the quality factor's initial concentration, C represents the quantity of the quality factor acquired or lost, K is the reaction rate, and t is the reaction time.

The rheological properties of food are significantly influenced by the reactive extrusion process. The viscosity of the food materials increases as the polymerization reaction progresses and decreases during depolymerization. The shear-thinning behavior of the non-Newtonian food products during reactive extrusion is defined by the Carreau–Yasuda equation, given by:

$$\eta = \eta_\infty + \left[(\eta_0 - \eta_\infty)\left(1 + (K\gamma)^a\right)\right]^{\frac{n-1}{a}} \tag{4.6}$$

where η_0 denotes the viscosity at zero shear rate, η_∞ denotes the viscosity at infinite shear rate, n is the power law index, and a and K are constants. The viscoelastic behavior of the food material is predicted by the above equation. Viscosity and viscoelastic properties directly reflect the chemical transformations occurring (Khalifeh and Clermont, 2005).

4.3 Physicochemical transformations in reactive extrusion

4.3.1 Polymerization

Polymerization is the process of producing polymers through the chemical conversion of one or more monomers. It involves linking a number of monomers in a covalent manner to form a polymer molecule or chain. Depending on the type, the monomers may require the presence of initiators or catalysts for the polymerization reaction (Li et al., 2021). Reactive extrusion is a powerful technique for developing numerous novel polymeric materials efficiently. It modifies the properties of existing polymers or is used to create new polymers (Gutiérrez and Alvarez, 2017). In reactive extrusion processing, the reactants are continuously sent along with the feed through the feed hopper. The extent of the polymerization reaction and reaction speed is determined by the temperature profile, use of activators, catalysts, and residence time of the material. The screw length and screw speed limit the degree of polymerization (Beyer, Hopmann and Cohnen, 2017). During the polymerization reaction, the viscosity increases with heating due to the rise in molecular weight throughout the process (Pietsch et al., 2017). Extrusion is used to modify starch by enhancing its functional properties, such as gelatinization behavior, water absorption capacity, and resistance to enzymatic degradation. In the reactive extrusion process for producing oxidized corn starch and polystyrene blends, the molecular weight increased due to crosslinking reactions in the presence of the catalyst zinc octanoate (Gutiérrez and Alvarez, 2017). Protein polymerization is another important aspect of reactive extrusion in food processing. For instance, in wheat gluten, raising the temperature above 75°C causes polymerization reactions through the formation of disulfide bonds between glutenin subunits. These reactions are induced by thermomechanical treatment, and the polymerization rate increases with temperature, ultimately resulting in increased complex viscosity (Pietsch et al., 2018). This process is particularly important in the production of high-protein extrudates. Reactive extrusion also facilitates the polymerization of sugars and other carbohydrates. Tremaine et al. (2014) investigated the polymerization of lactose using citric acid as the acid catalyst to produce oligosaccharides. An acid catalyst, along with heat, induces the polymerization of sugars. By precisely controlling the extrusion parameters and incorporating appropriate catalysts or initiators, food scientists can tailor the molecular structure of food ingredients to meet specific needs and consumer preferences.

4.3.2 Esterification

Esterification is a chemical reaction that forms esters as the primary product. In typical esterification, an organic acid (RCOOH) reacts with an alcohol (ROH), resulting in the formation of an ester (RCOOR) and water as products. This reaction is often catalyzed by acid catalysts to accelerate the process. The addition of hydrophilic ester groups (octyl succinic and acetyl groups) led to a decrease in resistance to melting and thermal transition by allowing water to more easily enter the interior of starch granules, further disrupting their orderly structure. Chemical modification of starch by esterification was carried out using citric acid, involving the reaction of starch with citric acid. The hydroxyl groups on starch molecules react with citric acid to form an anhydride, which is produced during the esterification process in reactive extrusion, to create citrate ester adducts. This esterification promotes crosslinking reactions, enhancing the structural integrity and digestibility profile of the starch (Ye et al., 2019). The extrusion process of sorghum starch opens up the starch molecule's structure due to the high temperature and pressure. This increases the contact between hydroxyl (OH) groups and other reactants, such as phosphate and acetyl groups, thereby enhancing the esterification rate of starch with the addition of acetic acid (García-Gurrola et al., 2019). The wood–plastic composites were formed by esterification through reactive extrusion. Functional properties and targeted modifications of the wood composites, including sander dust, cellulose, and hemicellulose, were achieved (Vaidya et al., 2016).

4.3.3 Gelatinization

During extrusion, gelatinization of starch occurs, which breaks down the strong intramolecular forces and causes the highly crystalline structure of starch to become disrupted and less ordered. This state ensures the proper blending of starch with other substances in the mixture (Wang et al., 2007; Singh and Muthukumarappan, 2016). Starch from different botanical origins has distinct structures in the starch molecules and double-helical crystalline structures organized inside the granules, leading to different gelatinization properties when heated in the presence of water. It also displays characteristic gelation temperatures, gelling ability, paste viscosity, clarity, and retrogradation rate (Ai and Jane, 2018). Extrusion generally takes place at a temperature of 40°C, and at this temperature, the initial melting of starch starts, which follows zeroth-order reaction kinetics, and then gelatinization occurs, following first-order reaction kinetics. Hence, starch gelatinization follows more than one reaction mechanism (Zhao et al., 2011).

4.3.4 Retrogradation

Retrogradation is greatly influenced by the structure of starch molecules. It is a process of gradual recombination of disordered amylose and amylopectin molecules into ordered structural molecules, leading to changes in the physical and functional properties of extruded products (Zhang et al., 2024). It is affected by the starch type, extrusion conditions, additives, or cross-linking agents. Changes such as texture, viscosity, and solubility are common in the extrudates. The texture changes can be either beneficial or undesirable; for instance, crispiness in snack food products is desirable, while some soft chew formation in the extrudates is undesirable. The changes in viscosity and solubility also affect the functional properties and final product applications. Native rice starch was gelatinized using improved extrusion cooking technology and retrograded after low-temperature storage (4°C). The retrograded starch showed a highly compact morphology (Zhang et al., 2014).

4.3.5 Hydrolysis

Hydrolysis in reactive extrusion processing refers to chemical reactions where molecules are broken down into smaller parts by the addition of water. The hydrolysis actions are defined as follows: acid-catalyzed hydrolysis, enzymatic hydrolysis, and mechanical hydrolysis processes. In acid-catalyzed hydrolysis processes, carboxylic acids such as ascorbic acid and citric acid are used to catalyze the hydrolysis process. In enzymatic processes, enzymes such as cellulase are used to catalyze the reactions. Mechanical forces, such as shear forces developed due to screw rotation and speed, also influence the hydrolysis process. Through hydrolysis, starch modification, such as an increase in molecular weight and the creation of materials with specific functionalities, can be achieved (Xie et al., 2006). In addition, polymer degradation processes, such as breaking down polylactic acid into smaller molecules, and the conversion of bio mass into constituent components like sugars and lignin for use in biofuel and other applications, are also facilitated by hydrolysis (Guiao et al., 2022; Wang et al., 2007).

4.3.6 *Crystallization*

Crystallization in reactive extrusion involves the formation of crystalline structures in the extruded material through the solidification and cooling of starch, fats, and sugar molecules. This crystallization process determines the texture, stability, and final quality of food products. During starch retrogradation, recrystallization of amylose and amylopectin molecules occurs, which increases digestibility, water-holding capacity, stability, and the molecular weight of the modified starch product through crosslinking and other chemical reactions (Huang et al., 2022). The crystallization process is influenced by the temperature profile of the extruder, which affects the textural properties of the product. Shear forces can also break molecular structures, forming new nucleation sites for crystallization. In addition, residence time and the composition of nucleating agents, modifiers, and stabilizers are used to control the crystallization process (Windhab, 1999).

4.4 Effect of critical parameters and processing variables on extruded product

The extruder parameters influencing the quality and properties of the extruded products are classified into two categories: machine parameters and material parameters. The machine parameters include screw speed, residence time, die geometry, screw and die geometry, and temperature, while the material parameters are composition and type of raw material, feed rate, moisture content, and particle size. These parameters significantly influence the chemical reactions occurring inside the extruder barrel during the reactive extrusion process (Singh and Muthukumarappan, 2014; Singh et al., 2019). The unique geometry of the screw extruder, such as screw length (L)-to-diameter (D) ratio and channel depth (H)-to-screw diameter (D) ratio, affects the pressure and velocity profiles and also impacts the shear rate of the feed material (Zhuang et al., 2023). Varying these parameters gives the extruder a unique ability to melt polymers and handle highly viscous systems (Li et al., 2021). These parameters collectively determine the textural properties of the extrudate, such as crispiness and hardness, which are crucial for consumer acceptance (Harper, 2019).

The temperature of the material fed into the extruder is adjusted by means of heaters applying heat to the barrel and a combination of screw revolution rate, screw configuration, and material viscosity, which apply shear to the feed material (Tremaine et al., 2014). High temperatures and shear forces enhance several chemical reactions and physicochemical transformations in food components. For example, the quaternary and tertiary structures of biopolymers change during the process (Harper, 1988). The temperature influences the polymerization reactions in the extrusion process. This was demonstrated by Pietsch et al. (2017), who found that raising the temperature from 90 to 160°C caused notable modifications in the polymerization reactions, including the creation of disulfide linkages between gliadins and glutenins. The extrudate's structural integrity and functional qualities are influenced by the kind and composition of raw materials used. Lower moisture levels increase the extrudate expansion, while higher moisture content decreases the expansion ratio, resulting in denser products (Thymi et al., 2005). In addition, the particle size of the feed material affects the textural properties of the extrudates, with finer granule particles leading to softer textures (Huber and Rokey, 1990).

In the extrusion process's design and operation, the specific mechanical energy (SME), is one of the crucial factors. The mechanical work that is done on the material's unit mass as it goes through the extruder is referred to as SME. Higher SME indicates more intense mechanical processing, leading to greater product transformations, mixing, shearing, and heat generation (Onwulata et al., 2001). Understanding and controlling SME is essential to achieve the desired product characteristics.

4.5 Changes in food components during reactive extrusion

4.5.1 Starch

Starch is composed of two major polysaccharides: amylose and amylopectin. Amylose is a linear polysaccharide that primarily consists of α-1,4-linked D-glucopyranose connections, with a small amount of α-1,6 links in branches. Amylopectin is a highly branched polysaccharide composed of D-glucopyranose links at α-1,4 and various lengths of linear chains joined by around 5% α-1,6 branch connections. These two components have distinct properties (Ai and Jane, 2018). In the food industry, starch is a natural, biodegradable polymer and a primary carbohydrate source in staple foods like rice, noodles, and bread, which are crucial for human diets. Reducing starch digestibility to lower the glycemic index is desirable and actively researched. Industrially, this can be achieved by modifying the starch structure through enzymatic, physical, or chemical methods (Cai et al., 2019).

As starch in its native form lacks properties such as solubility, stability, and functionality, extrusion offers a solution through modification via several physical and chemical reactions (Xie et al., 2006). During extrusion, significant changes such as gelatinization, melting, and fragmentation of starch occur. When starch undergoes gelatinization, water penetrates the granules, causing them to swell and lose their structure. As temperature increases, water breaks down crystalline regions, allowing starch molecules to become more mobile. This leads to the progressive melting of the crystalline regions, ultimately transforming the starch structure during extrusion. The mechanical energy involved in extrusion results in the fragmentation of starch molecules, inducing structural changes and enabling chemical reactions to take place (Lai and Kokini, 1991). Changes in starch are achieved through different reactions such as esterification, crosslinking, acetylation, hydroxypropylation, and dual modification. The presence of a large number of –OH groups affords hydrophilicity to the starch and aids in moisture affinity, which provides a pathway for reactive blending of starch with synthetic polymers (Maliger and Halley, 2014). The melting temperature increases during the reactive extrusion processing of chemically modified rice starch due to crosslinking reactions. It has been demonstrated that crosslinking strengthens the chemical bonds that are part of the intermolecular crosslinking. As a result, crystalline quality is improved, and the cross-linked bonds separating starch molecules are confirmed (Cai et al., 2019).

4.5.2 Protein

Proteins are made up of amino acids linked together with peptide bonds (Damodaran, 2017). Proteins undergo a number of conformational changes during the extrusion process that result in texturization, including hydrogen bonding, hydrophobic interactions, and the interconversion of thiol-disulfide linkages. Proteins' complex tertiary and quaternary structures are disrupted during the extrusion process by an increase in temperature and pressure. An increase in thiol content results from the breaking of hydrogen and disulfide bonds caused by elevated temperatures during the extrusion process, which opens up the protein structure. In addition, the formation of isopeptides takes place at high temperatures, resulting in intermolecular disulfide bond formation, which causes an increase in molecular weight. Moisture content also modifies the protein's conformational structure by altering its mobility. High moisture increases protein accessibility, allowing them to orient and align more during shear. Shear affects the temperature and pressure of the die, the moisture content and temperature of the extruder, and plays a major role in the texturization of protein. The extrusion die influences protein texturization and structural alterations by offering crosslinking sites for the exposed links to generate disulfide bridges, thereby facilitating protein denaturation (Arora et al., 2020).

4.5.3 Lipids

During reactive extrusion, lipids undergo significant changes due to the high temperatures and pressures involved in the process. Extrusion reduces lipid oxidation, which is a major concern, and increases the nutritional and sensory quality of foods. It forms lipid–amylose complexes and releases endogenous antioxidants during extrusion, which contributes to the delay of lipid oxidation. The feed moisture content significantly influences lipid oxidation. An increase in moisture content reduces the expansion of extrudates, leading to a lower lipid oxidation rate. The available surface area is an important factor in determining the rate of lipid oxidation. As the degree of expansion increases, it creates more surface area available for oxidation. In addition, larger cells and thinner cell walls make extrudates more susceptible to oxidation (Lin et al., 1998). Low water content in rye bran during extrusion promotes the Maillard reaction, thereby achieving lipid stabilization through an increase in radical scavenging activity. The presence of a high amount of lipids also affects the extrusion process. The excess lipid content acts as a lubricant, reduces expansion and slip, and makes the extrusion process difficult (Moisio et al., 2015).

4.5.4 Dietary fiber

Dietary fiber consists of plant parts that are edible or similar carbohydrates that undergo partial or complete fermentation in the large intestine but are indigestible and nonabsorbable in the human small intestine. It includes polysaccharides, oligosaccharides, lignin, and associated plant substances. Dietary fibers are recognized for their role in promoting various beneficial physiological effects, including gastrointestinal health, the reduction of blood cholesterol, and the regulation of blood glucose levels (Redgwell and Fischer, 2005). The availability of dietary fiber in the diet is increased by extrusion. The molecular mass of dietary fiber components, such as beta-glucan and arabinoxylan, increases during whole meal flour barley extrusion. The extrusion process converts insoluble dietary fiber into soluble dietary fiber, with parameters such as screw speed, moisture content, and temperature exerting significant influence on this conversion. Enzyme treatment significantly enhances the dietary fiber content in the extruded product. This treatment improves the pentosan content during the extrusion processing of rice bran. In addition, in the presence of citric acid and glucose, the polymerization reaction is enhanced, leading to the conversion of monomer carbohydrates into polysaccharides. This reaction achieves a dietary fiber yield of up to 60% (Arora et al., 2020).

4.5.5 Alteration of bioactive components

Antioxidants, particularly polyphenolics, are used to increase the shelf life of food and prevent oxidation, which can be influenced by variations in extrusion process parameters. High temperatures during extrusion result in a loss of polyphenolic content. However, the mechanical actions involved in the extrusion process release bound phenolics from foods, making them biologically available. This increase in polyphenolic content is attributed to the breakdown of complex polyphenols into low molecular weight phenolic compounds, the formation of Maillard reaction products, and the interaction of proteins and phenolics, which enhances stability (Arora et al., 2020). Extrusion increased phenolics, flavonoids, ferric reducing antioxidant power, and oxygen radical absorption capacity in brown rice extrudates, whereas it decreased these components in polished rice and brown rice extrudates (Zhang et al., 2018). An increase in polyphenolic content after extrusion was observed in rice bran, apple pomace, and corn-based extrudates (Arora et al., 2020; Zhang et al., 2018).

4.6 Future scope and applications

Reactive extrusion offers significant opportunities for developing functional ingredients from nutrient- and bioactive component-rich agro-waste products. By-products from the primary production industries of cereal and legume processing, which are high in dietary fibers, phytochemicals, vitamins, minerals, and proteins, can be value-added for human consumption through extrusion (Orozco-Angelino et al., 2023). In addition to enhancing food products, reactive extrusion can replace traditional packaging materials with biodegradable polymers, thereby reducing environmental impact and offering sustainable solutions for the food industry. This technology drives innovation by enabling the creation of novel food products and processing techniques that align with consumer preferences and market demands. Reactive extrusion is an effective technology for producing texturized starch products with improved digestibility and stability. This process not only enhances nutritional profiles but also allows for advancements in food packaging, texture, and shape control. Moreover, it facilitates the production of customized food products, advancing food technology through extrusion.

4.7 Summary

Reactive extrusion facilitates chemical reactions through the addition of reactants, which modify and change the structural and functional properties, producing a wide range of food products. The food material undergoes many physicochemical reactions such as gelatinization, polymerization, oxidation, crystallization, retrogradation, and denaturation. These chemical reactions are influenced by extruder parameters such as screw speed, screw length, temperature, residence time, and die geometry. The high temperature and shear caused by the mechanical forces induce many structural changes and modify the properties of the products through chemical reactions. Through reactive extrusion, novel polymer products are made by the addition of catalysts to induce polymerization reactions in the high-temperature and shear region. Starch modification and texturized starch molecules are produced by reactive extrusion, which also enhances the functionality of other components like proteins, lipids, and dietary fibers. For instance, proteins can be restructured to produce nutrient-rich products and their solubility can be increased, lipids can be stabilized, and dietary fibers can be modified to improve water-holding capacity, benefiting gut health.

4.8 Conclusion

Enhanced reaction efficiency and quality products compared to traditional reaction methodologies make it a cutting-edge technology in food processing. Reactive extrusion offers immense opportunities for producing value-added products from biopolymers through structural modification and chemical reactions. Chemical reactions that are not possible under normal atmospheric conditions are facilitated easily by extrusion, as it provides a continuous reaction medium along with high temperature and mechanical forces. Nutrient-rich functional foods and food fortification with high-nutrient ingredients, coupled with the ability to develop customized food products, make this technology versatile and highly adaptable. Furthermore, the potential for innovation in texture, shape, and nutritional enhancement ensures that reactive extrusion will continue to be a valuable tool in the advancement of food technology. Its ability to produce high-quality products drives both economic and health benefits on a global scale.

References

Ai, Y., Jane, J.L., 2018. Understanding Starch Structure and Functionality Starch in Food: Structure, Function and Applications, second ed. Elsevier Inc., Canada, pp. 151–178. Available from: http://www.sciencedirect.com/science/book/9780081008683. doi:10.1016/B978-0-08-100868-3.00003-2.

Arora, B., Yoon, A., Sriram, M., Singha, P., Rizvi, S.S.H., 2020. Reactive extrusion: a review of the physicochemical changes in food systems. Innov. Food Sci. Emerg. Technol. 14668564. 64. Available from: http://www.elsevier.com/locate/ifset. doi:10.1016/j.ifset.2020.102429.

Beyer, G., Hopmann, C., Cohnen, A., 2017. Introduction to Reactive Extrusion Reactive Extrusion: Principles and Applications. Wiley, pp. 1–10. doi:10.1002/9783527801541.ch1.

Cai, C., Wei, B., Tian, Y., Rongrong, M., Chen, L., Qiu, L., Jin, Z., 2019. Structural changes of chemically modified rice starch by one-step reactive extrusion. Food Chem. 03088146. 288, 354–360. doi:10.1016/j.foodchem.2019.03.017.

Damodaran, S., 2017. Amino Acids, Peptides, and Proteins Fennema's Food Chemistry. CRC Press, United States, pp. 235–356. Available from: http://www.tandfebooks.com/doi/book/10.1201/9781315372914. doi:10.1201/9781315372914.

De Graaf, R.A., Janssen, L.P.B.M., 2000. The production of a new partially biodegradable starch plastic by reactive extrusion. Polym. Eng. Sci. 0032-3888. 40 (9), 2086–2094. doi:10.1002/pen.11340.

Dey, D., Richter, J.K., Ek, P., Gu, B.J., Ganjyal, G.M., 2021. Utilization of Food Processing By-products in Extrusion Processing: a Review. Front. Sustain. Food Syst. 2571581X. 4. Available from: www.frontiersin.org/journals/sustainable-food-systems#. doi:10.3389/fsufs.2020.603751.

Dhar, P., Gaur, S.S., Soundararajan, N., Gupta, A., Bhasney, S.M., Milli, M., Kumar, A., Katiyar, V., 2017. Reactive Extrusion of Polylactic Acid/Cellulose Nanocrystal Films for Food Packaging Applications: influence of Filler Type on Thermomechanical, Rheological, and Barrier Properties. Ind. Eng. Chem. Res. 15205045. 56 (16), 4718–4735. Available from: http://pubs.acs.org/journal/iecred. doi:10.1021/acs.iecr.6b04699.

García-Gurrola, A., Rincón, S., Escobar-Puentes, A.A., Zepeda, A., Martínez-Bustos, F., 2019. Microencapsulation of red sorghum phenolic compounds with esterified sorghum starch as encapsulant materials by spray drying. Food Technol. Biotechnol. 13342606. 57 (3), 341–349. Available from: http://www.ftb.com.hr/. doi:10.17113/ftb.57.03.19.6146.

Guiao, K.S., Gupta, A., Tzoganakis, C., Mekonnen, T.H., 2022. Reactive extrusion as a sustainable alternative for the processing and valorization of biomass components. J. Clean. Prod. 09596526. 355. Available from: https://www.journals.elsevier.com/journal-of-cleaner-production. doi:10.1016/j.jclepro.2022.131840.

Gutiérrez, T.J., Alvarez, V.A., 2017. Properties of native and oxidized corn starch/polystyrene blends under conditions of reactive extrusion using zinc octanoate as a catalyst. React. Funct. Polym. 13815148. 112, 33–44. doi:10.1016/j.reactfunctpolym.2017.01.002.

Harper, J.M., 1988. Effects of Extrusion Processing on Nutrients. Sci. Bus. Media LLC, 365–391. doi:10.1007/978-94-011-7030-7_14.

Harper, J.M., 2019. Extrusion of Foods, CRC press, CRC press.

Hernandez-Hernandez, O., Julio-Gonzalez, L.C., Doyagüez, E.G., Gutiérrez, T.J., 2022. Structure-digestibility relationship from noodles based on organocatalytically esterified regular and waxy corn starch obtained by reactive extrusion using sodium propionate.

Food Hydrocoll. 0268005X. 131. Available from: https://www.journals.elsevier.com/food-hydrocolloids. doi:10.1016/j.foodhyd.2022.107825.

Huang, X., Liu, H., Yue, M., Mai, S., Cheng, L., 2022. Effects of Extrusion on Starch Molecular Degradation, Order–Disorder Structural Transition and Digestibility—A Review. Foods. 2304-8158. 11 (16), 2538. doi:10.3390/foods11162538.

Huber, G.R., Rokey, G.J., 1990. Extruded Snacks. Springer Sci. Bus. Media LLC, 107–138. doi:10.1007/978-1-4613-1477-6_7.

Khalifeh, A., Clermont, J.R., 2005. Numerical simulations of non-isothermal three-dimensional flows in an extruder by a finite-volume method. J. Non-Newtonian Fluid Mech. 03770257. 126 (1), 7–22. doi:10.1016/j.jnnfm.2004.12.002.

Lai, L.S., Kokini, J.L., 1991. Physicochemical Changes and Rheological Properties of Starch during Extrusion (A Review). Biotechnol. Prog. 8756-7938. 7 (3), 251–266. doi:10.1021/bp00009a009.

Li, T.T., Feng, L.F., Gu, X.P., Zhang, C.L., Wang, P., Hu, G.H., 2021. Intensification of Polymerization Processes by Reactive Extrusion. Ind. Eng. Chem. Res. 15205045. 60 (7), 2791–2806. Available from: http://pubs.acs.org/journal/iecred. doi:10.1021/acs.iecr.0c05078.

Lin, S., Hsieh, F., Huff, H.E., 1998. Effects of lipids and processing conditions on lipid oxidation of extruded dry pet food during storage. Anim. Feed Sci. Technol. 03778401. 71 (3-4), 283–294. doi:10.1016/s0377-8401(97)00157-0.

Liu, W., Mohanty, A.K., Askeland, P., Drzal, L.T., Misra, M., 2008. Modification of soy protein plastic with functional monomer with reactive extrusion. J. Polym. Environ. 15662543. 16 (3), 177–182. doi:10.1007/s10924-008-0095-6.

Maliger, R.B., Halley, P.J., 2014. Reactive Extrusion for Thermoplastic Starch-Polymer Blends Starch Polymers: From Genetic Engineering to Green Applications. Elsevier B.V., Australia, pp. 291–317. Available from: http://www.sciencedirect.com/science/book/9780444537300. doi:10.1016/B978-0-444-53730-0.00030-0.

Merci, A., Urbano, A., Grossmann, M.V.E., Tischer, C.A., Mali, S., 2015. Properties of microcrystalline cellulose extracted from soybean hulls by reactive extrusion. Food Res. Int. 09639969. 73, 38–43. Available from: www.elsevier.com/inca/publications/store/4/2/2/9/7/0. doi:10.1016/j.foodres.2015.03.020.

Moad, G., 2011. Chemical modification of starch by reactive extrusion. Prog. Polym. Sci. 00796700. 36 (2), 218–237. doi:10.1016/j.progpolymsci.2010.11.002.

Moisio, T., Damerau, A., Lampi, A.M., Partanen, R., Forssell, P., Piironen, V., 2015. Effect of extrusion processing on lipid stability of rye bran. Eur. Food Res. Technol. 14382385. 241 (1), 49–60. Available from: http://link.springer.de/link/service/journals/00217/index.htm. doi:10.1007/s00217-015-2433-y.

Onwulata, C.I., Konstance, R.P., Smith, P.W., Holsinger, V.H., 2001. Co-extrusion of Dietary Fiber and Milk Proteins in Expanded Corn Products. LWT. 00236438. 34 (7), 424–429. doi:10.1006/fstl.2000.0742.

Orozco-Angelino, X., Espinosa-Ramírez, J., Serna-Saldívar, S.O., 2023. Extrusion as a tool to enhance the nutritional and bioactive potential of cereal and legume by-products. Food Res. Int. 09639969. 169, 112889. doi:10.1016/j.foodres.2023.112889.

Pavani, M., Singha, P., Rajamanickam, D.T, Singh, S.K., 2024. Impact of extrusion processing on bioactive compound enriched plant-based extrudates: a comprehensive study and optimization using RSM and ANN-GA. Future Foods 9, 100286. doi: 10.1016/j.fufo.2023.100286.

Pietsch, V.L., Emin, M.A., Schuchmann, H.P., 2017. Process conditions influencing wheat gluten polymerization during high moisture extrusion of meat analog products. J. Food Eng. 02608774. 198, 28–35. Available from: http://www.sciencedirect.com/science/journal/02608774. doi:10.1016/j.jfoodeng.2016.10.027.

Pietsch, V.L., Karbstein, H.P., Emin, M.A., 2018. Kinetics of wheat gluten polymerization at extrusion-like conditions relevant for the production of meat analog products. Food Hydrocoll. 0268005X. 85, 102–109. Available from: https://www.sciencedirect.com/science/journal/0268005X. doi:10.1016/j.foodhyd.2018.07.008.

Raquez, J.M., Nabar, Y., Srinivasan, M., Shin, B.Y., Narayan, R., Dubois, P., 2008. Maleated thermoplastic starch by reactive extrusion. Carbohydr. Polym. 01448617. 74 (2), 159–169. doi:10.1016/j.carbpol.2008.01.027.

Redgwell, R.J., Fischer, M., 2005. Dietary fiber as a versatile food component: an industrial perspective. Mol. Nutr. Food Res. 49 (6), 521–535. doi:10.1002/mnfr.200500028.

Singh, S.K., Muthukumarappan, K., 2014. Effect of different extrusion processing parameters on physical properties of soy white flakes and high protein distillers dried grains-based extruded aquafeeds. J. Food Process Eng. 3 (6), 107. doi:10.5539/jfr.v3n6p107.

Singh, S.K., Muthukumarappan, K., 2016. Effect of feed moisture, extrusion temperature and screw speed on properties of soy white flakes based aquafeed: a response surface analysis. J. Sci. Food Agric. 96 (6), 2220–2229. doi:10.1002/jsfa.7339.

Singh, S.K., Muthukumarappan, K., 2017. Rheological characterization and CFD simulation of soy white flakes based dough in a single screw extruder. J. Food Process Eng. 40 (2), doi: 10.1111/jfpe.12368.

Singh, S.K., Singha, P., Muthukumarappan, K., 2019. Modeling and optimizing the effect of extrusion processing parameters on nutritional properties of soy white flakes-based extrudates using response surface methodology. Anim. Feed Sci. Technol. 254, 114197. doi:10.1016/j.anifeedsci.2019.06.001.

Singh, S.K, Muthukumarappan, K., 2017. A viscosity model for soy white flakes-based aquafeed dough in a single screw extruder. J. Food Process Eng. 40 (2). doi:10.1111/jfpe.12357.

Thymi, S., Krokida, M.K., Pappa, A., Maroulis, Z.B., 2005. Structural properties of extruded corn starch. J. Food Eng. 02608774. 68 (4), 519–526. doi:10.1016/j.jfoodeng.2004.07.002.

Tremaine, A.J., Reid, E.M., Tyl, C.E., Schoenfuss, T.C., 2014. Polymerization of lactose by twin-screw extrusion to produce indigestible oligosaccharides. Int. Dairy J. 09586946. 36 (1), 74–81. doi:10.1016/j.idairyj.2013.12.013.

Vaidya, A.A., Gaugler, M., Smith, D.A., 2016. Green route to modification of wood waste, cellulose and hemicellulose using reactive extrusion. Carbohydr. Polym. 01448617. 136, 1238–1250. Available from: http://www.elsevier.com/wps/find/journaldescription.cws_home/405871/description#description. doi:10.1016/j.carbpol.2015.10.033.

Wang, N., Yu, J., Ma, X., 2007. Preparation and characterization of thermoplastic starch/PLA blends by one-step reactive extrusion. Polym. Int. 0959-8103. 56 (11), 1440–1447. doi:10.1002/pi.2302.

Willett, J.L., Finkenstadt, V.L., 2003. Preparation of strach-graft-polyacrylamide copolymers by reactive extrusion. Polym. Eng. Sci. 0032-3888. 43 (10), 1666–1674. doi:10.1002/pen.10140.

Windhab, E.J., 1999. New Developments in Crystallization Processing. J. Therm. Anal. Calorim. 1388-6150. 57 (1), 171–180. doi:10.1023/a:1010103105425.

Xie, F., Long, Y., Liu, H., Chen, L., 2006. Starch Modification Using Reactive Extrusion. Starch-Stärke. 0038-9056. 58 (3-4), 131–139. doi:10.1002/star.200500407.

Xu, E., Campanella, O.H., Ye, X., Jin, Z., Liu, D., BeMiller, J.N., 2020. Advances in conversion of natural biopolymers: a reactive extrusion (REX)–enzyme-combined strategy for starch/protein-based food processing. Trends Food Sci. Technol. 09242244. 99, 167–180. Available from: http://www.elsevier.com/wps/find/journaldescription.cws_home/601278/description#description. doi:10.1016/j.tifs.2020.02.018.

Xu, Y., Miladinov, V., Hanna, M.A., 2004. Synthesis and characterization of starch acetates with high substitution. Cereal Chem. 00090352. 81 (6), 735–740. Available from: http://onlinelibrary.wiley.com/journal/10.1002/(ISSN)1943-3638/issues. doi:10.1094/CCHEM.2004.81.6.735.

Ye, J., Luo, S., Huang, A., Chen, J., Liu, C., McClements, D.J., 2019. Synthesis and characterization of citric acid esterified rice starch by reactive extrusion: a new method of producing resistant starch. Food Hydrocoll. 0268005X. 92, 135–142. Available from: https://www.journals.elsevier.com/food-hydrocolloids. doi:10.1016/j.foodhyd.2019.01.064.

Zhang, L., Zhao, J., Fei, L., Jiao, X., Zhang, Y., Yang, B., Quanhong, L., 2024. Insight to starch retrogradation through fine structure models: a review. Int. J. Biol. Macromol. 01418130. 273 (Pt 1), 132765. doi:10.1016/j.ijbiomac.2024.132765.

Zhang, R., Khan, S.A., Chi, J., Wei, Z., Zhang, Y., Deng, Y., Liu, L., Zhang, M., 2018. Different effects of extrusion on the phenolic profiles and antioxidant activity in milled fractions of brown rice. LWT. 00236438. 88, 64–70. Available from: https://www.journals.elsevier.com/lwt. doi:10.1016/j.lwt.2017.09.042.

Zhang, Y., Liu, W., Liu, C., Luo, S., Li, T., Liu, Y., Wu, D., Zuo, Y., 2014. Retrogradation behaviour of high-amylose rice starch prepared by improved extrusion cooking technology. Food Chem. 03088146. 158, 255–261. doi:10.1016/j.foodchem.2014.02.072.

Zhao, X., Wei, Y., Wang, Z., Chen, F., Ojokoh, A.O., 2011. Reaction kinetics in food extrusion: methods and results. Crit. Rev. Food Sci. Nutr. 15497852. 51 (9), 835–854. doi:10.1080/10408398.2010.483023.

Zhuang, Y., Saadatkhah, N., Morgani, M.S., Xu, T., Martin, C., Patience, G.S., Ajji, A., 2023. Experimental methods in chemical engineering: reactive extrusion. Can. J. Chem. Eng. 1939019X. 101 (1), 59–77. Available from: http://onlinelibrary.wiley.com/journal/10.1002/(ISSN)1939-019X. doi:10.1002/cjce.24538.

Extrusion-based three-dimensional printing

5

Soubhagya Tripathy[a], *Srutee Rout*[a], *Sushil Koirala*[b], *Bhesh Bhandari*[b], *Saphal Ghimire*[c], *and Prem Prakash Srivastav*[a]

[a]Department of Agricultural and Food Engineering, Indian Institute of Technology Kharagpur, Kharagpur, India

[b]School of Agriculture and Food Sustainability, The University of Queensland, Brisbane, QLD, Australia

[c]Department of Food, Agriculture, and Bioresources, Asian Institute of Technology, Thailand

5.1 Introduction

Three-dimensional (3D) printing, or additive manufacturing, has become a groundbreaking tool across various sectors, including aerospace, healthcare, construction, and, more recently, the food industry (Bănică et al., 2024). Its applications in food are particularly valuable for individuals with swallowing disorders (Lorenz et al., 2022) or children lacking adequate nutrient intake (Derossi et al., 2018). In addition, researchers are exploring its use in space missions, where food supply is constrained and requires innovation (Enfield et al., 2023). Among the different methods available, extrusion-based 3D printing is especially popular in food production because of its ease of use, material compatibility, and adaptability (Thorakkattu et al., 2025). This method works by depositing layers of semisolid or paste-like food materials, called "inks," to construct intricate, customized designs with minimal waste. As the need for sustainable food systems, personalized nutrition, and functional food increases, extrusion-based 3D printing is emerging as a promising approach for modern food production (Mudau and Adebo, 2024). A diverse range of materials has been successfully printed using this technique, such as chocolate (Mantihal et al., 2019), food emulsions like chocolate ganache (Kim et al., 2022) and custard cream (Cai et al., 2022), various pastes (Zhu et al., 2019), sugar-based confections (Kim et al., 2021), and hydrocolloid-enriched food inks (Gholamipour-Shirazi et al., 2019). This technology is especially beneficial for crafting customized products tailored to individual dietary requirements, textures, and visual appeal in food applications. Unlike conventional food production methods, which often rely on static molds and batch processes, 3D printing allows for on-demand customization, enabling the design of unique textures, shapes, and multilayered

Extrusion Cooking for Food and Feed. https://doi.org/10.1016/B978-0-443-29987-2.00005-1

food items (Waseem et al., 2024). This flexibility is particularly useful in fields such as clinical nutrition, geriatric food design, and food provision for space exploration.

Extrusion-based 3D food printing fundamentally relies on the careful design and optimization of "food inks," specialized formulations that must exhibit suitable rheological, thermal, and mechanical behaviors to enable smooth extrusion, structural stability, and postprint integrity (Hodhodi et al., 2025). These inks typically contain diverse food components such as proteins, carbohydrates, lipids, dietary fibers, and bioactive molecules. However, not all food matrices naturally lend themselves to 3D printing. As a result, pretreatment processes are often employed to enhance the functional and structural properties necessary for printability (Sohel et al., 2025). Common approaches to pretreatment include thermal, enzymatic, ultrasonic, cold plasma (CP), and microwave techniques, each offering distinct advantages; for instance, thermal treatment may promote starch gelatinization or protein denaturation for better extrusion; enzymatic methods can break down large molecules to improve flow; ultrasound can assist in particle size reduction and homogeneity; CP can modify surface properties to enhance compatibility; and microwaves can enable selective heating and improve the flow behavior of inks (Shen et al., 2023). In practical applications, ultrasound-assisted thawing of surimi has been shown to accelerate thawing and protect myofibrillar protein quality while improving the precision of 3D printing (Chen et al., 2021a). In addition, microwave-assisted printing has enabled rapid modulation of sensory attributes; for example, Guo et al. (2021) used microwave heating to control the release of color and flavor compounds in peach-based printed gels within 1–4 min.

Quality deterioration during food processing or storage can result from microbial activity, metabolite formation, or various adverse physical and chemical reactions (Shen et al., 2023; Liu et al., 2020). Low-moisture conditions generally inhibit microbial growth and reduce the risk of such reactions. Thus, postprinting treatments such as drying and sterilization are essential to ensure microbial safety, extend shelf-life, and meet consumer expectations (Liu et al., 2021; Nakra et al., 2025). Techniques like hot air, vacuum, and freeze drying reduce water content and maintain structural integrity. Meanwhile, thermal or nonthermal sterilization helps eliminate pathogens without significantly altering the nutritional or sensory profile of the printed food (Shen et al., 2023). The success of these treatments depends heavily on their compatibility with the printed food's geometry and material composition.

This chapter aims to examine extrusion-based 3D food printing in-depth by outlining its fundamental principles, essential components, strategies for ink development, pretreatment methodologies, and necessary postprocessing steps. It emphasizes the promising role of physical field-based pretreatments in enhancing print performance. Furthermore, the chapter explores spontaneous transformations occurring during printing, recent technological advancements, and the emerging potential of this platform to reshape the future of food by enabling customized, health-oriented, and sustainable innovations.

5.2 Fundamentals of 3D printing

3D food printing represents an interdisciplinary blend of engineering, materials science, and culinary innovation. Depositing edible materials layer by layer creates intricate and customized food structures. This technology holds transformative potential for personalized nutrition, creative culinary presentations, and more efficient food production systems (Sohel et al., 2025). Among the various 3D printing methods, extrusion-based printing is particularly versatile and can accommodate various food materials (Fig. 5.1). The process includes four essential stages: model design, material pretreatment,

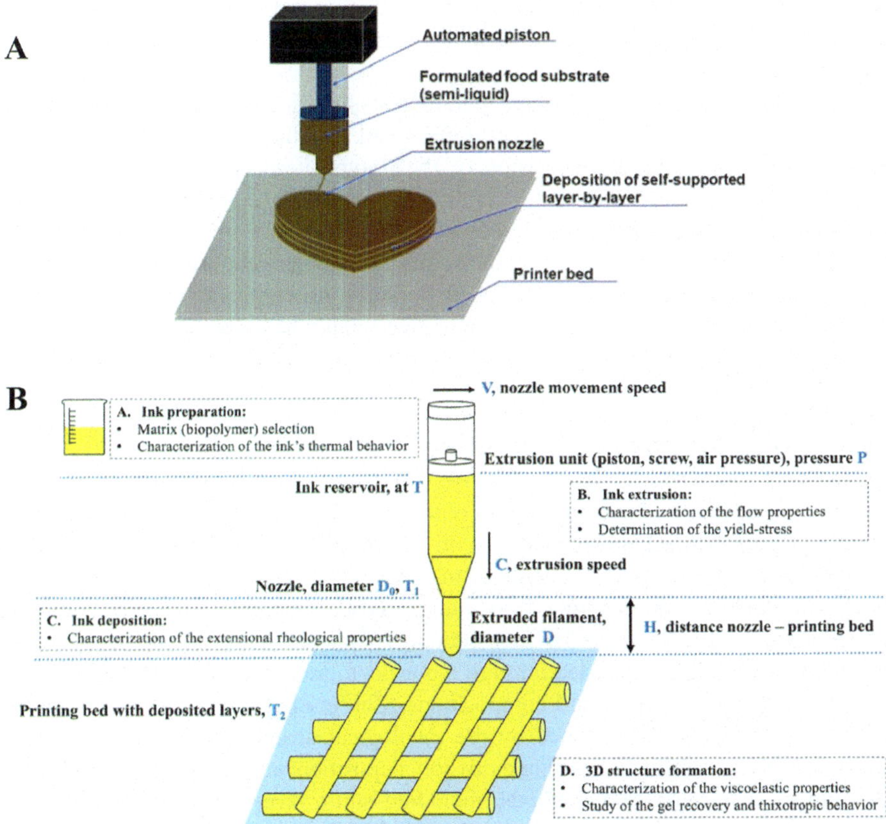

Fig. 5.1 (A) Different parts of an extrusion-based 3D printing (Adopted from Mantihal et al., 2020 with permission; License No: 6022820060545 dated May 05, 2025) and (B)working principle of extrusion-based 3D printing (Adopted from Outrequin et al., 2023 with permission; License No: 6025971024880 dated May 11, 2025).
From Outrequin, T.C.R., Gamonpilas, C., Siriwatwechakul, W., Sreearunothai, P., 2023. Extrusion-based 3D printing of food biopolymers: A highlight on the important rheological parameters to reach printability. J. Food Eng. 342. https://doi.org/10.1016/j.jfoodeng.2022.111371

printing, and postprocessing. In the model development phase, two primary approaches are used: forward and reverse design. Forward design, facilitated by computer-aided design (CAD), involves converting digital models into executable printing paths (Jayan and Wani, 2025), while reverse or bionic design relies on 3D scanning technology to convert physical objects into printable digital formats (Alemzadeh, 2025).

Proper material preparation is crucial, as successful printing depends heavily on the materials' flow properties and mechanical strength. To improve printability, ingredients often undergo pretreatment to enhance their rheological and structural characteristics (Shoel et al., 2025). The success of extrusion printing relies on ingredient viscosity, rheological behavior, and surface tension. These properties influence layer formation, structural fidelity, and adhesion between printed layers. Moreover, controlling thermal processes such as melting, solidification, and cooking is essential for achieving the final product's desired texture, structure, and flavor (Waseem et al., 2024; Rinshana et al., 2025).

During printing, various physical parameters significantly affect output quality. Factors like nozzle height, diameter, printing speed, and the speed at which the nozzle moves directly influence the precision of the printed item (Wang et al., 2025). For instance, smaller nozzle diameters generally improve surface smoothness and dimensional accuracy, while larger ones may reduce quality due to increased roughness (Shen et al., 2023). Shi et al. (2021) highlighted the influence of temperature on the print fidelity of potato starch protein matrices mixed with beeswax, noting that moderate temperature increases can enhance print accuracy. Hence, fine-tuning these operational parameters is essential for producing high-quality, structurally sound printed food products.

5.3 Extrusion-based 3D printing

Extrusion-based 3D food printing has become the predominant method in food customization and personalized nutrition owing to its versatility, precision, and capacity to accommodate diverse food materials (Liu et al., 2017; Hussain et al., 2022). This novel method enables the accurate deposition of edible components in layers, commonly known as food inks, to create 3D food products. This technique entails the sequential deposition of food inks, extruded through a nozzle, to fabricate intricate structures with customized shapes, textures, and compositions (Zhu et al., 2023). The efficacy of this process is significantly influenced by the rheological characteristics of the food ink, which dictate its performance during extrusion, deposition, and subsequent stabilization. Rheological characterization methods, including strain sweep, frequency sweep, and flow sweep, are essential for comprehending and enhancing these properties for printability (Bugday et al., 2024; Ghimire et al., 2025; Oh et al., 2024).

5.3.1 Principle of extrusion-based 3D printing

Extrusion-based 3D food printing is an additive manufacturing method that constructs 3D food structures through the sequential deposition of material layers. The procedure commences with a CAD model that delineates the geometry of the final product

(Ma et al., 2023). Food inks, including pastes, gels, or semisolid substances, are contained within a cartridge and extruded via a nozzle. The extrusion mechanism may be powered by pneumatic pressure, a screw system, or a piston, contingent upon the material characteristics and application specifications (Bugday et al., 2024). The material traverses the nozzle during extrusion and is deposited along a specified trajectory in the X- and Y-axes. Following the deposition of each layer, the build platform modifies its position along the Z-axis to facilitate the stacking of subsequent layers. The deposited substance solidifies via cooling (e.g., chocolate), curing (e.g., ultraviolet light or chemical crosslinking), or phase transition (e.g., gelation of hydrocolloids) to create stable layers. This methodical layer-by-layer technique facilitates exact control over geometry and permits the fabrication of complex designs (Agunbiade et al., 2022; Outrequin et al., 2023; Hussain et al., 2022). The following is a comprehensive schematic explanation of the procedure and its elements:

- Material feeding system: The food ink, which may consist of a paste, gel, or semisolid substance, is contained within a cartridge or hopper. This system guarantees uniform feeding of the material into the extrusion mechanism. Hydrocolloid gels and chocolate are frequently utilized because of their compatibility with extrusion systems (Outrequin et al., 2023; Hussain et al., 2022).
- Extrusion mechanism: The extrusion process may be propelled by pneumatic pressure, screw-driven systems, or pistons, contingent upon the rheological characteristics of the food ink. Pneumatic extrusion is especially efficient for pliable substances such as processed cheese or cookie dough. The extrusion mechanism regulates the flow rate and pressure to guarantee seamless deposition (Agunbiade et al., 2022; Outrequin et al., 2023; Hussain et al., 2022).
- The nozzle configures and applies the material onto the build platform. The diameter of the nozzle is essential for establishing resolution and accuracy, with standard diameters varying from 0.108 mm to 0.356 mm for food applications. Advanced systems employ computer vision to enhance nozzle movement and prevent under- or over-extrusion (Ma et al., 2023; Tan et al., 2018).
- A motion control system coordinates the nozzle's movement along the X–Y axes and adjusts the build platform along the Z-axis for layer accumulation. Feedforward motion control systems enhance accuracy to nearly 100%, ensuring precise layer deposition (Ma et al., 2023).
- Solidification mechanism: Upon deposition, the material solidifies via cooling (e.g., chocolate), curing (e.g., ultraviolet light or chemical crosslinking), or phase transitions (e.g., gelation of hydrocolloids). This step ensures the stabilization of each layer before adding subsequent layers (Tan et al., 2018; Zhu et al., 2023).

This flow illustrates the role of each component in ensuring accurate deposition and structural integrity during the printing process. Motion control systems guarantee synchronization between extrusion rate and nozzle movement to avert defects, while solidification mechanisms preserve structural integrity (Ma et al., 2023). Collectively, these components facilitate the development of complex designs customized for nutritional or esthetic needs. The basic working principle of extrusion-based 3D printing is presented in Fig. 5.1.

Extrusion-based 3D food printing accommodates diverse food materials, such as chocolate, doughs, hydrocolloids, plant-derived proteins, and composite formulations. Multimaterial extrusion systems augment versatility by incorporating various ingredients within a singular structure, thereby permitting the customization of texture, flavor, and nutritional composition (Agunbiade et al., 2022; Tan et al., 2018). For instance, alginate–pectin hydrogels have been effectively printed into beads with exact geometries utilizing pneumatic extrusion systems (Ma and Zhang, 2022; Rysenaer et al., 2023).

5.3.2 Advantages of extrusion-based 3D printing

In comparison with other 3D food printing techniques, such as inkjet printing or binder jetting, extrusion provides greater versatility in material choice, improved control over internal architecture, and enhanced dimensional precision for intricate geometries (Agunbiade et al., 2022). Due to their numerous distinct advantages, extrusion-based techniques have emerged as the predominant method for 3D food printing:

- Extrusion can accommodate a broad range of food viscosities and compositions, from low-viscosity batters to high-viscosity pastes, enabling the printing of complex multi-ingredient formulations (Agunbiade et al., 2022).
- Structural complexity: The layer-by-layer deposition technique facilitates texture customization by enabling the creation of intricate internal structures with diverse densities (10%–70%), including honeycomb, rectilinear, and Hilbert curve patterns (Demei et al., 2022; Derossi, 2021).
- Extrusion parameters, such as layer height, nozzle diameter, and deposition speed, can be precisely controlled to optimize printing outcomes. Typically established at 80% of the nozzle diameter, layer heights facilitate optimal shape fidelity (Agunbiade et al., 2022; Derossi, 2021).
- Extrusion-printed structures readily facilitate various postprocessing techniques, such as baking, steaming, and freezing, broadening the spectrum of potential textures and sensory attributes (Agunbiade et al., 2022).
- On-demand production capabilities streamline supply chains by reducing the necessity for extensive manufacturing infrastructure, potentially transforming paradigms of food distribution and personalization (Agunbiade et al., 2022).

5.4 Extrusion of food ink

A crucial element in 3D food printing is "food ink," a specially formulated, food-grade material. These materials must possess specific characteristics that allow them to be smoothly extruded through a printer nozzle while retaining their form after deposition (Table 5.1). Controlling the viscosity and taste of food inks is achievable through adjustments to the formulation (Domżalska and Jakubczyk, 2025). Food inks have been tailored to enable customization in 3D-printed foods, supporting innovations in texture, flavor, satiety, and visual appeal, thereby influencing dietary behaviors. The evolution of this technology is primarily driven by consumer demand and emerging printing technologies (Ma and Zhang, 2022). Some materials naturally exhibit sufficient extrudability and structural integrity without additional flow modifiers. However, when unconventional materials are used, modifications to physical or rheological properties may be required to achieve optimal printability. Incorporating

Table 5.1 Compositions of ink during extrusion-based 3D printing.

Ink ingredients	Printing speed	Temperature (°C)	Nozzle diameter (mm)	Nozzle height (mm)	Product	Rheology of ink	References
Textured soy protein, onion, bread crumbs powder, xanthan gum, gluten powder, salt, spices, and sunflower oil	5–10 mL/s	25±1	4	4	Plant-based burger	Non-Newtonian and pseudoplastic	Abedini et al. (2024)
Rice protein, xanthan gum, and water	50 mm/s	25.5±0.5	1.2	–	–	Non-Newtonian and pseudoplastic	Barrios-Rodríguez et al. (2024)
Egg white protein, tea polyphenols	15 mm/s	25	1.2	–	EWP-TP gels	Non-Newtonian and pseudoplastic	Liu et al. (2024)
Minced chicken breast, salt	25 mm/s	25-45	0.8-2	–	Chicken meat	Non-Newtonian and pseudoplastic	Yang et al. (2022)
Whey protein powder, peanut protein powder, hydrolyzed wheat protein, pea protein powder, spinach powder and purple sweet potato powder, potato flour, lemon powder, sodium alginate	2 mm/s	25±1	0.8–1.5	0.8–1.5	Fruit and vegetable gel	Non-Newtonian and pseudoplastic	Chen et al. (2021a)
Garden pea, carrot and bok choy, xanthan gum, kappa carrageenan, locust bean gum	25 mm/s	25±1	0.84	0.5	Printed vegetable for dysphagic patients	Non-Newtonian and pseudoplastic	Pant et al. (2021)
Mashed potato/purple sweet potato	24 mm/s	25	1.2	1.2	–	Non-Newtonian and thixotropic	He et al. (2020)
Cadbury dark chocolate, dark chocolate bittersweet flavor	70 mm/s	32	0.78	–	Dark chocolate	Non-Newtonian and pseudoplastic	Mantihal et al. (2019)
Dehydrated potato puree, whole milk	2 mm/s	10–30	2	2	3D printed potato	Non-Newtonian and pseudoplastic	Martínez-Monzó et al. (2019)

novel components, such as dietary fibers or proteins from alternative sources, can enhance the functionality of food inks for specific applications (Kim et al., 2019).

Rheology plays a central role in determining print quality. The yield stress (τ_{yield}) is an important parameter that indicates the minimum force needed to initiate flow. This value is influenced by the temperature, type, and concentration of biopolymers and other additives (Outrequin et al., 2023). For example, increasing protein content, such as using a blend of refined flour and pea protein isolate, can elevate the yield stress, improving print performance (Maldonado-Rosas et al., 2022). Zhou et al. (2023) formulated a successful printing ink using *Aronia melanocarpa* pulp with methylcellulose, hyaluronic acid, and pea albumin in a 100:14:1:1 ratio, which showed high precision and excellent rheological behavior. Despite the range of materials explored, only a few food inks meet both extrusion ease and shape retention requirements, and even fewer fall within the optimal yield stress window defined for ideal 3D printing performance (Liu et al., 2019b). For extrusion-based methods, avoiding shear thickening behavior is critical, as an increase in viscosity with higher shear rates can impede smooth extrusion, which is undesirable in printing applications (Wei et al., 2022). Therefore, preassessing the rheological profile, such as via a shear rate sweep between 1 and 1000 s^{-1} at the processing temperature is essential to ensure effective printing. Notably, most food inks mentioned thus far do not incorporate particulate matter. However, adding solids or powders may enhance the printed product's nutritional value or sensory attributes (Outrequin et al., 2023).

5.5 Various pretreatment methods for ink in 3D food printing

To ensure successful 3D printing, materials must exhibit suitable rheological traits such as proper viscosity, consistency, and elasticity. However, many natural food substances do not naturally possess these characteristics and are often unsuitable for direct printing (Shen et al., 2023). In recent years, various physical pretreatment techniques, such as thermal processing, enzymatic treatment, CP, ultrasound, and microwave methods, have gained significant interest in enhancing the printability of such materials.

5.5.1 *Thermal pretreatment*

Thermal pretreatment is widely employed in 3D food printing to alter the rheological behavior of printing inks, thereby improving their printability, structural strength, and postprinting integrity. Applying heat to ingredients such as proteins, starches, and polysaccharides can trigger molecular changes like denaturation, gelatinization, or cross-linking, all of which enhance gel formation and viscoelastic properties essential for maintaining the shape of printed constructs. For instance, Zheng et al. (2024) explored the use of dry heat treatment (DHT) at 70°C for 20 min as a preconditioning step to influence starch–lipid interactions in hot-extrusion 3D printing. Their findings revealed that DHT facilitated the complex formation between wheat starch and either oleic acid or corn oil, thereby increasing the resistant starch yield, offering valuable insights for formulating nutritious starch-based printed foods.

In addition, thermoplastic polymer nanocomposites, like PLA reinforced with nanoclay (NC) or graphene nanoplatelets (GNPs), can undergo thermal degradation at elevated temperatures, which impairs their flow behavior during printing. Ozdemir et al. (2024) noted that incorporating GNP, as opposed to NC, led to enhanced thermal stability of the nanocomposites. In a separate study, Liu et al. (2019a) developed a 3D-printable complex system by blending egg white protein with gelatin, corn starch, and sucrose, heating the mixture at 55°C for 10 min, and then conditioning it at 40°C for 20 min before printing. Their results showed that including egg white protein significantly enhanced the gel's firmness and elasticity, contributing to better printing accuracy and shape retention. Similarly, Liu et al., (2019b) prepared a composite gel by dissolving κ-carrageenan, xanthan gum, and potato starch in water, heating it at 90°C for 30 min, homogenizing it, and cooling it to 4°C. This thermal treatment improved the gelling capability, reduced setting time, and enhanced print quality and fidelity. However, precise thermal control is critical, as excessive heating can degrade components and diminish printing performance.

5.5.2 Enzymatic pretreatment

Enzymatic pretreatment provides a sustainable and selective method to modify the properties of food inks, enabling functional improvements without the need for harsh processing conditions. For instance, transglutaminase (TGase) is commonly employed to induce protein cross-linking, which supports enhanced structural integrity and network development in 3D-printed food products. In a study by Kangas et al. (2019), enzymatic treatment was applied to produce nanocellulose from unbleached and delignified pulps, with the latter showing greater responsiveness to the enzymatic process. A subsequent fluidization step was necessary to achieve nanocellulose with appropriate rheological and morphological characteristics suitable for microextrusion-based 3D printing. Their findings also confirmed the noncytotoxic nature of the resulting material, making it viable for constructing stable 3D-printed forms. Similarly, research by Schutyser et al. (2018) involved using TGase to cross-link sodium caseinate, improving its printability. The enzyme treatment raised the gelation temperature, enabling the formulation of printable structures even at relatively low protein concentrations (25% w/w).

5.5.3 CP pretreatment

CP pretreatment is gaining attention as an innovative, nonthermal approach for modifying food substrates' surface properties and functionalities to improve their performance in 3D food printing (Tripathy and Srivastav, 2023; Kumar et al., 2024). This technique facilitates chemical transformations such as cross-linking, oxidation, and partial polymerization of biomolecules, contributing to better ink adhesion and enhanced mechanical characteristics (Hoque et al., 2022; Tripathy and

Srivastav, 2023; Rout et al., 2024). Typically generated through gas discharge processes, CP systems used in the food sector include dielectric barrier discharge, radiofrequency (RF) discharge, and glow discharge (Tripathy and Srivastav, 2024). Studies have shown that RF air cold plasma treatment can alter the structural attributes of tapioca starch at multiple scales by promoting etching, oxidation, and depolymerization. In addition, CP-modified starch has demonstrated a higher storage modulus (G′) and a reduced loss tangent (tan δ), reflecting a more elastic and stable gel network (Sudheesh et al., 2023). In 3D printing, CP-treated starch has improved print precision, rising from 75.88% to 89.16% (Lu et al., 2024). This enhancement is likely linked to increased amylose concentration and the formation of hydrophilic groups such as carboxyls during CP-induced depolymerization and oxidation, contributing to a stronger and more compact starch gel matrix (Lu et al., 2024). Furthermore, inks subjected to CP treatment demonstrated superior printability compared to untreated counterparts (Fig. 5.2A), likely due to increased mechanical integrity and better self-supporting capacity of the emulsified printing ink (Liu et al., 2024).

5.5.4 Ultrasound pretreatment

Ultrasound is recognized as an environmentally friendly processing technique, and it involves pressure waves operating at frequencies above 20 kHz (Cui and Zhu, 2021). When propagated through a medium, ultrasound induces elastic vibrations that trigger cavitation and mechanical and thermal effects (Ojha et al., 2017). This pretreatment technique can decrease the viscosity of starch-based gels, enhancing the smooth extrusion of materials, which is especially useful in 3D printing. Simultaneously, it can enhance the gel's viscosity and swelling properties, contributing to the long-term structural stability of printed objects (Fig. 5.2B). At an ultrasonic power of 390 W, citrus pectin demonstrated superior textural attributes, such as enhanced hardness, elasticity, and viscoelastic behavior, compared to other variants, thereby improving the structural integrity and workability of pea protein systems in 3D printing applications (Gu et al., 2024).

Furthermore, ultrasound has effectively assisted in thawing surimi-based printing materials. Specifically, ultrasound-assisted thawing at 80 and 100 kHz not only reduced thawing time by 13.5%–40.4% but also minimized structural damage to myofibrillar proteins, thereby enhancing the 3D printability of surimi (Chen et al., 2021a). With its distinct texture, rich nutrient profile, and excellent gelling behavior of muscle proteins, surimi presents itself as a promising candidate for functional 3D food printing (Okita et al., 2021).

5.5.5 Microwave pretreatment

Microwave technology is widely used in food processing for operations such as thawing, drying, sterilizing, and baking because it is rapid, cost-effective, and energy-efficient (Verma et al., 2024). When applied as a pretreatment, microwaves enhance starch dispersion in water, promote starch pasting, and increase dough firmness (Fig. 5.2C). This rise in hardness strengthens the material's structural integrity, while improved elasticity supports better shape retention during processing (Cai et al., 2024).

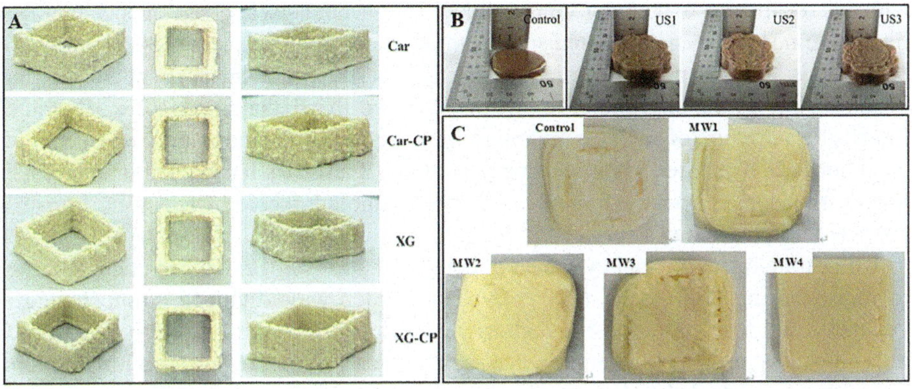

Fig. 5.2 Schematic diagram showing (A) Cold Plasma (Adopted from Liu et al., 2024 with permission; License No: 6021520759794 dated May 03, 2025), (B) Ultrasound (Gao et al., 2023) and (C) Microwave pretreatment (Adopted from Sun, Zhang, & Chen, 2020 with permission; License No: 6021521002083 dated May 03, 2025) for the Ink. Abbreviations: Car: Carrageenan; XG: Xanthan gum; CP: Cold Plasma, US: Ultrasound; MW: Microwave.
Car, Carrageenan; *XG,* Xanthan gum; *CP*, Cold Plasma; *US*, Ultrasound; *MW*, Microwave
From Liu, L., Xie, T., Cheng, W., Ding, Y., Xu, B., 2024. Characterization and mechanism of thermally induced tea polyphenols and egg white proteins gel system and its 3D printing. LWT 205, 116488 with permission; License No: 6021520759794 dated May 03, 2025; Gao, R., Song, R., Shen, L., Zhao, X., Xue, L., Li, J., Zheng, X., 2023. Evaluation of 3D printability of blueberry powder gel system under ultrasound pretreatment. LWT 190, 115577; Sun, Y., Zhang, M., Chen, H., 2020. LF-NMR intelligent evaluation of rheology and printability for 3D printing of cookie dough pretreated by microwave. LWT 132. doi:10.1016/j.lwt.2020.109752 with permission; License No: 6021521002083 dated May 03, 2025.

Furthermore, the energy from microwave photons can alter the configuration of starch molecules by affecting the arrangement of chemical bonds and the distribution of electron clouds within the molecular groups. Studies have shown that a brief microwave treatment (30 s) significantly boosts the printability of buckwheat starch combined with high-methoxy pectin, compared to samples subjected to 30-min steaming (Guo et al., 2021). In addition, using a low microwave power setting (70 W) enhances ink performance, conserves energy, and improves product appeal. The incorporation of salt also enhances the formability and structural stability of 3D-printed constructs (Fan et al., 2020).

5.6 Spontaneous changes in 3D food printing

Four-dimensional (4D) printing in food technology refers to 3D-printed products that can change shape, color, or flavor when subjected to external stimuli. Integrating efficient physical fields enables these printed items to transform more rapidly and precisely in response to specific triggers (Shen et al., 2023). As an additive manufacturing method, 3D food printing allows for precise control over the distribution of ingredients within a food matrix, making it possible to modulate flavor characteristics intentionally. For instance, Guo et al. (2022) created a gel of yellow peach and buckwheat starch embedded with microcapsules containing cinnamaldehyde essential oil and red pepper pigment. Electronic nose analysis indicated significant increases in esters, terpenes, and aromatic compounds when exposed to microwave energy (700 W) at intervals of 30, 60, and 90 s. Similarly, Phuhongsung et al. (2020) explored the flavor response of 3D-printed food using a mixture of soybean protein isolate, beetroot, and pumpkin under various pH conditions (2–10). After 5 h at alkaline pH values (8 and 10), notable shifts in aromatic hydrocarbons and terpene esters were observed, although these changes were primarily restricted to the product's surface.

Color transformation, a key aspect of consumer appeal, has also been demonstrated in food printing. Guo et al. (2022) evaluated microwave-induced color change in vegetable and fruit-based pastes such as yam, purple sweet potato, and yellow peach buckwheat. Within 15 s of microwave exposure at 700 W, samples with 5% microcapsules turned red due to capsule rupture and pigment release. Due to the pH-sensitive curcumin component, turmeric-infused sago starch matrices changed from yellow to reddish-orange in alkaline conditions (Shanthamma et al., 2021). Likewise, red cabbage juice containing anthocyanins enabled printed items to exhibit a range of hues from red to green–yellow under varying pH, enhancing visual cues related to freshness and flavor (Ghazal et al., 2021). Microwave-assisted discoloration in lotus root gel printed with curcumin emulsion was also observed. Upon heating to 65°C–75°C, the microcapsules broke, curcumin was deprotonated in an alkaline medium, and the color shifted to red within 1–3 min (Chen et al., 2021a). Furthermore, Guo et al. (2021) studied color changes in buckwheat dough mixed with yellow peach and microencapsulated pigments. Microwave treatment over 0–4 min disrupted capsule structures, leading to pigment, oil release, and visible yellowing of the printed food.

Spontaneous deformation in 3D-printed foods primarily results from dehydration. This occurs due to a moisture gradient within the printed material, generating internal stress and leading to physical distortion (Chen et al., 2021b). Rapid dehydration through microwave or infrared (IR) methods promotes swift deformation, although high temperatures can form an outer crust that hinders flexibility. Therefore, low-temperature dehydration is preferable for maintaining malleability (Liu et al., 2021). In one experiment, yellow peach-buckwheat paste petals bent up to 56.5° under microwave heating (150 W for 120 s), eliminating the need for support structures in printing, as verified using Rhino software (Guo et al., 2022). Teng et al. (2022) demonstrated that incorporating 10% soybean insoluble dietary fiber into whole potato flour helped preserve both form and texture in microwave-stimulated printed samples.

5.7 Drying of extrusion-based 3D-printed food products

Extrusion-based 3D printing significantly depends on postprocessing to attain structural integrity, texture, and functionality in produced items. The ink utilized in extrusion-based 3D printing has elevated moisture content to ensure printability and flow through nozzles (Domżalska and Jakubczyk, 2025). The printing method offers exceptional control over the shape and composition of food, yet the first printed product often retains significant moisture, requiring more postprocessing to attain the necessary final attributes.

Drying, the regulated extraction of moisture from a substance, is an essential postprocessing procedure for food products produced using extrusion-based 3D printing due to numerous critical factors that impact the safety, shelf life, palatability, and overall quality of the 3D-printed meals (Domżalska and Jakubczyk, 2025). The essential requirement for drying in extrusion 3D-printed food goods is to (Li et al., 2023; Martekha and Kaverina, 2023; Nachal et al., 2019):

- Enhance structural integrity and mechanical properties.
- Ensuring food safety and inhibiting microbial proliferation.
- Prolonging the shelf life and averting degradation.
- Enhancing palatability and texture.

5.7.1 Drying methods for extrusion-based 3D-printed food products

5.7.1.1 Air drying

Air drying in extrusion-based 3D food printing stabilizes the printed structure by reducing moisture and improving mechanical properties (Table 5.2). Locally heating the extruded material with hot air or IR drying helps form a crust and maintain internal moisture diffusion (Koirala et al., 2023). In 3D-printed pasta, hot air drying reduces the initial mass by 15%–35% in 0.5–2.5 min, depending on temperature and processing time, while IR drying creates a drier, harder surface that improves texture and reduces tackiness between layers (Martekha and Kaverina, 2023). Under harsh conditions, excessive crust formation can cause internal gas buildup and filament deformation. Air drying is an essential postprocessing step in extrusion-based 3D food printing because it improves structural integrity and customizes textures (Teng et al., 2022).

5.7.1.2 Oven drying

Extrusion-based 3D food printing uses oven drying to stabilize printed structures by reducing moisture and improving mechanical integrity (Table 5.2). The printed food structures are placed in a controlled-temperature oven at 40°C–80°C, depending on material properties and desired results (Agunbiade et al., 2022; Tan et al., 2018). Hydrocolloid-based formulations are more structurally stable after oven drying due to uniform moisture evaporation and surface hardening (Ma and Zhang, 2022). The stable outer crust and gradual internal moisture diffusion help maintain the printed layers' shape and prevent deformation during handling or storage. Protein-rich and fiber-enriched formulations benefit from oven drying for printability and sensory qualities (Guo et al., 2022; Rysenaer et al., 2023). Avoiding overheating, which can crack or shrink the printed structure, requires precise temperature control. Oven

Table 5.2 Summary of recent drying methods used for extrusion-based 3D-printed food products.

Drying method	Principle of drying	Research focus	Extruded 3D-printed food product	Typical application	Conclusion	References
Air drying	Evaporation of moisture in ambient or regulated airflow conditions	Systematic moisture extraction; enhancing airflow velocity and temperature (25°C–40°C)	Dough-based formulations, fruit purees	Economical drying technique for basic printed structures or prototypes	Appropriate for low-energy applications, yet slower than alternative methods; poses a risk of deformation in fragile structures	Agunbiade et al. (2022) and Martekha and Kaverina (2023)
Oven drying	Heat transfer via convection to remove moisture uniformly at controlled temperatures	Mitigating deformation risks in printed layers at temperatures ranging from 40°C to 80°C	Hydrocolloid gels and protein-enriched formulations	Sequential cooking and dehydrating for multifunctional food products, such as protein bars or snacks	Provides consistent drying; however, it may induce shrinkage if parameters are not precisely optimized	Tan et al. (2018)
Freeze drying (lyophilization)	Sublimation of ice under vacuum conditions to maintain structure and porosity	Retention of shape and formation of a porous matrix; optimization of freezing temperature (−20°C to − 80°C)	Hydrogels, yam gels, and fruit-derived confections	Lightweight foods; products favorable for rehydration; functional foods, especially for people with dysphagia	Maintains nutritional integrity and texture, yet is resource-intensive and time-consuming	Ma and Zhang (2022), Tan et al. (2018), and Kang et al. (2023)
Microwave drying	Microwave radiation facilitates volumetric heating, allowing for swift moisture extraction	Surface hardening and internal moisture retention; optimizing power levels (450–900 W) and duration (2–5 min)	Snacks derived from potatoes, vegetable gels	Convenient snacks; instant foods characterized by crunchy exteriors and tender interiors	Effective drying with improved texture necessitates meticulous optimization to avoid inconsistent heating	Li et al. (2023) and Chen et al. (2023)
Infrared (IR) drying	Infrared radiation facilitates localized surface hardening through heat transfer	Formation of surface crust; optimizing the distance of the infrared lamp (~1.77 cm) and heating power (~200 W)	Pasta dough, multilayered foods	Multitextured foods; crust development for enhanced layer adhesion	Accelerated drying with minimal shrinkage; suitable for multilayered constructions necessitating solid exteriors	Martekha and Kaverina (2023)

Vacuum drying	Extraction of moisture under diminished pressure to mitigate thermal damage at lower temperatures	Steady drying at moderate temperatures (40°C–60°C) to improve the shelf stability of printed foods	Mango jelly, fruit-derived formulations	Shelf-stable snacks, lightweight aerated products; functional foods for medical applications	Effective moisture extraction with minimal shrinkage yet necessitates specialized apparatus	Xie et al. (2023) and Yang et al. (2019)
Microwave + infrared drying	Combines volumetric microwave heating with localized infrared surface hardening for consistent drying	Improving surface texture while preserving internal moisture equilibrium in multilayered products	Snacks derived from potatoes, pasta formulations	Multitextured, ready-to-eat meals; instant meals requiring precision crust formation	Enhances drying efficiency and texture uniformity, yet necessitates coordinated parameter control	Li et al. (2023) and Martekha and Kaverina (2023)
Microwave + freeze drying	Integrates microwave heating with freeze drying to expedite sublimation and morphological alterations	Developing dynamic alterations in printed food (e.g., expansion effects in yam gels)	Yam gels, plant-derived snacks	Applications of 4D printing; functional foods featuring interactive textures or shapes	Facilitates swift drying while maintaining porous structures; optimal for 4D-printed food designs	Kang et al. (2023) and Li et al. (2023)

drying is essential for structural integrity and texture in extrusion-based 3D food printing (Guo et al., 2022).

5.7.1.3 Freeze drying (lyophilization)

Lyophilization, or freeze drying, is a common postprocessing method in extrusion-based 3D food printing to stabilize printed structures by removing moisture and preserving shape and structure (Table 5.2) (Jiang et al., 2021; Rysenaer et al., 2023; Sousa et al., 2022). Freezing the printed material at low temperatures (−20°C to −80°C) forms ice crystals, which are then sublimated under low vacuum conditions (below 2.6×10^4 Pa) to remove water vapor without a liquid phase. Studies show that lyophilized hydrogels retain their shape with <5%-dimensional reduction, making this method ideal for porous structures with high stability and low shrinkage (Jiang et al., 2021; Kuo et al., 2021).

Ice crystal nucleation and removal during lyophilization create a rigid matrix that strengthens printed food products. After freeze drying, extruded carboxymethyl cellulose-based hydrogels maintained layer adhesion and structural stability without splitting or separation (Jiang et al., 2021). Freeze-dried food products are ideal for texture-modified dysphagia diets and lightweight snacks due to their long shelf life, reduced weight, and preserved nutritional content (Lorenz et al., 2022). Lower freezing temperatures produce finer pores, which improve water uptake and rehydration in lyophilized structures. Lyophilization is time-consuming and resource-intensive, generally taking up to 24 h at−40°C or lower (Jiang et al., 2021; Wang and Florczyk, 2020). Despite these challenges, the technique is useful for creating lightweight, porous, shelf-stable, and functional 3D-printed food products.

5.7.1.4 Microwave drying

Microwave drying stabilizes extrusion-based 3D-printed food products by reducing moisture quickly and efficiently (Table 5.2). Microwaves heat material volumetrically from the inside out, making moisture evaporation faster and drying time shorter (Demei et al., 2022; Martekha and Kaverina, 2023). Studies have shown that microwave drying reduces the deformation of printed layers better than other methods (Feng et al., 2021). Microwave drying parameters, such as 450–900 W power and 2–5 min processing time, create a hardened surface layer while preserving internal softness, thereby improving texture (Dick et al., 2021; Martekha and Kaverina, 2023). Under controlled microwave drying, potato-based 3D-printed foods were crispy on the outside and tender on the inside. Microwave drying can also reduce the initial mass by 35% while maintaining the printed structure's shape and mechanical stability (Lee et al., 2025). However, excessive exposure may cause uneven heating or localized swelling, requiring precise parameter optimization. Ready-to-eat snacks and instant foods benefit from microwave drying for quick stabilization and tailored textures (Demei et al., 2022; Feng et al., 2021). Its fast drying and energy efficiency make it highly useful in extrusion-based 3D food printing.

5.7.1.5 IR drying

In extrusion-based 3D food printing, localized and efficient IR drying stabilizes printed structures by reducing moisture and increasing surface hardness (Table 5.2). In IR drying, IR radiation (1.5–8 μm) directly transfers heat to the printed material, allowing for rapid drying without significant deformation, unlike conventional methods (Lv et al., 2024; Martekha and Kaverina, 2023). This method makes thin extruded dough layers drier and harder, reducing tackiness and improving structural stability (Lee et al., 2025).

Optimal IR drying parameters include 200 W heating power, 1.5 min processing time, and 1.77 cm IR lamp-to-printed material distance (Martekha and Kaverina, 2023). These conditions ensure uniform crust formation and internal moisture diffusion, preventing cracking and shrinkage. Under IR drying, extrusion-based pasta dough exhibited better texture and less deformation compared to other methods (Lv et al., 2024). In addition, simultaneous IR drying during printing improves layer adhesion and allows multimaterial integration by precisely controlling heat distribution. IR drying is ideal for printed food products with crispy exteriors and soft interiors. Its rapid and controlled drying makes it essential for precise heat treatment applications, such as ready-to-eat meals and multilayered snacks (Hertafeld et al., 2018).

5.7.1.6 Vacuum drying

Reducing the boiling point of water and removing moisture at lower temperatures with vacuum drying is an effective post-processing method for extrusion-based 3D food printing (Table 5.2). This method protects the structural integrity and nutritional value. Vacuum drying improved the texture of 3D-printed mango juice jelly and reduced storage deformation. At pressures below 2.6×10^4 Pa and temperatures between 40°C and 60°C, moisture removal is uniform without affecting the shape of the printed layers (Yang et al., 2019).

Recent studies have shown that vacuum drying can reduce drying time by up to 50% for moisture-sensitive formulations like fruit or vegetable-based food inks. Combining vacuum drying with desiccants improves moisture removal and prevents storage reabsorption. This method creates a stable matrix with minimal shrinkage or deformation, making it ideal for lightweight, porous foods like aerated snacks or texture-modified foods for dysphagia patients (Tan et al., 2018; Yang et al., 2019). By preserving texture, shape, and nutritional content, vacuum drying can improve the quality and shelf life of extrusion-based 3D-printed food products.

5.7.1.7 Integrated drying approaches

To improve structural stability, texture, and functionality, integrated methods using microwave, IR, and freeze drying techniques for extrusion-based 3D food printing have been studied (Table 5.2). These methods often use 4D printing, where heat or moisture causes structures to change over time. They are suitable for 4D printing applications by allowing controlled food product transformations. 4D-printed foods utilize microwave energy or humidity to facilitate planned morphological transformations (Chen et al., 2023). For instance, the microwave-induced expansion of vegetable-based gels, such as potato-carrot composites, generates volumetric textures by regulating groove depths. Moisture-induced transformation in starch-protein matrices facilitates the formation of self-assembling structures during the drying process (Ma and Zhang, 2022). 4D food printing allows food structures to change over time in response to temperature, pH, and moisture (Koirala et al., 2025). Yam gels swell under microwave-induced heating due to localized water evaporation and structural deformation. Integrated freeze drying creates porous structures that rehydrate differently, creating interactive textures in vegetable-based snacks (Chen et al., 2024). Cricket-based snacks printed with purple potato puree show color change stimuli from anthocyanin redistribution under pH stimulation (Kang et al., 2023). By tailoring infill patterns and drying methods, chocolate-coated protein bars have customized flavor profiles that improve crunchiness and taste uniformity. These advances demonstrate 4D food printing's ability to create functional and interactive food products for personalized nutrition and sensory innovation. These technologies underscore the transition to precision drying for tailored textures, nutritional preservation, and structural advancement in 3D/4D food printing.

Microwave and IR drying are often used together for fast, localized heating. IR drying hardens the printed product while maintaining internal softness, making it suitable for multilayered foods like pasta dough. The optimal IR parameters are two units of heating power, 1.5 min of processing time, and 1.77 cm from the IR source to minimize deformation and improve texture (Martekha and Kaverina, 2023). Volumetric heating from microwave drying accelerates moisture removal. At 450–900 W, microwave-treated potato-based foods had crispy exteriors and uniform internal moisture distribution (Li et al., 2023). This combination improves product quality and drying efficiency.

Microwave-assisted freeze drying uses rapid microwave heating and vacuum sublimation to dry printed foods quickly while preserving their porous structure. This method works well for light snacks and texture-modified foods like yam gels. Due to localized swelling, microwave-induced heating in freeze-dried yam gels caused shape transformations ("spreading wings") in 30 s (Chen et al., 2023). It preserves nutritional content and improves rehydration, making it ideal for functional foods.

5.7.2 Key factors significantly influencing the drying process

Various drying techniques can be utilized for extrusion-based 3D-printed food items, each functioning on distinct principles and producing specific consequences on the final product attributes. The choice of the optimal drying method is contingent upon factors, including the type of food ink utilized, the intended final texture and moisture content, the printed structure (shape, size, porosity), required ambient temperature, humidity levels, the nutritional profile to be preserved, and overall cost and time considerations. Diverse methods exist, each possessing distinct advantages and disadvantages. Challenges such as deformation, cracking, microbial proliferation, and nutrient depletion necessitate mitigation (Feng et al., 2021). Moreover, the ideal drying method for extrusion 3D-printed food is contingent upon the material's composition and characteristics; specifically,

- For gels with elevated moisture content, freeze drying and microwave vacuum drying ensure excellent form retention while preserving structure and porosity (Feng et al., 2021).
- Pastes, such as pasta dough, are customarily dried using hot air or IR technologies, which regulate the desired textures, including hardness (Martekha and Kaverina, 2023).
- For purees with elevated water content, freeze drying and microwave vacuum drying are employed to preserve structure and nutrients (Feng et al., 2021).

Comprehension and regulation of drying are crucial to producing high-quality, safe, and visually appealing 3D-printed food items. The subsequent tables summarize the prevalent drying procedures employed for extrusion-based 3D-printed objects in recent years, together with their research focus and conclusions.

5.8 Sterilization of extrusion-based 3D-printed food products

During 3D food printing, the material (or "ink") often comes into contact with components such as the printer's extrusion piston and internal piping, and the printed item is typically exposed to ambient air throughout the process. These conditions can contribute to microbial contamination, making hygiene and microbial safety-critical considerations in 3D food printing (Shen et al., 2023; Liu et al., 2025). For instance, Severini et al. (2018) demonstrated the creation of smoothies using fruit and vegetable-based inks, and microbial testing indicated that even after thoroughly sanitizing all materials, contamination still occurred due to contact with printer surfaces and prolonged air exposure. Microbial proliferation is closely tied to water activity (aw); levels below 0.6 generally inhibit microbial growth, while values above 0.85 significantly increase the risk (Shen et al., 2023). Previous research involving mushroom-based inks reported water activity values of 0.60 before and 0.66 after printing, with the increase attributed to water becoming entrapped within the fiber matrix during formulation (Keerthana et al., 2020; Demei et al., 2022).

In recent developments, nonthermal sterilization technologies such as CP, pulsed electric fields, ultraviolet light, pulsed light, and ultrasound have been explored for their effectiveness in ensuring microbial safety (Fig. 5.3). These approaches offer advantages like rapid processing and sterilizing packaged products, thus minimizing recontamination (Shyam and Palaniappan, 2023). Sharma et al. (2020) applied air plasma sterilization for 30 s to polyvinyl alcohol microspheres loaded with guggulsterone before incorporating them into a fibrin-based bioink designed for 3D bioprinting neural tissues. However, they did not evaluate the sterilization's impact on cell viability or print performance. Similarly, Hodder et al. (2019) studied four sterilization techniques on alginate/methylcellulose (alg/MC) bioinks containing bovine chondrocytes. They found that γ-irradiation at 25 kGy significantly compromised the bioink's viscosity and structural stability, making it unsuitable for bioprinting applications. Despite these

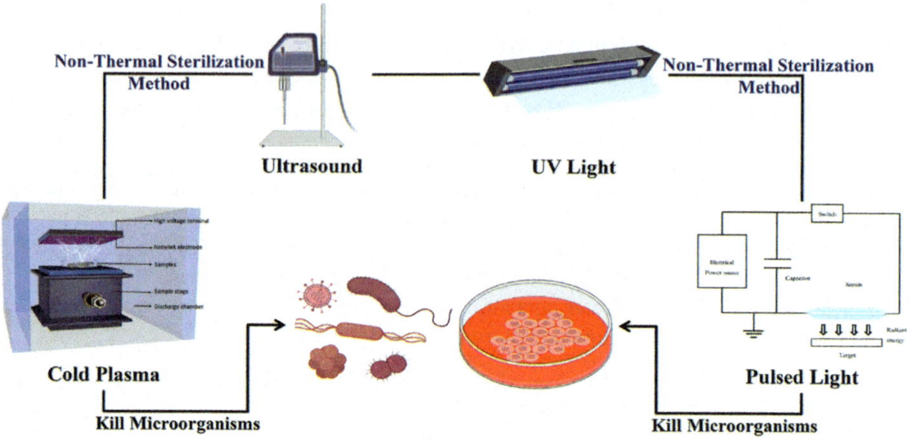

Fig. 5.3 Sterilization process to reduce the microbial load in the 3D printed food products (Mandal et al., 2020; Wu et al., 2023).
From Mandal, R., Mohammadi, X., Wiktor, A., Singh, A., Pratap Singh, A., 2020. Applications of pulsed light decontamination technology in food processing: an overview. Appl. Sci. 10 (10). doi:10.3390/app10103606; Wu, Y., Feng, X., Zhu, Y., Li, S., Hu, Y., Yao, Y., Zhou, N., 2023. The effect of atmospheric dielectric barrier discharge cold plasma treatment on the nutritional and physicochemical characteristics of various legumes. Foods 12 (17), 3260.

studies, there remains limited discussion about hygiene protocols specific to 3D printers and food printing workflows in the literature. In addition, the absence of robust data on microbial loads and printer sterilization reliability underlines the need for more focused research on decontamination strategies tailored to 3D food printing systems.

5.9 Advances and innovations in 3D food printing

Recent progress in 3D food printing, especially extrusion-based techniques, has significantly changed the production of personalized and functional foods. Advancements in nozzle design, material formulations, and digital modeling techniques have made it possible to precisely layer various food components with tailored shapes, textures, and nutritional properties. The development of multimaterial printing enables the concurrent extrusion of ingredients with different rheological characteristics, increasing the complexity and functionality of the printed foods. In addition, the integration of artificial intelligence and real-time monitoring has led to improvements in printing precision, consistency, and scalability. These innovations are beneficial in healthcare, providing solutions for personalized nutrition for patients with specific dietary requirements, as well as in the culinary field for creating intricate designs that are difficult to achieve using traditional methods. As research continues to overcome challenges related to material compatibility, printability, and postprocessing, extrusion-based 3D food printing is set to play a significant role in enabling sustainable, customizable, and intelligent food production.

5.10 Future scope and opportunities

The potential of extrusion-based 3D food printing to transform personalized nutrition, food sustainability, and on-demand food production is vast. Ongoing developments in materials science offer opportunities to create innovative food inks that are nutrient-dense and tailored to individual dietary preferences. Incorporating real-time sensors and AI technologies will allow for precise manipulation of texture, structure, and nutritional composition during printing. In addition, combining this technology with advanced pretreatment methods, such as CP, ultrasound, or enzymatic treatments, could significantly improve the bioavailability and stability of bioactive compounds in printed foods. As the technology evolves, its use in clinical nutrition, space exploration, military rations, and elderly care will likely expand. Future research should focus on creating scalable, cost-efficient printing systems, standardizing printable food-grade materials, and enhancing consumer acceptance to move 3D-printed foods from research laboratories to the marketplace. Cross-disciplinary collaborations will be key in driving advancements and establishing extrusion-based 3D food printing as a mainstream food processing method.

5.11 Conclusion

Extrusion-based 3D printing has become a groundbreaking technology in the food industry, providing exceptional levels of customization, accuracy, and the ability to create personalized nutrition. This review explores the core concepts of extrusion printing, emphasizing the significance of optimizing food ink formulations and various pretreatment techniques, such as thermal, enzymatic, CP, ultrasound, and microwave treatments, to improve printability and the quality of the final product. It also discusses the natural physicochemical changes that occur during and after printing, as well as the critical role of postprocessing techniques, such as drying and sterilization, to ensure the stability and safety of the product. With ongoing advancements and cross-disciplinary research, extrusion-based 3D food printing presents exciting prospects for the development of functional foods, sustainable food design, and customized dietary solutions, positioning it as a key tool in the future of food manufacturing.

References

Abedini, A., Hosseini, H., Shariatifar, N., Molaee-aghaee, E., Sadighara, P., 2024. Studying the impact of 3d printing technology on safety indicators of plant-based burger. Food Chem. X. 25901575. 22, 101489. doi:10.1016/j.fochx.2024.101489.

Agunbiade, A.O., Song, L., Agunbiade, O.J., Ofoedu, C.E., Chacha, J.S., Duguma, H.T., Hossaini, S.M., Rasaq, W.A., Shorstkii, I., Osuji, C.M., Owuamanam, C.I., Okpala, C.O.R., Korzeniowska, M., Guine, R.P.F., 2022. Potentials of 3D extrusion-based printing in resolving food processing challenges: a perspective review. J. Food

Process Eng. 17454530. 45 (4). Available from: http://onlinelibrary.wiley.com/journal/10.1111/(ISSN)1745-4530. doi:10.1111/jfpe.13996.

Alemzadeh, K., 2025. Innovative bionics product life-cycle management methodology framework with built-in reverse biomimetics: from inception to clinical validation. Biomimetics. 2313-7673. 10 (3), 158. doi:10.3390/biomimetics10030158.

Bănică, C.-F., Sover, A., Anghel, D.-C., 2024. Printing the future layer by layer: a comprehensive exploration of additive manufacturing in the era of industry 4.0. Appl. Sci. 2076-3417. 14 (21), 9919. doi:10.3390/app14219919.

Barrios-Rodríguez, Y.F., Igual, M., Martínez-Monzó, J., García-Segovia, P., 2024. Exploration of changes in rheological and spectral properties of rice protein inks before and after 3D printing. LWT. 00236438. 210, 116808. doi:10.1016/j.lwt.2024.116808.

Bugday, Z.Y., Venkatachalam, A., Anderson, P.D., van der Sman, R.G.M., 2024. Rheology of paste-like food inks for 3D printing: effects of nutrient and water content. Curr. Res. Food Sci. 26659271. 9, 100847. doi:10.1016/j.crfs.2024.100847.

Cai, M., Zhang, Y., Cao, H., Sen, L., Zhang, Y., Huang, K., Song, H., Guan, X., 2024. Exploring the remarkable effects of microwave treatment on starch modification: from structural evolution to changed physicochemical and digestive properties. Carbohydr. Polym. 01448617. 343, 122412. doi:10.1016/j.carbpol.2024.122412.

Cai, Q., Zhong, Y., Xu, M., Huang, Q., Lu, X., 2022. 3D printed high oil custard cream: effects of whey protein isolate, hydroxypropylated starch and carrageenan on physicochemical properties and printing performance. LWT. 00236438. 156, 113039. doi:10.1016/j.lwt.2021.113039.

Chen, F., Zhang, M., Liu, Z., Bhandari, B., 2021b. 4D deformation based on double-layer structure of the pumpkin/paper. Food Struct. 27, 100168.

Chen, X., Zhang, M., Tang, T., 2023. Microwave-induced rapid shape change of 4D printed vegetable-based food. Foods 12 (11). doi:10.3390/foods12112158.

Chen, X., Zhang, M., Tang, T., Dongxing, Y., 2024. The color/shape/flavor of yam gel with double emulsified microcapsules changed synchronously in 4D printing induced by microwave. Int. J. Biol. Macromol. 01418130. 260 (Pt 2), 129631. doi:10.1016/j.ijbiomac.2024.129631.

Chen, Y., Zhang, M., Phuhongsung, P., 2021a. 3D printing of protein-based composite fruit and vegetable gel system. LWT. 00236438. 141, 110978. doi:10.1016/j.lwt.2021.110978.

Cui, R., Zhu, F., 2021. Ultrasound modified polysaccharides: a review of structure, physicochemical properties, biological activities and food applications. Trends Food Sci. Technol. 107, 491–508.

Demei, K., Zhang, M., Phuhongsung, P., Mujumdar, A.S., 2022. 3D food printing: controlling characteristics and improving technological effect during food processing. Food Res. Int. 09639969. 156, 111120. doi:10.1016/j.foodres.2022.111120.

Derossi, A., 2021. 3D Food printing: opportunities, principles, limitations, and new ways in food production. In: IUFoST Scientific Information Bulletins (SIBs).

Derossi, A., Caporizzi, R., Azzollini, D., Severini, C., 2018. Application of 3D printing for customized food. A case on the development of a fruit-based snack for children. J. Food Eng. 02608774. 220, 65–75. doi:10.1016/j.jfoodeng.2017.05.015.

Dick, A., Gao, Y., Bhandari, B., Prakash, S., 2021. Influence of drying method and 3D design on the 4D morphing of beef products. Appl. Food Res. 27725022. 1 (2), 100017. doi:10.1016/j.afres.2021.100017.

Domżalska, Z., Jakubczyk, E., 2025. Characteristics of food printing inks and their impact on selected product properties. Foods. 2304-8158. 14 (3), 393. doi:10.3390/foods14030393.

Enfield, R.E., Pandya, J.K., Lu, J., McClements, D.J., Kinchla, A.J., 2023. The future of 3D food printing: opportunities for space applications. Crit. Rev. Food Sci. Nutr. 63 (29), 10079–10092.

Fan, H., Zhang, M., Liu, Z., Yufen, Y., 2020. Effect of microwave-salt synergetic pre-treatment on the 3D printing performance of SPI-strawberry ink system. LWT. 00236438. 122, 109004. doi:10.1016/j.lwt.2019.109004.

Feng, C., Zhang, M., Liu, Z., Mujumdar, A., Wang, Y., Chang, L., 2021. Effect of drying method on post-processing stability and quality of 3D printed rose-yam paste. Dry. Technol. 15322300. 39 (9), 1196–1204. Available from: https://www.tandf.co.uk/journals/titles/07373937.asp. doi:10.1080/07373937.2020.1851708.

Ghazal, A.F., Zhang, M., Bhandari, B., Chen, H., 2021. Investigation on spontaneous 4D changes in color and flavor of healthy 3D printed food materials over time in response to external or internal pH stimulus. Food Res. Int. 09639969. 142, 110215. doi:10.1016/j.foodres.2021.110215.

Ghimire, S., Gamonpilas, C., Umar, M., Anal, A.K., 2025. Production of 3D-printed meat analogues using pea, fava, and mung bean proteins: a comparison study. Food Struct. 22133291. 44, 100419. doi:10.1016/j.foostr.2025.100419.

Gholamipour-Shirazi, A., Norton, I.T., Mills, T., 2019. Designing hydrocolloid based food-ink formulations for extrusion 3D printing. Food Hydrocoll. 95, 161–167.

Guo, C., Zhang, M., Bhandari, B., Devahastin, S., 2022. Investigation on simultaneous change of deformation, color and aroma of 4D printed starch-based pastes from fruit and vegetable as induced by microwave. Food Res. Int. 09639969. 157, 111214. doi:10.1016/j.foodres.2022.111214.

Guo, C., Zhang, M., Devahastin, S., 2021. Color/aroma changes of 3D-Printed buckwheat dough with yellow flesh peach as triggered by microwave heating of gelatin-gum Arabic complex coacervates. Food Hydrocoll. 0268005X. 112 (106358). doi:10.1016/j.foodhyd.2020.106358.

He, C., Zhang, M., Guo, C., 2020. 4D printing of mashed potato/purple sweet potato puree with spontaneous color change. Innov. Food Sci. Emerg. Technol. 14668564. 59, 102250. doi:10.1016/j.ifset.2019.102250.

Hertafeld, E., Zhang, C., Jin, Z., Jakub, A., Russell, K., Lakehal, Y., Andreyeva, K., Bangalore, S., J., Mezquita, Blutinger, J., Lipson, H., 2018. Multi-material three-dimensional food printing with simultaneous infrared cooking. 3D Print. Addit. Manuf. 6. doi:10.1089/3dp.2018.0042.

Hodder, E., Duin, S., Kilian, D., Ahlfeld, T., Seidel, J., Nachtigall, C., Bush, P., Covill, D., Gelinsky, M., Lode, A., 2019. Investigating the effect of sterilisation methods on the physical properties and cytocompatibility of methyl cellulose used in combination with alginate for 3D-bioplotting of chondrocytes. J. Mater. Sci. Mater. Med. 0957-4530. 30 (1). doi:10.1007/s10856-018-6211-9.

Hodhodi, A., Rahimabadi, E.Z., Simorgh, S., Safari, P., 2025. 3D Food Printing: A Promising Approach to Meet Food Security. Springer Science and Business Media LLC, pp. 1–43. doi:10.1007/978-3-031-71164-0_52-1.

Hoque, M., McDonagh, C., Tiwari, B.K., Kerry, J.P., Pathania, S., 2022. Effect of cold plasma treatment on the packaging properties of biopolymer-based films: a review. Appl. Sci. 2076-3417. 12 (3), 1346. doi:10.3390/app12031346.

Hussain, S., Malakar, S., Arora, V.K., 2022. Extrusion-based 3D food printing: technological approaches, material characteristics, printing stability, and post-processing. Food Eng. Rev. 14 (1), 100–119. doi:10.1007/s12393-021-09293-w.

Jayan, L.S., Wani, K.M., 2025. Utilization of industrial fruits and vegetables waste streams for the production of 3D printed and synbiotic foods. J. Food Meas. Charact. 21934134. 19 (6), 3867–3887. Available from: https://www.springer.com/journal/11694. doi:10.1007/s11694-025-03263-0.

Jiang, X., Huang, Y., Cheng, Y., Zhang, Z., Shi, X., Qin, H., 2021. Effects of lyophilization on the release profiles of 3d printed delivery systems fabricated with carboxymethyl cellulose hydrogel. Polymers. 20734360. 13 (5), 1–11. Available from: https://www.mdpi.com/2073-4360/13/5/749/pdf. doi:10.3390/polym13050749.

Kang, Z., Wang, Z., Zhang, W., Fang, Z., 2023. 4D printing of edible insect snacks: color changes with external pH stimulation. Food Biosci. 22124292. 55, 102971. doi:10.1016/j.fbio.2023.102971.

Kangas, H., Felissia, F.E., Filgueira, D., Ehman, N.V., Vallejos, M.E., Imlauer, C.M., Lahtinen, P., Area, M.C., Chinga-Carrasco, G., 2019. 3D printing high-consistency enzymatic nanocellulose obtained from a soda-ethanol-O_2 pine sawdust pulp. Bioengineering. 23065354. 6 (3). Available from: https://www.mdpi.com/2306-5354/6/3/60/pdf. doi:10.3390/bioengineering6030060.

Keerthana, K., Anukiruthika, T., Moses, J.A., Anandharamakrishnan, C., 2020. Development of fiber-enriched 3D printed snacks from alternative foods: a study on button mushroom. J. Food Eng. 02608774. 287, 110116. doi:10.1016/j.jfoodeng.2020.110116.

Kim, H.B., Kim, S.W., Ko, J.B., Yang, Y.J., Oh, G.Y., Kim, Y.W., Kim, H.C., 2021. Optimization of extrusion process for 3D hard candy manufacture. J. Manuf. Process. 15266125. 71, 580–588. Available from: http://www.elsevier.com/wps/find/journaldescription.cws_home/620379/description#description. doi:10.1016/j.jmapro.2021.09.035.

Kim, S.M., Woo, J.H., Kim, H.W., Park, H.J., 2022. Formulation and evaluation of cold-extruded chocolate ganache for three-dimensional food printing. J. Food Eng. 02608774. 314. Available from: http://www.sciencedirect.com/science/journal/02608774. doi:10.1016/j.jfoodeng.2021.110785.

Kim, T.K., Yong, H.I., Kim, Y.B., Kim, H.W., Choi, Y.S., 2019. Edible insects as a protein source: a review of public perception, processing technology, and research trends. Food Sci. Anim. Resour. 39 (4), 521.

Koirala, S., Prakash, S., Karim, A., Bhandari, B., 2023. Shape morphing of foods: mechanism, strategies, and applications. Trends Food Sci. Technol. 09242244. 139, 104135. doi:10.1016/j.tifs.2023.104135.

Koirala, S., Prakash, S., Karim, A., Bhandari, B., 2025. 4D shape morphing in 3D-printed pea protein structures through designed surface grooves under drying and frying conditions. Innov. Food Sci. Emerg. Technol. 14668564. 100, 103924. doi:10.1016/j.ifset.2025.103924.

Kumar, N.S., Dar, A.H., Dash, K.K., Kaur, B., Pandey, V.K., Singh, A., et al., 2024. Recent advances in cold plasma technology for modifications of proteins: a comprehensive review. J. Agric. Food Res. 16, 101177.

Kuo, C.-C., Qin, H., Cheng, Y., Jiang, X., Shi, X., 2021. An integrated manufacturing strategy to fabricate delivery system using gelatin/alginate hybrid hydrogels: 3D printing and freeze-drying. Food Hydrocoll. 0268005X. 111, 106262. doi:10.1016/j.foodhyd.2020.106262.

Lee, C.K.W., Xu, Y., Yuan, Q., Chan, Y.H., Poon, W.Y., Zhong, H., Chen, S., Li, M.G., 2025. Advanced 3D food printing with simultaneous cooking and generative AI design. Adv. Mater. 15214095. 37 (13). Available from: http://onlinelibrary.wiley.com/journal/10.1002/(ISSN)1521-4095. doi:10.1002/adma.202408282.

Li, N., Li, H., Liu, Z., Lv, S., Xie, S., Shi, C., Wu, Y., 2023. Preparation and product characterization of microwaveable food using lentinus edodes protein through 3D printing. Polymers. 2073-4360. 15 (18), 3736. doi:10.3390/polym15183736.

Liu, H., Changchun, W., Lin, S., Lam, J., Ning, X., Chen, Y., 2025. Advances in 3D and 4D printing of soft robotics and their applications. Adv. Intell. Syst. 2640-4567. 7 (6). doi:10.1002/aisy.202400699.

Liu, L., Meng, Y., Dai, X., Chen, K., Zhu, Y., 2019a. 3D printing complex egg white protein objects: properties and optimization. Food Bioprocess Technol. 12, 267–279.

Liu, L., Xie, T., Cheng, W., Ding, Y., Xu, B., 2024. Characterization and mechanism of thermally induced tea polyphenols and egg white proteins gel system and its 3D printing. Lwt 205,116488.

Liu, Y., Yu, Y., Liu, C., Regenstein, J.M., Liu, X., Zhou, P., 2019b. Rheological and mechanical behavior of milk protein composite gel for extrusion-based 3D food printing. LWT. 00236438. 102, 338–346. Available from: https://www.journals.elsevier.com/lwt. doi:10.1016/j.lwt.2018.12.053.

Liu, Z., Chang, H., Guo, C., Chen, F., Bhandari, B., Zhang, M., 2021. Dehydration-triggered shape transformation of 4D printed edible gel structure affected by material property and heating mechanism. Food Hydrocoll. 0268005X. 115, 106608. doi:10.1016/j.foodhyd.2021.106608.

Liu, Z., Zhang, M., Bhandari, B., Wang, Y., 2017. 3D printing: printing precision and application in food sector. Trends Food Sci. Technol. 09242244. 69, 83–94. doi:10.1016/j.tifs.2017.08.018.

Lorenz, T., Iskandar, M.M., Baeghbali, V., Ngadi, M.O., Kubow, S., 2022. 3D food printing applications related to dysphagia: a narrative review. Foods 11 (12), 1789. Available from: https://www.mdpi.com/2304-8158/11/12/1789.

Lu, K., Du, Z., Chen, J., Kong, Y., Huang, Q., 2024. Radio frequency air cold plasma pretreatment to enhance 3D printing performance of tapioca starch by altering its rheological behavior and water migration. Food Hydrocoll. 0268005X. 153, 110032. doi:10.1016/j.foodhyd.2024.110032.

Lv, Y., Wang, B., Cheng, Y., Lv, W., Zeng, S., Xiao, H., 2024. Printing characteristics and microwave infrared-induced 4D printing of chestnut powder composite paste. J. Food Eng. 02608774. 382, 112197. doi:10.1016/j.jfoodeng.2024.112197.

Ma, Y., Potappel, J., Chauhan, A., Schutyser, M.A.I., Boom, R.M., Zhang, L., 2023. Improving 3D food printing performance using computer vision and feedforward nozzle motion control. J. Food Eng. 02608774. 339. Available from: http://www.sciencedirect.com/science/journal/02608774. doi:10.1016/j.jfoodeng.2022.111277.

Ma, Y., Zhang, L., 2022. Formulated food inks for extrusion-based 3D printing of personalized foods: a mini review. Curr. Opin. Food Sci. 22147993. 44, 100803. doi:10.1016/j.cofs.2021.12.012.

Maldonado-Rosas, R., Tejada-Ortigoza, V., Cuan-Urquizo, E., Mendoza-Cachú, D., Morales-de la Peña, M., Alvarado-Orozco, J.M., Campanella, O.H., 2022. Evaluation of rheology and printability of 3D printing nutritious food with complex formulations. Addit. Manuf. 22148604. 58. Available from: https://www.journals.elsevier.com/additive-manufacturing. doi:10.1016/j.addma.2022.103030.

Mantihal, S., Prakash, S., Godoi, F.C., Bhandari, B., 2019. Effect of additives on thermal, rheological and tribological properties of 3D printed dark chocolate. Food Res. Int. 18737145. 119, 161–169. Available from: https://www.elsevier.com/inca/publications/store/4/2/2/9/7/0. doi:10.1016/j.foodres.2019.01.056.

Martekha, A.N., Kaverina, Y.E., 2023. Kinetic evaluation and optimization of the drying process of 3D printed pasta: key factors influence on the finished product quality. Food Sci. Technol. 1678-457X. 43. doi:10.1590/fst.067522.

Martínez-Monzó, J., Cárdenas, J., García-Segovia, P., 2019. Effect of temperature on 3D printing of commercial potato puree. Food Biophys. 1557-1858. 14 (3), 225–234. doi:10.1007/s11483-019-09576-0.

Mudau, M., Adebo, O.A., 2024. Three dimensional (3D)-printed foods: a review of recent advances in their ingredients, printing techniques, food printers, post-processing methods, consumer acceptance and safety. J. Food Process Eng. 17454530. 47 (5). Available from: http://onlinelibrary.wiley.com/journal/10.1111/(ISSN)1745-4530. doi:10.1111/jfpe.14621.

Nachal, N., Moses, J.A., Karthik, P., Anandharamakrishnan, C., 2019. Applications of 3D printing in food processing. Food Eng. Rev. 1866-7910. 11 (3), 123–141. doi:10.1007/s12393-019-09199-8.

Nakra, S., Tripathy, S., Srivastav, P.P., 2025. Drying as a preservation strategy for medicinal plants: physicochemical and functional outcomes for food and human health. Phytomed. Plus 5 (2), 100762.

Oh, Y., Lee, S., Lee, N.K., Rhee, J.K., 2024. Improving the three-dimensional printability of potato starch loaded onto food ink. J. Microbiol. Biotechnol. 17388872. 34 (4), 891–901. Available from: https://www.jmb.or.kr/journal/view.html?uid=6402&vmd=Full. doi:10.4014/jmb.2311.11040.

Ojha, K.S., Mason, T.J., O'Donnell, C.P., Kerry, J.P., Tiwari, B.K., 2017. Ultrasound technology for food fermentation applications. Ultrason. Sonochem. 18732828. 34, 410–417. Available from: https://www.elsevier.com/inca/publications/store/5/2/5/4/5/1. doi:10.1016/j.ultsonch.2016.06.001.

Okita, A., Takahashi, K., Itakura, M., Horio, A., Yamamoto, R., Nakamura, Y., Osako, K., 2021. A novel soft surimi gel with functionality prepared using alcalase for people suffering from dysphagia. Food Chem. 344, 128641.

Outrequin, T.C.R., Gamonpilas, C., Siriwatwechakul, W., Sreearunothai, P., 2023. Extrusion-based 3D printing of food biopolymers: a highlight on the important rheological parameters to reach printability. J. Food Eng. 02608774. 342. Available from: http://www.sciencedirect.com/science/journal/02608774. doi:10.1016/j.jfoodeng.2022.111371.

Ozdemir, B., Hernández-del-Valle, M., Gaunt, M., Schenk, C., Echevarría-Pastrana, L., Fernández-Blázquez, J.P., Wang, D.Y., Haranczyk, M., 2024. Toward 3D printability prediction for thermoplastic polymer nanocomposites: insights from extrusion printing of PLA-based systems. Addit. Manuf. 22148604. 95. Available from: https://www.sciencedirect.com/science/journal/22148604. doi:10.1016/j.addma.2024.104533.

Pant, A., Lee, A.Y., Karyappa, R., Lee, C.P., An, J., Hashimoto, M., Tan, U.X., Wong, G., Chua, C.K., Zhang, Y., 2021. 3D food printing of fresh vegetables using food hydrocolloids for dysphagic patients. Food Hydrocoll. 0268005X. 114. Available from: https://www.sciencedirect.com/science/journal/0268005X. doi:10.1016/j.foodhyd.2020.106546.

Phuhongsung, P., Zhang, M., Devahastin, S., 2020. Influence of surface pH on color, texture and flavor of 3D printed composite mixture of soy protein isolate, pumpkin, and beetroot. Food Bioprocess Technol. 13 (9), 1600–1610.

Rinshana, P.F., Murugesan, B., Kim, Y.H., Alaguthevar, R., Rhim, J.W., 2025. Advances in 3D food printing technology: innovation and applications in the food industry. Food Sci. Biotechnol. 20926456. 34 (2), 403–421. Available from: https://www.springer.com/journal/10068. doi:10.1007/s10068-024-01779-7.

Rout, S., Tripathy, S., Srivastav, P.P., 2024. Effect of cold plasma for modulating macromolecules and bioactive composition of food: unveiling mechanisms and synergies with other emerging techniques. Food Biosci. 61, 104545.

Rysenaer, V.B.J., Ahmadzadeh, S., Van Bockstaele, F., Ubeyitogullari, A., 2023. An extrusion-based 3D food printing approach for generating alginate-pectin particles. Curr. Res. Food Sci. 26659271. 6. Available from: https://www.journals.elsevier.com/current-research-in-food-science. doi:10.1016/j.crfs.2022.11.023.

Schutyser, M.A.I., Houlder, S., de Wit, M., Buijsse, C.A.P., Alting, A.C., 2018. Fused deposition modelling of sodium caseinate dispersions. J. Food Eng. 02608774. 220, 49–55. doi:10.1016/j.jfoodeng.2017.02.004.

Severini, C., Derossi, A., Ricci, I., Caporizzi, R., Fiore, A., 2018. Printing a blend of fruit and vegetables. New advances on critical variables and shelf life of 3D edible objects. J. Food Eng. 02608774. 220, 89–100. doi:10.1016/j.jfoodeng.2017.08.025.

Shanthamma, S., Preethi, R., Moses, J.A., Anandharamakrishnan, C., 2021. 4D printing of sago starch with turmeric blends: a study on pH-triggered spontaneous color transformation. ACS Food Sci. Technol. 2692-1944. 1 (4), 669–679. doi:10.1021/acsfoodscitech.0c00151.

Sharma, R., Smits, I.P.M., De La Vega, L., Lee, C., Willerth, S.M., 2020. 3D Bioprinting pluripotent stem cell derived neural tissues using a novel fibrin bioink containing drug releasing microspheres. Front. Bioeng. Biotechnol. 22964185. 8. Available from: http://journal.frontiersin.org/journal/bioengineering-and-biotechnology#archive. doi:10.3389/fbioe.2020.00057.

Shen, D., Zhang, M., Mujumdar, A.S., Li, J., 2023. Advances and application of efficient physical fields in extrusion based 3D food printing technology. Trends in Food Science and Technology. 09242244. 131, 104–117. Available from: http://www.elsevier.com/wps/find/journaldescription.cws_home/601278/description#description. doi:10.1016/j.tifs.2022.11.017.

Shi, Y., Zhang, M., Bhandari, B., 2021. Effect of addition of beeswax based oleogel on 3D printing of potato starch-protein system. Food Struct. 22133291. 27, 100176. doi:10.1016/j.foostr.2021.100176.

Shyam, R., Palaniappan, A., 2023. Effect of sterilization techniques on biomaterial inks' properties and 3D bioprinting parameters. Bioprinting. 24058866. 33, e00294. doi:10.1016/j.bprint.2023.e00294.

Sohel, A., Sahu, S., Mitchell, G.R., Patel, M.K., 2025. 3D Food printing: a comprehensive review and critical analysis on technologies, food materials, applications, challenges, and future prospects. Food Eng. Rev. 18667929. 17 (2), 220–248. Available from: https://www.springer.com/journal/12393. doi:10.1007/s12393-025-09400-1.

Sousa, T., Kajave, N., Dong, P., Gu, L., Florczyk, S., Kishore, V., 2022. Optimization of freeze-fresh methodology for 3D printing of microporous collagen constructs. 3D Print. Addit. Manuf. 2329-7662. 9 (5), 411–424. doi:10.1089/3dp.2020.0311.

Sudheesh, C., Sunooj, K.V., Aaliya, B., Navaf, M., Akhila, P.P., Mir, S.A., Sabu, S., Sasidharan, A., Sudheer, K.P., Sinha, S.K., Sajeevkumar, V.A., George, J., 2023. Effect of energetic neutrals on the kithul starch retrogradation; Potential utilization for improving mechanical and barrier properties of films. Food Chem. 03088146. 398, 133881. doi:10.1016/j.foodchem.2022.133881.

Tan, C., Toh, W.Y., Wong, G., Li, L., 2018. Extrusion-based 3D food printing – materials and machines. Int. J. Bioprint. 24248002. 4 (2). Available from: https://accscience.com/journal/IJB/volumes_and_issues. doi:10.18063/IJB.v4i2.143.

Teng, X., Li, C., Mujumdar, A.S., Zhang, M., 2022. Progress in extrusion-based food printing technology for enhanced printability and printing efficiency of typical personalized foods: a review. Foods. 23048158. 11 (24). Available from: https://www.mdpi.com/journal/foods. doi:10.3390/foods11244111.

Thorakkattu, P., Awasti, N., Babu, K.S., Khanashyam, A.C., Deliephan, A., Shah, K., Singh, P., Pandiselvam, R., Nirmal, N.P., 2025. 3D printing: trends and approaches toward achieving long-term sustainability in the food industry. Crit. Rev. Biotechnol. 15497801. 45 (1), 48–68. Available from: http://www.tandfonline.com/loi/ibty20. doi:10.1080/07388551.2024.2344577.

Tripathy, S., Srivastav, P.P., 2023. Effect of dielectric barrier discharge (DBD) cold plasma-activated water pre-treatment on the drying properties, kinetic parameters, and physico-chemical and functional properties of Centella asiatica leaves. Chemosphere. 00456535. 332, 138901. doi:10.1016/j.chemosphere.2023.138901.

Tripathy, S., Srivastav, P.P., 2024. Synergistic effects of dielectric barrier discharge (DBD) cold plasma pretreatment combined with microwave drying on the physicochemical and functional properties of centella asiatica leaves. Food Bioproc. Technol. 19355149. 17 (11), 3979–3998. Available from: https://www.springer.com/journal/11947. doi:10.1007/s11947-024-03364-6.

Verma, D.K., Tripathy, S., Srivastav, P.P., 2024. Microwave heating in rice and its influence on quality and techno-functional parameters of rice compositional components. J. Food Compos. Anal. 128, 106030.

Wang, J., Hu, H., Liu, Z., Shi, Y., Huang, Y., 2025. Optimization of 3D printing nozzle parameters and the optimal combination of 3D printer process parameters for engineering plastics with high melting points and large thermal expansion coefficients. Materials. 1996-1944. 18 (3), 500. doi:10.3390/ma18030500.

Wang, Z., Florczyk, S.J., 2020. Freeze-fresh: a 3D printing technique to produce biomaterial scaffolds with hierarchical porosity. Materials. 19961944. 13 (2). Available from: https://res.mdpi.com/d_attachment/materials/materials-13-00354/article_deploy/materials-13-00354.pdf. doi:10.3390/ma13020354.

Waseem, M., Tahir, A.U., Majeed, Y., 2024. Printing the future of food: the physics perspective on 3D food printing. Food Phys. 29500699. 1. Available from: https://www.sciencedirect.com/journal/food-physics/about/aims-and-scope. doi:10.1016/j.foodp.2023.100003.

Wei, M., Lin, K., Sun, L., 2022. Shear thickening fluids and their applications. Mater. Des. 216, 110570.

Xie, Y., Liu, Q., Zhang, W., Yang, F., Zhao, K., Dong, X., Prakash, S., Yuan, Y., 2023. Advances in the potential application of 3D food printing to enhance elderly nutritional dietary intake. Foods. 2304-8158. 12 (9), 1842. doi:10.3390/foods12091842.

Yang, F., Min, Z., Liu, Y., 2019. Effect of post-treatment microwave vacuum drying on the quality of 3D-printed mango juice gel. Dry. Technol. 37 (14), 1757–1765. doi:10.1080/07373937.2018.1536884.

Yang, G., Tao, Y., Wang, P., Xu, X., Zhu, X., 2022. Optimizing 3D printing of chicken meat by response surface methodology and genetic algorithm: feasibility study of 3D printed chicken product. LWT. 00236438. 154, 112693. doi:10.1016/j.lwt.2021.112693.

Zheng, B., Qiu, Z., Liu, Z., Chen, L., 2024. Pre-dry heat treatment alters the structure and ultimate in vitro digestibility of wheat starch-lipids complex in hot-extrusion 3D printing. Carbohydr. Polym. 01448617. 334, 122026. doi:10.1016/j.carbpol.2024.122026.

Zhou, Q., Nan, X., Zhang, S., Zhang, L., Chen, J., Jiayi, L., Wang, H., Ruan, Z., 2023. Effect of 3D food printing processing on polyphenol system of loaded aronia melanocarpa and post-processing evaluation of 3D printing products. Foods. 2304-8158. 12 (10), 2068. doi:10.3390/foods12102068.

Zhu, S., Stieger, M.A., van der Goot, A.J., Schutyser, M.A.I., 2019. Extrusion-based 3D printing of food pastes: correlating rheological properties with printing behaviour. Innovative Food Science and Emerging Technologies. 14668564. 58. Available from: http://www.elsevier.com/locate/ifset. doi:10.1016/j.ifset.2019.102214.

Zhu, W., Iskandar, M.M., Baeghbali, V., Kubow, S., 2023. Three-dimensional printing of foods: a critical review of the present state in healthcare applications, and potential risks and benefits. Foods. 23048158. 12 (17). Available from: https://www.mdpi.com/journal/foods. doi:10.3390/foods12173287.

Further Reading

Ming, G., Huiwen, G., Raghavan, V., Wang, J., 2024. Ultrasound treatment improved the physicochemical properties of pea protein with pectin ink used for 3D printing. Future Foods. 26668335. 9, 100377. doi:10.1016/j.fufo.2024.100377.

Bioextrusion

6

Beijia Wan[a], *Siyu Yao*[a], *Zhengzong Wu*[b], *and Enbo Xu*[a]
[a]College of Biosystems Engineering and Food Science, Zhejiang University, Hangzhou, China
[b]Qilu University of Technology, Hangzhou, China

6.1 Introduction

To meet the demands for diverse and high-quality products as the food and feed industries evolve, interdisciplinary processing technology has been continuously developed. Among these technologies, extrusion—a high-temperature, short-time, continuous, multifunctional thermal process (even combined with biocatalysis, as detailed below)—is widely used in the fields of food and feed (Bouvier and Campanella, 2014; Li et al., 2019). It integrates material mixing, stirring, cutting, heating, molding, and other operations to achieve high productivity. As early as the mid-1930s, extrusion was applied to pasta and cereal processing. However, due to the distinctive structures of natural polymers, conventional extrusion often shows limited capacity for deep processing. To address this limitation, enzymes can be purposefully introduced before, during, or after the extrusion process, assisting extrusion actions to enhance the conversion of substrates. Notably, in the 1980s, an enzyme was added to extrusion with a starch-based substrate, demonstrating that thermostable α-amylase (TαA) could accelerate the liquefaction of starch. This pivotal development marked the inception of enzymatic reactive extrusion (eREX, referred to as such only in this book chapter). Since enzymes function as biocatalysts, this technique of simultaneously introducing exogenous enzymes into the extrusion process can also be defined as "bioextrusion" (Xu et al., 2020). Subsequently, researchers have added enzymes from various sources to both food and nonfood extrusion processes. To date, the range of bioextruded materials has expanded from the initial starch-based foods to other biopolymers, such as proteins and cellulose.

Bioextrusion technology not only opens the enclosed structure of raw materials through physical actions but also achieves accurate conversion at the molecular level through the addition of specific enzymes (Xu et al., 2020). This innovative approach integrates thermomechanical extrusion with biocatalysis, providing a novel solution strategy for processing natural biological macromolecules. It is expected to be implemented in industrial scenarios or potentially used for economic and environmental value in the fields of saccharification and fermentation (Dang and Vasanthan, 2019). Bioextrusion has demonstrated substantial advantages in the bioconversion of natural biopolymers, such as starch-, protein-, and biomass-based materials (Gatt et al., 2019;

Extrusion Cooking for Food and Feed. https://doi.org/10.1016/B978-0-443-29987-2.00006-3

Xu et al., 2020). Thus in this book chapter, we focus on the enzymatic reactions that break down complex biopolymer structures, enhance reaction selectivity and efficiency, and enable specific modifications of edible biopolymers during bioextrusion. This process may improve the physical and chemical properties of extrudates, enhance the sensory quality of edible substrates, increase the stability of endogenous/exogenous bioactive compounds, or convert (agro)waste resources into valuable products. Overall, the application prospects of bioextrusion as an enzyme-assisted thermomechanical technology are becoming increasingly vast in fields such as food, feed, pharmaceuticals, and biomaterials.

6.2 General processing features of bioextrusion

Enzymes play crucial roles in the extrusion process as bioreactors to achieve the desired processing goals (Xu et al., 2020). Enzymatic action, offering good selectivity for naturally polymeric substrates, is safer and faster than other chemical reactions (Shokri et al., 2022; Zhang et al., 2015). Extensive research on enzymes has led to the extraction, purification, and characterization of a wide range of enzymes with improved stability and catalytic efficiency, providing a fundamental development of enzymatic extrusion, i.e., bioextrusion.

In high-efficiency enzymatic reactions, it is generally preferred to maintain high moisture (with substrate concentration less than around 10%), low-shear mixing, and optimal batch conditions with moderate pH. However, enzymes struggle to act on the raw material matrices directly due to their high structural order, strength, and degrees of polymerization and crystallinity, especially when some of the materials are surrounded by resistant outer layers (Alvira et al., 2010; Shokri et al., 2022; Zhang et al., 2015). Therefore specific mechanical or chemical pretreatment is required to expose substrates, and extrusion can break down the naturally organized matrix composition of food/feed through shear, pressure, heating, and expansion for subsequent enzyme reactions.

Furthermore, both single- and twin-screw extruders have been widely used as enzymatic reactors. However, due to the high enzyme consumption (≥5%) in single-screw extrusion systems, researchers often choose twin-screw extruders as enzyme reactors (Choudhury and Gautam, 1998; Vergnes and Berzin, 2006; Xu, 2019; Xu et al., 2015c). In addition, twin-screw extruders are preferred for their versatility, stability, flexibility, and notably their strong micromixing capabilities at the mesoscale and molecular levels, thereby creating more favorable conditions for enzymatic catalytic reactions (Bouvier and Campanella, 2014; Cheng and Manas-Zloczower, 1997; Grafelman and Meagher, 1995; Zheng and Rehmann, 2014).

6.2.1 Bioextrusion-induced breaking of material structure

The structure, composition, and properties of raw materials have an important influence on bioextrusion, which affects the flow behavior, mixing uniformity, and enzymatic hydrolysis efficiency of materials simultaneously. Most raw materials are composed of natural biological macromolecules (such as polysaccharides and proteins) (Bouvier and Campanella, 2014; Xu et al., 2020; Zhang et al., 2015), which often hinder enzyme activity. Without physical or chemical pretreatment assistance, enzymes themselves rarely open the integrated structure of natural biopolymers with high degrees of polymerization, crystallinity, and physical strength (Zhang et al., 2012). That is, structural destruction or pretreatment, such as extrusion of raw materials, facilitates the loosening and depolymerization of complex starch-based substrates with proteins, lipids, or by-products such as bran and husk, making it easier for enzymes to attach to and degrade substrates (Cui et al., 2011; Ditudompo and Takhar, 2015; Manoi and Rizvi, 2010; Zheng et al., 2006).

The natural intact tissue (or tightly connected structure) of raw materials can act as a physical barrier to enzymes (Alvira et al., 2010; Zhang et al., 2015), and the presence of natural chemical components or cell walls may also affect the contact and reaction efficiency of enzymatic hydrolysis. The breaking of this barrier is associated with changes in the condensed structure, microstructure size, and even the exposure of functional group bonding via bioextrusion. For example, the natural and ordered structure of starch particles can be destroyed and degraded by enzymatic attack (Sun et al., 2010), especially during extrusion processing (Xu et al., 2015c).

Specifically, it has been shown that high-amylose maize starch granules are robust and exhibit little damage and gelatinization under most extrusion conditions, resulting in low digestibility of the extrudate (Ye et al., 2018). Compared with α-(1,4/6)-linked starch chains, β-(1,4)-linked cellulose is more difficult to digest by enzymes due to its strong structure and high crystallinity, as well as the presence of hemicellulose and lignin. Enzyme-resistant extrudates can also be formed during extrusion due to various factors, including retrogradation, reorganization, and thermomechanically induced chemical modifications such as amylose–lipid binding (Duque et al., 2013; Martínez et al., 2014; Zhang et al., 2015).

It is worth noting that the potential reactions and property changes of the complex raw materials have various effects on the stability of the bioextrusion process with high enzyme activity. For example, condensation and repolymerization in cellulose–hemicellulose matrices aggravate enzyme resistance (Duque et al., 2013; Jeunink and Cheftel, 1979), decreasing the breaking effect of bioextrusion; similarly, excess aggregation occurs in proteins (Jeunink and Cheftel, 1979; Liu and Hsieh, 2008; Zheng et al., 2006). During extrusion, some fat components in the materials are hydrolyzed by their own lipase to produce free fatty acids, which form complexes with protein and starch under high temperature and high pressure (Gulati et al., 2020).

6.2.2 Bioextrusion-induced synergistic effects of physical and biological reactions

The functional zones in an intermeshing corotating twin-screw extruder consist of the feeding throat/preconditioner, power drive, screw-barrel-die assembly, and control system. During the bioextrusion process, the material (usually with exogenous enzymes) undergoes physical, mechanical, and thermal effects. Particularly in the mixing section, a series of chemical reactions, such as polymerization, degradation,

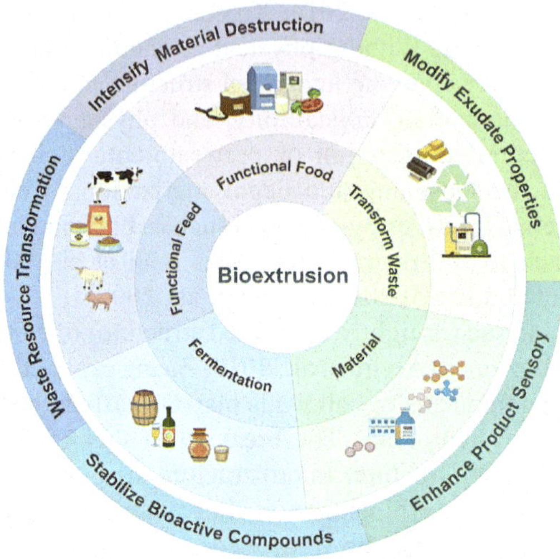

Fig. 6.1 Application and material conversion of bioextrusion.

modification, and the Maillard reaction, also occur in different components within the materials under enzymatic action.

The general mechanism of enzyme-induced bioextrusion synergy involves the actions of physical forces and biological reactions. The extrusion process creates a looser hierarchical or expansive structure of the substrate for enzyme contact; enzymes, in turn, break down the complex biopolymer structures and enable specific conversions or modifications of biopolymers. Exogenous enzymes can be incorporated into the feed prior to the extrusion process or injected at different locations in the extruder (Fig. 6.1). The bioextrusion process (Fig. 6.1) subjects the enzyme and substrate to strong mechanical actions (shear forces, barrel/screw friction, and pressure build-up) and thermal effects (heat input/generation, transfer, and accumulation) (Sun et al., 2010). In the bioextrusion microenvironment, the enzyme exhibits a radial action path from the center to the outer region of the substrate, enabling a more expansive interaction with the biopolymer (Xu et al., 2020).

6.3 Material conversion of bioextrusion

6.3.1 *Intensification of the destruction of material structure*

The combination of enzymatic hydrolysis and extrusion exacerbates the change in material structure, inducing a looser state that facilitates shear fracture, enhances enzyme accessibility, and ultimately boosts reaction efficiency. During bioextrusion, the cell structure of the raw material is destroyed under the action of friction and shear force, as shown in Fig. 6.2. The starch, protein, and other contents flow out (with glycosidic bonds, peptide bonds, or disulfide bonds enzymatically breaking), which is

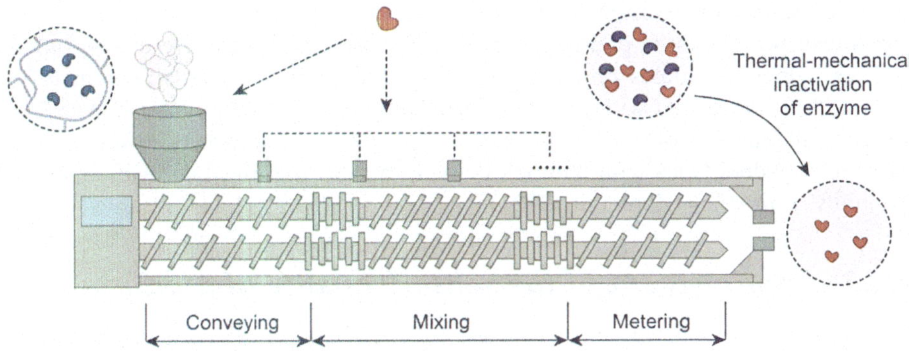

Fig. 6.2 General processing features of bioextrusion.

conducive to water entering the matrix interior and heat transfer, thereby speeding up its ripening. Starch, protein, and their degradation products can also form complexes with lipids, vitamins, phytic acid, etc., which affect the expansion, digestion, and in vivo absorption of materials (Zhou et al., 2022).

Wu et al. (2020) found that starch was extruded and simultaneously treated by enzymatic hydrolysis, achieving higher crystallinity, a more uniform pore structure, and a higher gelatinization temperature than those of native porous starch. Pea protein, after bioextrusion, exhibited greater porosity, a more uneven surface, a larger specific surface area, and a looser, more open, and extended structure compared to its conventionally extruded counterpart. Bioextrusion treatment had significant effects on the secondary structure of proteins in pea flour, such as a reduction or even disappearance of the alpha-helix, a decrease in beta-sheet content, and an increase in beta-turn and random coil structures (Li et al., 2022; Qi et al., 2022).

6.3.2 *Changes in the physical and chemical properties of exudates*

Bioextrusion has a significant effect on the rheological, thermal, and other physicochemical properties of materials. Compared with untreated pea powder, the increase in amylose content of bioextruded pea powder improved its fluidity, making it easier to rearrange, tangle, and aggregate to form a stronger gel structure (Qi et al., 2022). The energy storage modulus and loss modulus of bioextruded pea powder were higher than those of raw and conventionally extruded pea powders. The former exhibited good viscoelasticity and temperature stability. A similar result was also observed in the changes in wheat flour after bioextrusion (Zhao et al., 2022). The enthalpic value, gelatinization temperature, peak viscosity, and relative crystallinity of wheat flour decreased significantly, which promoted starch gelatinization.

Bioextrusion is widely used for protein modification and processing pretreatment. Protein foaming and stability are mainly related to its structure and molecular mass, while bioextrusion can alter these properties to enable proteins to quickly adsorb, expand, or rearrange at the air–water interface, rapidly forming air–water interface films for improved foaming and stability (Jiang et al., 2020). It has also been confirmed that extrusion pretreatment destroys the natural structure of soy protein isolates, which facilitates protease hydrolysis and improves emulsification (Chen et al., 2011). Compared with pure extrudates, protein degradation after bioextrusion increases the number of dissociable groups. This results in an enhanced static charge number of hydrolyzates with hydrophilicity, as well as improved emulsification and solubility of proteins (Zhu et al., 2021).

6.3.3 Improvement of the sensory quality of products

By adding xylanase during extrusion, the water absorption capacity, viscoelasticity, and flexibility of corn flour were significantly increased (Platt-lucero et al., 2013). Tortillas prepared from this flour had decreased hardness and increased flexibility, being 15% softer than the control (without xylanase). This improved texture makes tortillas easier to chew and digest. Similarly, when employing bioextrusion with TαA, particularly at concentrations ranging from 0.05% to 0.10%, the porous structure of the noodles facilitates enhanced rehydration properties, allowing for superior water absorption and retention. This ultimately enhanced the overall taste and palatability of the noodles, rendering them softer and more pliable than traditional extruded noodles (Li et al., 2018a). This improvement is primarily attributed to the effect of TαA on starch granules during the high-intensity extrusion process, which promotes the internal breakdown of α-1,4-glucosidic bonds and the generation of water-soluble degradation products (oligosaccharides and dextrins). The addition of medium-temperature α-amylase (MαA) likewise exhibited positive impacts on noodle rehydration and palatability (Li et al., 2018b, 2019). Optimized enzyme concentrations (e.g., TαA and MαA at 0.8‰ and 1.6‰, respectively) facilitated the formation of a well-developed porous structure, whereas excessively high enzyme concentrations may exhibit inhibitory effects (Li et al., 2018a, 2018b).

Bioextrusion has a significant effect on a product's color and flavor. The deep degradation products of starch and protein, such as small molecular sugars, peptides, and amino acids, can serve as precursors for the Maillard reaction, imparting special color and flavor to food. Bioextrusion alters the composition and content of volatile compounds in the product by influencing the Maillard reaction and lipid degradation processes, thereby affecting the product's flavor (Xu et al., 2015e; Zhang et al., 2015). The content of volatile flavor compounds, such as aldehydes, alcohols, ketones, and heterocycles, increased, accompanied by a small amount of fat degradation and oxidation, which enhanced the product's aroma and significantly improved its sensory score (Zhang and Peterson, 2018). Compared with traditional steam-processed rice wine, Chinese rice wine processed through enzyme extrusion exhibited a higher fermentation rate and efficiency and contained more flavor precursors, particularly reducing sugars and amino acids (Li et al., 2013). The brightness value of corn flour after bioextrusion was higher than that achieved through standard extrusion, and the color difference between corn flour and raw materials was reduced. This indicates that the combined use of enzymatic hydrolysis and extrusion can mitigate the excessive browning phenomenon observed during the extrusion process (Zhao et al., 2019).

Bioextrusion also promotes the pigments of the material itself (such as anthocyanins and proanthocyanidins) to be released during enzymatic hydrolysis or decomposed under high temperature and high pressure, leading to changes in the extrudate color (Pismag et al., 2023; Zhao et al., 2019). Bioextrusion affects the reducing sugar content of quinoa by regulating the temperature and α-amylase concentration, which in turn affects the Maillard reaction and ultimately changes the color (Pismag et al., 2023). As the temperature and enzyme concentration increase, the luminosity of the product decreases and the browning index increases. Therefore bioextrusion may further improve the taste and nutritional quality and increase consumers' eating desire.

6.3.4 Increase in the stability of bioactive compounds

Generally, the extrusion process decreases the content of bioactive components and antioxidant capacities. The extrusion process can disrupt the polymeric matrix, cell wall, and covalent bonds in biomaterials, improving the bioaccessibility of bioactives like phenols; however, it also causes the destruction or decarboxylation of free phenols or the polymerization of certain phenols (such as tannins) (Repo-Carrasco-Valencia et al., 2009). The combined effects usually lead to a significant loss of total phenolic content (Brennan et al., 2011). For example, low retention rates (20%–40%) were observed for phenols in extruded corn (Wang and Ryu, 2013), potato (Nayak et al., 2011), pea (Nayak et al., 2011), banana (Sarawong et al., 2014), etc.

Bioextrusion can release and protect certain bioactive substances in materials (Chen et al., 2022), with the content of phytic acid, tannin, and other anti-nutrients reduced, leading to increased nutrient absorption. By introducing α-amylase, the retention rate of phenolic acids and flavonoids in grains and soybeans can be increased, likely through an indirect protection mechanism (Chen et al., 2022; Sarawong et al., 2014; Xu et al., 2015d). Specifically, these bioactives, as nonsubstrates of amylase, are able to maintain their structures in the mild reaction environment of bioextrusion. Similarly, phenolic retention has also been achieved in the bioextrusion of brown rice catalyzed by TαA/MαA (Zeng et al., 2017, 2018). As previously mentioned, bioextrusion can effectively retain phenols during quinoa flour processing, resulting in a high content of extractable and hydrolysable phenolic compounds (Pismag et al., 2023). It also influences the release and stability of nonextractable proanthocyanidins. By introducing TαA in the extrusion process, starch–ferulic acid complexes with a high retention rate of ferulic acid (>70%) and excellent release performance in vitro can be formed (Ma et al., 2022). It is speculated that the starch hydrolyzate produced by bioextrusion forms a flexible protective layer around phenolic molecules. Bioextrusion with MαA significantly improves the phenol retention rate in brown rice, especially when the enzyme concentration reaches 1000 U/g, with the phenol retention rate being twice that of traditional extrusion methods (Qin et al., 2023).

For starch-based materials, methods that increase the total retention content (70%–100%) of endogenous bioactives in products may be more economical than methods that add auxiliary materials rich in bioactive ingredients (Chiu et al., 2012). An in-depth study of the protection mechanisms of bioextrusion may help further optimize processing parameters to minimize the loss of bioactive compounds and enhance their bioaccessibility in the final products.

6.3.5 Transformation of waste resources

6.3.5.1 Enzyme-containing by-products

A variety of by-products or raw material scraps (containing enzymes) have been used as biomaterials for extrusion. Enzyme-containing side products [such as brewer's spent grain (BSG)] were added to the starch extrusion process (Stojceska et al., 2008; Sobukola et al., 2013). BSG, consisting of the husks and outer layers of barley malts, has an annual world production of~30 million tons (Rommi et al., 2018). The mixture of rice flour and BSG has significant application potential as a raw material for the production of extruded snacks with physical properties similar to rice extrudates. (Nascimento et al., 2017) observed that the apparent density of bioextrudates increased with the addition of BSG. It is noteworthy that in addition to the composition, texture, and properties of BSG, the residual activity of amylase, protease, and other enzymes may also have a positive impact on the final product. Steinmacher et al. (2012) used cellulase to catalyze BSG through extrusion for subsequent bread making. Extruded beer waste grains can also be used to produce anti-digestive, low-sugar biscuits (Gutierrez-Barrutia et al., 2023). Furthermore, other untreated by-products, such as Chinese rice wine grain, were also mixed and extruded with raw materials for the cyclic utilization of both residues and enzymes (mainly amylase) in them (Xu et al., 2015a). This strategy of waste bioextrusion not only reduces the cost of the product but also promotes the reuse of food/feed industrial waste, resulting in positive environmental benefits.

6.3.5.2 Agricultural residues

The materials of agricultural residues, such as soybean and corn husks, are rich in complex components like starch, protein, and fiber, which can be depolymerized and loosened by extrusion to enhance enzyme digestibility and subsequent saccharification (Cui et al., 2011; Qu et al., 2017). Through the application of advanced bioextrusion, these bioresources can be efficiently transformed to produce diverse and high-value-added products, such as soluble fibers and soluble arabinoxylan. For example, bioextrusion can be used to produce soluble fiber from soybean dregs (Qu et al., 2017). The use of bioextrusion to treat rice bran can effectively disrupt the structure of the bran cell wall. The conversion of insoluble dietary fiber into soluble components in rice bran is significantly enhanced. Simultaneous extrusion–xylanase treatments are equally as efficient as extrusion alone in improving fiber solubility (Dang and Vasanthan, 2019). Bioextrusion allows for the use of whole cereals, reducing process costs and contaminant effluents (e.g., cooking liquor), which contributes to environmental protection and sustainable development (Platt-Lucero et al., 2013).

6.3.5.3 Lignocellulosic biomass

Bioextrusion also shows potential in the treatment of lignocellulosic biomass such as straw and bagasse (Vandenbossche et al., 2014). By adding cellulase to the extruder's inner cavity, the decomposition of biomass was achieved with high-efficiency enzymatic saccharification. In one study, a biomimetic flexible extrusion reactor was developed to enhance mass transfer by squeezing different parts, which significantly improved the effect of enzymatic hydrolysis of wheat straw (Deng et al., 2022). Most pretreatments, including extrusion for lignocellulose degradation, have a very low solid content (1%–2%) during subsequent enzymatic saccharification. The extrusion of barley straw was conducted with cellulase and hemicellulase and produced an extrudate with 35% (w/w) of solids in soluble form (Duque et al., 2014). A high solid loading level of 30% could then be used for sugar recovery.

A twin-screw extruder is used to combine soda and enzyme-catalyzed extrusion on the biomass to improve the availability of cellulose and glucose yields for the production of bioethanol (Duque et al., 2020). Although the residence time is too short to achieve complete saccharification, bioconversion of considerable glucose (16.4%–49.5%) and xylose (9.6%–19.4%) yields is obtained from enzymatically extruded material (Vandenbossche et al., 2014). Similarly, 15% of C6 sugar was found in extruded blue agave bagasse mixed with an enzymatic cocktail (exoglucanase, endoglucanase, hemicellulase, xylanase, β-glucosidase), and it is favorable for bioethanol fermentation (Montiel et al., 2016).

In addition, bioextrusion is used to process cellulose fibers into nanoscale functional materials that can be utilized in energy storage, barrier layers, and pharmaceuticals (Banvillet et al., 2021; Razzak et al., 2022; Rol et al., 2019). Other studies focus on the enzymatic hydrolysis, saccharification, and functionalization of cellulose fibers (Gatt et al., 2019; Leal et al., 2021; Li et al., 2022). Thus the bioextrusion technology of biomass materials demonstrates significant application prospects and development potential for conversion into bioethanol, sugar, or dietary fiber.

6.4 Applications of bioextrusion

6.4.1 Saccharification and fermentation

The use of bioextrusion simplifies and accelerates the process of converting starch into sugars, which can subsequently be used for fermentation (or wine brewing). Instead of traditional starch gelatinization by high-water heating, bioextrusion employs an extruder to complete the gelatinization process, with amylase added to rapidly degrade starch. For instance, extrusion enzymolysis (i.e., bioextrusion) was used to investigate the extrusion liquefaction and glycation process of grain starch (Ye and Shen, 2019). Bioextrusion technology has significant advantages in the application of starch to produce sugar, as it avoids common problems such as excessively high viscosity, low solubility, and rapid retrogradation (or recrystallization) (Hakulin et al., 1983; Likimani et al., 1991; Hakulin et al., 1983). This improves the conversion efficiency of starch, optimizes product quality, and provides an efficient and environmentally friendly production method for industries.

For wine making, bioextrusion has shown great potential, especially in improving fermentation efficiency, optimizing wine quality, and promoting the degradation and transformation of raw materials. For example, barley flour was modified by bioextrusion for beer fermentation, which significantly increased the alcohol content of finished beer (5.41% vol) (Zhao et al., 2022). α-Amylase was added during sorghum extrusion to produce Xiaoqu liquor (Pan et al., 2019). It was also used as a pretreatment method for yellow rice wine, which had an alcohol content (9.4%), total acid (7.15 g/L), total sugar (2.7 g/L), high nonsugar soluble solids (50.02 g/L), and gamma-aminobutyric acid (0.47 mg/L), showing good nutritional value and unique flavor (Bian et al., 2021). In addition, bioextrusion increased the phenolic content and antioxidant properties (DPPH, ABTS, and FRAP values of 75.87, 76.21, and 203.32 mg Trolox equivalent/L, respectively) of the wine samples (Bian et al., 2022). Furthermore, the degradation degree of glutinous rice flour by enzymatic puffing and its correlation with rice wine quality indexes were also investigated (Guan et al., 2022).

Thus the application of bioextrusion as a pretreatment in saccharification, fermentation, and winemaking optimized the physical and chemical properties of raw materials, increased fermentation efficiency, and improved the quality characteristics of final products. This approach may provide a theoretical basis for controlling the quality indices of products by regulating the enzyme content in bioextrusion.

6.4.2 Functional food design

As aforementioned, bioextrusion can optimize the quality, functional performance, sensory characteristics, and stability of bioactive substances in functional foods. For example, in the high-solid environment of bioextrusion, the supramolecular structure modification of resistant starch (RS) is generated by adjusting different ratios of amylose to amylopectin and varying pullulanase levels (Tang et al., 2022). Through the bioextrusion–debranching strategy of high starch concentration, the formation of resistant type III starch is improved. Similar to the twice-extrusion method, it gelatinizes and partially degrades the starch, and then the treated starch is further extruded and mixed with pullulanase for debranching (Liu et al., 2023). Different amylose content and resistant starch forms can be obtained by adjusting the amount of amylose. Moreover, through the extrusion–debranching–complexing extrusion strategy, amylose–lipid complexes can be effectively debranched and complexed, and resistant type V starch with high thermal stability and low digestibility can be produced (RS content of enzymatically debranched extruded corn starch–lauric acid complexes reaches 38.34%), which is suitable for the production of low glycemic index foods (Liu et al., 2024).

Whole corn flour and tortillas can be utilized by processing whole grains to retain their nutritional benefits. The addition of xylanase to the extruded flour leads to the depolymerization of arabinoxylan, which improves the handling properties of the dough and enhances the texture of the tortilla (Platt-Lucero et al., 2013). The soluble fiber content of corn bran was increased through extrusion–enzyme treatment, which was applied to the production of functional cakes (Haghighi-Manesh and Azizi, 2018). By using different concentrations of amylase, moisture content, and screw speed, maltodextrins with specific properties (such as dextrose equivalent, DE and maltooligosaccharide compositions) can be utilized as adhesives, expanders, flavor encapsulators, fat substitutes, and other functional ingredients in foods (Ansharullah et al., 2020).

In addition, bioextrusion was used to reduce or neutralize the effect of gluten immunogenicity to prepare gluten-free or very low-gluten ready-to-eat snack products (Sharikov et al., 2020). Bioextrusion also improved oat milk production by altering its physicochemical properties and enhancing stability and quality. Some researchers (Liu et al., 2023; Ren et al., 2023) found that α-amylase reduced oat starch molecular weight, increased water solubility, reduced sugar content, and achieved nearly complete gelatinization. This led to lower viscosity and more regular fluid characteristics, improving oat milk stability and taste.

6.4.3 Plant-based meat production

Extrusion can be divided into low-moisture (20%–40%) and high-moisture (40%–80%) modes for treating materials (Chen et al., 2021). High-moisture extrusion is regarded as the most promising new plant-based meat processing technology because of its advantages, including low energy consumption and minimal nutrient loss, with no need for subsequent processing (Zhang et al., 2022). The high-pressure and high-temperature environment during extrusion provides ideal conditions for reactions such as stretching, cross-linking, and aggregation of plant protein molecular chains (Zhang et al., 2019). This process allows plant proteins to form an ordered, compact reticular fiber structure, which significantly enhances the taste and texture of plant-based meat. The combination of enzymatic modification and extrusion technology further improves the modification effect of plant proteins.

It is worth mentioning that transglutaminase (TGase) plays a key role in the bioextrusion of plant-based proteins. As a crosslinking enzyme, TGase has the ability to bind glutamine residues with lysine, creating ε-(γ-glutamyl) lysine bonds, thereby altering the functional characteristics of proteins and improving the properties of plant proteins, such as viscosity, gelation, and water-holding capacity (Qi et al., 2022). For plant-based meat, the introduction of TGase helps to form a tighter fiber structure, making it more similar to real animal meat in taste and texture. By applying a combination of TGase and protein–glutaminase, bioextrusion significantly enhances the formation of the fiber structure of oat protein during high-moisture extrusion. The positive effects of adding TGase (0.1%–0.2%) to extruded peanut protein on fiber structure formation and tensile strength were observed, while an excessive amount (0.3%) of TGase had an adverse effect (Zhang et al., 2021). By using high-moisture extrusion and TGase modification, the extrusion properties and fiber structure changes of plant proteins from different sources during the organizing process were investigated (Zhang et al., 2023). In addition, bioextrusion could destroy antinutrients and allergens, improving digestibility and reducing sensitization (Faisal et al., 2022; Kaleda et al., 2020). Furthermore, plant proteins of different sources and contents show different structural changes under thermomechanical treatment. For instance, soybean protein responded significantly during extrusion, while rice protein showed relatively poor extrudability. Overall, this technology combines enzymatic modification with extrusion to successfully simulate the fiber structure and texture of animal meat through fine regulation of plant proteins, supporting the development of plant-based meat products.

6.4.4 Feed manufacture

Existing studies have demonstrated the promoting effects of bioextrusion in feed manufacture. For instance, for dog feed and other feeds, the addition of TαA can significantly alter extrusion parameters, increasing production and decreasing electrical energy consumption (Carciofi et al., 2012). Experimental data show that the addition of amylases can increase extruder output by 28%–43% while reducing electrical energy consumption by 22%–29%. Specifically, it promotes the breakdown of starch chains, reducing the viscosity and resistance of the dough within the extruder, resulting in a smoother extrusion process. In addition, it does not dramatically affect the appearance and quality of the feed and has no adverse effects on the animal's nutritional absorption or appetite (Carciofi et al., 2012).

Beyond improving production efficiency and energy conservation, bioextrusion also positively impacts feed quality and animal production performance. The addition of enzyme mixtures (one-third glucoamylase and two-thirds endoamylase) was investigated prior to the extrusion processing of dairy cow feed (Mendowski et al., 2020). Compared to treatment groups without additives and those with added reducing sugars, the treatment group with added enzyme mixtures exhibited lower furan content and moderate levels of Maillard compounds in the feed. Cows fed bioextruded feed produced milk with higher fat content and lower urea, had decreased ruminal ammonia levels, and showed a tendency for increased plasma essential amino acid concentrations.

It can also enhance the nutritional value of feeds and maintain intestinal health in animals. The addition of enzymatic formulations (amylase) promotes the digestion and absorption of nutrients in the feed, increasing the content of small molecular sugars and improving intestinal morphology and barrier function (Mendowski et al., 2020). These changes contribute to the improvement of feed digestibility and utilization, reducing nutrient waste. In addition, bioextrudate can regulate the intestinal microbiome, reducing polysaccharide fermentation and volatile fatty acid production in the gut microbiota, maintaining intestinal microbial balance, and promoting healthy animal growth (Sá et al., 2013).

Moreover, bioextrusion broadens the range of feed raw materials (Mendowski et al., 2020; Sá et al., 2013). The bioextruded dog food, using agricultural by-products such as wheat bran, is not only environmentally friendly and cost-effective but also leverages the beneficial effects of dietary fiber in wheat bran on animal intestinal health. In the processing of specific feeds, bioextrusion also exhibits unique advantages. For instance, the use of TGase in the twin-screw extrusion of extruded fish feed significantly improves the durability, hardness, and elasticity of fish feed pellets (Wolska et al., 2015). This technology not only enhances the physical quality of the feed but also reduces losses and waste during transportation and storage. In addition, it has successfully transformed feathers, which were previously regarded as waste, into valuable feed ingredients that can replace part of the fish meal for white shrimp farming. This substitution could result in a reduction of approximately 55% of fish meal without compromising the growth or nutrient conversion efficiency of shrimp (Mendoza et al., 2001).

While research on bioextrusion may be relatively limited currently, it is believed that this technology will make significant contributions to the sustainable development of the feed industry in the future (Carciofi et al., 2012). We can further explore the application effects of bioextrusion technology in different feed types, animal species, and production environments, determining the optimal balance between enzyme costs, improvements in food production rates, and energy cost savings.

6.4.5 Other applications

6.4.5.1 Polyester polymerization for food packaging

Bioextrusion technology has great potential and promising application prospects in polyester synthesis. By embedding the enzyme into the polyester, the degradation, modification, and introduction of specific functions of the polyester can be realized, offering advantages such as high efficiency, environmental protection, and sustainable development. By adding enzymes during extrusion, microlipase is embedded in the polyester, which could provide the polyester with a self-degradation function without altering the mechanical properties of the material (Huang et al., 2021). This approach offers a novel solution to address the environmental problems associated with food-related plastic waste.

Bioextrusion has made significant progress in continuous enzyme-catalyzed polymerization and processing. By optimizing extrusion parameters and catalyst selection, polymer synthesis and processing with high efficiency and high monomer conversion can be achieved (Li et al., 2020; Spinella et al., 2015). In addition, bioextrusion enables the specific functionalization of polyester by incorporating other additives or functional molecules into the polyester matrix. For example, chitosan polycaprolactone (Cs-g-PCL) grafted copolymer with antibacterial properties was synthesized in a twin-screw extruder using a lipase (CALB and NovoCor) catalytic reaction extrusion method (Vaidya et al., 2019). This method eliminates the need for harmful solvents or protection–deprotection chemical steps, making it a green and environmentally friendly synthesis technology. It is conceivable that this solvent-free and green enzyme-catalyzed grafting process could be extended to other polymer–enzyme combinations for applications in food/feed packaging.

6.4.5.2 Porous biomaterial fabrication

Porous starch, as a common type of modified material, has shown significant potential in applications such as adsorption and separation due to its unique porous sponge-like structure and large specific surface area. There are many methods for preparing porous starch, including physical, chemical, and biological methods (Dura et al., 2014). Among these, the enzymatic method using α-amylase and glucoamylase is considered a simple, safe, and industrialized production approach (Wu et al., 2020). However, due to the low specific surface area and high crystalline region of natural starch, the enzymatic process usually requires assistance from methods such as thermal–mechanical pretreatment (Tatsumi and Katano, 2005; Xie et al., 2019; Xu et al., 2019, 2015b).

In recent years, the combination of extrusion and enzymatic hydrolysis has become a new method for preparing porous starch (Xu et al., 2018, 2015b). Compared to that from a single-mode process, this biologically extruded and enzymatically hydrolyzed porous starch exhibited more uniform and controllable pore structures, significantly improving its adsorption capacity, especially for water, oil, and dyes (Wu et al., 2020). Porous starch-based materials were successfully fabricated through the interaction of metal ions and starch enzymes during bioextrusion (Xu et al., 2018). Further methods for modifying porous starch-based materials were achieved by adding zinc ions and TαA–magnesium complexes as biocatalysts. Increasing the concentration of starch enzymes enhanced the average pore number of extrudates while maintaining the pore size constant. In addition, enzyme-assisted bioextrusion allows for the preparation of porous biomaterials with efficient adsorption and magnetic selectivity, such as the creation of zero-valent iron particles (Fe° Ps) embedded in a three-dimensional zinc-modified starch framework using bioextrusion, demonstrating a hierarchical porous structure (including mesopores and macropores) (Xu, 2019). They have potential uses in the food- and feed-related industries as fat adsorbents and moisture retainers to improve the quality and texture of products, or as separators to treat food/feed processing wastewater for by-products.

6.4.5.3 Microencapsulation

Microencapsulation is an advanced encapsulation technique that encapsulates and protects active core substances inside a protective shell. Starch was used as the wall material, with resveratrol as the core material, and TαA added as a release improvement agent for coextrusion to prepare the microencapsulation particles (Chen et al., 2020). The results showed that coextrusion increased the release rate (86.7%) of resveratrol from bioextruded particles and improved the photostability of resveratrol compared with traditional extrusion (58.2%). The overall antioxidant activity of microcapsules could be enhanced by combining chitosan coating with bioextrusion, providing additional health benefits (Song et al., 2021). Thus bioextrusion microcapsule technology has advantages in improving release speed, reducing particle size, and enhancing stability and protection, making it promising for a variety of applications such as food additives, pharmaceuticals, and cosmetics.

6.4.5.4 *Synthetic oligopeptide*

Recently, a remarkable breakthrough was achieved in the oligomerization of L-amino acids through eREX (i.e., bioextrusion) catalyzed by papain. This method, pioneered by Ardila-Fierro et al. (2018), has opened up new avenues for the synthesis of oligopeptides with controlled properties. Moreover, oligopeptides synthesized using papain-catalyzed extrusion possess unique biocatalytic properties. For instance, oligo (L-Leu) can be utilized as a biocatalyst in the Juliá–Colonna enantioselective epoxidation of chalcone derivatives (Allen et al., 1998). This capability also highlights the potential of these oligopeptides in chiral synthesis and other biocatalytic applications (Bérubé et al., 2017), offering unprecedented opportunities in peptide engineering, which could lead to novel applications in functional foods, as well as biotechnology and biomedicine.

6.5 Summary and perspectives

Bioextrusion technology combines thermomechanical actions with enzymatic catalysis to achieve efficient conversion of natural biological macromolecules, which have been widely used in food and feed applications. The general mechanism of bioextrusion involves the synergistic effects of physical forces and biological reactions: the extrusion process creates a looser substrate structure, enhancing enzymatic interaction; enzymes, in turn, break down complex biopolymer structures, enhance reaction selectivity and efficiency, and enable specific modifications of biopolymers. In addition, bioextrusion intensifies the destruction of the raw material structure, facilitating enzymatic access and degradation of substrates.

Factors such as enzyme concentration, temperature, moisture content, and screw speed can affect enzyme activity and stability during bioextrusion. Understanding and controlling these factors are crucial for optimizing enzyme performance in the high-solid reaction microenvironment of extrusion. Mathematical modeling and optimization techniques are useful tools for analyzing the complex system of bioextrusion.

Applications of bioextrusion include improving the physical and chemical properties of products, enhancing the sensory quality of the product, increasing the stability of bioactive compounds, converting waste resources, as well as functional product manufacture, brewing, and producing future foods and feeds. Each application scenario requires specific adaptations of process parameters and enzymes to achieve the objectives.

In conclusion, bioextrusion is a promising technology that provides a novel solution for processing natural biological macromolecules into edible products. It holds great potential for innovation and development in the food and feed industries, leading to a high-quality, functional, and sustainable production model.

The future research points of bioextrusion could involve many aspects, including structural formation mechanisms, functionalization, biopolymerization reactions, process simulation and modeling, equipment and machinery optimization, pretreatment and extrusion correlation analyses, as well as enzyme stability and activity investigations, cost reduction, new product development, and safety evaluation.

(1) Improve the stability and activity of the enzymes used in the bioextrusion process and enhance the compatibility of the enzymes with substrates and solvents under a range of conditions (especially temperature and pH). Attention should be paid to the correlation analysis in bioextrusion systems to determine the most appropriate treatment scheme. In addition, to achieve the industrial application of the bioextrusion technique, it is necessary to reduce the cost of the enzyme (or increase its reuse cycle) with a modified extruder for biomachine updates.

(2) In-depth understanding of the mechanism of the formation process of food/feed tissue structure, exploring the functionalization of plant and animal proteins, and the functional modification of other biological polymeric materials. This will help optimize the quality and taste of products (expanded food, ready-to-eat food, meat substitutes, etc.) and promote the development of novel edible ingredients with specific nutrients and functions.

(3) Conduct process simulation and modeling of the transportation process and rheological behavior in the bioextrusion process, optimize process parameters, and improve production efficiency. In terms of equipment and machinery, optimize the performance of the extruder, especially the design of die components, to improve production efficiency and product quality. In the future, with the further development of AI technology, it is expected to achieve the intelligent control and optimization of the bioextrusion process.

(4) Research in new product development includes topics such as prototyping and modeling, risk analysis and safety studies, sensory assessment and consumer acceptance studies, and upgrading of food by-products and waste. Together, these studies will drive the wide applications of bioextrusion technology in food/feed and other fields and contribute to their sustainable development.

Acknowledgments

The authors thank the National Natural Science Foundation of China (32272464) and the Young Elite Scientists Sponsorship Program by CAST (2022QNRC001) for research funding.

References

Allen, J.V., Bergeron, S., Griffiths, M.J., Mukherjee, S., Roberts, S.M., Williamson, N.M., Wu, L.E., 1998. Juliá-Colonna asymmetric epoxidation reactions under non-aqueous conditions: rapid, highly regio- and stereo-selective transformations using a cheap, recyclable catalyst. J. Chem. Soc., Perkin Trans. 0300922X. 1 (19), 3171–3179. doi:10.1039/a805407j.

Alvira, P., Tomás-Pejó, E., Ballesteros, M., Negro, M.J., 2010. Pretreatment technologies for an efficient bioethanol production process based on enzymatic hydrolysis: a review. Bioresour. Technol. 09608524. 101 (13), 4851–4861. doi:10.1016/j.biortech.2009.11.093.

Ansharullah, A., Rianda, L., Wahyuni, S., Tamrin,, 2020. Optimization in preparing sago based maltodextrin in a twin screw extruder. J. Appl. Life Sci. Int. 2394-1103. 45–54. doi:10.9734/jalsi/2020/v23i930186.

Ardila-Fierro, K.J., Crawford, D.E., Körner, A., James, S.L., Bolm, C., Hernández, J.G., 2018. Papain-catalysed mechanochemical synthesis of oligopeptides by milling and twin-screw extrusion: application in the Juliá-Colonna enantioselective epoxidation. Green Chem. 14639270. 20 (6), 1262–1269. Available from: http://pubs.rsc.org/en/journals/journal/gc. doi:10.1039/c7gc03205f.

Banvillet, G., Gatt, E., Belgacem, N., Bras, J., 2021. Cellulose fibers deconstruction by twin-screw extrusion with in situ enzymatic hydrolysis via bioextrusion. Bioresour. Technol. 09608524. 327, 124819. doi:10.1016/j.biortech.2021.124819.

Bérubé, C., Barbeau, X., Lagüe, P., Voyer, N., 2017. Revisiting the Juliá–Colonna enantioselective epoxidation: supramolecular catalysis in water. Chem. Commun. 1359-7345. 53 (37), 5099–5102. doi:10.1039/C7CC01168G.

Bian, S., Xu, E., Fu, X., Jin, Z., Jiao, A., 2021. Comparison of different thermal treatments on the physicochemical properties of *Apios fortunei* used for yellow wine fermentation. LWT. 00236438. 145. Available from: https://www.journals.elsevier.com/lwt. doi:10.1016/j.lwt.2021.111518.

Bian, S., Zhang, R., Liu, Q., Guan, Z., Jin, Z., Zhu, K., Jiao, A., 2022. Effects of the addition of thermostable α-amylase on the physicochemical and antioxidant properties of extrusion-pretreated *Apios fortunei* used for yellow wine fermentation. LWT. 00236438. 154. Available from: https://www.journals.elsevier.com/lwt. doi:10.1016/j.lwt.2021.112845.

Bouvier, J.M., Campanella, O.H., 2014. Extrusion processing technology. Food Non-Food Biomaterials. Wiley Blackwell, p. 544. Available from: https://www.wiley.com/en-us/Extrusion+Processing+Technology%3A+Food+and+Non-Food+Biomaterials-p-9781444338119.

Brennan, C., Brennan, M., Derbyshire, E., Tiwari, B.K., 2011. Effects of extrusion on the polyphenols, vitamins and antioxidant activity of foods. Trends Food Sci. Technol. 09242244. 22 (10), 570–575. doi:10.1016/j.tifs.2011.05.007.

Carciofi, A.C., Palagiano, C., Sá, F.C., Martins, M.S., Gonçalves, K.N.V., Bazolli, R.S., Souza, D.F., Vasconcellos, R.S., 2012. Amylase utilization for the extrusion of dog diets. Anim. Feed Sci. Technol. 03778401. 177 (3-4), 211–217. doi:10.1016/j.anifeedsci.2012.08.017.

Chen, L., Chen, J., Ren, J., Zhao, M., 2011. Modifications of soy protein isolates using combined extrusion pre-treatment and controlled enzymatic hydrolysis for improved emulsifying properties. Food Hydrocoll. 0268005X. 25 (5), 887–897. Available from: https://www.journals.elsevier.com/food-hydrocolloids. doi:10.1016/j.foodhyd.2010.08.013.

Chen, Q., Zhang, J., Liu, L., Wang, Q., 2021. Research advances on macromolecular interactions during the high moisture extrusion process. J. Chin. Inst. Food Sci. Technol. 10097848. 21 (8), 350–359. Available from: http://www.oriprobe.com/journals/zgspxb.html. doi:10.16429/j.1009-7848.2021.08.039.

Chen, S., Zong, J., Jiang, L., Ma, C., Li, H., Zhang, D., 2020. Improvement of resveratrol release performance and stability in extruded microparticle by the α-amylase incorporation. J. Food Eng. 02608774. 274. Available from: http://www.sciencedirect.com/science/journal/02608774. doi:10.1016/j.jfoodeng.2019.109842.

Chen, Y., Chen, L., Xiao, Z., Gao, L., 2022. Effects of enzymolysis and fermentation on the antioxidant activity and functional components of a coarse cereal compound powder based on principal component analysis and microstructure study. J. Food Sci. 17503841. 87 (8), 3573–3587. Available from: http://onlinelibrary.wiley.com/journal/10.1111/(ISSN)1750-3841. doi:10.1111/1750-3841.16217.

Cheng, H., Manas-Zloczower, I., 1997. Study of mixing efficiency in kneading discs of co-rotating twin-screw extruders. Polym. Eng. Sci. 00323888. 37 (6), 1082–1090. Available from: http://onlinelibrary.wiley.com/journal/10.1002/(ISSN)1548-2634. doi:10.1002/pen.11753.

Chiu, H.W., Peng, J.C., Tsai, S.J., Lui, W.B., 2012. Effect of extrusion processing on antioxidant activities of corn extrudates fortified with various Chinese yams (*Dioscorea* sp.). Food Bioprocess Technol. 19355149. 5 (6), 2462–2473. doi:10.1007/s11947-011-0675-7.

Choudhury, G.S., Gautam, A., 1998. Comparative study of mixing elements during twin-screw extrusion of rice flour. Food Res. Int. 09639969. 31 (1), 7–17. doi:10.1016/s0963-9969(98)00053-2.

Cui, C., Zhao, H., Zhao, M., Chai, H., 2011. Effects of extrusion treatment on enzymatic hydrolysis properties of wheat gluten. J. Food Process Eng. 17454530. 34 (2), 187–203. doi:10.1111/j.1745-4530.2008.00348.x.

Dang, T.T., Vasanthan, T., 2019. Modification of rice bran dietary fiber concentrates using enzyme and extrusion cooking. Food Hydrocoll. 0268005X. 89, 773–782. Available from: https://www.journals.elsevier.com/food-hydrocolloids. doi:10.1016/j.foodhyd.2018.11.024.

Deng, Z., Liao, Q., Xia, A., Huang, Y., Zhu, X., Qiu, S., Zhu, X., 2022. A bio-inspired flexible squeezing reactor for efficient enzymatic hydrolysis of lignocellulosic biomass for bioenergy production. Renew. Energy. 18790682. 191, 92–100. Available from: http://www.journals.elsevier.com/renewable-and-sustainable-energy-reviews/. doi:10.1016/j.renene.2022.03.114.

Ditudompo, S., Takhar, P.S., 2015. Hybrid mixture theory based modeling of transport mechanisms and expansion-thermomechanics of starch during extrusion. AIChE J. 15475905. 61 (12), 4517–4532. Available from: http://onlinelibrary.wiley.com/journal/10.1002/(ISSN)1547-5905. doi:10.1002/aic.14936.

Duque, A., Doménech, P., Álvarez, C., Ballesteros, M., Manzanares, P., 2020. Study of the bioprocess conditions to produce bioethanol from barley straw pretreated by combined soda and enzyme-catalyzed extrusion. Renew. Energy. 18790682. 158, 263–270. Available from: http://www.journals.elsevier.com/renewable-and-sustainable-energy-reviews/. doi:10.1016/j.renene.2020.05.130.

Duque, A., Manzanares, P., Ballesteros, I., Negro, M.J., Oliva, J.M., González, A., Ballesteros, M., 2014. Sugar production from barley straw biomass pretreated by combined alkali and enzymatic extrusion. Bioresour. Technol. 09608524. 158, 262–268. doi:10.1016/j.biortech.2014.02.041.

Duque, A., Manzanares, P., Ballesteros, I., Negro, M.J., Oliva, J.M., Saez, F., Ballesteros, M., 2013. Optimization of integrated alkaline–extrusion pretreatment of barley straw for sugar production by enzymatic hydrolysis. Process Biochem. 13595113. 48 (5-6), 775–781. doi:10.1016/j.procbio.2013.03.003.

Dura, A., Błaszczak, W., Rosell, C., 2014. Functionality of porous starch obtained by amylase or amyloglucosidase treatments. Carbohyd. Polym. 101, 837–845. Available from: https://www.sciencedirect.com/science/article/pii/S0144861713010242. doi:10.1016/j.carbpol.2013.10.013.

Faisal, S., Zhang, J., Meng, S., Shi, A., Li, L., Wang, Q., Maleki, S.J., Adhikari, B., 2022. Effect of high-moisture extrusion and addition of transglutaminase on major peanut allergens content extracted by three step sequential method. Food Chem. 18737072. 385. Available from: https://www.elsevier.com/locate/foodchem. doi:10.1016/j.foodchem.2022.132569.

Gatt, E., Khatri, V., Bley, J., Barnabé, S., Vandenbossche, V., Beauregard, M., 2019. Enzymatic hydrolysis of corn crop residues with high solid loadings: new insights into the impact of bioextrusion on biomass deconstruction using carbohydrate-binding modules. Bioresour. Technol. 09608524. 282, 398–406. doi:10.1016/j.biortech.2019.03.045.

Grafelman, D.D., Meagher, M.M., 1995. Liquefaction of starch by a single-screw extruder and post-extrusion static-mixer reactor. J. Food Eng. 02608774. 24 (4), 529–542. doi:10.1016/0260-8774(95)90768-7.

Guan, Z., Zhang, Z., Ren, X., Bian, S., Xu, E., Jin, Z., Jiao, A., 2022. Study on the relationship between the degradation degrees of glutinous rice starch extruded with different α-amylases and the qualities of Chinese rice wine. Int. J. Food Sci. Technol. 13652621. 57 (8), 4792–4803. Available from: http://onlinelibrary.wiley.com/journal/10.1111/(ISSN)1365-2621. doi:10.1111/ijfs.15660.

Gulati, P., Brahma, S., Rose D., 2020. Chapter 13 - impacts of extrusion processing on nutritional components in cereals and legumes: carbohydrates, proteins, lipids, vitamins, and minerals. In: Extrusion Cooking, 2nd ed. In: Cereal Grains Processing. pp. 415–443. Available from: https://www.sciencedirect.com/science/chapter/edited-volume/pii/B9780128153604000134. doi:10.1016/B978-0-12-815360-4.00013-4.

Gutierrez-Barrutia, M.B., Cozzano, S., Arcia, P., Del Castillo, M.D., 2023. Assessment of in vitro digestion of reduced sugar biscuits with extruded brewers' spent grain. Food Res. Int. 18737145. 172. Available from: https://www.elsevier.com/inca/publications/store/4/2/2/9/7/0. doi:10.1016/j.foodres.2023.113160.

Haghighi-Manesh, S., Azizi, M.H., 2018. Integrated extrusion-enzymatic treatment of corn bran for production of functional cake. Food Sci. Nutr. 2048-7177. 6 (7), 1870–1878. doi:10.1002/fsn3.738.

Hakulin, S., Linko, Y.-.Y., Linko, P., Seiler, K., Seibel, W., 1983. Enzymatic conversion of starch in twin-screw HTST-extruder. Starch-Stärke. 0038-9056. 35 (12), 411–414. doi:10.1002/star.19830351203.

Huang, Q., Kimura S., Iwata T., 2021. Development of self-degradable aliphatic polyesters by embedding lipases via melt extrusion. Polym. Degrad. Stabil. 190, 109647. Available from: https://www.sciencedirect.com/science/article/pii/S0141391021001671. doi:10.1016/j.polymdegradstab.2021.109647.

Jeunink, J., Cheftel, J.C., 1979. Chemical and physicochemical changes in field bean and soybean proteins texturized by extrusion. J. Food Sci. 0022-1147. 44 (5), 1322–1325. doi:10.1111/j.1365-2621.1979.tb06430.x.

Jiang, Y., Jia, J., Xiong, D., Xu, X., Yang, Y., Liu, X., Duan, X., 2020. Effects of short-term fermentation with lactic acid bacteria on egg white: characterization, rheological and foaming activities. Food Hydrocoll. 0268005X. 101. Available from: https://www.journals.elsevier.com/food-hydrocolloids. doi:10.1016/j.foodhyd.2019.105507.

Kaleda, A., Talvistu, K., Tamm, M., Viirma, M., Rosend, J., Tanilas, K., Kriisa, M., Part, N., Tammik, M.L., 2020. Impact of fermentation and phytase treatment of pea-oat protein blend on physicochemical, sensory, and nutritional properties of extruded meat analogs. Foods. 23048158. 9 (8). Available from: https://www.mdpi.com/2304-8158/9/8/1059. doi:10.3390/foods9081059.

Leal, F.H.P.N., Senna, C.D.A., Kupski, L., Mendes, G.D.R.L., Badiale-Furlong, E., 2021. Enzymatic and extrusion pretreatments of defatted rice bran to improve functional properties of protein concentrates. Int. J. Food Sci. Technol. 13652621. 56 (11), 5445–5451. Available from: http://onlinelibrary.wiley.com/journal/10.1111/(ISSN)1365-2621. doi:10.1111/ijfs.15017.

Li, C., Pan, S., Xu, W., Lu, Y., Wang, P., Zhang, F., Gross, R.A., 2020. Lipase-catalyzed ring-opening copolymerization of ω-pentadecalactone and δ-valerolactone by reactive extrusion. Green Chem. 14639270. 22(3), 662–668. Available from http://pubs.rsc.org/en/journals/journal/gc. doi:10.1039/c9gc04111g.

Li, H., Jiao, A., Xu, X., Wu, C., Wei, B., Hu, X., Jin, Z., Tian, Y., 2013. Simultaneous saccharification and fermentation of broken rice: an enzymatic extrusion liquefaction pretreatment for Chinese rice wine production. Bioprocess Biosyst. Eng. 16157605. 36 (8), 1141–1148. doi:10.1007/s00449-012-0868-0.

Li, J., Jiao, A., Deng, L., Rashed, M.M.A., Jin, Z., 2018b. Porous-structured extruded instant noodles induced by the medium temperature α-amylase and its effect on selected cooking properties and sensory characteristics. Int. J. Food Sci. Technol. 13652621. 53 (10), 2265–2272. Available from: http://onlinelibrary.wiley.com/journal/10.1111/(ISSN)1365-2621. doi:10.1111/ijfs.13815.

Li, J., Jiao, A., Rashed, M.M.A., Deng, L., Xu, X., Jin, Z., 2018a. Effect of thermostable α-amylase addition on producing the porous-structured noodles using extrusion treatment. J. Food Sci. 17503841. 83 (2), 332–339. Available from: http://www3.interscience.wiley.com/journal/118509799/issueyear?year=2008. doi:10.1111/1750-3841.14010.

Li, J., Rashed, M.M.A., Deng, L., Jin, Z., Jiao, A., 2019. Thermostable and mesophilic α-amylase: effects on wheat starch physicochemical properties and their applications in extruded noodles. J. Cereal Sci. 10959963. 87, 248–257. Available from: http://www.elsevier.com/inca/publications/store/6/2/2/8/5/9/index.htt. doi:10.1016/j.jcs.2019.04.013.

Li, S., Hu, N., Zhu, J., Zheng, M., Liu, H., Liu, J., 2022. Influence of modification methods on physicochemical and structural properties of soluble dietary fiber from corn bran. Food Chem. X. 25901575. (14). Available from https://www.journals.elsevier.com/food-chemistry-x. doi:10.1016/j.fochx.2022.100298.

Likimani, T.A., Sofos, J.N., Maga, J.A., Harper, J.M., 1991. Extrusion cooking of corn soybean mix in presence of thermostable alpha-amylase. J. Food Sci. 56 (1), 99–105. Available from: https://ift.onlinelibrary.wiley.com/doi/10.1111/j.1365-2621.1991.tb07985.x. doi:10.1111/j.1365-2621.1991.tb07985.x.

Liu, K.S., Hsieh, F.H., 2008. Protein-protein interactions during high-moisture extrusion for fibrous meat analogues and comparison of protein solubility methods using different solvent systems. J. Agric. Food Chem. 00218561. 56 (8), 2681–2687. doi:10.1021/jf073343q.

Liu, Q., Guan, H., Guo, Y., Wang, D., Yang, Y., Ji, H., Jiao, A., Jin, Z., 2024. Structure and in vitro digestibility of amylose–lipid complexes formed by an extrusion-debranching-complexing strategy. Food Chem. 18737072. 437. Available from: https://www.elsevier.com/locate/foodchem. doi:10.1016/j.foodchem.2023.137950.

Liu, Q., Shi, J., Jin, Z., Jiao, A., 2023. Development and characterization of resistant starch produced by an extrusion–debranching strategy with a high starch concentration. Food Hydrocoll. 0268005X. 136. Available from: https://www.sciencedirect.com/science/journal/0268005X. doi:10.1016/j.foodhyd.2022.108276.

Ma, S., Zhu, Q., Yao, S., Niu, R., Liu, Y., Qin, Y., Zheng, Y., Tian, J., Li, D., Wang, W., Liu, D., Xu, E., 2022. Efficient retention and complexation of exogenous ferulic acid in starch: could controllable bioextrusion be the answer? J. Agric. Food Chem. 15205118. 70 (47), 14919–14930. Available from: http://pubs.acs.org/journal/jafcau. doi:10.1021/acs.jafc.2c04261.

Manoi, K., Rizvi, S.S.H., 2010. Physicochemical characteristics of phosphorylated cross-linked starch produced by reactive supercritical fluid extrusion. Carbohydr. Polym. 01448617. 81 (3), 687–694. doi:10.1016/j.carbpol.2010.03.042.

Martínez, M.M., Calviño, A., Rosell, C.M., Gómez, M., 2014. Effect of different extrusion treatments and particle size distribution on the physicochemical properties of rice flour. Food Bioprocess Technol. 19355149. 7 (9), 2657–2665. Available from: http://www.springer.com/life+sci/food+science/journal/11947. doi:10.1007/s11947-014-1252-7.

Mendowski, S., Chapoutot, P., Chesneau, G., Ferlay, A., Enjalbert, F., Cantalapiedra-Hijar, G., Germain, A., Nozière, P., 2020. Effects of pretreatment with reducing sugars or an enzymatic cocktail before extrusion of fava bean on nitrogen metabolism and performance of dairy cows. J. Dairy Sci. 00220302. 103 (1), 396–409. doi:10.3168/jds.2019-17286.

Mendoza,, Dios, D., Vazquez,, Cruz,, Ricque,, Aguilera,, Montemayor,, 2001. Fishmeal replacement with feather-enzymatic hydrolyzates co-extruded with soya-bean meal in practical diets for the Pacific white shrimp (*Litopenaeus vannamei*). Aquac. Nutr. 13535773. 7 (3), 143–151. doi:10.1046/j.1365-2095.2001.00164.x.

Montiel, C., Hernández-Meléndez, O., Vivaldo-Lima, E., Hernández-Luna, M., Bárzana, E., 2016. Enhanced bioethanol production from blue agave bagasse in a combined extrusion–saccharification process. Bioenerg. Res. 1939-1234. 9 (4), 1005–1014. doi:10.1007/s12155-016-9747-x.

Nascimento T., Calado V., Carvalho C., 2017. Effect of Brewer's spent grain and temperature on physical properties of expanded extrudates from rice. LWT-Food Sci. Technol. 79, 145–151. Available from: https://www.sciencedirect.com/science/article/pii/S0023643817300361. doi:10.1016/j.lwt.2017.01.035.

Nayak, B., Liu, R.H., Berrios, J.D.J., Tang, J., Derito, C., 2011. Bioactivity of antioxidants in extruded products prepared from purple potato and dry pea flours. J. Agric. Food Chem. 15205118. 59 (15), 8233–8243. doi:10.1021/jf200732p.

Pan, X., Jiao, A., Long, J., Min, D., Jin, Z., 2019. Design and optimization of an efficient enzymatic extrusion pretreatment for xiaoqu liquor fermentation. J. Food Sci. Biotechnol. 16731689. 38 (9), 27–32. Available from: http://spyswjs.cnjournals.com/spyswjs/issue/browser. doi:10.3969/j.issn.1673-1689.2019.09.004.

Pismag, R., Pico, J., Fernández, A., Hoyos, J.L., Martinez, M.M., 2023. α-Amylase reactive extrusion enhances the protein digestibility of saponin-free quinoa flour while preserving its total phenolic content. Innov. Food Sci. Emerg. Technol. 14668564. 88. Available from: http://www.elsevier.com/locate/ifset. doi:10.1016/j.ifset.2023.103448.

Platt-lucero, L., Ramírez-wong, B., Torres-cháVvez, P., LÓpez-cervantes, J., Sánchez-machado, D., Carvajal-millan, E., Martínez-bustos F., Quintero-ramos, A., Morales-rosas I., 2013. Effect of xylanase on extruded nixtamalized corn flour and tortilla: physicochemical and rheological characteristics. J. Food Process Eng. 36 (2), 179–186. Available from: https://onlinelibrary.wiley.com/doi/10.1111/j.1745-4530.2011.00667.x. doi:10.1111/j.1745-4530.2011.00667.x.

Qi, M., Peng, H., Song, J., Zhang, J., Wang, S., Ma, C., 2022. Effects of extrusion and enzymatic hydrolysis on the in vitro starch digestibility, protein structure and rheological properties of pea flour. Shipin Kexue/Food Science. 10026630. 43 (1), 76–82. Available from: http://www.spkx.net.cn/CN/article/showTenYearVolumnDetail.do?nian=2018. doi:10.7506/spkx1002-6630-20201120-216.

Qin, Y., Ma, S., Zhou, J., Li, D., Chen, J., Wang, W., Cheng, H., Wu, Z., Tian, J., Xu, E., Liu, D., 2023. Conversion of endogenous phenolic acid in brown rice by bioextrusion of mesophilic A-amylase. SSRN. 15565068. Available from: https://www.ssrn.com/index.cfm/en/. doi:10.2139/ssrn.4354817.

Qu, M., Liu, Y., Zhang, G., Xue, S.J., Fan, H., Wang, Y., Yu, S., 2017. Identification and characterization of soybean dreg soluble dietary fibre by combination of extrusion pretreatment and enzymatic modification. Food Qual. Saf. 23991402. 1 (2), 153–160. Available from: academic.oup.com/fqs . doi:10.1093/fqsafe/fyx014.

Razzak, A., Khiari, R., Moussaoui, Y., Belgacem, M.N., 2022. Cellulose nanofibers from schinus molle: preparation and characterization. Molecules. 14203049. 27 (19). Available from: http://www.mdpi.com/journal/molecules. doi:10.3390/molecules27196738.

Ren, X., Yang, Y., Liu, Q., Wang, Y., Jin, Z., Jiao, A., 2023. Effects of enzymatic extrusion on the structure and physicochemical properties of oat flour and its application in oat milk production. Int. J. Food Sci. Technol. 13652621. 58 (9), 4638–4651. Available from: http://onlinelibrary.wiley.com/journal/10.1111/(ISSN)1365-2621. doi:10.1111/ijfs.16568.

Repo-Carrasco-Valencia, R., Peña, J., Kallio, H., Salminen, S., 2009. Dietary fiber and other functional components in two varieties of crude and extruded kiwicha (*Amaranthus caudatus*). J. Cereal Sci. 07335210. 49 (2), 219–224. doi:10.1016/j.jcs.2008.10.003.

Rol, F., Saini, S., Meyer, V., Petit-Conil, M., Bras, J., 2019. Production of cationic nanofibrils of cellulose by twin-screw extrusion. Ind. Crops Prod. 09266690. 137, 81–88. doi:10.1016/j.indcrop.2019.04.031.

Rommi, K., Niemi, P., Kemppainen, K., Kruus, K., 2018. Impact of thermochemical pre-treatment and carbohydrate and protein hydrolyzing enzyme treatment on fractionation of protein and lignin from brewer's spent grain. J. Cereal Sci. 07335210. 79, 168–173. doi:10.1016/j.jcs.2017.10.005.

Sá, F.C., Vasconcellos, R.S., Brunetto, M.A., Filho, F.O.R., Gomes, M.O.S., Carciofi, A.C., 2013. Enzyme use in kibble diets formulated with wheat bran for dogs: effects on processing and digestibility. J. Anim. Physiol. Anim. Nutr. 0931-2439. 97 (s1), 51–59. doi:10.1111/jpn.12047.

Sarawong, C., Schoenlechner, R., Sekiguchi, K., Berghofer, E., Ng, P.K.W., 2014. Effect of extrusion cooking on the physicochemical properties, resistant starch, phenolic content and antioxidant capacities of green banana flour. Food Chem. 18737072. 143, 33–39. Available from: https://www.elsevier.com/locate/foodchem. doi:10.1016/j.foodchem.2013.07.081.

Sharikov, A.Y., Serba, E.M., Sokolova, E.N., Borsheva, Y.A., Rimareva, L.V., 2020. Towards the development of wheat extruded snacks with reduced gluten toxicity. IOP Conf. Ser.: Earth Environ. Sci. 17551315. 548 (8), 082060. https://iopscience.iop.org/journal/1755-1315. doi:10.1088/1755-1315/548/8/082060.

Shokri, Z., Seidi, F., Saeb, M.R., Jin, Y., Li, C., Xiao, H., 2022. Elucidating the impact of enzymatic modifications on the structure, properties, and applications of cellulose, chitosan, starch and their derivatives: a review. Mater. Today Chem. 24685194. 24. Available from: https://www.journals.elsevier.com/materials-today-chemistry/. doi:10.1016/j.mtchem.2022.100780.

Sobukola, O., Babajide J., Ogunsade O., 2013. Effect of brewers spent grain addition and extrusion parameters on some properties of extruded yam starch-based pasta. J. Food Process Preserv. 37 (5), 734–743. Available from: https://onlinelibrary.wiley.com/doi/10.1111/j.1745-4549.2012.00711.x. doi:10.1111/j.1745-4549.2012.00711.x.

Song, J., Zong, J., Ma, C., Chen, S., Li, H., Zhang, D., 2021. Microparticle prepared by chitosan coating on the extruded mixture of corn starch, resveratrol, and α-amylase controlled the resveratrol release. Int. J. Biol. Macromol. 18790003. 185, 773–781. Available from: https://www.elsevier.com/locate/ijbiomac. doi:10.1016/j.ijbiomac.2021.06.154.

Spinella, S., Ganesh, M., Lo Re, G., Zhang, S., Raquez, J.-M., Dubois, P., Gross, R.A., 2015. Enzymatic reactive extrusion: moving towards continuous enzyme-catalysed polyester polymerisation and processing. Green Chem. 1463-9262. 17 (8), 4146–4150. doi:10.1039/C5GC00992H.

Steinmacher, N.C., Honna, F.A., Gasparetto, A.V., Anibal, D., Grossmann, M.V.E., 2012. Bioconversion of brewer's spent grains by reactive extrusion and their application in bread-making. LWT-Food Sci. Technol. 00236438. 46 (2), 542–547. doi:10.1016/j.lwt.2011.11.011.

Stojceska, V., Ainsworth, P., Plunkett, A., Ibanoğlu, S., 2008. The recycling of brewer's processing by-product into ready-to-eat snacks using extrusion technology. J. Cereal Sci. 10959963. 47 (3), 469–479. doi:10.1016/j.jcs.2007.05.016.

Sun, H., Zhao, P., Ge, X., Xia, Y., Hao, Z., Liu, J., Peng, M., 2010. Recent advances in microbial raw starch degrading enzymes. Appl. Biochem. Biotechnol. 02732289. 160 (4), 988–1003. doi:10.1007/s12010-009-8579-y.

Tang, J., Zhou, J., Zhou, X., Li, D., Wu, Z., Tian, J., Xu, E., Liu, D., 2022. Rearranged supramolecular structure of resistant starch with polymorphic microcrystals prepared in high-solid enzymatic system. Food Hydrocoll. 0268005X. 124. Available from: https://www.journals.elsevier.com/food-hydrocolloids. doi:10.1016/j.foodhyd.2021.107215.

Tatsumi, H., Katano, H., 2005. Kinetics of the surface hydrolysis of raw starch by glucoamylase. J. Agric. Food Chem. 0021-8561. 53 (21), 8123–8127. doi:10.1021/jf050934c.

Vaidya, A.A., Hussain, I., Gaugler, M., Smith, D.A., 2019. Synthesis of graft copolymers of chitosan-poly(caprolactone) by lipase catalysed reactive extrusion. Carbohydr. Polym. 01448617. 217, 98–109. Available from: http://www.elsevier.com/wps/find/journaldescription.cws_home/405871/description#description. doi:10.1016/j.carbpol.2019.03.081.

Vandenbossche, V., Brault, J., Vilarem, G., Hernández-Meléndez, O., Vivaldo-Lima, E., Hernándezz-Luna, M., Barzana, E., Duque, A., Manzanares, P., Ballesteros, M., Mata, J., Castellón, E., Rigal, L., 2014. A new lignocellulosic biomass deconstruction process combining thermo-mechano chemical action and bio-catalytic enzymatic hydrolysis in a twin-screw extruder. Ind. Crops Prod. 09266690. 55, 258–266. doi:10.1016/j.indcrop.2014.02.022.

Vergnes, B., Berzin, F., 2006. Modeling of reactive systems in twin-screw extrusion: challenges and applications. C. R. Chim. 1878-1543. 9 (11-12), 1409–1418. doi:10.1016/j.crci.2006.07.006.

Wang, Y.Y., Ryu, G.H., 2013. Physicochemical and antioxidant properties of extruded corn grits with corn fiber by CO_2 injection extrusion process. J. Cereal Sci. 10959963. 58 (1), 110–116. doi:10.1016/j.jcs.2013.03.013.

Wolska, J., Jonkers, J., Holst, O., Adlercreutz, P., 2015. The addition of transglutaminase improves the physical quality of extruded fish feed. Biotechnol. Lett. 0141-5492. 37 (11), 2265–2270. doi:10.1007/s10529-015-1911-4.

Wu, W., Jiao, A., Xu, E., Chen, Y., Jin, Z., 2020. Effects of extrusion technology combined with enzymatic hydrolysis on the structural and physicochemical properties of porous corn starch. Food Bioprocess Technol. 19355149. 13 (3), 442–451. Available from: http://www.springer.com/life+sci/food+science/journal/11947. doi:10.1007/s11947-020-02404-1.

Xie, Y., Li, M.N., Chen, H.Q., Zhang, B., 2019. Effects of the combination of repeated heat-moisture treatment and compound enzymes hydrolysis on the structural and physicochemical properties of porous wheat starch. Food Chem. 18737072. 274, 351–359. Available from: https://www.elsevier.com/locate/foodchem. doi:10.1016/j.foodchem.2018.09.034.

Xu, E.B., 2019. Regulation mechanism of enzymatically extruded rice quality and its application in Chinese rice wine.

Xu, E., Campanella, O.H., Ye, X., Jin, Z., Liu, D., BeMiller, J.N., 2020. Advances in conversion of natural biopolymers: a reactive extrusion (REX)–enzyme-combined strategy for starch/protein-based food processing. Trends Food Sci. Technol. 09242244. 99, 167–180. Available from: http://www.elsevier.com/wps/find/journaldescription.cws_home/601278/description#description. doi:10.1016/j.tifs.2020.02.018.

Xu, E., Li, H., Wu, Z., Wang, F., Xu, X., Jin, Z., Jiao, A., 2015e. Influence of enzymatic extrusion liquefaction pretreatment for Chinese rice wine on the volatiles generated from extruded rice. J. Food Sci. 17503841. 80 (1), C29. Available from: http://www3.interscience.wiley.com/journal/118509799/issueyear?year=2008. doi:10.1111/1750-3841.12719.

Xu, E., Long, J., Wu, Z., Li, H., Wang, F., Xu, X., Jin, Z., Jiao, A., 2015a. Characterization of volatile flavor compounds in Chinese rice wine fermented from enzymatic extruded rice. J. Food Sci. 17503841. 80 (7), C1476. Available from: http://www3.interscience.wiley.com/journal/118509799/issueyear?year=2008. doi:10.1111/1750-3841.12935.

Xu, E., Wu, Z., Ding, T., Ye, X., Jin, Z., Liu, D., 2019. Magnetic (Zn-St)10Fe0n (n = 1, 2, 3, 4) framework of macro-mesoporous biomaterial prepared via green enzymatic reactive extrusion for dye pollutants removal. ACS Appl. Mater. Interfaces. 19448252. 11 (46), 43553–43562. Available from: http://pubs.acs.org/journal/aamick. doi:10.1021/acsami.9b14750.

Xu, E., Wu, Z., Long, J., Jiao, A., Jin, Z., 2018. Porous starch-based material prepared by bioextrusion in the presence of zinc and amylase-magnesium complex. ACS Sustain. Chem. Eng. 21680485. 6 (8), 9572–9578. Available from: http://pubs.acs.org/journal/ascecg. doi:10.1021/acssuschemeng.8b02128.

Xu, E., Wu, Z., Long, J., Wang, F., Pan, X., Xu, X., Jin, Z., Jiao, A., 2015b. Effect of thermostable α-amylase addition on the physicochemical properties, free/bound phenolics and antioxidant capacities of extruded hulled and whole rice. Food Bioprocess Technol. 19355149. 8 (9), 1958–1973. Available from: http://www.springer.com/life+sci/food+science/journal/11947. doi:10.1007/s11947-015-1552-6.

Xu, E., Wu, Z., Long, J., Wang, F., Xu, X., Jin, Z., Jiao, A., 2015d. Improved bioaccessibility of phenolics and antioxidant activity of glutinous rice and its fermented Chinese rice wine by simultaneous extrusion and enzymatic hydrolysis. J. Funct. Foods. 17564646. 17, 214–226. Available from: http://www.elsevier.com/wps/find/journaldescription.cws_home/717426/description#description. doi:10.1016/j.jff.2015.05.032.

Xu, E., Wu, Z., Wang, F., Li, H., Xu, X., Jin, Z., Jiao, A., 2015c. Impact of high-shear extrusion combined with enzymatic hydrolysis on rice properties and Chinese rice wine fermentation. Food Bioprocess Technol. 19355149. 8 (3), 589–604. Available from: http://www.springer.com/life+sci/food+science/journal/11947. doi:10.1007/s11947-014-1429-0.

Ye, J., Hu, X., Luo, S., Liu, W., Chen, J., Zeng, Z., Liu, C., 2018. Properties of starch after extrusion: a review. Starch/Staerke. 1521379X. 70, 11–12. Available from: http://onlinelibrary.wiley.com/journal/10.1002/(ISSN)1521-379X. doi:10.1002/star.201700110.

Ye, X., Shen, D., 2019. Mechanism of starch degradation of corn grist degermed by extruded at low-temperature and multienzyme synergistic degradation. Nongye Gongcheng Xuebao/Transactions of the Chinese Society of Agricultural Engineering. 10026819. 35 (14), 291–299. Available from: http://www.tcsae.org/nygcxb/ch/index.aspx. doi:10.11975/j.issn.1002-6819.2019.14.036.

Zeng, Z., Li, Y., Yang, R., Liu, C., Hu, X., Luo, S., Gong, E., Ye, J., 2017. The relationship between reducing sugars and phenolic retention of brown rice after enzymatic extrusion. J. Cereal Sci. 10959963. 74, 244–249. Available from: http://www.elsevier.com/inca/publications/store/6/2/2/8/5/9/index.htt. doi:10.1016/j.jcs.2017.02.016.

Zeng, Z., Luo, S., Liu, C., Hu, X., Gong, E., Miao, J., 2018. Phenolic retention of brown rice after extrusion with mesophilic α-amylase. Food Biosci. 22124306. 21, 8–13. Available from: http://www.journals.elsevier.com/food-bioscience/. doi:10.1016/j.fbio.2017.10.008.

Zhang, B., Dhital, S., Flanagan, B.M., Luckman, P., Halley, P.J., Gidley, M.J., 2015. Extrusion induced low-order starch matrices: enzymic hydrolysis and structure. Carbohydr. Polym. 01448617. 134, 485–496. Available from: http://www.elsevier.com/wps/find/journaldescription.cws_home/405871/description#description. doi:10.1016/j.carbpol.2015.07.095.

Zhang, J., Chen, Q., Kaplan, D.L., Wang, Q., 2022. High-moisture extruded protein fiber formation toward plant-based meat substitutes applications: science, technology, and prospect. Trends Food Sci. Technol. 09242244. 128, 202–216. Available from:

http://www.elsevier.com/wps/find/journaldescription.cws_home/601278/description#description. doi:10.1016/j.tifs.2022.08.008.

Zhang, J., Chen, Q., Liu, L., Zhang, Y., He, N., Wang, Q., 2021. High-moisture extrusion process of transglutaminase-modified peanut protein: effect of transglutaminase on the mechanics of the process forming a fibrous structure. Food Hydrocoll. 0268005X. 112. Available from: https://www.journals.elsevier.com/food-hydrocolloids. doi:10.1016/j.foodhyd.2020.106346.

Zhang, J., Li, T., Chen, Q., Liu, H., Kaplan, D.L., Wang, Q., 2023. Application of transglutaminase modifications for improving protein fibrous structures from different sources by high-moisture extruding. Food Res. Int. 18737145. 166. Available from: https://www.elsevier.com/inca/publications/store/4/2/2/9/7/0. doi:10.1016/j.foodres.2023.112623.

Zhang, J., Liu, L., Liu, H., Yoon, A., Rizvi, S.S.H., Wang, Q., 2019. Changes in conformation and quality of vegetable protein during texturization process by extrusion. Crit. Rev. Food Sci. Nutr. 15497852. 59 (20), 3267–3280. Available from: https://www.tandfonline.com/loi/bfsn20. doi:10.1080/10408398.2018.1487383.

Zhang, L., Peterson, D.G., 2018. Identification of bitter compounds in extruded corn puffed products. Food Chem. 18737072. 254, 185–192. Available from: https://www.elsevier.com/locate/foodchem. doi:10.1016/j.foodchem.2018.01.161.

Zhang, S., Xu, Y., Hanna M., 2012. Pretreatment of corn stover with twin-screw extrusion followed by enzymatic saccharification. Appl. Biochem. Biotechnol. 166 (2), 458–469. Available from: https://link.springer.com/article/10.1007/s12010-011-9441-6. doi:10.1007/s12010-011-9441-6.

Zhao, S., Jiao, A., Yang, Y., Liu, Q., Wu, W., Jin, Z., 2022. Modification of physicochemical properties and degradation of barley flour upon enzymatic extrusion. Food Biosci. 22124306. 45. Available from: http://www.journals.elsevier.com/food-bioscience/. doi:10.1016/j.fbio.2021.101243.

Zhao, Z.H., Liu, L., Zhang, M.W., Zhang, R.F., Xiao, J., Wei, Z.C., Ma, Y.X., 2019. Combined effect of enzymatic pretreatment and extrusion on quality properties of brown rice flour. Food Sci. 40 (1), 108–116.

Zheng, J., Rehmann, L., 2014. Extrusion pretreatment of lignocellulosic biomass: a review. Int. J. Mol. Sci. 1422-0067. 15 (10), 18967–18984. doi:10.3390/ijms151018967.

Zheng, X.Q., Li, L.T., Liu, X.L., Wang, X.J., Lin, J., Li, D., 2006. Production of hydrolysate with antioxidative activity by enzymatic hydrolysis of extruded corn gluten. Appl. Microbiol. Biotechnol. 01757598. 73 (4), 763–770. doi:10.1007/s00253-006-0537-9.

Zhou, C., Wu, M., Sun, D., Wei, W., Yu, H., Zhang, T., 2022. Twin-screw extrusion of oat: evolutions of rheological behavior, thermal properties and structures of extruded oat in different extrusion zones. Foods. 23048158. 11 (15). Available from: https://www.mdpi.com/journal/foods. doi:10.3390/foods11152206.

Zhu, X., Liu, Y., Zhu, Y., Jiang, Y., Liu, L., 2021. Progress in understanding the emulsification properties of soy protein modified by extrusion pretreatment and enzymatic hydrolysis. Shipin Kexue/Food Science. 10026630. 42 (21), 365–371. Available from: http://www.spkx.net.cn/CN/article/showTenYearVolumnDetail.do?nian=2018. doi:10.7506/spkx1002-6630-20200819-245.

Hybrid extrusion

7

Carlos Alberto Avila-Orta[a], *Heidi Andrea Fonseca-Florido*[b], *Carlos Andrés Covarrubias-Gordillo*[c], *Florentino Soriano-Corral*[d], *Victor Javier Cruz-Delgado*[d], *and Victoria Graciano-de la Cruz*[e]
[a]Department of Advanced Materials, Research Center in Applied Chemistry, Saltillo, Mexico
[b]Department of Advanced Materials, SECIHTI, Research Center in Applied Chemistry, Saltillo, Mexico
[c]Faculty of Chemical Sciences, Autonomous University of Coahuila, Saltillo, Mexico
[d]Department of Plastics Transformation Processes, Research Center in Applied Chemistry, Saltillo, Mexico
[e]Department of Chemical Engineering and Biochemistry, Mexico Technology Institute/Durango Technology Institute., Durango, Mexico

7.1 Introduction

The extrusion process is a multifunctional process that integrates mixing, cooking, cutting, and shaping operations, which can eventually cause the expansion of food products. The desired product characteristics can be achieved by adjusting feed moisture, barrel temperature, screw speed, pressure, screw diameter, and other extrusion parameters. The advantages of extrusion technology include its speed, high productivity, excellent versatility, and low energy consumption. It can also combine various ingredients, thereby facilitating the development of new products, enabling continuous and cost-effective production, and ensuring high and consistent product quality through precise control. Another noteworthy aspect of this process is its classification as a clean technology, characterized by the absence of waste generation. The extrusion process has been applied for over 80 years in food industry applications. However, industry requirements have evolved over the years, shifting from focusing on higher product volumes to developing new textures and fulfilling emerging nutritional demands. A recent trend to address these challenges involves the integration of additive manufacturing (AM) with secondary production techniques; this articulation of techniques is known as hybrid processes. Hybrid AM solutions have the potential to enhance current production systems by utilizing the combined advantages of fused technologies. This has promoted the development and implementation of integrated techniques, particularly within the extrusion process, leading to the emergence of hybrid extrusion methods. Additive processes have been incorporated into food extrusion systems to address specific current needs. For example,

Extrusion Cooking for Food and Feed. https://doi.org/10.1016/B978-0-443-29987-2.00007-5

different extruder dies shape foods with high starch content, such as pasta, cereals, snacks, and pet foods. Another example is the use of specialized tooling to achieve the optimal texture of the final product. In addition, tools for coextruded foods are employed to create the desired light and crispy texture with the proper ingredient encapsulation.

In the last decade, there has been an emphasis on reducing and optimizing processes while preserving nutritional properties and sensory variables for the consumer. This entails the addition and variation of conditions, as well as oxidizing agents, during the extrusion process. This chapter provides an overview of the key AM processes in extrusion, specifically hybrid extrusion, as applied to food production. Hybrid processes are categorized into subtopics, each focusing on the additive variables of the extrusion process. The chapter presents and discusses how incorporating these additional variables into the extrusion process affects the characteristics of the final product. Exploring the relationship between product properties, nutritional profiles, and variables managed during hybrid extrusion can assist food manufacturers and researchers in broadening the scope of their applications. Recent studies from the literature have emphasized the need to conduct further research and enhance the potential uses of products in the extrusion process.

7.2 Thermal energy

Extrusion processing has emerged as a versatile and widely adopted technology in the food industry, enabling the production of a diverse range of products, from breakfast cereals to snacks and pet foods. Part of its versatility and acceptance in the industrial sector is due to the fact that extrusion cookers combine pumping, mixing, kneading, and heating operations in a single piece of equipment. However, these combined operations interact in a manner that renders them inseparable, although it is possible to adjust the parameters within this integrated framework. The physical properties of extruded foods reflect the efficiency of the extrusion process (optimization of the variables of the unit operations that develop the process) and the suitability of the ingredients used. One of the key variables in food extrusion processing is the applied thermal energy, as it can significantly influence the properties and characteristics of the final food products. Extrusion processing generally involves the application of either high or low temperatures, resulting in considerable structural disruption and alterations to the biopolymers in food materials. These changes can significantly affect the nutritional properties and the bioactive compounds present in the extruded products, such as vitamins, as well as the content of bioactive compounds and proteins. In the review presented by Hu et al. (2024), it is noted that heat application can cause structural modifications, resulting in protein denaturation, exposure of hydrophobic groups, and subsequent aggregation or flocculation. Understanding the influence of thermal energy on the properties of extruded foods is essential for optimizing the extrusion process and creating high-quality, nutritious, and consumer-desirable food products.

During the extrusion process, there are several ways to generate heating. An overview of the current state of four emerging thermal technologies: ohmic heating, microwave heating, radiofrequency (RF) heating, and infrared (IR) heating (Hu et al., 2024), is presented.

7.2.1 Ohmic

Ohmic heating is a technique in which the food material is heated by passing an electric current through it. In other words, in ohmic heating, energy is directly dissipated into the food, providing rapid heating while preserving its organoleptic qualities, such as color and nutritional value. Studies on this type of heating have shown that it could improve the quality and preservation of food and efficiently modify the functional properties of proteins, such as solubility, water/oil retention, and emulsification characteristics, aligning with global requirements for environmentally friendly policies promoting "green" and cost-effective processing technologies (Avelar et al., 2021; Kaur and Singh, 2016). This type of heating presents its variables, such as the particle size of the food (Zareifard et al., 2003), the increase in ionic mobility due to structural changes, and the electrical conductivity of foods (Darvishi et al., 2013).

7.2.2 Microwave

Microwave heating is achieved by applying an oscillating electric field that interacts with dipolar molecules, generating internal molecular friction and resulting in volumetric heating of the material (Zhou et al., 2023). The application of microwaves for the pretreatment or posttreatment of foods has become increasingly common, giving rise to a substantial body of literature on the subject. Therefore it is currently considered an additive operation to the extrusion process, and it is discussed in a dedicated section in this chapter.

7.2.3 Radiofrequency

RF heating operates on a principle similar to microwave heating, where polar molecules try to align with an external electric field, producing internal friction within the molecules. The main difference is that the electromagnetic waves propagate in a single direction between the two parallel plates in RF heating. In contrast, in microwave heating, the waves are generated by specialized oscillating tubes and are transmitted in all directions. It belongs to a part of the high-frequency electromagnetic spectrum and occupies a range between 1 and 300 MHz. However, to avoid disruptions in communication systems, the US Federal Communications Commission permits only 13.56, 27.12, and 40.68 MHz for industrial, scientific, and medical applications (Ling et al., 2020).

The thermal treatment induced by RF is primarily influenced by factors associated with the sample and the equipment. For example, key factors include its dielectric and thermal properties, while for the equipment, the influencing factors are mainly the frequency, output power, and electrode gap (Gao et al., 2023).

RF heating has emerged as a prominent alternative to traditional methods used in extrusion, owing to its significant depth penetration and volumetric heating capabilities. Recently, RF heating has also been applied in pasteurization processes to improve food safety (Cui et al., 2023).

7.2.4 Infrared

IR thermal energy has shorter wavelengths than microwaves and RF, enabling more direct and localized heating of materials. Heat transfer occurs through radiation without a medium, transferring energy between two surfaces at different temperatures. In the case of food, IR radiation is utilized not only to heat food surfaces and dehydrate agricultural products but also to decontaminate and disinfect foods (Wiriyaumpaiwong et al., 2004; Andrejko et al., 2011). IR is commonly employed in food processing under specific heat conditions, offering several advantages, including higher heat transfer, immediate efficiency, and precise temperature regulation. This enables a controlled, compact, and homogeneous process (Jun and Irudayaraj, 2008). The application of this heating method has been investigated by various researchers across different food types, demonstrating its effectiveness in achieving drying, disinfection, disinfestation, and enhancing the nutritional quality and stabilization of foods. In addition, several authors highlight that this technology could offer substantial economic and environmental benefits by reducing processing costs and eliminating the need for chemicals in the process (Andrejko et al., 2011; Pan, 2021; Liu et al., 2022).

It is important to note that most studies and industry practices do not integrate extrusion with the aforementioned heating methods in a single-step process. Instead, these heating methods are primarily used as pretreatments or posttreatments for food extrusion. Consequently, the authors describe extrusion heating as a form of synergistic heating, in which heat variables, cutting, and pressure exerted in the barrel collectively influence the properties of processed foods. Below is a compilation of the key studies where thermal energy was the primary variable examined in the extrusion process, along with the main findings regarding its impact on the final properties of the foods.

7.2.5 Key studies of heating mechanisms coupled to extrusion systems

Baller et al. (2021) evaluated the influence of specific thermal energy (STE) applications on extruder preconditioners in cat dry food. Six STE applications were tested with mass temperatures of 45°C, 55°C, 65°C, 75°C, 85°C, and 95°C, and extrusion parameters, starch gelatinization, and kibble formation were evaluated. The increase in STE induced more significant kibble expansion and starch gelatinization, while lower STE and starch gelatinization induced higher butyrate and total volatile fatty acid (VFA) contents in feces. Cats showed the most significant preference for the formulation with the highest STE. Finally, the authors concluded that STE application is essential for enhancing kibble expansion, starch gelatinization, cat preferences for food, extruder productivity, and reducing specific mechanical energy applications. Foods with lower starch gelatinization lead to increased fecal VFA, possibly affecting gut health.

Similarly, Pacheco et al. (2018) evaluated the effects of STE at the same temperatures in dry food for dogs. Kibble expansion increased (bulk density, expansion rate, specific length), and hardness decreased with increasing STE. In comparison, according to the STE application, nutrient digestibility and food palatability did not change. Feces dry matter increased, but pH and fermentation product content remained unchanged.

Schmid et al. (2021) emphasized that in the extrusion process, the heat supplied to the mass is linked to mechanical stress, making it difficult to measure or calculate its effect on the outcomes (influence on structural changes and functional properties of foods). To mitigate the issue mentioned above, the authors worked with specific stress in a suitable chamber (called a closed-cavity rheometer) to control mechanical stress and heat in the processing of apple pulp. The results showed that with increasing temperature and mechanical stress, soluble and low-molecular-weight soluble dietary fiber contents increased, whereas insoluble fiber contents decreased. On the contrary, no changes in the water solubility index (WSI) and the water absorption index (WAI) up to 90°C were observed. Nevertheless, WSI and WAI decreased significantly at 120°C. Finally, at 160°C, WSI and WAI increased.

Sinaki et al. (2022) evaluated the effect of thermal energy generated by microwaves on the physical and thermal properties of cellular snacks made from yellow pea and red lentil flours. The results showed that with increasing temperature and moisture content during extrusion, the products became darker, crunchier, and crispier, with a higher expansion index. In addition, the authors reported that extrusion followed by microwave heating resulted in extensive damage to the starch granular structure and complete denaturation of proteins.

Jafari et al. (2024) studied the impact of extrusion cooking temperature (60°C, 85°C, and 110°C) on the particle size distribution of curcumin nanoemulsions stabilized with native whey protein concentrate and extruded whey protein concentrate. It was found that when the temperature was raised to 85°C, no significant changes were observed between the extrudates. However, the temperature of 110°C and its interaction with pH resulted in protein denaturation, leading to a considerable increase in particle size distribution. Larger particle sizes reflect more significant protein denaturation and aggregation. The authors mentioned weak van der Waals bonds that stabilize the aggregates created below 85°C. In contrast, more rigid and dense aggregates were formed when the samples were extruded at 110°C.

Köllmann et al. (2024) hypothesize that the intensity of the thermal treatment can explain a large part of the physicochemical changes that occur during extrusion. For this reason, near-infrared spectroscopy was used as a novel method to quantify the thermal process intensity during high-moisture extrusion (HME). The authors found that the degree of texturing of the extrudates was correlated with the equivalent temperature of the cutting cell. The knowledge generated could contribute to further optimization of processing conditions to achieve the desired meat analog products.

In the work of Žilić et al. (2014), a study was conducted to determine the indicators of the Maillard reaction: furosine, hydroxymethylfurfural (HMF), acrylamide, and color induced during soybean extrusion with heating induced by an IR and a microwave source. The results showed that soy proteins were susceptible to the Maillard reaction, and furosine formed rapidly under mild heating conditions during extrusion with IR heating.

However, the formation of acrylamide significantly increased with decreasing moisture content. After a short extrusion time, IR heating at 140°C, and microwave heating at 135°C for 5 min, HMF concentrations increased to 11.34, 26.21, and 34.97 $\mu g\ g^{-1}$, respectively.

The review by Singh et al. (2007) mentions that temperature is one of the main variables affecting protein digestibility in wheat. The increase in extrusion temperature (100°C–140°C) improves the degree of inactivation of protease inhibitors in wheat, and consequently, protein digestibility is increased. Finally, the authors mention that extrusion, even at its highest temperature, does not have the opposite effect on protein digestibility.

van den Einde et al. (2004) investigated the reduction in the molecular weight of corn starch generated from two different thermal treatments, 90°C–160°C and 90°C–140°C. During the first 5 min of treatment, a decrease in molecular weight was observed, measured by intrinsic viscosity. After this time interval, no significant effect was observed. Furthermore, the authors mentioned that at higher temperatures, thermomechanical decomposition could be divided into a mechanical part, depending on the maximum shear stress, and a thermal decomposition part, which again depended only on time at shorter time scales.

In the review by Dalbhagat et al. (2019), the authors mention that in rice processing, it is necessary to consider that the temperature of the food mainly depends on the barrel temperature; however, the food temperature is higher than the barrel temperature due to the friction between the barrel surface and the feed granules as they pass through the screws. The authors compile various works on rice and rice-based products, where it has been observed that the combined effect of temperature, moisture content, and screw speed reduces the carbohydrate content from 53% (in native flour) to 1%–8% in the final extruded products due to the breakdown of amylopectin into intermediate molecular weight material.

Finally, Pietsch et al. (2017) investigated the influence of process conditions on the polymerization behavior of wheat gluten during HME processing. The authors found that extruder temperatures ranging from 90°C to 160°C caused a significant change in polymerization behavior, which was reflected in the visual appearance of the samples.

Heat is an indispensable variable in food extrusion, essential for achieving, preserving, and adapting the quality and functional properties of food products. Current research emphasizes the development of "green" and cost-effective processing technologies, aligning with global requirements for environmentally friendly policies. However, the specific method of heat application in extrusion is highly dependent on the food being processed, considering its chemical, rheological, and textural properties.

Studies in this field demonstrate significant advancements, particularly highlighting the intricate relationship between operating temperatures, residence times, and the core properties of extruded foods. These properties include carbohydrate content, intrinsic viscosity, degradation, and overall chemical composition. Consequently, heat energy is recognized as a priority variable with a substantial impact on the food extrusion process.

7.3 Hybrid wave-assisted extrusion

This section explores the use of ultrasound waves, microwaves, and extrusion in food processing. The initial objective was to investigate whether wave-based processes, such as those involving ultrasound, could facilitate extrusion in a hybrid manner, similar to their application in synthetic polymer processing. In such cases, ultrasound probes are integrated into extruders or applied to the material as it flows through the system. The probes are subjected to microwave heating, resulting in synergistic enhancements to the final material properties. However, the literature on this topic remains limited, as most studies indicate that wave-based processes are predominantly employed in pretreatments, posttreatments, or as alternative methods to extrusion. This section provides an overview of wave physics and their interaction with matter. Then, the setup configurations for ultrasound-assisted and microwave-assisted extrusion, as reported in the literature, are presented.

7.3.1 *Waves*

Waves transfer energy through matter or space by propagating a dynamic disturbance—defined as a deviation from equilibrium—of one or more physical quantities, where the disturbance oscillates periodically. Oscillatory waves propagate through periodic variations characterized by crests and troughs. Their key characteristics include amplitude (the maximum displacement from the equilibrium, measured at the crest), wavelength (the distance of one complete oscillation, measured from crest to crest or trough to trough), period (the time required for one complete oscillation), and frequency (the number of complete oscillations per second, measured in Hertz). Waves can be longitudinal or transverse. In the former, the oscillations occur parallel to the direction of energy transfer, creating alternating compressions and rarefactions within a medium. In the latter, oscillations are perpendicular to the direction of energy transfer and propagate without needing a medium. Common waves, such as mechanical sound waves, are generally longitudinal when propagating in gases, liquids, and solids but can also be transverse when propagating through solids. In contrast, electromagnetic waves, characterized by the interplay of electric and magnetic fields forming an electromagnetic field, are inherently transverse. Electromagnetic waves can propagate through gases, liquids, solids, and even in a vacuum, traveling at the speed of light. In all cases, waves are primarily characterized by their wavelength, measured in meters (m), or their frequency, measured in Hertz (Hz). Sound waves can be classified as infrasonic (frequencies below 20 Hz, inaudible to humans), audible (frequencies ranging from 20 Hz to 20 kHz), or ultrasonic (frequencies above 20 kHz, extending to several gigahertz, and also inaudible to humans). Similarly, electromagnetic waves can be categorized into distinct bands based on their frequencies: radio waves (<3 × 10 Hz) (Ling et al., 2020), microwaves (3 × 10 Hz) (Ling et al., 2020) (–3 × 10 Hz) (Cui et al., 2023), IR (3 × 10 Hz) (Cui et al., 2023) (–4 × 10 Hz) (Jun and Irudayaraj, 2008), visible light (4 × 10 Hz) (Jun and Irudayaraj, 2008) (–7.5 × 10 Hz) (Cui et al., 2023), ultraviolet (7.5 × 10 Hz) (Jun and Irudayaraj, 2008) (–3 × 10 Hz) (Liu et al., 2022), X-rays (3 × 10 Hz) (Liu et al., 2022) (–3 × 10 Hz) (Schmid et al., 2021), and gamma rays (>3 × 10 Hz) (Schmid et al., 2021). Ultrasound waves and microwaves are particularly interesting; they possess high energy levels and can induce chemical changes or initiate chemical reactions when propagating through liquids.

7.3.1.1 Ultrasound waves

High-frequency ultrasound refers to low-amplitude sound (low-power) and is primarily associated with the physical interaction of the wave with the medium. It typically operates within a frequency range of 2–10 MHz. High-frequency ultrasound waves are widely utilized in medical imaging and chemical analysis. In contrast, low-frequency ultrasound waves, ranging from 20 to 100 kHz, are employed in cleaning, plastic welding, and sonochemistry applications. With advancements in high-performance equipment, sonochemistry applications can now achieve frequencies as high as 2 MHz (Mason and Lorimer, 2003).

7.3.1.2 Interaction of ultrasound waves with matter

In low-viscosity liquid systems, such as Newtonian fluids, the chemical and physical effects of ultrasound are attributed to the phenomenon of acoustic cavitation (Mason, 1999). Acoustic cavitation refers to the rapid formation and subsequent collapse of vapor bubbles within the liquid in a medium. Acoustic cavitation occurs in response to a reduction in pressure caused by the propagation of an acoustic wave, wherein the static pressure falls below the vapor pressure of the liquid. Bubbles form, grow, and ultimately collapse explosively during the characteristic rarefaction and compression of ultrasound waves. The collapse generates intense localized heating and produces extremely high pressures, reaching up to 1000 atm (Suslick and Price, 1999).

The strict definition of cavitation is based on the isothermal transition of the liquid-vapor phase boundary within a fluid due to a pressure decrease (Ashokkumar et al., 2016; Brennen, 2013; Trevena, 1984). However, this phenomenon does not apply to low-fluidity macromolecules in molten or non-Newtonian fluids. Consequently, this mechanism alone cannot fully account for the effects induced by ultrasound in such systems (Ávila-Orta et al., 2013). Therefore alternative mechanisms have been proposed, primarily involving the molecular relaxation of chain segments and the fatigue effects induced by the compression and rarefaction of the aforementioned longitudinal waves. These mechanisms are evaluated through response variables, which are quantified using thermal–mechanical characterization techniques and rheological analysis of the resulting materials (Yuntao et al., 2005; Espinoza-Gonzalez et al., 2016).

7.3.2 Microwaves

Microwaves represent a category of electromagnetic radiation characterized by frequencies ranging from 300 GHz to 300 MHz, corresponding to wavelengths between 1 mm and 1 m. Commonly used household microwave ovens and microwave reactors operate at a standardized frequency of 2.45 GHz (12.24 cm wavelength) to minimize interference with telecommunications frequencies. Within this electromagnetic energy spectrum, molecular rotation is selectively influenced without impacting molecular structures, as the energy associated with microwaves is insufficient to disrupt molecular bonds (Hayes, 2002).

7.3.2.1 Interaction of microwaves with matter

Microwaves can interact with matter through three primary mechanisms: (1) reflection, which occurs on conductive surfaces such as metals and graphite, preventing microwave penetration; (2) transmission, where microwaves penetrate insulating materials without significant energy absorption; and (3) absorption, which takes place in dielectric materials, resulting in the conversion of microwave energy into heat and, therefore, causing rapid heating of the medium (Bogdal et al., 2003).

When microwaves interact with a substance, energy transfer leading to rapid heating occurs primarily through two mechanisms: dipole rotation and ionic conduction. In the case of dipole rotation, polar molecules attempt to align with the electric field of the microwave, which changes direction constantly at a high frequency. This continuous reorientation induces rotational motion, generating friction within the substance. The resulting frictional energy is subsequently converted into heat. In contrast, ionic conduction occurs when ionic species or free ions are present within the substance exposed to microwave radiation. The oscillating electric field induces movement of these ions as they attempt to align with the rapidly changing field. This motion leads to collisions between the ions, which in turn generate heat (de la Hoz and Loupy, 2013). Consequently, the continuously oscillating electric field is the underlying mechanism responsible for microwave heating, also referred to as dielectric heating.

At a frequency of 2.45 GHz, the electric field oscillates 4.9×10 times per second (Ling et al., 2020). This rapid oscillation induces intense agitation of molecules through cyclic orientation, resulting in internal heating, as previously explained.

Microwave heating relies on the capacity of liquids or solids to absorb electromagnetic energy and convert it into heat. The efficiency of this energy conversion is governed by the loss tangent ($\tan \delta$), which represents the most critical property in microwave processing. The loss tangent quantifies the material's ability to polarize in response to the electric field and efficiently transform absorbed electromagnetic energy into thermal energy. The rate at which a substance heats under microwave radiation is directly proportional to its loss tangent (tan δ), which is defined as the ratio of the dielectric loss (ε″) to the dielectric constant (ε″): tan δ=ε/ε″.

Materials with a tan δ value equal to or greater than 0.1 are considered effective absorbers of microwave radiation, as they efficiently convert electromagnetic energy into heat. Conversely, materials with a tan δ value equal to or <0.0001 are regarded as poor absorbers, as they exhibit minimal energy conversion and are relatively transparent to microwave radiation (Komorowska-Durka et al., 2015).

7.3.3 Microwaves heating vs conventional heating

Microwave heating offers several distinct advantages over conventional heating methods, including instantaneous and direct heating, homogeneous heat distribution at high temperatures, selective heating, and energy savings. One of the key benefits of microwave heating is its ability to rapidly reach the target temperature, thereby minimizing the formation of secondary reaction products (Hayes, 2002; Kappe, 2008). This capability is particularly advantageous in chemical reactions, as it leads to improved reaction yields by preventing side reactions. These advantages make microwave radiation an increasingly popular energy source in chemical synthesis, as the direct interaction between microwave radiation and the molecules in reaction mixtures facilitates a rapid increase in temperature. In contrast, conventional heating traditionally relies on conduction from an external heat source, such as an oil bath. This method is less efficient at transferring energy to the system, since it depends on the material's conductivity and results in a temperature profile where the walls of the reaction vessel are hotter than the reaction mixture itself. Conversely, microwave radiation provides a more efficient form of heating. The energy is transferred directly to the molecules in the reaction mixture, enabling uniform heating. Furthermore, reaction vessels are made from materials such as quartz, polytetrafluoroethylene (PTFE), or borosilicate glass, which are transparent to microwave radiation and allow for the formation of an inverted temperature gradient compared to that obtained by conventional heating, where the temperature of the walls often exceeds that of the mixture.

7.3.4 Experimental setups

The integration of ultrasound probes and microwave technology within extruders is referred to as hybrid extrusion. These processes are usually referred to as ultrasound-assisted extrusion and microwave-assisted extrusion. Experimental setups for both ultrasound-assisted extrusion and microwave-assisted extrusion have been reported in the literature. No documented schemes or photographs for ultrasound-assisted extrusion were found in the scientific literature related to food processing. However, existing studies claim that extrusion and ultrasound techniques are utilized in combination (Chemat et al., 2011; Knorr et al., 2004; Patist and Bates, 2008; Singla and Sit,

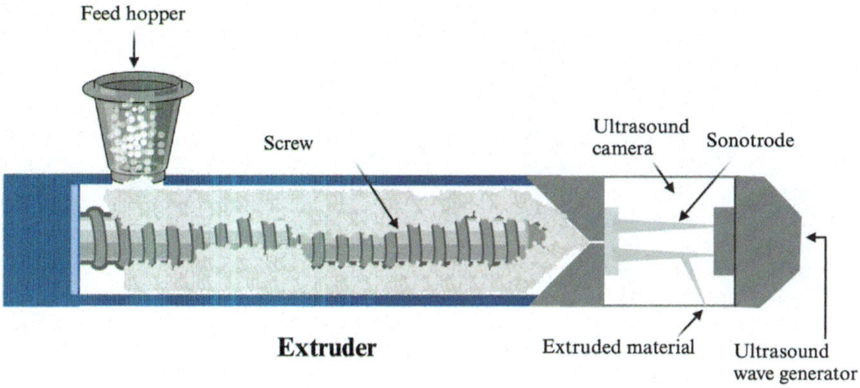

Fig. 7.1 Scheme of an experimental setup for ultrasound-assisted extrusion. Modified from Ávila-Orta, C.A., González-Morones, P., Agüero- Valdez, D., González-S ánchez, A., Martínez-Colunga, J.G., Mata-Padilla, J.M., Cruz-Delgado, V.J., 2019. Ultrasound-assisted melt extrusion of polymer nanocomposites. In: Subbarayan, S. (Ed.), Nanocomposites - Recent Evolutions. IntechOpen, London, p. 167.

2021). In contrast, this is not the case for synthetic polymers, where ultrasound probes have been integrated into extruders, either perpendicular to the barrel or in a counter-flow arrangement at the exit die or nozzle. This integration aims to reduce viscosity, improve mixing, and enhance the dispersion of nanoparticles in thermoplastic polymers (Ávila-Orta et al., 2019; Isayev et al., 2009).

Fig. 7.1 illustrates a schematic representation of an ultrasound probe integrated with the exit die nozzle of an extruder. The ultrasound system consists of an ultrasound wave generator, a sonotrode, an expansion chamber that facilitates the expansion of the extruded material, and an exit nozzle through which the ultrasonically processed material is discharged.

While these hybrid systems have primarily been employed for the extrusion of synthetic thermoplastic polymers (synthetic macromolecules), they can be readily adapted for applications in food processing, where biomacromolecules are the main materials being processed (Ávila-Orta et al., 2019).

Conversely, to the best of our knowledge, the recent integration of microwave technology within an extrusion line has been reported, building upon a previous microwave system utilizing a waveguide (Graf et al., 2020). Rapid heating, enhanced energy efficiency, and improved product quality are among the anticipated benefits of microwave-assisted extrusion.

Fig. 7.2 presents a schematic representation of a microwave system integrated into an extruder (Schmidt et al., 2022). This system is positioned perpendicularly to the barrel and comprises three main components: a microwave generator, a waveguide, and a tuner.

It is noteworthy that the extruder utilized is a prototype of a single extruder (Kern et al., 2019), modified with a PTFE tube and screw. PTFE was selected due to its exceptional transparency to microwaves.

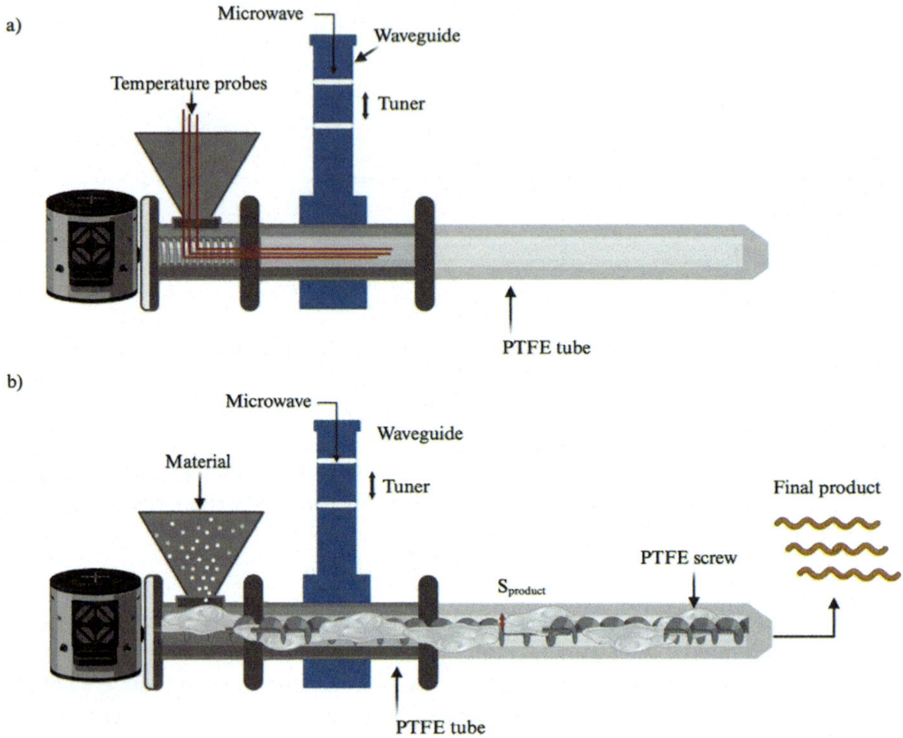

Fig. 7.2 Scheme of an experimental setup for microwave-assisted extrusion. Schematic representation of a microwave system integrated into an extruder (Schmidt et al., 2022). Modified from Schmidt, F., Graf, B., Hinrichs, J., Kern, C., 2022. Continuous microwave-assisted extrusion for high moisture texturized foods: a feasibility study. Innov. Food Sci. Emerg. Techn. 78, 3. doi:10.1016/j.ifset.2022.103020.

7.3.5 Applications and case studies

7.3.5.1 Ultrasound-assisted extrusion

The application of high-energy ultrasound waves at correct angles to the metal barrel induces radial vibration, thereby reducing drag friction and enhancing throughput (Chemat et al., 2011; Knorr et al., 2004; Patist and Bates, 2008). This phenomenon has also been researched through numerical simulation studies (Mousavi et al., 2007). The authors reported that the application of ultrasound waves results in the reduction of both extrusion force and material flow stress. Although the reduction of drag resistance through the use of ultrasound waves in food processing has been mentioned, no scientific reference to a case study on ultrasound hybrid extrusion was found. However, the combination of ultrasound with extrusion technology in food processing has been extensively reported and reviewed in current literature (Singla and Sit, 2021). The integration of these techniques typically occurs in a sequential manner, employed for pretreatment or posttreatment processes. A pertinent example is the pretreatment of rice hulls, which involved an initial extrusion process followed by ultrasound treatment (Zhang et al., 2020) to improve enzymatic hydrolysis. The authors of this study concluded that the combination of extrusion and subsequent ultrasound resulted in a substantial enhancement of reduced sugar yields. To conclude, the sequential application of extrusion and ultrasound presents a promising chemical-free pretreatment method for augmenting the enzymatic hydrolysis of rice hulls, thereby facilitating the production of fermentable sugars. This approach offers notable advantages in terms of environmental sustainability, process efficiency, and the enhancement of product yield.

7.3.5.2 Microwave-assisted extrusion

To the best of our knowledge, only a single report has been published regarding microwave-assisted extrusion. In this study, the feasibility of utilizing microwave-assisted continuous extrusion to produce high-moisture textured foods was explored (Schmidt et al., 2022). Traditional extrusion methods often face limitations in achieving the desired textural properties and operational efficiency. The integration of microwave heating, which seeks to enhance the process through uniform heating, improved energy efficiency, and superior product quality, presents a potential solution to overcome these limitations. In this study, the authors demonstrate that microwave-assisted extrusion enhances the production of high-moisture textured foods. Furthermore, the ability to precisely control heating facilitates more effective management of texture properties, a crucial factor in the development of plant-based meat analogs. Moreover, the process demonstrates considerable potential for energy savings and scalability, resulting in a promising approach for the food industry. In conclusion, the authors affirm that microwave-assisted extrusion is both feasible and efficient for producing high-moisture textured foods.

Hybrid extrusion, assisted by ultrasound or microwaves, represents an emerging technological frontier, with a limited number of studies reported to date. Nevertheless, existing experimental setups have already been implemented, enabling the use of ultrasound and microwaves in the extrusion processing of food for both pets and humans. Although no comprehensive reports on the hybrid ultrasound-assisted extrusion of food have been found, exploring the potential benefits and disadvantages of utilizing this technology for food processing would be of considerable interest.

7.3.6 Future directions and research opportunities

Ultrasound-assisted extrusion, wherein extrusion is followed by ultrasound treatment, emerges as a promising chemical-free pretreatment method for enhancing the enzymatic hydrolysis of rice hulls in the production of fermentable sugars. This approach presents significant advantages in terms of environmental sustainability, process efficiency, and product yield, thereby positioning it as a viable option for the biofuel and bioproduct industries. Future studies should prioritize the optimization of combined pretreatment parameters and the scaling up of the process for industrial applications. In addition, economic analyses and life cycle assessments would be essential to comprehensively evaluate the feasibility and environmental impact of this method in large-scale operations. A key question that arises is whether hybrid ultrasound-assisted extrusion could serve as an alternative to the traditional sequential step process.

Microwave-assisted extrusion, which integrates microwave heating, offers notable advantages in terms of product quality and process efficiency, indicating a potential transformation in extrusion technology for food production. However, further investigation is required to optimize process parameters and assess the scalability of this technology for industrial applications. In addition, sensory evaluations and consumer acceptance studies would be crucial in determining the commercial viability of the products produced using this method.

7.4 AM technology

This section underscores the importance of AM and its influence on the fabrication and design of edible food products through extrusion-based processes, specifically extrusion three-dimensional (3D) printing. To comprehend the 3D printing process, it is essential to understand the conventional food extrusion process. This process involves feeding a semi-solid material, such as resin, dough, paste, puree, or other viscous substances, into an extruder. Within the extruder, the raw material is transported through a barrel by one or more screws toward an extrusion die or nozzle. At this stage, the material acquires a solid or foamed shape, after which it is subjected to drying or cooking.

AM, in this context, refers to the manufacturing of 3D food structures guided by computer-aided design technology (Sanchez et al., 2019). AM encompasses a range of technologies, including extrusion 3D printing, binder jetting, selective laser sintering (SLS), ball-back extrusion for assessing texture characteristics, and an innovative extrusion posttreatment method known as roller embossing. These technologies facilitate the formation of unique food structures, but most importantly, each of them provides different nutritional properties.

One of the primary advantages of food products processed via AM technology is the preservation of their nutritional characteristics. Furthermore, AM enables the customization of food products.

7.4.1 Extrusion 3D printing

An extrusion 3D printer is a system designed to fabricate 3D objects by depositing successive layers on top of a fluid medium or surface. This process involves physical transformation as layers are deposited to construct the desired 3D structure (Hull, 1986). Initially, extrusion 3D printing was associated with lithography and mainly employed for the creation of prototypes or artistic objects. Nowadays, extrusion 3D printing has evolved significantly and is now utilized to produce diverse items, including but not limited to bone prostheses, lightweight components, and food products with customized properties. Food extrusion 3D printers using fermented and malted materials are emerging as a promising bridge for the potential fabrication of animal and human tissues or organs.

In the context of AM, the term "printing" refers to a process in which the raw material is often called ink. This terminology extends to extrusion-based 3D printing of food, where the material used is also referred to as "food ink," associated with lithography processes. Fig. 7.3 illustrates a schematic representation of a 3D printer.

Recently, two primary methods dominate the extrusion 3D printing process: SLS and fused deposition modeling (FDM). SLS is predominantly focused on powder-based materials, such as sugar or sugar-rich powders, which are sintered and bonded layer by layer through the application of heat (Godoi et al., 2016). In contrast, FDM is employed for semi-solid materials, including dough, puree, paste, or other viscous substances. The process may be conducted at room temperature or elevated temperatures, depending on the melting characteristics of the food ink, to ensure proper adhesion between layers.

A notable advantage of FDM-based 3D food printing is the ability to use several deposition units (two or more extruders) simultaneously. This capability enables the production of complex, multicomponent food items with different formulations, such as candies, burgers, pizza, cookies, and other edible products. Consequently, extrusion 3D printing is often synonymous with FDM due to its widespread application.

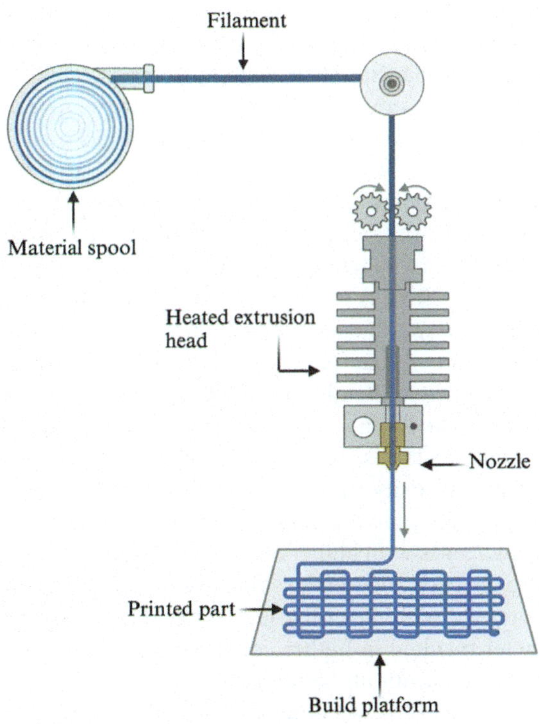

Fig. 7.3 Schematic representation of the 3D printing process.

The process of extrusion 3D printing is inherently complex, as it is influenced by numerous factors. Among these, the most critical parameters that significantly affect the accuracy and high quality of 3D-printed food products are given in the following (Derossi et al., 2018).

The rheological behavior of food ink during extrusion—characterized by properties such as pseudoplasticity, shear thinning, shear thickening, thixotropy, Bingham behavior, and others—is associated with various processing conditions. These conditions include die dimensions, temperature, extrusion speed, nozzle movement speed, layer height, feed percentage, and, most importantly, the composition of the food ink.

Each ingredient type and its concentration within the formulation plays a critical role in altering the rheological behavior of the ink. These parameters must be carefully optimized to ensure stable flow, layer-by-layer adhesivity, and the structural integrity of the final 3D-printed structure.

Temperature, as previously discussed, plays an important role in influencing the rheology of food raw materials. Based on this factor, the extrusion 3D printing process can be classified into three categories: (1) room-temperature extrusion (soft material extrusion), (2) hot-melt extrusion, and (3) hydrogel-foaming extrusion.

Printing time is also intrinsically linked to the rheological behavior of food ink, which exhibits pseudoplastic and shear-thinning characteristics during the extrusion process. These rheological properties are influenced by the temperature and composition of the formulation.

Postprocessing following deposition is essential to ensuring the quality of 3D-printed food. It preserves the nutritional properties, structural integrity, and final geometry of the product, enhancing its palatability, sensory appeal, and shelf life.

Several researchers have investigated the parameters governing extrusion 3D printing, mainly those related to rheological behavior and its association with printing conditions.

Zhu et al. (2019) conducted a study to examine the correlation between the rheological behavior and printability of several food inks using tomato puree as a base material; the authors subjected the puree to varying centrifugation forces (1000–10,000 g force for 20 min) to obtain tomato pastes for evaluating rheological and printable properties. Experiments were conducted at room temperature, with the samples printed into a hollow square column measuring 30 × 30 mm, a wall thickness of 2.4 mm, and a printing layer height of 0.6 mm.

The food inks were printed using two nozzle diameters (1.2 and 0.8 mm) at printing speeds of 25 mm/s and 18 mm/s, respectively. The study established a correlation between zero-shear rate viscosity and collapsed stress (the total sample weight divided by the collapsed height acting on the bottom sample). The results indicated that zero-shear rate viscosity increases with higher centrifugation forces due to the packaging of tomato particles (broken cells and tomato skin in the puree dispersion). In addition, the collapsed stress increased with the collapsed height. The findings also noted that the larger nozzle diameter resulted in higher collapsed stress and storage modulus (G'), leading to improved printability of the tomato paste.

Similarly, Zheng et al. (2021) conducted a study focusing on the rheological behavior of food inks and its correlation with extrusion 3D printing parameters, including extrusion rate (mm/s) (Kaur and Singh, 2016), nozzle movement speed (mm/s), and fill rate (%). The study examined several food ink formulations in the form of pastes or doughs, such as Sidney pulp/potato starch (SP/PS), sorghum flour/edible soybean oil (SF/ESBO), spinach/pumpkin powders (S/PP), whey protein/carrot powder (WP/CP), and other blends. All formulations were prepared in water with salt, and some included xanthan gum as an additive.

Two models—a cone and a cylinder—were printed with diameters and heights of 30 mm, and the collapse height was recorded for both models. The authors reported the stability and printability of each food ink by measuring the bottom diameter (mm) and the collapsed height (%) of the cone model. The printability behavior of the food ink blends was observed in the following order: whey protein/carrot powder (WP/CP) > Sidney pulp/potato starch (SP/PS) > spinach/pumpkin powders (S/PP) > sorghum flour/edible soybean oil (SF/ESBO). This trend was directly correlated with the viscosity of the food ink, as measured using a plate-plate rheometer. Higher viscosity enhanced stability by minimizing the diameter variation and increasing the collapsed height, both critical factors in achieving printability.

In a study by Liu et al. (2019) a system based on sodium caseinate (NaCN) and milk protein concentrate (MPC) compounds was examined. The authors found a strong correlation between the MPC content in the NaCN/MPC mixtures and the rheological behavior and textural properties of the resulting materials. These properties, including hardness, adhesiveness, gumminess, springiness, and chewiness, were reported to change with varying concentrations.

The viscosity of the compounds increased with higher MPC content, with significant changes in rheological behavior observed between 350 and 500 g/L MPC concentrations. At 300 g/L MPC, the mixture exhibited shear-thinning fluid behavior, while at 400–500 g/L MPC, it transitioned to a thixotropic behavior. The authors suggest that materials exhibiting appropriate thixotropy are advantageous for extrusion 3D printing. In such materials, viscosity decreases under shear rate (a beneficial property); still, the material rethickens and recovers its structure once the shear stress is removed, and this process may lead to a decay in the textural properties.

As several authors have highlighted, there is a clear relationship between the rheological properties of food inks and the composition of various formulations, with interactions between components significantly influencing rheological behavior. Consequently, the extrusion 3D printing process becomes increasingly complex when designing customized foods, considering nutritional content, shape, color, texture, shelf life, and digestibility parameters.

In this context, Liu et al. (2020) explored the use of different rice starches, specifically those with high and low amylopectin content, in combination with an extrusion 3D printing process and steaming method. The authors stated that the rheological properties of rice doughs were influenced qualitatively—though not quantitatively—by the composition, and these properties were correlated with printability. They found that rice dough with high amylopectin content exhibited excellent printability; however, when subjected to steam, the printed products experienced significant deformation and rapid digestion. In contrast, rice doughs with lower amylopectin content (i.e., higher amylose content) demonstrated enhanced printability and maintained structural stability during the steaming process, contributing to improved nutritional properties and extended shelf life.

The use of various starches in food manufacturing is well-established due to their excellent rheological properties. Starch acts as a fluid thickener in juices, purees, pulps, and other processed foods. In the context of extrusion-based 3D processes, starch serves a dual purpose: first, to enhance the nutritional content and, second, to improve the rheological properties and printability of materials that are otherwise unsuitable for printing, such as fruits and vegetables. For instance, it is well known that fruits and vegetables are rich in numerous health-beneficial compounds. Markovinović et al. (2023) reported a study in this regard, using corn and wheat starches as thickeners in a strawberry-based compound. Strawberries are a functional fruit notable for their high antioxidant content and other nutritional elements. However, their rheological behavior renders them unsuitable for use in 3D printing processes, as is the case with most vegetables or fruits. The authors observed that incorporating starch significantly enhances the printability of strawberry-based food inks while ensuring that the dimensional and structural stability of the printed product remains consistent, irrespective of the type of starch used. However, the study revealed that wheat starch preserves the functional properties of strawberries. In addition to starches, other polysaccharides, such as xanthan gum (XG) and guar gum (GG), have been employed as thickeners in various emulsions, including those based on soy protein isolate (SPI), to enhance rheological properties and improve extrusion 3D printability. A critical consideration for these gums is their physical interaction with the food matrix. For example, SPI–GG emulsions with a 0.5% GG content exhibit a low dimensional printing deviation, excellent structural support, and a smooth texture. In contrast, SPI/XG emulsions tend to result in hardened textures and surface roughness (Jie et al., 2022).

The development of new food formulations is widely investigated and approached from different perspectives. These include preserving nutritional properties to produce healthier and more edible food, controlling and optimizing rheological properties to enhance food ink printability, adjusting texture properties for customer acceptance, and incorporating natural bioactive agents to extend shelf life. It can be stated that nearly all critical challenges in formulation have been addressed. However, significant challenges remain within the domain of extrusion 3D printing technology.

One notable issue is the bottleneck created by using small-capacity cartridges for food ink, which limits the accessibility of such technology for broader applications.

While using larger cartridges could address this issue, it introduces another challenge: increased cleaning time for the extrusion 3D printer; thus the challenge remains.

7.4.2 Ball back extrusion

As highlighted in the previous section, the texture properties of food inks—hardness, adhesiveness, gumminess, springiness, and chewiness—are fundamental factors in the food ink itself and the final product manufactured through the extrusion 3D printing process. An important aspect to consider regarding food texture is its handling during eating and ingestion. Different textures require specific oral manipulations and bolus formation to ensure safe and effective swallowing (Lillford, 2018). Another significant factor is the target customer group for whom the food is designed: infants, school-age children, adults, or older people. Equally important is addressing the needs of individuals with specific conditions, such as dysphagia (Prabha et al., 2021). In alignment with the previously mentioned considerations, obtaining an accurate texture measurement is critical for designing extrusion 3D-printed food. Several empirical methods have been employed to assess texture, all based on shear/stress deformation for fluid or soft solid foods and strain/stress fracture for solid foods. Commonly used techniques, such as syringe flow, mimic analysis, and texture profile analysis, are popular in quality control due to their low cost. However, these methods are limited by their qualitative nature and lack of accuracy in texture measurement (Chen et al., 2021).

The plate back extrusion (PBE) technique has also been utilized for texture evaluation, focusing on flow and deformation properties (rheology), as represented in Fig. 7.4. This method involves a cylindrical cup and a smaller-diameter cylindrical probe. The food ink or soft food is fed into the cup, and the probe, connected to a load cell, is inserted. The fluid is compressed and squeezed, creating flow through the narrow gap between the cup and the probe walls. The stress required for squeezing is recorded and correlated with texture properties. However, due to the sharp angle of the probe causing flow instability, the texture measurements obtained through this method lack accuracy.

Chen et al. (2021) introduced a novel method called ball back extrusion (BBE). This method comprises a cylindrical cup maintained at a controlled temperature and a spherical probe designed to avoid flow instabilities observed in the PBE method.

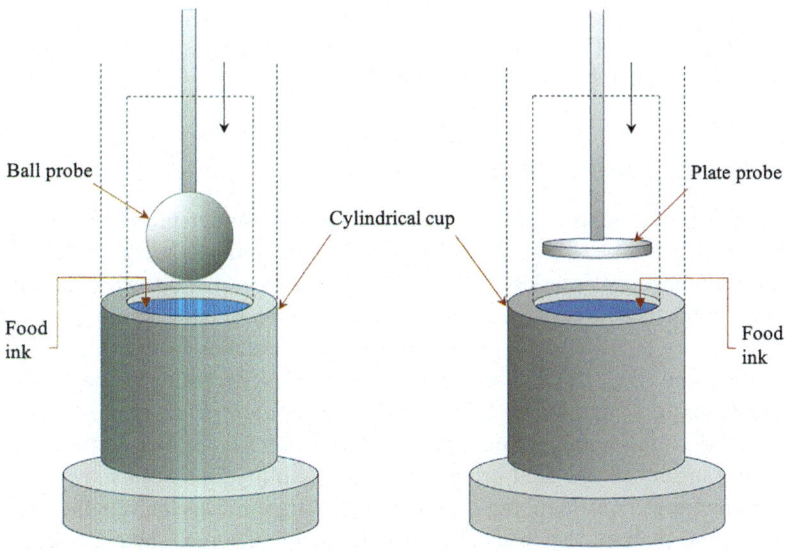

Fig. 7.4 Conceptual representation of the ball-and-plate probe diagram.

The BBE technique enables accurate flow measurements. Consequently, different aspect ratios—defined as the ratio of the diameter of the spherical probe to the inner diameter of the cylindrical cup—can be utilized.

In the BBE method, the spherical probe is connected to a load cell, facilitating precise movement control and force measurement. Compared to the PBE method, the spherical geometry of the BBE probe avoids the flow instabilities observed in PBE. It prevents bubble formation and accumulation, a common issue with the PBE technique. Both BBE and PBE methods are suitable for quality control within industrial environments and applicable to Newtonian and non-Newtonian fluids. By varying the aspect ratio and shear rate, these methods yield multiple apparent stress measurements, which can be associated with the material's texture properties.

The BBE method provides a detailed assessment of texture properties by analyzing specific points and areas of the stress-deformation curve, independent of whether the fluid exhibits Newtonian or non-Newtonian behavior. Kaewsorn et al. (2023) reported using BBE to evaluate the texture properties of cooked rice grains, correlating specific measurements with different texture attributes. The relationships are outlined as follows:

1. Hardness: Hardness is determined by the maximum compression force (N) recorded at the point where the spherical probe, in its compression movement, reaches 1 mm from the bottom of the cup.
2. Toughness: Toughness is quantified as the area under the stress–deformation curve (Nmm) from the start of the probe's compression movement to the point where the probe is fully retracted. This measurement reflects the energy required to deform the sample.
3. Stickiness: Stickiness is measured as the maximum negative force (N) observed during the retraction of the spherical probe. This occurs when the sample adheres to the probe, creating a pulling force as the probe moves upward.
4. Adhesiveness: Adhesiveness is defined as the area under the curve (Nmm) during the probe's retraction phase, extending from the point of maximum negative force to the point of zero deformation. This represents the total energy required to overcome the adhesive forces between the probe and the sample note.

7.4.3 Roller embossing

The determination of texture properties extends beyond understanding the rheological properties required for extrusion 3D printing, such as hardness, adhesiveness, springiness, and thickness, or assessing the sensory properties of food, such as chewiness, toughness, and stickiness, which are essential for ensuring adequate swallowing. Texture attributes, such as springiness, hardness, and toughness, also play a significant role in the embossing process of 3D-printed food. The roller embossing process involves stamping a 3D pattern onto a laminated dough, paste, or gum using either a hot or cold roller with a specifically desired pattern. These texture properties significantly influence the final product's shape, stability, and appearance.

Critical parameters influencing the food embossing process include the rheological behavior of the material, such as shear thinning, thixotropic, shear thickening, and Newtonian fluid characteristics. If the rheological properties are unsuitable for the roller embossing process, textural parameters become increasingly important. In this context, the laminate dough must achieve an optimal balance of hardness, springiness, and toughness to ensure that the embossed pattern is preserved throughout the process.

AM has garnered significant interest due to its inherent practicality. Nonetheless, 3D printing extrusion is a multifactorial process, incorporating key elements such as rheological properties, formulation composition, digestibility, nutritional attributes (e.g., enhanced health and edibility), extended shelf-life, and effective swallowing. Maintaining a balance among these interdependent factors poses a considerable challenge. Varying the ink in food printing can significantly alter the final product's composition, overall nutritional content, shelf life, ease of digestion, and critical textural attributes. The development of novel optimization models, potentially employing artificial intelligence, presents a feasible short-term strategy to address this complexity.

7.5 Atmosphere- and agent-assisted extrusion

Extruded food represents a viable alternative for producing nutritionally dense snacks rich in vitamins, minerals, proteins, and fibers, while offering desirable textures that cater to the growing trend of on-the-go consumption. Extrusion technology is also extensively utilized in pet food and animal feed production, enhancing their nutritional quality and textural attributes such as crispness, hardness, and crunchiness.

Key extrusion parameters, including water content, profile temperature, screw speed, die dimensions, and feed rate, have a significant impact on the final quality characteristics of extruded food, such as expansion ratio, density, color, texture, and microstructure. These factors are associated with the consumption attributes of the product. In addition, the integration of assisted extrusion technologies, such as blowing agents, supercritical fluids, and vacuum processes, can further enhance the nutritional and sensory properties of extruded foods.

The processing of protein-rich and high-fiber-content foods often presents a challenge, as these ingredients can inhibit the expansion process, leading to denser and less crunchy products. Water serves as the sole blowing agent.

However, the use of atmosphere-controlled and agent-assisted extrusion technologies offers significant advantages. These methods can enhance key textural and appearance attributes, including more substantial expansion, reduced density, a more desirable crunchy texture, and a more homogeneous microstructure.

7.5.1 Blowing agent injection-assisted extrusion

Human and pet food are complex systems composed of starch, proteins, lipids, dietary fiber, bioactive components, and additives. These interact during the extrusion process and influence the final product's appearance and sensory quality.

Starch and protein play crucial roles among the primary macromolecules in extruded foods. Starch is essential in regulating the expansion ratio, density, and cell morphology of extruded food. These characteristics are influenced by the high temperatures, pressure, and shear forces encountered during extrusion, which lead to changes in the starch's structure, molecular order, and chain disentanglement (Dircio-Morales et al., 2023). These structural alterations, in turn, affect the starch's viscous properties, enhancing its ability to interact with other components, including proteins, lipids, polyphenols, dietary fiber, and nonpolysaccharide substances (Mohamed, 2023).

Water serves as the primary driving force for the expansion of extruded foods. The generation of water vapor creates higher pressure within the dough, and when the dough passes through the extruder die, a pressure drop causes the expansion. The water content is crucial in developing the appropriate dough viscosity and facilitating cell growth for product expansion. However, relying solely on water as the forming agent or blowing agent can lead to a lack of control over microstructure expansion, potentially resulting in an undesirable textural appearance in the extruded food.

Extruded foods, including puffed, blasted, light snacks, and other extruded products, are typically high in starch. Various proteins, dietary fibers, and bioactive components are incorporated to enhance their nutritional profile and promote health benefits. However, including these ingredients could alter the quality and sensory properties of the final product, which, in turn, may affect consumer acceptance. For instance, extruded rice starch snacks enriched with legume protein, such as pea protein, exhibited a nonuniform structure when the protein content reached 30% (w/w). This was due to a delay in the solidification process after the dough exited the die, which decreased the glass transition temperature. At higher protein levels of 50% (w/w), the extrusion process resulted in low expansion (Philipp et al., 2017).

The expansion process of extruded food is influenced by the combination of raw materials, extrusion conditions, and blowing agents, which determine the final product's physical and textural characteristics. A blowing agent is an additive that facilitates the formation of a cellular microstructure during the expansion process, similar to the foaming effect in polymers and expanded foods. Blowing agents can be classified into two categories: chemical and physical.

Chemical blowing agents, such as azodicarbonamide, isocyanate, crosslinkers, and hydrazides, generate exothermic or endothermic reactions during extrusion, which can enhance the physical appearance of the final product. However, these agents are not regularly used in food production due to potentially undesirable effects on health and safety. Physical blowing agents, including supercritical fluids and gases, offer a promising alternative for extrusion-assisted technologies (Chauvet et al., 2017).

Raw materials with relatively low starch content and high protein, or formulations with high moisture content, typically result in extrudates that collapse upon exiting the die, failing to expand properly. To address this issue, a valve system that injects pressurized gases, such as supercritical fluids, into the barrel of the extruder is employed. This injection creates nucleation sites and promotes the growth of bubbles, facilitating the expansion process. Carbon dioxide (CO_2) is widely used in food extrusion due to its solubility in water and feed material. Other gases, such as nitrogen (N_2) and air, have also been incorporated into extrusion processes to achieve the desired texture and expansion properties (Koksel and Masatcioglu, 2018; Chan et al., 2019; Luo et al., 2020; Sinaki et al., 2023).

Gases have been injected into various food matrices in food extrusion processes to enhance product characteristics. In a study on air injection-assisted extrusion, pea flour with different particle sizes (small and large) was extruded using a twin-screw extruder under different temperature profiles (120°C, 140°C, and 160°C) and air injection pressures (0 kPa, 150 kPa, 300 kPa). The results showed that extrudates produced with air injection at temperatures of 140°C and 160°C exhibited higher emulsion capacity and stability, making them more suitable as ingredients for emulsified foods. The authors concluded that air injection-assisted extrusion, when combined with optimal particle size distribution and temperature profile, is a promising technique for enhancing the techno-functional properties of pulse flours. This process has the potential to expand the applications of pulse flours in the food industry, improving their utility in a range of food formulations (Sinaki et al., 2023).

Nitrogen, an affordable, abundant, and environmentally friendly gas, has gained attention as a potential physical blowing agent in the extrusion process. Nitrogen gas was injected at two pressure levels (100 and 200 kPa) to produce cereal-based snack foods through gas-assisted extrusion. The study compared conventional and nitrogen gas-assisted extrusion at these two injection pressures. The results showed that nitrogen injection led to significantly more expanded extrudates, evidenced by decreased extrudate density and increased expansion index. Furthermore, nitrogen injection at 200 kPa enhanced the extrudate's crunchiness, resulting in a more homogeneous and smoother surface than conventional extrusion. Nitrogen injection-assisted extrusion significantly increased the cold viscosity of extrudates compared to traditional extrusion. This process also resulted in extrudates with a less yellow and red appearance while being brighter overall, which brought their color closer to that of raw wheat flour. This indicates that less intensive browning reactions occurred during nitrogen injection-assisted extrusion. Regarding functional properties, the water solubility of the extrudates was enhanced at a nitrogen injection pressure of 200 kPa. However, no significant difference was observed in their water-binding capacity (Li et al., 2019).

To promote the consumption of nutritionally dense crops, such as lentils, extrusion cooking has been identified as a versatile and efficient method for producing puffed cereal and pulse-based snack foods that appeal to consumers. Nitrogen (N_2) injection during extrusion can potentially manipulate the expansion, microstructure, texture, and color of red lentil extrudates. In one study, a corotating twin-screw extruder with a screw length-to-diameter ratio of 25:1, a circular die with a diameter of 2.3 mm, and five temperature control zones (set at 60, 90, 120, 140, and 150°C from feed-to-die sections) was used to extrude red lentil flour. Nitrogen gas was employed as a physical blowing agent at three different pressures: 300, 400, and 500 kPa. The results indicated that nitrogen injection significantly influenced the extrudates' properties, including their expansion and texture, while also impacting their visual appearance, thus demonstrating the potential of nitrogen-assisted extrusion to improve the quality and appeal of pulse-based snack foods. Nitrogen injection at 300 kPa N_2 significantly decreased the extrudate density from 0.25–0.41 kg/m^3 to 0.12–0.26 kg/m

(Kaur and Singh, 2016) across all feed moisture levels and screw speeds studied. This injection pressure also substantially altered the microstructure of the extrudates, resulting in maximum expansion, reduced density, and a more uniform distribution of smaller cells. These changes in microstructure contributed to improved textural properties, such as increased crispiness. However, higher feed moisture content was found to increase the hardness of the extrudates while simultaneously decreasing crispiness (Luo et al., 2020).

The growing demand for plant-based meat alternatives and the desire for these meat analogs to closely replicate the textural quality and taste of traditional meat products have led to the development of gas-assisted HME for soy-based meat analogs. In this process, a gas injection assembly was used to inject nitrogen gas at varying pressures of 0 bar (conventional extrusion), 1 bar, and 2.5 bar. The cooling die temperature, combined with increased gas injection pressure, resulted in significant changes in the textural properties of the meat analogs, including lower density, reduced hardness, chewiness, and gumminess, as well as the formation of gas bubbles that altered the microstructure.

Despite these changes, the novel gas-assisted HME technique did not affect the degree of protein texturization, indicating that it holds great potential for manipulating the end product's texture and color. The entrapped gas bubbles within the protein matrix during extrusion significantly influenced the physical properties of the meat analogs, enhancing their potential as plant-based substitutes for traditional meat products (Ghanghas et al., 2024).

Pulses, including legumes rich in protein, fiber, and micronutrients, are increasingly used as raw materials for snack foods. In a study exploring pulse flours, two types of flours—red lentils and yellow peas—were utilized for extrusion using a laboratory-scale twin-screw extruder. The extrusion process was carried out at 200 rpm, with a die temperature of 150 °C and a die diameter of 2.3 mm. Two physical blowing agents, nitrogen (N_2) and carbon dioxide (CO_2), were injected into the extruder barrel through an injector valve positioned 22.9 cm before the die exit at a fixed injection pressure of 300 kPa.

The study demonstrated that physical blowing agents significantly impacted the physical properties of extrudates. Nitrogen was found to be more effective for red lentil flour in its dry form, while carbon dioxide was more suitable for yellow pea flour in its wet form. Both blowing agents successfully influenced the expansion, microstructure, texture, and color of extruded snacks, showing potential benefits for developing healthier, plant-based snack foods and enhancing their potential in the health-conscious food market (Chan et al., 2019). In summary, gas injection-assisted extrusion offers significant advantages with the growing global demand for snack foods and the increasing trend to incorporate protein-rich ingredients for healthier and more nutritious options or to develop meat analogs. This technique improves the nutritional profile, texture quality, appearance, and taste of snack foods by generating nucleation sites and microbubbles. As a result, gas-assisted extrusion offers an effective method for enhancing physical properties, making them more appealing to consumers seeking healthier alternatives and expanding the potential for plant-based meat substitutes.

7.5.2 Supercritical fluid-assisted extrusion

In supercritical fluid extrusion (SCFX), a supercritical blowing agent is injected into the extruder, acting both as an expander and a plasticizer to reduce the viscosity of the material. Fig. 7.5 represents a supercritical-assisted extrusion process. Unlike conventional steam extrusion, SCFX typically uses supercritical CO_2, the most commonly employed blowing agent in this process. SCFX is carried out at low temperatures (below 100°C) and under moderate shear conditions, making it particularly suitable for processing sensitive raw materials (Zhou et al., 2021; Panak-Balentic et al., 2017). The supercritical CO_2 enhances solubilization by increasing pressure or decreasing temperature, creating a single-phase system due to the reduction in surface tension (Panak-Balentic et al., 2017; Liu et al., 2018).

As previously mentioned, the structure and properties of proteins pose challenges during extrusion, particularly in higher protein-content foods. The high temperatures and shear forces typically used in conventional extrusion processes can hinder the expansion of proteins, affecting the final product's quality and appearance. In contrast, SCFX offers a solution by enabling the production of a diverse range of extruded food products at lower temperatures compared to conventional extrusion. This results in energy savings and better preservation of heat-sensitive nutrients. In addition, SCFX helps prevent the formation of undesirable compounds that can arise due to excessive heat, making it a more favorable option (Zhou et al., 2021; Iqbal and Rizvi, 2023; Gopirajah et al., 2020).

An extrusion process was employed to compare steam and supercritical CO_2 (SC-CO_2) extrusion using MPC. The injection rate and pressure for SC-CO_2 were 1.7 kg/h and 15.2 MPa, respectively, with a water injection rate of 15.7 kg/h. Both

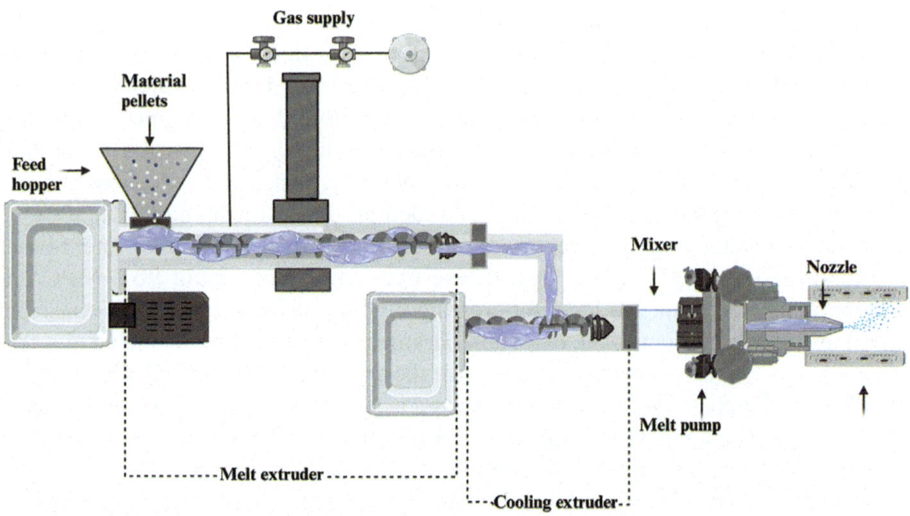

Fig. 7.5 Supercritical fluid-assisted extrusion scheme. Diagram of a SCFX process, showing material flow through extruders with strategic supercritical fluid injection. This setup enables enhanced product expansion and modification.

processes were operated at 100 rpm. The extrudates produced by steam extrusion exhibited a more yellow color, accompanied by the highest color change, indicating extensive Maillard browning. In contrast, the SCFX extrudates exhibited higher lightness and whiteness, the highest radial expansion ratio, and lower bulk density. These results were attributed to the lower processing temperatures and increased nucleation rate associated with the SCFX process. The WSI of SCFX extrudates increased sevenfold compared to steam-expanded extrudates, highlighting the advantages of SCFX in enhancing solubility and product quality (Yoon et al., 2021).

Another example of SCFX application is the incorporation of grape and apple pomace powders, which are rich in dietary fiber and nutrients, into formulations containing MPC with over 81% milk protein. This incorporation aimed to improve the nutritional value of the extruded products while reducing food waste. The goal was to produce puffed products and understand the dynamics of cell nucleation and expansion characteristics concerning the pressure drop rate and SC-CO_2 injection rate. The results indicated that, for optimal expansion and microstructure characteristics, SCFX required a combination of an appropriate SC-CO_2 saturation level and a high-pressure drop rate, reaching up to 141.04 MPa/s. This combination was critical for achieving the desired puffing effect and improving the product's overall texture and structure.

Pomace-based extruded products achieved an impressive expansion ratio of up to 12.3, with a microcellular structure featuring cells ranging in size from 278 to 540 μm and a cell density of 10 (Zareifard et al., 2003) − 10 (Marra et al., 2009) cells/cm. The use of SC-CO_2 up to its saturation limit is essential for diffusing into the melt under high pressure, which facilitates significant expansion and contributes to the formation of a well-defined cellular structure in the extrudates. Extrudates with similar expansion ratios can exhibit varying hardness levels depending on their cellular structure, including average cell size and nucleation density. Specifically, a higher nucleation density, smaller cell sizes, and thinner wall thickness result in less rigid extruded products. SCFX enables the production of puffed products with a high expansion ratio, favorable microcellular structure, and improved cell density, which can be fine-tuned by adjusting pressure drop rates and the amount of dissolved supercritical fluid during the extrusion process. The supercritical fluid-assisted extrusion technology provides precise control over the cellular morphology, expansion characteristics, and physicochemical properties of the final extruded products (Iqbal and Rizvi, 2023).

Identical results have been reported in developing novel high milk protein-based extrudates using SCFX technology. These extrudates, containing 90% MPC and incorporating rice flour, are ideal for creating highly nutritious and health-conscious cereal and snack products. The extrusion parameters for this process included a feed rate of 35 kg/h, a water injection of 24.4% of the dry feed flow rate, and a screw speed of 108 rpm. The SC-CO_2 injection rate was set at 0.44 kg/h, with an injection pressure of 25.5 MPa. The temperature profile in the first two zones was maintained between 20°C and 30°C, while the last three were kept below 50°C, with a die pressure ranging from 10 to 13 MPa. These optimized conditions helped achieve a desirable product with superior nutritional quality and textural properties. The results demonstrated that increasing the amount of MPC in the extrudates led to higher radial and overall expansions and improvements in crispness, porosity, and water activity. However, higher MPC content in the extrudates reduced axial expansion, hardness, piece density, apparent density, and bulk density. The increased MPC levels also raised the viscosity of the extrudates, contributing to the formation of stable, porous structures. This enhanced viscosity facilitated the development of a more robust gel network within the extrudate dispersion, resulting in higher stress required to break the gel and flow. These findings suggest that the optimized extrusion process is well-suited for developing highly nutritious and healthy cereal and snack products (Zhou et al., 2021).

In the study of SCFX-processed extrusion products composed of MPC along with carbohydrates, the objective was to examine the impact of sugar, maltodextrin, and cornstarch on the physical and textural properties of the resulting puffed MPC products in comparison to commercial puffed products. The CO_2 injection rate was set at 0.27 kg/h. The results showed that the highest concentration of MPC resulted in the highest expansion ratio and porosity, as well as the lowest bulk density. In contrast, adding starch, maltodextrin, or sugar reduced the expansion ratio and increased the bulk density. At 5% concentrations, starch, maltodextrin, or sugar decreased the hardness of the extrudates. However, hardness increased when the concentrations were raised from 5% to 10%. These findings suggest that adding carbohydrates to MPC can enhance the textural properties (such as hardness, brittleness, and crispness) of puffed products, making them comparable to commercial puffed cereal products (Liu et al., 2018).

Functionalized milk protein concentrate (f-MPC) emulsifying properties were compared to MCP without extrusion and commercial sodium caseinate. The extrusion process was conducted with die pressures ranging from 1.200 to 1.900 psi, allowing continuous supercritical CO_2 flow into the protein-polymer melt with a CO_2 flow rate of 1.2 kg/h. Several emulsion properties were evaluated, including the emulsion activity index, creaming index, viscosity, droplet size distribution, and microstructure. The results showed that the f-MPC exhibited superior emulsion activity, a reduced creaming index, and viscosity comparable to commercial sodium caseinate. These enhanced emulsifying properties of f-MPC suggest that it could be a viable alternative to nondairy sodium caseinate for use in various food formulations (Gopirajah et al., 2020).

SCFX is an innovative process that integrates traditional extrusion with supercritical fluid technology, enabling the development of commercially viable, nutritionally dense, and healthy puffed products. This process is particularly beneficial for applications involving protein concentrates. Supercritical CO_2 (SC-CO_2) acts as a plasticizer in the molten dough, reducing its viscosity, lowering the melting point, and decreasing the glass transition temperature. These changes improve the extrusion process and broaden its potential applications, resulting in enhanced product quality, texture, and opportunities for creating healthier, more functional food products with improved physical properties and consumer appeal.

7.5.3 Vacuum-assisted extrusion

A hybrid extrusion cooking system aligns with the growing demand for texturally rich snacks and convenient, on-the-go products while incorporating high bioactive compounds such as anthocyanins and phenolic acids. Vacuum pressure plays a crucial role in modifying the physical characteristics of air, creating an airless void within a confined space. This process enhances expansion by minimizing thermal energy transfer through convection, which reduces heat loss. As a result, the system helps to preserve the bioactive compounds, thereby enhancing both the sensory and nutritional qualities of the final product (Shahrum et al., 2024).

Despite its numerous benefits, vacuum-assisted extrusion in food processing remains underexplored. To assess its potential, a study was conducted using white and brown rice flour blends to evaluate the impact of vacuum extrusion cooking on expansion and phytochemical properties. The parameters, such as screw speed (ranging from 200 to 400 rpm), barrel temperature (70–100°C), and vacuum pressure (0–400 mmHg), were varied during the extrusion process. A vacuum chamber was integrated into the extruder to facilitate the experiment. Analysis of the expanded samples revealed that increasing the vacuum level reduced bulk density, highlighting the potential of vacuum-assisted extrusion to improve product texture and expandability.

At lower temperatures, vacuum-assisted extrusion can promote water evaporation, which increases the expansion of the extrudates. The pressure difference at the die exit, enhanced by the vacuum level, leads to water evaporation and the formation of bubbles, facilitating product expansion. In addition, the formation of a porous structure increases water absorption. One of the key benefits of vacuum-assisted extrusion is its ability to reduce the loss of phenolic compounds during the process. For instance, the final concentrations of the major anthocyanin species (C3G) and peonidin-3-glucoside (P3G) were significantly higher in vacuum-assisted extrusion than in conventional extrusion. Specifically, the content of C3G was 0.021 μm/L and 3.21 μm/L (P3G) in vacuum-assisted extrusion, while conventional extrusion resulted in lower levels, with C3G at 0.008 μm/L and P3G at 1.91 μm/L. The researchers concluded that vacuum-assisted extrusion is an effective method for producing extrudates from pigmented rice at lower temperatures while preserving a higher concentration of phytochemicals (Ramchiary and Das, 2020).

The integration of gas injection-assisted extrusion, particularly utilizing nitrogen, and the innovative approach of SCFX represent substantial advancements in food processing technologies. These methodologies furnish a robust repertoire for addressing the growing global demand for organoleptically desirable and nutritionally reliable snack foods, concurrently with the increasing scholarly and industrial interest in plant-based meat alternatives. Through the strategic modulation of the extrusion environment via gas injection and the exploitation of the unique physicochemical properties of supercritical fluids, such as CO_2, manufacturers can meticulously control the physical attributes of their products. This precise control enables the production of products characterized by reduced density, enhanced textural quality, and superior visual appeal, often accompanied by improved nutritional profiles derived from protein-rich constituents such as lentils.

SCFX, distinguished by its operational parameters of lower temperatures and moderate shear, further enhances processing capabilities by enabling the valorization of delicate raw materials while concurrently improving product quality and textural integrity. The plasticizing effect attributed to supercritical CO_2 optimizes the extrusion process, creating novel avenues for the development of innovative food products with enhanced functionality and improved consumer acceptance. Ultimately, these sophisticated extrusion techniques offer effective strategies for creating healthier, more sustainable, and commercially successful food options that cater to evolving consumer preferences and emerging dietary trends.

7.6 Concluding remarks

The integration of AM technologies with secondary production techniques has led to the development of hybrid extrusion methods. These methods combine the advantages of fused technologies to enhance current production systems. Thermal energy applied during the extrusion process can significantly influence the properties and characteristics of the final food products. Various emerging thermal technologies, such as ohmic heating, microwave heating, RF heating, and IR heating, have been explored. The application of ultrasonic and microwave waves in food extrusion has shown improvements in the quality of the final product. However, the literature on this topic is limited, and further studies are needed to fully understand the benefits and drawbacks of these technologies. AM, specifically extrusion 3D printing, has proven to be a promising technology for producing personalized food. This technology allows for the creation of unique food structures with specific nutritional properties. The injection of physical agents, such as gases and supercritical fluids, during the extrusion process can improve the textural and nutritional properties of extruded foods. These techniques offer significant advantages in terms of product quality and process efficiency. In summary, the importance of hybrid extrusion and agent-assisted technologies in food production is emphasized. These innovations have the potential to improve the quality, efficiency, and nutritional properties of food products, aligning with current trends in sustainability and health.

Hybrid extrusion represents a significant advancement in food processing, strategically integrating AM techniques and auxiliary technologies to increase product quality, nutritional value, and sustainability. This chapter focuses on the essential role of various additive variables, including thermal energy, wave-based technologies, and physical blowing agents, in manipulating the extrusion process and final product characteristics.

Heat, as a decisive variable, is fundamental for achieving and preserving desirable food properties, with emerging thermal technologies offering precise control over structural and compositional changes. Furthermore, extrusion 3D printing appears as a transformative AM solution, enabling the creation of customized food with tailored nutritional and textural attributes. The rheological behavior of food inks is critical for successful printing, supported by advanced textural assessment methods, such as BBE.

The integration of gas injection-assisted extrusion (e.g., nitrogen) and SCFX represents a significant advancement in producing expanded products with enhanced textural and nutritional profiles, particularly for protein-rich materials. SCFX, operating at lower temperatures, offers distinct advantages for preserving sensitive components and improving sustainability. While some hybrid approaches are still developing, their collective potential is immense, providing efficacious strategies for healthier, more sustainable, and commercially successful food options that align with evolving consumer demands. Continued research in these synergistic technologies is necessary to fully unlock their transformative capacity within the global food industry.

References

Andrejko, D., Grochowicz, J., Goździewska, M., Kobus, Z., 2011. Influence of infrared treatment on mechanical strength and structure of wheat grains. Food Bioprocess Technol. 1935-5130. 4 (8), 1367–1375. doi:10.1007/s11947-009-0238-3.

Ashokkumar, M., Cavalieri, F., Chemat, F., Okitsu, K., Sambandam, A., Yasui, K., Zisu, B., 2016. Handbook of Ultrasonics and Sonochemistry. Springer Singapore, Australia, pp. 1–1487. doi:10.1007/978-981-287-278-4.

Avelar, Z., Vicente, A.A., Saraiva, J.A., Rodrigues, R.M., 2021. The role of emergent processing technologies in tailoring plant protein functionality: new insights. Trends Food Sci. Technol. 09242244. 113, 219–231. Available from: http://www.elsevier.com/wps/find/journaldescription.cws_home/601278/description#description. doi:10.1016/j.tifs.2021.05.004.

Ávila-Orta, C.A., González-Morones, P., Agüero- Valdez, D., González-Sánchez, A., Martínez-Colunga, J.G., Mata-Padilla, J.M., Cruz-Delgado, V.J., 2019. Ultrasound-assisted melt extrusion of polymer nanocomposites. IntechOpen. doi:10.5772/intechopen.80216.

Ávila-Orta, C., Espinoza-González, C., Martínez-Colunga, G., Bueno-Baqués, D., Maffezzoli, A., Lionetto, F., 2013. An overview of progress and current challenges in ultrasonic treatment of polymer melts. Adv. Polym. Technol. 10982329. 32 (1), E582. doi:10.1002/adv.21303.

Baller, M.A., Pacheco, P.D.G., Vitta-Takahashi, A., Putarov, T.C., Vasconcellos, R.S., Carciofi, A.C., 2021. Effects of thermal energy on extrusion characteristics, digestibility and palatability of a dry pet food for cats. J. Anim. Physiol. Anim. Nutr. 14390396. 105 (1), 76–90. Available from: http://onlinelibrary.wiley.com/journal/10.1111/(ISSN)1439-0396. doi:10.1111/jpn.13606.

Bogdal, D., Lukasiewicz, M., Pielichowski, J., Miciak, A., Bednarz, S., 2003. Microwave-assisted oxidation of alcohols using Magtrieve™. Tetrahedron. 00404020. 59 (5), 649–653. doi:10.1016/s0040-4020(02)01533-8.

Brennen, C.E., 2013. Cavitation and Bubble Dynamics Cavitation and Bubble Dynamics. Cambridge University Press, United States, pp. 1–249. doi:10.1017/CBO9781107338760.

Chan, E., Masatcioglu, T.M., Koksel, F., 2019. Effects of different blowing agents on physical properties of extruded puffed snacks made from yellow pea and red lentil flours. J. Food Process Eng. 17454530. 42 (3). Available from: http://onlinelibrary.wiley.com/journal/10.1111/(ISSN)1745-4530. doi:10.1111/jfpe.12989.

Chauvet, M., Sauceau, M., Fages, J., 2017. Extrusion assisted by supercritical CO2: a review on its application to biopolymers. J. Supercrit. Fluids. 08968446. 120, 408–420. doi:10.1016/j.supflu.2016.05.043.

Chemat, F., Zill-E-Huma, Khan, M.K., 2011. Applications of ultrasound in food technology: processing, preservation and extraction. Ultrason. Sonochem. 13504177. 18 (4), 813–835. www.elsevier.com/inca/publications/store/5/2/5/4/5/1. doi:10.1016/j.ultsonch.2010.11.023.

Chen, W., Hadde, E.K., Chen, J., 2021. Development of a ball back extrusion technique for texture analysis of fluid food. J. Texture Stud. 17454603. 52 (4), 461–469. Available from: http://onlinelibrary.wiley.com/journal/10.1111/(ISSN)1745-4603. doi:10.1111/jtxs.12613.

Cui, B., Wang, K., Hu, N., Sun, Y., Mao, C., Ye, P., Pang, H., Wang, Y., Fu, H., Chen, X., Wang, Y., 2023. Combination of radio frequency heating and antibacterial agents for microorganism control in the production of packaged tofu. Food Control. 09567135. 154, 110015. doi:10.1016/j.foodcont.2023.110015.

Dalbhagat, C.G., Mahato, D.K., Mishra, H.N., 2019. Effect of extrusion processing on physicochemical, functional and nutritional characteristics of rice and rice-based products: a review. Trends Food Sci. Technol. 09242244. 85, 226–240. Available from: http://www.elsevier.com/wps/find/journaldescription.cws_home/601278/description#description. doi:10.1016/j.tifs.2019.01.001.

Darvishi, H., Khostaghaza, M.H., Najafi, G., 2013. Ohmic heating of pomegranate juice: electrical conductivity and pH change. J. Saudi Soc. Agric. Sci. 1658077X. 12 (2), 101–108. Available from: https://www.journals.elsevier.com/journal-of-the-saudi-society-of-agricultural-sciences/. doi:10.1016/j.jssas.2012.08.003.

de la Hoz, A., Loupy, A., 2013. Microwave–materials interactions and dielectric properties: from molecules and macromolecules to solids and colloidal suspensions. Microw. Org. Synth. 1, 1–56. doi:10.1002/9783527651313.ch1.

Derossi, A., Caporizzi, R., Azzollini, D., Severini, C., 2018. Application of 3D printing for customized food. A case on the development of a fruit-based snack for children. J. Food Eng. 02608774. 220, 65–75. doi:10.1016/j.jfoodeng.2017.05.015.

Dircio-Morales, M.A., Fonseca-Florido, H.A., Velazquez, G., Ávila-Orta, C.A., Ramos-De Valle, L.F., Hernández-Gámez, F., Rivera-Salinas, J.E., Soriano-Corral, F., 2023. Relationship among extrusion conditions, cell morphology, and properties of starch-based foams—a review. Starch/Staerke. 1521379X. 75, 1–2. Available from: http://onlinelibrary.wiley.com/journal/10.1002/(ISSN)1521-379X. doi:10.1002/star.202200103.

Espinoza-Gonzalez, C., Avila-Orta, C., Martinez-Colunga, G., Lionetto, F., Maffezzoli, A., 2016. A measure of CNTs dispersion in polymers with branched molecular architectures by UDMA. IEEE Trans. Nanotechnol. 1536-125X. 15 (5), 731–737. doi:10.1109/TNANO.2016.2530697.

Gao, J., Mingtai, W., Du, S., Zhang, H., Wang, S., Ling, B., 2023. Recent advances in food processing by radio frequency heating techniques: a review of equipment aspects. J. Food Eng. 02608774. 357, 111609. doi:10.1016/j.jfoodeng.2023.111609.

Ghanghas, N., Nadimi, M., Paliwal, J., Koksel, F., 2024. Gas-assisted high-moisture extrusion of soy-based meat analogues: impacts of nitrogen pressure and cooling die temperature on density, texture and microstructure. Innov. Food Sci. Emerg. Technol. 14668564. 92, 103557. doi:10.1016/j.ifset.2023.103557.

Godoi, F.C., Prakash, S., Bhandari, B.R., 2016. 3d printing technologies applied for food design: status and prospects. J. Food Eng. 02608774. 179, 44–54. Available from: http://www.sciencedirect.com/science/journal/02608774. doi:10.1016/j.jfoodeng.2016.01.025.

Gopirajah, R., Singha, P., Javad, S., Rizvi, S.S.H., 2020. Emulsifying properties of milk protein concentrate functionalized by supercritical fluid extrusion. J. Food Process. Preserv. 17454549. 44 (10). Available from: http://onlinelibrary.wiley.com/journal/10.1111/(ISSN)1745-4549. doi:10.1111/jfpp.14754.

Graf, B., Kapfer, T., Ostertag, F., Hinrichs, J., 2020. New experimental set-up for testing microwave technology to continuously heat fouling-sensitive food products like milk concentrates. Innov. Food Sci. Emerg. Technol. 14668564. 65, 102453. doi:10.1016/j.ifset. 2020.102453.

Hayes, B.L., 2002. Microwave Synthesis: chemistry at the Speed of Light.

Hu, J., Xu, H., Shi, R., Gantumur, M.-A., Jiang, Z., Hou, J., 2024. Emerging thermal modifying methods in milk protein: a review. Trends in Food Science & Technology. 09242244. 146, 104407. doi:10.1016/j.tifs.2024.104407.

Hull, C.W., 1986. Apparatus for production of three-dimensional objects by stereolithography.

Iqbal, A., Rizvi, S.S.H., 2023. Cell nucleation dynamic and expansion characteristics of milk protein puffs containing fruit pomace made by supercritical fluid extrusion. Food Bioprocess Technol. 19355149. 16 (8), 1746–1756. Available from: https://www.springer.com/journal/11947. doi:10.1007/s11947-023-03018-z.

Isayev, A.I., Kumar, R., Lewis, T.M., 2009. Ultrasound assisted twin screw extrusion of polymer-nanocomposites containing carbon nanotubes. Polymer. 00323861. 50 (1), 250–260. Available from: http://www.journals.elsevier.com/polymer/. doi:10.1016/j.polymer.2008.10.052.

Jafari, M., Parastouei, K., Abbaszadeh, S., 2024. Development of curcumin-loaded nanoemulsion stabilized with texturized whey protein concentrate: characterization, stability and in vitro digestibility. Food Sci. Nutr. 2048-7177. 12 (3), 1655–1672. doi:10.1002/fsn3.3860.

Jie, Y., Wang, X.-Y., Dong, L., Wang, L.-J., Wang, Y., 2022. Development of soy protein isolate emulsion gels as extrusion-based 3D food printing inks: effect of polysaccharides incorporation. Food Hydrocoll. 0268005X. 131, 107824. doi:10.1016/j.foodhyd.2022.107824.

Jun, S., Irudayaraj, J.M., 2008. Food Processing Operations Modeling: Design and Analysis, Second edn. CRC Press, United States, pp. 1–357. Available from: https://www.taylorfrancis.com/books/e/9781420055542.

Kaewsorn, K., Maichoon, P., Pornchaloempong, P., Krusong, W., Sirisomboon, P., Tanaka, M., Kojima, T., 2023. Evaluation of precision and sensitivity of back extrusion test for measuring textural qualities of cooked germinated brown rice in production process. Foods. 2304-8158. 12 (16), 3090. doi:10.3390/foods12163090.

Kappe, C.O., 2008. Microwave dielectric heating in synthetic organic chemistry. Chem. Soc. Rev. 14604744. 37 (6), 1127–1139. doi:10.1039/b803001b.

Kaur, N., Singh, A.K., 2016. Ohmic heating: concept and applications—a review. Crit. Rev. Food Sci. Nutr. 15497852. 56 (14), 2338–2351. Available from: www.tandf.co.uk/journals/titles/10408398.asp. doi:10.1080/10408398.2013.835303.

Kern, C., Bähler, B., Hinrichs, J., Nöbel, S., 2019. Waterless single screw extrusion of pasta-filata cheese: process design based on thermo-rheological material properties. J. Food Eng. 02608774. 260, 58–69. doi:10.1016/j.jfoodeng.2019.04.015.

Knorr, D., Zenker, M., Heinz, V., Lee, D.U., 2004. Applications and potential of ultrasonics in food processing. Trends Food Sci. Technol. 09242244. 15 (5), 261–266. Available from: http://www.elsevier.com/wps/find/journaldescription.cws_home/601278/description#description. doi:10.1016/j.tifs.2003.12.001.

Koksel, F., Masatcioglu, M.T., 2018. Physical properties of puffed yellow pea snacks produced by nitrogen gas assisted extrusion cooking. LWT. 00236438. 93, 592–598. Available from: https://www.journals.elsevier.com/lwt. doi:10.1016/j.lwt.2018.04.011.

Köllmann, N., Vringer, R., Mishra, P., Zhang, L., van der Goot, A.J., 2024. Near-infrared spectroscopy to quantify overall thermal process intensity during high-moisture extrusion of soy protein concentrate. Food Res. Int. 18737145. 186. Available from: https://www.sciencedirect.com/science/journal/09639969. doi:10.1016/j.foodres.2024.114320.

Komorowska-Durka, M., Dimitrakis, G., Bogdał, D., Stankiewicz, A.I., Stefanidis, G.D., 2015. A concise review on microwave-assisted polycondensation reactions and curing of polycondensation polymers with focus on the effect of process conditions. Chem. Eng. J. 13858947. 264, 633–644. Available from: www.elsevier.com/inca/publications/store/6/0/1/2/7/3/index.htt. doi:10.1016/j.cej.2014.11.087.

Li, X., Masatcioglu, M.T., Koksel, F., 2019. Physical and functional properties of wheat flour extrudates produced by nitrogen injection assisted extrusion cooking. J. Cereal Sci. 07335210. 89, 102811. doi:10.1016/j.jcs.2019.102811.

Lillford, P., 2018. Texture and breakdown in the mouth: an industrial research approach. J. Texture Stud. 0022-4901. 49 (2), 213–218. doi:10.1111/jtxs.12279.

Ling, B., Cheng, T., Wang, S., 2020. Recent developments in applications of radio frequency heating for improving safety and quality of food grains and their products: a review. Crit. Rev. Food Sci. Nutr. 1040-8398. 60 (15), 2622–2642. doi:10.1080/10408398.2019.1651690.

Liu, H., Hebb, R.L., Putri, N., Rizvi, S.S.H., 2018. Physical properties of supercritical fluid extrusion products composed of milk protein concentrate with carbohydrates. Int. J. Food Sci. Technol. 13652621. 53 (3), 847–856. doi:10.1111/ijfs.13624.

Liu, S., Ren, Y., Yin, H., Nickerson, M., Pickard, M., Yongfeng, A., 2022. Improvement of the nutritional quality of lentil flours by infrared heating of seeds varying in size. Food Chem. 03088146. 396, 133649. doi:10.1016/j.foodchem.2022.133649.

Liu, Y., Tang, T., Duan, S., Qin, Z., Zhao, H., Wang, M., Li, C., Zhang, Z., Liu, A., Han, G., Dingtao, W., Chen, H., Shen, L., Wenjuan, W., Dong, Y., 2020. Applicability of rice doughs as promising food materials in extrusion-based 3D printing. Food Bioprocess Technol. 1935-5130. 13 (3), 548–563. doi:10.1007/s11947-020-02415-y.

Liu, Y., Yun, Y., Liu, C., Regenstein, J.M., Liu, X., Zhou, P., 2019. Rheological and mechanical behavior of milk protein composite gel for extrusion-based 3D food printing. LWT. 00236438. 102, 338–346. doi:10.1016/j.lwt.2018.12.053.

Luo, S., Chan, E., Masatcioglu, M.T., Erkinbaev, C., Paliwal, J., Koksel, F., 2020. Effects of extrusion conditions and nitrogen injection on physical, mechanical, and microstructural properties of red lentil puffed snacks. Food Bioprod. Process. 09603085. 121, 143–153. doi:10.1016/j.fbp.2020.02.002.

Markovinović, A., et al., 2023. 3D Printing of functional strawberry snacks: food design, Texture, antioxidant bioactive compounds, and microbial stability. Antioxidants 12, 1–19.

Marra, F., Zell, M., Lyng, J.G., Morgan, D.J., Cronin, D.A., 2009. Analysis of heat transfer during ohmic processing of a solid food. J. Food Eng. 02608774. 91 (1), 56–63. doi:10.1016/j.jfoodeng.2008.08.015.

Mason, T.J., 1999. Advances in Sonochemistry, Vol. 5. Jai Press, Greenweech.

Mason, T.J., Lorimer, J.P., 2003. Front Matter and Subject Index Applied Sonochemistry: Uses of Power Ultrasound in Chemistry and Processing. Wiley, Wiley. 10.1002/352760054X.fmatter_indsub.

Mohamed, I.O., 2023. Interaction of starch with some food macromolecules during the extrusion process and its effect on modulating physicochemical and digestible properties. A review. Carbohydr. Polym. Technol. Appl. 26668939. 5, 100294. doi:10.1016/j.carpta.2023.100294.

Mousavi, S.A.A.A., Feizi, H., Madoliat, R. 2007. Investigations on the effects of ultrasonic vibrations in the extrusion process. J. Mater. Process. Technol. 09240136. (187-188), 657–661. doi: 10.1016/j.jmatprotec.2006.11.168.

Pacheco, P.D.G., Putarov, T.C., Baller, M.A., Peres, F.M., Loureiro, B.A., Carciofi, A.C., 2018. Thermal energy application on extrusion and nutritional characteristics of dog foods.

Anim. Feed Sci. Technol. 03778401. 243, 52–63. Available from: http://www.journals.elsevier.com/animal-feed-science-and-technology. doi:10.1016/j.anifeedsci.2018.07.003.

Pan, Z., 2021. Innovative infrared heating technologies for food and agricultural processing. Technol. Innov. 21, 1–16.

Panak-Balentic, J., Ackar, D., Jozinovic, A., Babic, J., Milicevic, B., Jokic, S., Pajin, B., Subaric, D., 2017. Application of supercritical carbon dioxide extrusion in food processing technology. Chem. Ind. 0367-598X. 71 (2), 127–134. doi:10.2298/hemind150629024p.

Patist, A., Bates, D., 2008. Ultrasonic innovations in the food industry: from the laboratory to commercial production. Innov. Food Sci. Emerg. Technol. 14668564. 9 (2), 147–154. doi:10.1016/j.ifset.2007.07.004.

Philipp, C., Oey, I., Silcock, P., Beck, S.M., Buckow, R., 2017. Impact of protein content on physical and microstructural properties of extruded rice starch-pea protein snacks. Journal of Food Engineering. 02608774. 212, 165–173. Available from: http://www.sciencedirect.com/science/journal/02608774. doi:10.1016/j.jfoodeng.2017.05.024.

Pietsch, V.L., Emin, M.A., Schuchmann, H.P., 2017. Process conditions influencing wheat gluten polymerization during high moisture extrusion of meat analog products. J. Food Eng. 02608774. 198, 28–35. Available from: http://www.sciencedirect.com/science/journal/02608774. doi:10.1016/j.jfoodeng.2016.10.027.

Prabha, K., Ghosh, P., Abdullah, S., Joseph, R.M., Krishnan, R., Rana, S.S., Pradhan, R.C., 2021. Recent development, challenges, and prospects of extrusion technology. Future Foods. 26668335. 3, 100019. doi:10.1016/j.fufo.2021.100019.

Ramchiary, M., Das, A.B., 2020. Vacuum-assisted extrusion of red rice (bao-dhan) flour: physical and phytochemical comparison with conventional extrusion. J. Food Process. Preserv. 17454549. 44 (8). Available from: http://onlinelibrary.wiley.com/journal/10.1111/(ISSN)1745-4549. doi:10.1111/jfpp.14570.

Sanchez, L.C., Beatrice, C.A.G., Lotti, C., Marini, J., Bettini, S.H.P., Costa, L.C., 2019. Rheological approach for an additive manufacturing printer based on material extrusion. Int. J. Adv. Manuf. Technol. 14333015. 105 (5-6), 2403–2414. Available from: http://www.springerlink.com/content/0268-3768. doi:10.1007/s00170-019-04376-9.

Schmid, V., Trabert, A., Keller, J., Bunzel, M., Karbstein, H.P., Emin, M.A., 2021. Defined shear and heat treatment of apple pomace: impact on dietary fiber structures and functional properties. Eur. Food Res. Technol. 1438-2377. 247 (8), 2109–2122. doi:10.1007/s00217-021-03776-0.

Schmidt, F., Graf, B., Hinrichs, J., Kern, C., 2022. Continuous microwave-assisted extrusion for high moisture texturized foods: a feasibility study. Innov. Food Sci. Emerg. Technol. 14668564. 78, 103020. doi:10.1016/j.ifset.2022.103020.

Shahrum, M.A., Rajendran, T.K., Maidin, S., Azri, M.A., Ismail, S., Kamarulnizam, M., 2024. Dimensional accuracy of acrylonitrile butadiene styrene and polylactic acid samples printed in vacuum-assisted material extrusion system. Inst. Phys. Malays. Eng. Res. Express. 26318695. 6 (2). Available from: https://iopscience.iop.org/journal/2631-8695. doi:10.1088/2631-8695/ad4306.

Sinaki, N.Y., Masatcioglu, M.T., Paliwal, J., Koksel, F., 2022. Development of cellular high-protein foods: third-generation yellow pea and red lentil puffed snacks. Foods. 23048158. 11 (1). Available from: https://www.mdpi.com/2304-8158/11/1/38/pdf. doi:10.3390/foods11010038.

Sinaki, N.Y., Paliwal, J., Koksel, F., 2023. Enhancing the Techno-Functionality of Pea Flour by Air Injection-Assisted Extrusion at Different Temperatures and Flour Particle Sizes.

Foods. 23048158. 12 (4). Available from: www.mdpi.com/journal/foods. doi:10.3390/foods12040889.

Singh, S., Gamlath, S., Wakeling, L., 2007. Nutritional aspects of food extrusion: a review. International Journal of Food Science & Technology. 0950-5423. 42 (8), 916–929. doi:10.1111/j.1365-2621.2006.01309.x.

Singla, M., Sit, N., 2021. Application of ultrasound in combination with other technologies in food processing: a review. Ultrason. Sonochem. 13504177. 73 (105506). doi:10.1016/j.ultsonch.2021.105506.

Suslick, K.S., Price, G.J., 1999. Applications of ultrasound to materials chemistry. Annu. Rev. Mater. Sci. 29, 295–326. doi:10.1146/annurev.matsci.29.1.295.

Trevena, D.H., 1984. Cavitation and the generation of tension in liquids. J. Phys. D Appl. Phys. 00223727. 17 (11), 2139–2164. doi:10.1088/0022-3727/17/11/003.

van den Einde, R.M., Akkermans, C., van der Goot, A.J., Boom, R.M., 2004. Molecular breakdown of corn starch by thermal and mechanical effects. Carbohydr. Polym. 01448617. 56 (4), 415–422. doi:10.1016/j.carbpol.2004.03.006.

Wiriyaumpaiwong, S., Somchart Soponronnarit, Somkiat Prachayawarakorn, 2004. Comparative study of heating processes for full-fat soybeans. J. Food Eng. 02608774. 65 (3), 371–382. doi:10.1016/j.jfoodeng.2004.01.036.

Yoon, A.K., Singha, P., Rizvi, S.S.H., 2021. Steam vs. SC–CO2–based extrusion: comparison of physical properties of milk protein concentrate extrudates. J. Food Eng. 292, 1–5.

Yuntao, L., Jiang, L., Guo, S., Huilin, L., 2005. Mechanochemical degradation kinetics of high-density polyethylene melt and its mechanism in the presence of ultrasonic irradiation. Ultrason. Sonochem. 13504177. 12 (3), 183–189. doi:10.1016/j.ultsonch.2003.10.011.

Zareifard, M.R., Ramaswamy, H.S., Trigui, M., Marcotte, M., 2003. Ohmic heating behaviour and electrical conductivity of two-phase food systems. Innov. Food Sci. Emerg. Technol. 14668564. 4 (1), 45–55. doi:10.1016/s1466-8564(02)00088-7.

Zhang, Y., Tingting, L., Shen, Y., Wang, L., Zhang, H., Qian, H., Xiguang, Q., 2020. Extrusion followed by ultrasound as a chemical-free pretreatment method to enhance enzymatic hydrolysis of rice hull for fermentable sugars production. Ind. Crops Prod. 09266690. 149, 112356. doi:10.1016/j.indcrop.2020.112356.

Zheng, Z., Zhang, M., Liu, Z., 2021. Investigation on evaluating the printable height and dimensional stability of food extrusion-based 3D printed foods. J. Food Eng. 02608774. 306, 110636. doi:10.1016/j.jfoodeng.2021.110636.

Zhou, R., Mitra, P., Melnychenko, A., Rizvi, S.S.H., 2021. Quality attributes and rheological properties of novel high milk protein-based extrudates made by supercritical fluid extrusion. Int. J. Food Sci. Technol. 13652621. 56 (8), 3866–3875. Available from: http://onlinelibrary.wiley.com/journal/10.1111/(ISSN)1365-2621. doi:10.1111/ijfs.15003.

Zhou, X., Tang, Z., Pedrow, P.D., Sablani, S.S., Tang, J., 2023. Microwave heating based on solid-state generators: new insights into heating pattern, uniformity, and energy absorption in foods. J. Food Eng. 02608774. 357. Available from: http://www.sciencedirect.com/science/journal/02608774. doi:10.1016/j.jfoodeng.2023.111650.

Zhu, S., Stieger, M.A., van der Goot, A.J., Schutyser, M.A.I., 2019. Extrusion-based 3D printing of food pastes: correlating rheological properties with printing behavior. Innov. Food Sci. Emerg. Technol. 14668564. 58. Available from: http://www.elsevier.com/locate/ifset. doi:10.1016/j.ifset.2019.102214.

Žilić, S., Ataç Mogol, B., Akıllıoğlu, G., Serpen, A., Delić, N., Gökmen, V., 2014. Effects of extrusion, infrared and microwave processing on Maillard reaction products and phenolic compounds in soybean. J. Sci. Food Agric. 0022-5142. 94 (1), 45–51. doi:10.1002/jsfa.6210.

Role of carbohydrates in extrusion

8

Nasser Al-Habsi[a], *Maha Al-Khalili*[a], *and Pankaj B. Pathare*[b]
[a]Department of Food Science and Nutrition, College of Agricultural & Marine Sciences, Sultan Qaboos University, Muscat, Oman
[b]Department of Soils, Water & Agricultural Engineering, College of Agricultural & Marine Sciences, Sultan Qaboos University, Muscat, Oman

1 Introduction

Extrusion is a processing technology that transforms raw materials into different food products by using heat, pressure, and mechanical shear (Choton et al., 2020). It is used to produce various food products such as breakfast cereals, pasta, and snack foods, along with other types of nonfood products like bioplastics (Giannakas, 2024; Jebalia et al.; Lin, 2022; Mohammed et al., 2024; Galanakis, 2022). Food extrusion allows the development of products with specific textures, shapes, tastes, enhanced digestibility, and extended shelf-life (Choton et al., 2020; Lazou, 2024; Qi et al., 2024; Sule et al., 2024), which has led to the progressive implementation of this technology in different sectors.

Protein-based ingredients from both animal (meat, dairy, eggs, etc.) and plant (wheat, legume, soy, pea, etc.) origins deliver functional properties such as texture, structure, and nutrition in extruded products. Extrusion plays a significant role in the functional and sensory properties of the protein, which makes it ideal for the development of protein-rich products and meat analogs (Mosibo et al., 2022).

Similarly, extrusion techniques could be used in the modification of proteins from ice to enhance their characteristics, such as solubility, water-holding capacity, and emulsion ability by altering the bonds' structure. This could enhance protein functionality, enabling broader applications in food and improving their use in various contexts (Gao et al., 2022). These proteins are essential and are used in many extruded products, such as meat analogs and protein-rich foods, which contribute to the overall mouthfeel and sensory characteristics (Plattner, 2020; Liu and Hsieh, 2007; Zhang et al., 2023).

Customers' awareness and demand are shifting toward more sustainable plant-origin options, making carbohydrates from plants a preferable choice for extrusion technology (Wibawa et al., 2023), which offers more nutritious, eco-friendly, and green alternatives. As consumer awareness continues to grow, the trend toward using sustainable sources increases interest in carbohydrate-based products with

Extrusion Cooking for Food and Feed. https://doi.org/10.1016/B978-0-443-29987-2.00008-7

functional versatility and environmental benefits. Moreover, the potential for innovation can be achieved due to the diverse sources of carbohydrates from plants, such as the inclusion of alternative flours and fibers in extruded products (Ek et al., 2020).

Carbohydrates could be either complex polysaccharides or simple sugars, depending on the number of monomers they contain (Lee et al., 2024). Based on their structural composition and origin, polysaccharides can be further categorized as starch-based or nonstarch-based (Li et al., 2023; Oumeddour et al., 2024).

Starch-based polylactides consist primarily of amylose and amylopectin, which occur in cereals, tubers, and legumes (Obadi and Bin, 2024; Rostamabadi et al., 2024). Nonstarch polysaccharides include less soluble cellulose, hemicellulose, and pectin, which are found in plant cell walls (Agbana et al., 2024; Chen et al., 2024). Simple sugars, which consist of mono- or disaccharides, are less complex (small) carbohydrate molecules such as glucose, fructose, and sucrose that are readily water-soluble and considered a rapid source of energy (Saitkulov et al., 2022; Xiao et al., 2024). In extrusion technology, carbohydrates play a crucial role in the functional and structural integrity, as well as the sensory attributes, of the extrudate. Both polysaccharides and simple sugars are employed to attain these characteristics, catering to diverse product requirements and consumer preferences. Most of the reviews conducted on carbohydrate extrusion have focused on individual carbohydrates with specific functions. The interaction and synergistic effects of different types of carbohydrates remain largely unexplored.

In addition, the influence of temperature, moisture content, screw speed, and extrusion configurations as major processing parameters requires further understanding to optimize product characteristics and sustainability. Therefore this chapter aims to briefly overview the role of carbohydrates in extrusion processing by focusing on their properties, exploring the impact of processing parameters, and examining their potential for innovation in food products. By addressing current knowledge gaps and exploring emerging trends, the goal is to enhance our understanding of how carbohydrates can contribute to sustainable and high-quality food production in extrusion manufacturing.

2 Types of carbohydrates used in extrusion

Carbohydrates play a primary role in the extrusion process due to their physicochemical properties, particularly their ability to undergo gelatinization. Gelatinization is the stage at which carbohydrates, mainly starches, absorb water and swell, leading to the disruption of the crystalline structure under high heat and moisture content (Li, 2022). This process is fundamental to the textural and structural developments required for many extruded food products. The gelatinized structure of the carbohydrates makes the mixture more stable and elastic, with a higher expansion ability in the final product (Dey et al., 2024). This capability makes carbohydrates a versatile component that enables the production of a wide variety of extruded products, ranging from puffy snacks to dense pasta (van der Sman et al., 2018; Bouasla and Wójtowicz, 2021).

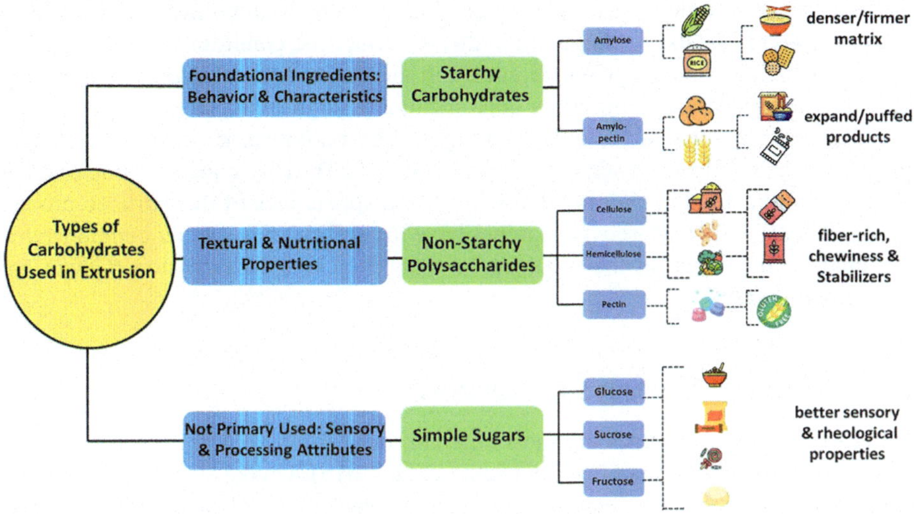

Fig. 8.1 Carbohydrates used in extrusion, classified based on their chemical composition and functional properties.

Carbohydrates incorporated in extrusion are typically grouped based on their basic chemical composition and functionality (Fig. 8.1). This classification contributes to the understanding of their behavior and reactions during extrusion and their influence on the final product's attributes. The major categories of carbohydrates are discussed below:

2.1 *Starchy carbohydrates*

Starchy carbohydrates, composed mainly of polysaccharides such as amylose and amylopectin, are fundamental ingredients in food extrusion. These two components influence the behavior of starch during extrusion and the characteristics of the final product due to their distinct structural and functional properties (Chen et al., 2022). Amylose is a glucose molecule arranged in linear chains that form strong intermolecular interactions, leading to the formation of firm, dense textures (Chen et al., 2024). On the contrary, amylopectin is highly branched in structure, which promotes rapid swelling with greater expansion under heating and extrusion (Vamadevan and Bertoft, 2020). The ratio of amylose to amylopectin in a starch source is a key determinant of its characteristics and influences its functionality in extrusion. For example, amylose-rich starches, like those derived from corn or rice, are less likely to expand and thus are more suitable for producing dense and cohesive products like noodles, pasta, and crackers (Obadi and Xu, 2021; Sánchez-Aldana Villarruel et al., 2021; Low et al., 2020). These starches contribute to the rigidity and structural integrity required in such applications. On the contrary, amylopectin-rich starches derived from sources such as waxy maize, barley, or potato starch contribute significantly to expansion, resulting in airy and porous textures. These characteristics make them ideal for the formulation of extruded products such as breakfast cereals and other puffed products (Okoronkwo et al., 2019 ; Singh et al., 2014). Under the conditions of the extrusion process, which involves high temperature and shear forces, the crystalline structure of the starches is disrupted, initiating gelatinization (Singh et al., 2014). This process requires water absorption, swelling, and molecular realignment, creating a complex gel-like matrix (Zhang et al., 2021; Cai et al., 2023). This matrix is important in forming texture and maintaining the stability and appearance of the extruded product (Schmid et al., 2022). Moreover, starches have the ability to bind other components in the matrix, thus creating the required compatibility in food extrusion (Gozzo and Glittenberg, 2009).

2.2 Nonstarchy polysaccharides

Nonstarchy polysaccharides are those that are less readily soluble in water, such as cellulose, hemicellulose, and pectin (Sethy et al., 2015). Unlike starchy polysaccharides, they do not undergo gelatinization. Instead, their main functions are to improve the structural, textural, and nutritional properties of extrudates (Ek et al., 2023; Schmid et al., 2020; Wolf, 2010).

In particular, cellulose and hemicellulose are commonly derived from products like cereal grains, legumes, and vegetables (Zhang et al., 2020). These are used to increase the dietary fiber content of extruded products. Therefore their incorporation aligns with consumer demand for health-promoting foods that are rich in fiber. In addition, these polysaccharides promote a dense or fibrous texture, which is desirable for high-fiber snacks, including cereal bars or meal replacements (Serna Saldívar and Hernández, 2020; Grasso, 2020). Another critical role of nonstarchy polysaccharides is their water-binding capacity (Zhao et al., 2021), which plays a role in the texture and stability of extruded products. Due to their ability to retain moisture, their inclusion leads to added benefits like better chewiness and resilience of the final product (Yi et al., 2022; Rough and Wilson, 2005). In addition, their ability to form networks within the matrix of the product minimizes collapse or deformation, which enhances the structural stability of the product (Chatterjee et al., 2018; El Awad; Azrak et al., 2020). Hemicellulose and pectin act as hydrocolloids, which means that they contribute to stabilizing the extruded product (Goff and Guo, 2019). The structural integrity of the formulated matrix is supported due to the interaction of these compounds with water and other ingredients. This functionality is especially essential in gluten-free or low-starch formulations (Foschia et al., 2017; Ahmed et al., 2024), since traditional binding agents may be absent.

2.3 Simple sugars

Simple sugars like glucose, sucrose, and fructose are primarily added to extrusion formulations due to their ability to caramelize or burn under high temperatures (Sasanam et al., 2023). However, their addition in small amounts can significantly boost the sensory attributes and processing behavior of extrudates (Dey et al., 2024;

Kraujutienė et al., 2019). One of the most notable contributions of simple sugars is their role in flavor and color development. These sugars undergo caramelization and Maillard reactions during the extrusion process, resulting in the formation of desirable flavors and golden-brown hues (Lohinova and Petrusha, 2023; Ostrikov et al., 2022). Through these changes, the sensory appeal of products is enhanced, as seen in breakfast cereals, snack bars, and confectionery extrudates (do Carmo et al., 2019; Meng et al., 2019b; Gunes et al., 2022). In addition, simple sugars influence the rheological properties of the dough during extrusion (Jeong et al., 2021). By modifying the plasticity of the mixture, small amounts of sugar can lower the extrusion temperature and reduce energy requirements (Huang et al., 2024). This effect is particularly useful when heat-sensitive ingredients are included in the formulation. In addition, sugar contributes to the mouthfeel and texture of the extruded product by providing a subtle sweetness and a smooth or glossy surface finish (Nussinovitch and Hirashima, 2023).

3 Functional roles of carbohydrates during extrusion

Carbohydrates are an added-value ingredient essential to the function of extrusion, influencing the texture, structure, and overall quality of extruded products. Table 8.1 provides a summary of these roles, highlighting their primary contributions.

3.1 Melt formation and viscosity

The formulation of the melt during extrusion is a critical stage at which carbohydrates play a dominant role in determining the rheological properties of the extruded material. The melt is the semifluid, viscous mixture of ingredients that undergoes thermal and mechanical processes as it passes through the extruder (Luo et al., 2023; Cotabarren et al., 2019). The characteristics of the melt are critical in determining the texture, structure, and overall quality of the final product. Carbohydrates, especially starches, are fundamental to melt formation due to their unique ability to gelatinize under elevated temperatures and shear forces (Donmez et al., 2021; Logié et al., 2018). The gelatinization stage includes the swelling and disruption of starch granules as they absorb water and undergo structural changes (Donmez et al., 2021). During this process, starch is transformed into a viscous, cohesive gel that contributes to the uniformity and flowability of the melt (Kumalasari et al., 2022).

There are several factors contributing to the viscosity of the melt, such as temperature, moisture content, and shear force (Qi et al., 2024; Yu et al., 2024). Higher temperatures and adequate moisture levels are required to facilitate starch gelatinization and enhance the melt to achieve the desired flow properties (Donmez et al., 2021). However, insufficient water content or insufficient thermal energy could negatively influence the gelatinization process, resulting in a melt that is either too rigid or prone to breakage (Schirmer et al., 2015; Fu et al., 2012).

In addition, nonstarchy polysaccharides, dietary fibers, and simple sugars can significantly interact with starch and thus influence melt viscosity (Mohamed, 2023;

Table 8.1 Functional roles of carbohydrates during extrusion.

Functional role	Key effect	Components	Key factors	Attributes	References
Melt formation and viscosity	Determines flow and uniformity	Starches, fibers, sugars	Gelatinization, moisture, temperature, shear	Stable texture, cohesive melt	(Luo et al., 2023; Cotabarren et al., 2019; Donmez et al., 2021; Kumalasari et al., 2022; Mohamed, 2023)
Texture and expansion	Controls expansion, texture	Amylose, amylopectin, fibers	Amylose-to-amylopectin ratio, pressure drop	Puffed snacks, dense products	(Vamadevan and Bertoft, 2020; Chen et al., 2024; Beech et al., 2022; Kristiawan et al., 2016; Alemayehu, 2017)
Binding and structure	Ensures cohesion	Starches, cellulose, hemicellulose	Gelatinization, gel-like networks	Prevents separation, maintains form	(Kristiawan et al., 2016; Lille et al., 2018; Trigui et al., 2020; Zhang et al., 2024; Delić et al., 2023)
Flavor and color	Improves sensory appeal	Reducing sugars	Maillard reaction, high extrusion temperature	Enhanced flavor, appealing brown color	(Sasanam et al., 2021; Shakoor et al., 2022)
Nutritional enhancement	Adds health benefits	Dietary fibers like cellulose, inulin	Fiber enrichment, prebiotic benefits	Digestive health support	(Li, 2022; Guarino et al., 2020)

Li et al., 2023). For example, specific fibers can enhance water retention and thereby increase the viscosity and stability of the melt (Yan et al., 2015). Similarly, the addition of sugars can modify viscosity by modulating the hydrogen bond network, which alters the melt's structural properties (Dey et al., 2024). These interactions highlight the complex interplay of ingredients that influence the rheological behavior of the melt during extrusion. Ultimately, understanding and controlling the properties of the melt are crucial for achieving consistent product quality with desired textural attributes in extruded foods. The ability of carbohydrates to form and manipulate the melt makes them indispensable in extrusion formulations.

3.2 Product texture and expansion

Carbohydrates, particularly starches, are fundamental in determining functional properties like the texture and expansion of extruded products. Under extrusion processing conditions (i.e., high temperature and high shear), starch gelatinization occurs, leading to the granules absorbing water, swelling, and consequently rupturing, releasing amylose and amylopectin to form a viscous gel (Vamadevan and Bertoft, 2020; Chen et al., 2024; Beech et al., 2022). When the mixture exits the extruder die and encounters a sudden pressure drop, the trapped water vaporizes, contributing to puffing or expansion of the formulation (Alavi and Rizvi, 2009; Aghajanzadeh et al., 2024). The degree of expansion depends on the starch composition, moisture content, temperature, and shear force during extrusion (Thymi et al., 2005; Ditudompo et al., 2016; Kristiawan et al., 2016; Zhihua et al., 2024).

The expansion behavior of starch during extrusion is heavily affected by the ratio of amylose to amylopectin (Chen et al., 2024). Amylose tends to create more rigid and less viscous melts due to its linear chain structure. The denser and firmer matrix is attributed to the ability of amylose molecules to interact and align more strongly within the mixture. These properties are considered advantageous for applications where a strong, cohesive structure is required, such as in pasta or dense snack products (Li et al., 2019). On the contrary, the molecular structure of amylopectin, which is highly branched, contributes to higher viscosity in the melt. The branched nature of amylopectin prevents close molecular packing, allowing for greater water absorption and expansion (Beech et al., 2022). This leads to the formation of a more fluid, pliable melt that can expand significantly during the extrusion process (Kallu et al., 2017). Products such as puffed snacks and breakfast cereals benefit from this attribute, as the high viscosity of the melt leads to enhanced expansion and aeration (Alemayehu, 2017).

Moisture content affects this process, as lower moisture promotes expansion; however, excessively low levels risk incomplete gelatinization (Meng et al., 2019a; Budi et al., 2016). Nonstarchy polysaccharides like cellulose and hemicellulose, which are poorly gelatinized, contribute to the structural integrity and fiber content, resulting in firmer textures and reducing brittleness or collapse (Bouvier and Campanella, 2014; Zhao et al., 2005). These compounds, along with other ingredients such as proteins and sugars, interact to formulate the final texture and expansion of the product (Singh et al., 2007; Guerrero et al., 2012), thus enabling a wide range of extruded foods designed based on consumer preferences.

3.3 Binding and structure

Carbohydrates play a fundamental role as building blocks in extrusion, ensuring that various ingredients adhere together to form a stable, cohesive product. Starches, in particular, are integral due to their ability to gelatinize, wherein the heat and moisture in the extruder cause starch granules to swell and release their contents, thus creating a viscoelastic matrix (Kristiawan et al., 2016). This matrix binds other ingredients such as proteins, fats, and fibers, leading to structural integrity that prevents separation during and after extrusion (Lille et al., 2018; Delić et al., 2023). Nonstarchy polysaccharides such as cellulose, hemicellulose, and pectin can contribute to binding by forming gel-like networks or stabilizing the structure (Trigui et al., 2020; Liao et al., 2024; Zhang et al., 2024), enhancing stiffness and preventing product collapse. These carbohydrate interactions ensure the extruded product's stability, texture, and overall quality, making them indispensable in complex formulations with diverse ingredients.

3.4 Other functional roles

Carbohydrates contribute to other functional roles in extrusion, such as the flavor, color, and nutritional value of the final product. The Maillard reaction occurs at high temperatures in the extrusion process, involving the interaction between reducing sugars and amino acids. This reaction enhances the product's flavor and develops an appealing brown color (Sasanam et al., 2021). It not only improves sensory attributes but also increases the intensity of the product's flavor profile (Shakoor et al., 2022). In addition, carbohydrates classified as dietary fibers, such as cellulose and inulin, increase the fiber content of extruded foods, improving their nutritional value by supporting digestive health, enhancing bowel regularity, and fostering the growth of beneficial gut bacteria (Li, 2022; Guarino et al., 2020). This multifunctionality makes carbohydrates essential for achieving both sensory and nutritional characteristics in extrusion-based food products.

4 Impact of processing parameters on carbohydrate functionality

The functionality of carbohydrates in extrusion is highly influenced by processing parameters such as screw speed, extrusion configurations (e.g., twin-screw systems), moisture content, and pressure. These factors affect the behavior of carbohydrates during extrusion and determine the texture, structure, and quality of the final product. Fig. 8.2 illustrates the impact of major processing parameters on carbohydrate extrusion.

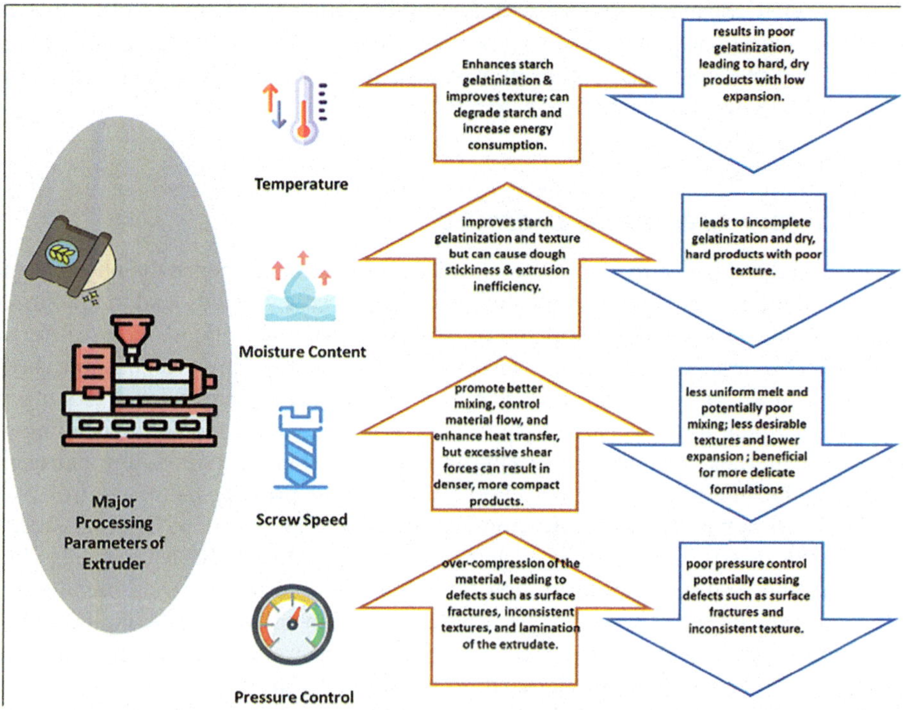

Fig. 8.2 The impact of processing parameters at extreme levels on carbohydrate extrusion.

4.1 Temperature and moisture content

Temperature and moisture are fundamental factors in determining the behavior of carbohydrates during extrusion, particularly for starch gelatinization. Adequate moisture (around 10%–30% by weight) (Berrios et al., 2010; Zambrano et al.; Lin, 2022) and heat (commonly in the range of 120°C–180°C) (Moscicki et al., 2013; Selling, 2010; Kaur et al., 2015; Arêas, 1992) enable starch granules to swell, gelatinize, and integrate into the melt, contributing to the desired texture and structure of the final product. Temperature plays a crucial role not only in the gelatinization of starch but also in enzyme activity (Wang et al., 2023). In addition, temperature influences the degree of starch gelatinization, the expansion of extrudates, and the development of their pore structure. As the temperature increases, the activity and plasticity of starch molecules are enhanced, leading to the formation of a more open and uniform porous structure. Rafiq et al. (2018) demonstrated that the degree of gelatinization is positively correlated with both temperature and feed moisture in twin-screw extrusion. In dry extrusion (i.e., low moisture content), poor gelatinization of the starch occurs, leading to the formation of a hard product with a low degree of expansion (Guo et al., 2020).

This can negatively affect the texture and taste of the extrudate. However, in the case of high moisture content, or "wet extrusion," starch gelatinization is improved and leads to better expansion, texture, and digestibility. Higher moisture (>40%) initiates starch particles to absorb water more easily, thus improving the overall quality of the product and reducing energy consumption and wear on extrusion equipment (Choi et al., 2021; Jongsutjarittam and Charoenrein, 2014).

4.2 Screw speed and pressure

Screw speed and pressure within the extruder significantly influence the physical and functional properties of the extruded product. Higher screw speeds lead to the formation of a more stable melt temperature, with minimal fluctuations across different materials and processing conditions (Abeykoon et al., 2016). However, in some cases, excessive shear at 500–700 rpm can reduce the degree of expansion, resulting in shrinkage, which forms denser products with a more compact texture, depending on the raw materials' characteristics (Philipp et al., 2018). In twin-screw extrusion systems, the intermeshing screws boost mixing and provide better control over material flow, consequently improving heat transfer (Martin, 2016; Qi et al., 2024). High screw speeds in twin-screw systems generate significant shear forces, resulting in a smoother and more uniform melt by efficiently deforming starch granules and other ingredients (Abeykoon, 2022). The combination of twin-screw configuration and high screw speed ensures better control over material flow, facilitating even processing and consistent product quality (Nandi et al., 2021).

de la Peña and Manthey (2017) found that higher hydration (30%–34%) reduces extrusion pressure. The formulation of the dough notably affects this pressure, with semolina dough exhibiting a smaller decrease in pressure compared to whole-meal semolina. The plasticizing effect of water in the dough enables its movement within the extruder, thereby reducing the extrusion pressure. Similarly, higher hydration levels (i.e., 32%–34%) reduce pressure in the extruder and are associated with a decrease in the diameter and density of the spaghetti. In contrast, at 30% hydration, the higher consistency and resistance to flow exert higher pressure on the Teflon die coatings, resulting in larger spaghetti diameters (de la Peña et al., 2014).

Extrusion pressure, typically measured using a capillary ram extruder, is an important factor reflecting the rheological properties of pastes, specifically when assessing small quantities of new advanced ceramic materials before scaling up production (Li and Bridgwater, 2000). The pressure drop along the die entry and die land is considered a key factor in the paste extrusion process. The flow behavior of the paste can be estimated when the pressure drop is determined (Akdogan, 1996). Understanding the flow characteristics in dies is crucial for controlling extruder performance and the quality of the extrudate (Padmanabhan and Bhattacharya, 1989). Poor control over extrusion pressure could lead to negative outcomes such as surface fractures, lamination of extrudates (Benbow et al., 1991; Person and Denn, 1997), liquid migration, and powder packing problems (Burbidge et al., 1995; Blackburn and Böhm, 1994). Therefore extrusion pressure is a crucial determinant in ensuring proper extrusion processes.

5 Modification of carbohydrates for extrusion

The modulation of carbohydrates is an effective approach to improving their functional properties for use in extrusion processes. These modifications can resolve specific challenges associated with native carbohydrates, including limited solubility, poor water absorption, or inconsistent textural outcomes (Olawoye et al., 2023), consequently broadening their applicability across various extruded food products. For instance, extrusion processing can enhance the water absorption and solubility of flours, which increases their application potential (Patil and Kaur, 2018).

Pregelatinization is a common method by which starches are precooked and dried before their use in the extrusion process (Jiang et al., 2022; Hong and Liu, 2023). This modification leads to the water absorption and dispersion of starch during processing, which results in smoother melts and consistent textural attributes in the final product. Pregelatinized starches are particularly effective in applications that demand rapid hydration or precise texture control, such as ready-to-eat soups, semiprepared sauces, instant noodles, ready-to-bake mixes, and instant flours (Carvalho et al., 2013; Cui et al., 2024).

Chemical modifications, such as alkaline/acidic digestion, esterification, and etherification, can enhance carbohydrate functionality by modifying their structural properties (Al-Khalili et al., 2022, 2021; Al-Mamari et al., 2022; Al-Harrasi et al., 2024; Ashogbon and Akintayo, 2014; Korma, 2016; Masina et al., 2017). These interventions can improve functional properties such as solubility, viscosity, or control of gelling behavior, making carbohydrates more versatile in extrusion. For instance, chemically treated fibers exhibit enhanced physicochemical properties, such as improved solubility, stability, and thermal resistance, which enhance their functionality in applications like thickening, gelling, and emulsification. In addition, they provide health benefits, such as improved digestion and the ability to encapsulate bioactive ingredients or deliver essential micronutrients (Gałkowska et al., 2023).

Another approach to pretreat carbohydrates is enzymatic treatment. Enzymes work to modify carbohydrates by breaking down the structure of starch, thereby changing its physical and chemical properties. Extrusion technology involves high temperature, mechanical energy, and moisture to deform starch molecular and granular structures, which enhances enzyme access and catalysis. For instance, this process cleaves α-(1→4) and α-(1→6) bonds and breaks down crystalline or ordered forms, such as retrograded starch or starch–lipid complexes (Lai and Kokini, 1991). In addition, enzymes can alter the degree of polymerization by selectively hydrolyzing carbohydrate molecules, thus improving their processability and functionality in extrusion (Xu et al., 2020; Gatt et al., 2018). For example, enzymatically treated starches have shown enhanced expansion properties (Bangar et al.; Lin, 2022) which help in the creation of puffed products with desirable textures.

The use of modulated carbohydrates not only improves the efficiency and versatility of the extrusion process, but also expands the production of innovative food products with specific textures, flavors, and nutritional profiles. This adaptability emphasizes the importance of carbohydrate modification in developing extrusion technology and meeting evolving consumer demands.

6 Future trends in carbohydrate utilization for extrusion

As the trend of sustainability and innovation increases in food manufacturing and processing, there is a growing interest in utilizing novel carbohydrate sources for extrusion. Alternative carbohydrates from pseudocereals, such as quinoa and amaranth (Mir et al., 2018; García-Salcedo et al., 2018), as well as unconventional flours from sources like legumes, peas, and other underutilized crops (Pedrosa et al., 2021; Navin Venketeish et al., 2024; Adebo, 2023), are considered exciting opportunities to enhance the nutritional profile of extruded foods. Furthermore, the augmentation of starches with fiber-rich materials from fruits and vegetables (Berrios et al., 2010; Onwulata et al., 2014; Patil et al., 2017) is becoming increasingly common to enhance texture, taste, and health benefits in extrusion technology. Fibers extracted from agrobyproducts such as seeds, pomace, and peels (Al-Khalili et al., 2023; Tedeschi et al., 2018; Tauferova et al., 2021; Schmid et al., 2020; Yang et al., 2021; Kharchoufi et al., 2018; de Moraes Crizel et al., 2018) offer potential to contribute to the circular economy, reduce environmental impact, and combat global food insecurity. In addition, carbohydrates sourced locally or derived from by-products of other food production processes show a promising avenue for advancing these goals.

7 Conclusion

Carbohydrates are an indispensable ingredient in the extrusion process, contributing to shaping the texture, structure, and overall quality of the extrudate. From their involvement in melt formation and viscosity control to their impact on product texture, expansion, and binding properties, carbohydrates are essential for achieving the desired product characteristics. Starches, in particular, are central to many of these functions due to their ability to gelatinize under heat and shear forces, while non-starchy polysaccharides contribute additional functionalities such as structural integrity and nutritional benefits. The interaction between carbohydrate type, processing parameters, and product characteristics is key to optimizing extrusion outcomes. Variations in parameters such as temperature, moisture content, pressure, screw speed, and configuration directly impact the behavior of carbohydrates, affecting processes such as gelatinization, viscosity, and expansion. Understanding these interactions allows for precise control over the extrusion process, enabling the production of a wide range of products tailored to consumer preferences and technological requirements. The future of carbohydrate innovation in extrusion technology is promising due to the growing interest in sustainability and nutrition. Exploring alternative carbohydrate sources, such as underutilized crops and agrobyproducts, could further support sustainable and health-conscious products. Ongoing advancements in modified carbohydrates and novel ingredients have the potential to promote product quality and environmental responsibility, ensuring extrusion technology remains a leader in food innovation.

References

Abeykoon, C., 2022. Sensing technologies for process monitoring in polymer extrusion: a comprehensive review on past, present and future aspects. Meas. Sens. 26659174. 22, 100381. doi:10.1016/j.measen.2022.100381.

Abeykoon, C., Kelly, A.L., Brown, E.C., Coates, P.D., 2016. The effect of materials, process settings and screw geometry on energy consumption and melt temperature in single screw extrusion. Appl. Energy. 03062619. 180, 880–894. Available from: http://www.elsevier.com/inca/publications/store/4/0/5/8/9/1/index.htt. doi:10.1016/j.apenergy.2016.07.014.

Adebo, J.A., 2023. A review on the potential food application of lima beans (Phaseolus lunatus L.), an underutilized crop. Appl. Sci. 13.

Agbana, M.R., Angeletti, B.S., Buecker, H.C., Tseng, Y.C., Davis, B.E., Schendel, R.R., 2024. Characterizing the non-starch polysaccharides of hempseed cell walls. Eur. Food Res. Technol. 14382385. 250 (9), 2405–2419. Available from: https://www.springer.com/journal/217. doi:10.1007/s00217-024-04548-2.

Aghajanzadeh, S., Sultana, A., Ziaiifar, A.M., Khalloufi, S., 2024. Formation of pores and bubbles and their impacts on the quality attributes of processed foods: a review. Food Res. Int. 09639969. 188, 114494. doi:10.1016/j.foodres.2024.114494.

Ahmed, M.W., Jothi, J.S., Saifullah, M., Hannan, M.A., Mohibbullah, M., 2024. Impact of drying temperature on textural, cooking quality, and microstructure of gluten-free pasta. In: Development of Gluten-Free Pasta. Elsevier, United States, 65–110. Available from: https://www.sciencedirect.com/book/9780443132384. doi:10.1016/B978-0-443-13238-4.00010-1.

Akdogan, H., 1996. Pressure, torque, and energy responses of a twin screw extruder at high moisture contents. Food Res. Int. 09639969. 29 (5-6), 423–429. doi:10.1016/s0963-9969(96)00036-1.

Al-Harrasi, K., Al-Habsi, N., Myint, M.T.Z., Rahman, M.S., 2024. Molecular and structural characterization of micronized lignocellulose from date-pits by alcoholic fractionations. Carbohydr. Polym. Technol. Appl. 26668939. 7. Available from: https://www.sciencedirect.com/science/journal/26668939. doi:10.1016/j.carpta.2024.100451.

Al-Khalili, M., Al-Habsi, N., Al-Alawi, A., Al-Subhi, L., Myint, M.T.Z., Al-Abri, M., Waly, M.I., Al-Harthi, S., Al-Mamun, A., Rahman, M.S., 2021. Structural characteristics of alkaline treated fibers from date-pits: residual and precipitated fibers at different pH. Bioact. Carbohydr. Diet. Fibre. 22126198. 25. Available from: http://www.journals.elsevier.com/bioactive-carbohydrates-and-dietary-fibre/. doi:10.1016/j.bcdf.2020.100251.

Al-Khalili, M., Al-Habsi, N., Al-Kindi, M., Rahman, M.S., 2022. Characteristics of crystalline and amorphous fractions of date-pits as treated by alcohol-water pressure cooking. Bioact. Carbohydr. Diet. Fibre. 22126198. 28, 100331. doi:10.1016/j.bcdf.2022.100331.

Al-Khalili, M., Al-Habsi, N., Rahman, M.S., 2023. Applications of date pits in foods to enhance their functionality and quality: a review. Front. Sustain. Food Syst. 6, 1101043.

Al-Mamari, A., Al-Habsi, N., Al-Khalili, M., Rahman, M.S., 2022. Extraction and characterization of residue fibers from defatted date-pits after alkaline-acid digestion: effects of different pretreatments. J. Therm. Anal. Calorim. 15882926. 147 (17), 9405–9416. Available from: http://www.springer.com/sgw/cda/frontpage/0,11855,1-40109-70-35752391-0,00.html. doi:10.1007/s10973-022-11237-3.

Alavi, S., Rizvi, S.S., Supercritical fluid extrusion—a novel method for producing microcellular structures in starch-based matrices. Novel food processing—effects on rheological and functional properties. (2009), 403–420

Alemayehu, H., Product Development and Quality Characterization of Pulse-based Extruded Ready to Eat Snacks. (2017).

Arêas, J.A.G., 1992. Extrusion of food proteins. Crit. Rev. Food Sci. Nutr. 1040-8398. 32 (4), 365–392. doi:10.1080/10408399209527604.

Ashogbon, A.O., Akintayo, E.T., 2014. Recent trend in the physical and chemical modification of starches from different botanical sources: a review. Starch-Stärke. 0038-9056. 66 (1-2), 41–57. doi:10.1002/star.201300106.

Azrak, S.M.E.A., Costakis, W.J., Moon, R.J., Schueneman, G.T., Youngblood, J.P., 2020. Continuous Processing of Cellulose Nanofibril Sheets through Conventional Single-Screw Extrusion. ACS Appl. Polym. Mater. 26376105. 2 (8), 3365–3377. Available from: pubs.acs.org/journal/aapmcd. doi:10.1021/acsapm.0c00477.

Beech, D., Beech, J., Gould, J., Hill, S., 2022. Effect of amylose/amylopectin ratio and extent of processing on the physical properties of expanded maize starches. Int. J. Food Sci. Technol. 0950-5423. 57 (4), 2298–2309. doi:10.1111/ijfs.15581.

Benbow, J., Jazayeri, S., Bridgwater, J., 1991. The flow of pastes through dies of complicated geometry. Powder Technol. 65, 393–401.

Berrios, J.D.J., Morales, P., Cámara, M., Sánchez-Mata, M.C., 2010. Carbohydrate composition of raw and extruded pulse flours. Food Res. Int. 09639969. 43 (2), 531–536. doi:10.1016/j.foodres.2009.09.035.

Blackburn, S., Böhm, H., 1994. The influence of powder packing on the rheology of fibre-loaded pastes. J. Mater. Sci. 0022-2461. 29 (16), 4157–4166. doi:10.1007/bf00414194.

Bouasla, A., Wójtowicz, A., 2021. Gluten-Free Rice Instant Pasta: effect of Extrusion-Cooking Parameters on Selected Quality Attributes and Microstructure. Processes. 2227-9717. 9 (4), 693. doi:10.3390/pr9040693.

Bouvier, J.M., Campanella, O.H., 2014. Extrusion Processing Technology: food and Non-Food Biomaterials. In: Extrusion Processing Technology: Food and Non-Food Biomaterials, pp. 1–518, Available from: http://www.wiley.com/remtitle.cgi?isbn=1444338110. doi:10.1002/9781118541685. 9781444338119.

Budi, F.S., Hariyadi, P., Budijanto, S., Syah, D., 2016. Effect of dough moisture content and extrusion temperature on degree of gelatinization and crystallinity of rice analogues. J. Dev. Sustain. Agric. 10, 91–100.

Burbidge, A., Bridgwater, J., Saracevic, Z., 1995. Liquid migration in paste extrusion: advanced materials. Chem. Eng. Res. Des. 73, 810–816.

Cai, Q., Li, X., Ding, X., Wang, H., Hu, X., 2023. Effects of quercetin and Ca $(OH)_2$ addition on gelatinization and retrogradation properties of Tartary buckwheat starch. Lwt 178, 114488.

Carvalho, A.V., Bassinello, P.Z., Rios, A.D.O., Ferreira, T.F., Carvalho, R.N., Koakuzu, S.N., 2013. Characterization of pre-gelatinized rice and bean flour. Ciencia E Tecnologia de Alimentos. 1678457X. 33 (2), 245–250. Available from: http://www.scielo.br/pdf/cta/v33n2/aop_cta_5804.pdf. doi:10.1590/S0101-20612013005000049Brazil.

Chatterjee, T., O'Donnell, K.P., Rickard, M.A., Nickless, B., Li, Y., Ginzburg, V.V., Sammler, R.L., 2018. Rheology of Cellulose Ether Excipients Designed for Hot Melt Extrusion. Biomacromolecules. 15264602. 19 (11), 4430–4441. Available from: http://pubs.acs.org/journal/bomaf6. doi:10.1021/acs.biomac.8b01306.

Chen, Q., Zhang, J., Zhang, Y., Kaplan, D.L., 2022. Qiang Wang, Protein-amylose/amylopectin molecular interactions during high-moisture extruded texturization toward plant-based meat substitutes applications. Food Hydrocoll. 0268005X. 127, 107559. doi:10.1016/j.foodhyd.2022.107559.

Chen, X., Zhu, L., Zhang, H., Gangcheng, W., Cheng, L., Zhang, Y., 2024. A review of endogenous non-starch components in cereal matrix: spatial distribution and mechanisms for inhibiting starch digestion. Crit. Rev. Food Sci. Nutr. 1040-8398. 1–16. doi:10.1080/10408398.2024.2370487.

Choi, O.-J., Zhao, C.-C., Ameer, K., Eun, J.-B., 2021. Effects of soy flour types and extrusion-cooking conditions on physicochemical, microstructural and sensory characteristics of puffed rice snack base. Int. J. Food Eng. 1556-3758. 17 (6), 473–483. doi:10.1515/ijfe-2019-0157.

Choton, S., Gupta, N., Bandral, J.D., 2020. Nadira Anjum, Ankita Choudary, Extrusion technology and its application in food processing: a review. Pharma Innov. 23498242. 9 (2), 162–168. doi:10.22271/tpi.2020.v9.i2d.4367.

Cotabarren, I.M., Cruces, S., Palla, C.A., 2019. Extrusion 3D printing of nutraceutical oral dosage forms formulated with monoglycerides oleogels and phytosterols mixtures. Food Res. Int. 126, 108676.

Cui, C., Li, D., Wang, L.J., Wang, Y., 2024. Microstructure of Extrusion-Cooked Whole Grain in Controlling Product Quality. Food Rev. Int. 15256103. 40 (6), 1621–1646. Available from: www.tandf.co.uk/journals/titles/87559129.asp. doi:10.1080/87559129.2023.2227981.

de la Peña, E., Manthey, F.A., 2017. Effect of Formulation and Dough Hydration Level on Extrusion, Physical and Cooked Qualities of Nontraditional Spaghetti. J. Food Process Eng. 17454530. 40 (1). Available from: http://www.blackwellpublishers.co.uk/journal_new.asp?site=1. doi:10.1111/jfpe.12301.

de la Peña, E., Manthey, F.A., Patel, B.K., Campanella, O.H., 2014. Rheological properties of pasta dough during pasta extrusion: effect of moisture and dough formulation. J. Cereal Sci. 10959963. 60 (2), 346–351. Available from: http://www.elsevier.com/inca/publications/store/6/2/2/8/5/9/index.htt. doi:10.1016/j.jcs.2014.05.013.

de Moraes Crizel, T., de Oliveira Rios, A., Alves, V.D., Bandarra, N., Moldão-Martins, M., Flôres, S.H., 2018. Biodegradable Films Based on Gelatin and Papaya Peel Microparticles with Antioxidant Properties. Food Bioprocess Technol. 19355149. 11 (3), 536–550. Available from: http://www.springer.com/life+sci/food+science/journal/11947. doi:10.1007/s11947-017-2030-0.

Delić, J., Ikonić, P., Jokanović, M., Peulić, T., Ikonić, B., Banjac, V., Vidosavljević, S., Stojkov, V., Hadnađev, M., 2023. Sustainable snack products: impact of protein- and fiber-rich ingredients addition on nutritive, textural, physical, pasting and color properties of extrudates. Innov. Food Sci. Emerg. Technol. 14668564. 87, 103419. doi:10.1016/j.ifset.2023.103419.

Dey, D., Gu, B.J., Ek, P., Ross, C.F., Saunders, S.R., Ganjyal, G.M., 2024. Sugar inclusion influences the expansion characteristics of corn starch extrudates. J. Food Sci. 17503841. 89 (11), 7760–7775. Available from: http://onlinelibrary.wiley.com/journal/10.1111/(ISSN)1750-3841. doi:10.1111/1750-3841.17403.

Ditudompo, S., Takhar, P.S., Ganjyal, G.M., Hanna, M.A., 2016. Effect of extrusion conditions on expansion behavior and selected physical characteristics of cornstarch extrudates. Trans. ASABE. 21510032. 59 (4), 969–983. Available from: http://elibrary.asabe.org/azdez.asp?JID=3&AID=47312&ConfID=t2016&v=59amp;i=4amp;T=2amp;redirType=. doi:10.13031/trans.59.11467.

do Carmo, C.S., Varela, P., Poudroux, C., Dessev, T., Myhrer, K., Rieder, A., Zobel, H., Sahlstrøm, S., Knutsen, S.H., 2019. The impact of extrusion parameters on physicochemical, nutritional and sensorial properties of expanded snacks from pea and oat fractions. LWT 112, 108252.

Donmez, D., Pinho, L., Patel, B., Desam, P., Campanella, O.H., 2021. Characterization of starch–water interactions and their effects on two key functional properties: starch gelatinization and retrogradation. Curr. Opin. Food Sci. 22147993. 39, 103–109. Available from: http://www.journals.elsevier.com/current-opinion-in-food-science/. doi:10.1016/j.cofs.2020.12.018.

Ek, P., Gu, B.J., Richter, J.K., Dey, D., Saunders, S.R., Ganjyal, G.M., 2023. High methoxyl pectin can improve the extrusion characteristics and increase the dietary fiber content of starch-cellulose extrudates. J. Food Sci. 17503841. 88 (10), 4156–4168. Available from: http://onlinelibrary.wiley.com/journal/10.1111/(ISSN)1750-3841. doi:10.1111/1750-3841.16742.

Ek, P., Kowalski, R.J., Ganjyal, G.M., 2020. Raw material behaviors in extrusion processing I (Carbohydrates). In: Ganjyal, G.M. (Ed.), Extrusion cooking. Woodhead Publishing, pp.119–152.

Foschia, M., Horstmann, S.W., Arendt, E.K., Zannini, E., 2017. Legumes as Functional Ingredients in Gluten-Free Bakery and Pasta Products. Annu. Rev. Food Sci. Technol. 19411421. 8, 75–96. Available from: http://www.annualreviews.org/journal/food. doi:10.1146/annurev-food-030216-030045.

Fu, Z.Q., Wang, L.J., Li, D., Adhikari, B., 2012. Effects of partial gelatinization on structure and thermal properties of corn starch after spray drying. Carbohydr. Polym. 01448617. 88 (4), 1319–1325. doi:10.1016/j.carbpol.2012.02.010.

Galanakis, C.M., 2022. Sustainable applications for the valorization of cereal processing by-products. Foods. 23048158. 11 (2). Available from: https://www.mdpi.com/2304-8158/11/2/241/pdf. doi:10.3390/foods11020241.

Gałkowska, D., Kapuśniak, K., Juszczak, L., 2023. Chemically modified starches as food additives. Molecules 28, 7543.

Gao, Y., Sun, Y., Zhang, Y., Sun, Y., Jin, T., 2022. Extrusion modification: effect of extrusion on the functional properties and structure of rice protein. Processes. 2227-9717. 10 (9), 1871. doi:10.3390/pr10091871.

García-Salcedo, Á.J., Torres-Vargas, O.L., Aeiza-Calderón, H., 2018. Physical-chemical characterization of quinoa (Chenopodium quinoa Willd.), amaranth (Amaranthus caudatus L.), and chia (Salvia hispanica L.) flours and seeds. Acta Agronómica 67, 215–222.

Gatt, E., Rigal, L., Vandenbossche, V., 2018. Biomass pretreatment with reactive extrusion using enzymes: a review. Ind. Crops Prod. 122, 329–339.

Giannakas, A.E., 2024. Extrusion of biopolymers for food applications. In: Advances in Biopolymers for Food Science and Technology. Elsevier, Greece, 137–169. Available from: https://www.sciencedirect.com/book/9780443190056. doi:10.1016/B978-0-443-19005-6.00007-4.

Goff, H.D., Guo, Q., 2019. The Role of Hydrocolloids in the Development of Food Structure. R. Soc. Chem. 1–28. doi:10.1039/9781788016155-00001.

Gozzo, A., Glittenberg, D., 2009. Starch: a Versatile Product from Renewable Resources for Industrial Applications. R. Soc. Chem. 238–263. doi:10.1039/9781847552686-00238.

Grasso, S., 2020. Extruded snacks from industrial by-products: a review. Trends Food Sci. Technol. 09242244. 99, 284–294. doi:10.1016/j.tifs.2020.03.012.

Guarino, M.P.L., Altomare, A., Emerenziani, S., Di Rosa, C., Ribolsi, M., Balestrieri, P., Iovino, P., Rocchi, G., Cicala, M., 2020. Mechanisms of action of prebiotics and their effects on gastro-intestinal disorders in adults. MDPI AG, Italy Nutrients. 20726643. 12 (4). Available from: https://www.mdpi.com/2072-6643/12/4/1037/pdf. doi:10.3390/nu12041037.

Guerrero, P., Beatty, E., Kerry, J.P., de la Caba, K., 2012. Extrusion of soy protein with gelatin and sugars at low moisture content. J. Food Eng. 02608774. 110 (1), 53–59. doi:10.1016/j.jfoodeng.2011.12.009.

Gunes, R., Palabiyik, I., Konar, N., Toker, O.S., 2022. Soft confectionery products: quality parameters, interactions with processing and ingredients. Food Chem. 03088146. 385, 132735. doi:10.1016/j.foodchem.2022.132735.

Guo, Z., Teng, F., Huang, Z., Lv, B., Xiqiao, L., Babich, O., Wenhua, Y., Yang, L., Wang, Z., Jiang, L., 2020. Effects of material characteristics on the structural characteristics and flavor substances retention of meat analogs. Food Hydrocoll. 0268-005X. 105, 105752. doi:10.1016/j.foodhyd.2020.105752.

Hong, Y., Liu, X., 2023. Pre-gelatinized Modification of Starch. Physical Modifications of Starch, Springer.

Huang, J., Zhang, M., Mujumdar, A.S., Chunli, L., 2024. Modulation of starch structure, swallowability and digestibility of 3D-printed diabetic-friendly food for the elderly by dry heating. Int. J. Biol. Macromol. 01418130. 264 (Pt 2), 130629. doi:10.1016/j.ijbiomac.2024.130629.

Jeong, S., Park, Y., Lee, S., 2021. Assessment of turanose as a sugar alternative in a frozen dough system: rheology, tomography, and baking performance. LWT 141, 110869.

Jiang, L., Mingming, Q., Deng, Y., Suo, W., Song, J., Zhang, M., Zheng, H., Zhang, D., Chen, S., Li, H., 2022. Extrusion-induced pre-gelatinization and hydrolyzation of rice adjunct contributed to the mashing performance. LWT. 00236438. 158, 113126. doi:10.1016/j.lwt.2022.113126.

Jongsutjarittam, O., Charoenrein, S., 2014. The effect of moisture content on physicochemical properties of extruded waxy and non-waxy rice flour. Carbohydr. Polym. 01448617. 114, 133–140. doi:10.1016/j.carbpol.2014.07.074.

Kallu, S., Kowalski, R.J., Ganjyal, G.M., 2017. Impacts of cellulose fiber particle size and starch type on expansion during extrusion processing. J. Food Sci. 82, 1647–1656.

Kaur, S., Sharma, S., Singh, B., Dar, B.N., 2015. Effect of extrusion variables (temperature, moisture) on the antinutrient components of cereal brans. J. Food Sci. Technol. 0022-1155. 52 (3), 1670–1676. doi:10.1007/s13197-013-1118-4.

Kharchoufi, S., Parafati, L., Licciardello, F., Muratore, G., Hamdi, M., Cirvilleri, G., Restuccia, C., 2018. Edible coatings incorporating pomegranate peel extract and biocontrol yeast to reduce Penicillium digitatum postharvest decay of oranges. Food Microbiol. 07400020. 74, 107–112. doi:10.1016/j.fm.2018.03.011.

Korma, S.A.E.H., 2016. Chemically Modified Starch and Utilization in Food Stuffs, Science Publishing Group.

Kraujutienė, I., Rasa, V., Rasa, P., 2019. Production of dragées from extruded corn: the product's effect on glucose level in blood. Ann. Food Sci. Technol. 20 (1), 105.

Kristiawan, M., Chaunier, L., Valle, G.D., Ndiaye, A., Vergnes, B., 2016. Modeling of starchy melts expansion by extrusion. Trends Food Sci. Technol. 09242244. 48, 13–26. doi:10.1016/j.tifs.2015.11.004.

Kumalasari, R., Sholichah, E., Haryanto, A., HANIFAH, U., Mayasti, N.K.I., YUNIAR, A.D., 2022. Evaluation of uniformity of physical and texture quality in manufacture of gluten-free noodles using single-screw extruders: a case study on local SMEs in Subang district-Indonesia. Food Sci. Technol. 1678-457X. 42. doi:10.1590/fst.82421.

Lai, L.S., Kokini, J.L., 1991. Physicochemical Changes and Rheological Properties of Starch during Extrusion (A Review). Biotechnol. Prog. 8756-7938. 7 (3), 251–266. doi:10.1021/bp00009a009.

Lazou, A.E., 2024. Food extrusion: an advanced process for innovation and novel product development. Crit. Rev. Food Sci. Nutr. 15497852. 64 (14), 4532–4560. Available from: https://www.tandfonline.com/loi/bfsn20. doi:10.1080/10408398.2022.2143474.

Lee, H., Song, J., Lee, B., Cha, J., 2024. Hyeyoung Lee, Food carbohydrates in the gut: structural diversity, microbial utilization, and analytical strategies. Food Sci. Biotechnol. 1226-7708. 33 (9), 2123–2140. doi:10.1007/s10068-024-01648-3.

Li, C., 2022. Recent progress in understanding starch gelatinization – An important property determining food quality. Carbohydr. Polym. 01448617. 293, 119735. doi:10.1016/j.carbpol.2022.119735.

Li, H., Gidley, M.J., Dhital, S., 2019. High-amylose starches to bridge the "Fiber Gap": development, structure, and nutritional functionality. Compr. Rev. Food Sci. Food Saf. 18, 362–379.

Li, S., Chen, W., Zongo, A.W.S., Chen, Y., Liang, H., Li, J., Li, B., 2023. Effects of non-starch polysaccharide on starch gelatinization and digestibility: a review. Food Innov. Adv. 2836774X. 2 (4), 302–312. Available from: https://www.maxapress.com/article/doi/10.48130/FIA-2023-0029. doi:10.48130/FIA-2023-0029.

Li, Y.Y., Bridgwater, J., 2000. Prediction of extrusion pressure using an artificial neural network. Powder Technol. 00325910. 108 (1), 65–73. doi:10.1016/s0032-5910(99)00254-5.

Liao, G., Sun, E., Kana, E.B.G., Huang, H., Sanusi, I.A., Qu, P., Jin, H., Liu, J., Shuai, L., 2024. Renewable hemicellulose-based materials for value-added applications. Carbohydr. Polym. 01448617. 341. Available from: https://www.sciencedirect.com/science/journal/01448617. doi:10.1016/j.carbpol.2024.122351.

Lille, M., Nurmela, A., Nordlund, E., Metsä-Kortelainen, S., Sozer, N., 2018. Applicability of protein and fiber-rich food materials in extrusion-based 3D printing. J. Food Eng. 02608774. 220, 20–27. doi:10.1016/j.jfoodeng.2017.04.034.

Lin, S., 2022. Dietary fiber in bakery products: source, processing, and function. Adv. Food Nutr. Res. 99, 37–100. Available from: http://www.elsevier.com/wps/find/bookdescription.cws_home/704624/description#description. doi:10.1016/bs.afnr.2021.12.001.

Liu, K.S., Hsieh, F., 2007. Protein–Protein Interactions in High Moisture-Extruded Meat Analogs and Heat-Induced Soy Protein Gels. J. Am. Oil Chem. Soc. 0003-021X. 84 (8), 741–748. doi:10.1007/s11746-007-1095-8.

Logié, N., Valle, G.D., Rolland-Sabaté, A., Descamps, N., Soulestin, J., 2018. How does temperature govern mechanisms of starch changes during extrusion? Carbohydr. Polym. 01448617. 184, 57–65. Available from: http://www.elsevier.com/wps/find/journaldescription.cws_home/405871/description#description. doi:10.1016/j.carbpol.2017.12.040.

Lohinova, A., Petrusha, O., 2023. Maillard reaction in food technologies. Ukrainian J. Food Sci. 23101008. 11 (2), 81–109. doi:10.24263/2310-1008-2023-11-2-4.

Low, Y., Effarizah, M., Cheng, L., 2020. Factors influencing rice noodles qualities. Food Rev. Int. 36, 781–794.

Luo, Y., Sun, W., Bao, M., Zhu, X., Ning, C., Zhang, W., Li, Y., Zhang, X., 2023. Process fundamentals and quality investigation in extrusion 3D printing of shear thinning materials: extrusion process based on Nishihara model. Int. J. Adv. Manuf. Technol. 0268-3768. 124 (1-2), 245–264. doi:10.1007/s00170-022-10506-7.

Martin, C., 2016. Twin Screw Extruders as Continuous Mixers for Thermal Processing: a Technical and Historical Perspective. AAPS PharmSciTech. 1530-9932. 17 (1), 3–19. doi:10.1208/s12249-016-0485-3.

Masina, N., Choonara, Y.E., Kumar, P., du Toit, L.C., Govender, M., Indermun, S., Pillay, V., 2017. A review of the chemical modification techniques of starch. Carbohydr. Polym. 01448617. 157, 1226–1236. Available from: http://www.elsevier.com/wps/find/journaldescription.cws_home/405871/description#description. doi:10.1016/j.carbpol.2016.09.094.

Meng, L., Liu, H., Yu, L., Duan, Q., Chen, L., Liu, F., Shao, Z., Shi, K., Lin, X., 2019a. How water acting as both blowing agent and plasticizer affect on starch-based foam. Ind. Crops Prod. 134, 43–49.

Meng, X., Ji, J., Qi, X., Nie, X., 2019b. Effect of anticaking agents on hardening and Maillard-induced protein aggregation in high-protein nutrition bars formulated with whey protein concentrate. LWT 108, 261–267.

Mir, N.A., Riar, C.S., Singh, S., 2018. Nutritional constituents of pseudo cereals and their potential use in food systems: a review. Ind. Crops Prod. 75, 170–180.

Mohamed, I.O., 2023. Interaction of starch with some food macromolecules during the extrusion process and its effect on modulating physicochemical and digestible properties. A review. Carbohydr. Polym. Technol. Appl. 26668939. 5, 100294. doi:10.1016/j.carpta.2023.100294.

Mohammed, S., Dubey, P.K., Mishra, A.A., Rahman, S., 2024. Valorisation of jackfruit seed flour in extrusion and bakery products: a review. Food Sci. Biotechnol. 20926456. 33 (14), 3167–3180. Available from: https://www.springer.com/journal/10068. doi:10.1007/s10068-024-01665-2.

Moscicki, L., Mitrus, M., Wojtowicz, A., Oniszczuk, T., Rejak, A., 2013. Extrusion-Cooking of Starch. InTech. doi:10.5772/52323.

Mosibo, O.K., Ferrentino, G., Alam, M.R., Morozova, K., Scampicchio, M., 2022. Extrusion cooking of protein-based products: potentials and challenges. Crit. Rev. Food Sci. Nutr. 15497852. 62 (9), 2526–2547. Available from: https://www.tandfonline.com/loi/bfsn20. doi:10.1080/10408398.2020.1854674.

Nandi, U., Trivedi, V., Ross, S.A., Douroumis, D., 2021. Advances in twin-screw granulation processing. Pharmaceutics. 19994923. 13 (5). Available from: https://www.mdpi.com/1999-4923/13/5/624/pdf. doi:10.3390/pharmaceutics13050624.

Navin Venketeish, K., Govindarajan, N., Pandiselvam, R., Mousavi Khaneghah, A., 2024. Unlocking the potential of underutilized legumes: nutritional benefits, anti-nutritional challenges, and mitigation strategies. J. Food Meas. Charact. 1–15.

Nussinovitch, A., Hirashima, M., 2023. Use of Hydrocolloids to Control Food Appearance, Flavor, Texture, and Nutrition. Wiley. doi:10.1002/9781119702139.

Obadi, M., Bin, X., 2024. A review of the effects of physical processing techniques on the characteristics of legume starches and their application in low-glycemic index foods. Int. J. Biol. Macromol. 01418130. 279 (Pt 1), 135124. doi:10.1016/j.ijbiomac.2024.135124.

Obadi, M., Xu, B., 2021. Review on the physicochemical properties, modifications, and applications of starches and its common modified forms used in noodle products. Food Hydrocoll. 0268005X. 112, 106286. doi:10.1016/j.foodhyd.2020.106286.

Okoronkwo, N., Mgbakogu, C., Mbaeyi-Nwaoha, I., 2019. Production and evaluation of breakfast cereals from rice, African yam-bean and orange-fleshed sweet potato. Asian Food Sci. J. 11, 1–17.

Olawoye, B., Jolayemi, O.S., Akinyemi, T.Y., Nwaogu, M., Oluwajuyitan, T.D., Popoola-Akinola, O.O., Fagbohun, O.F., Akanbi, C.T., 2023. Starch: Advances in Modifications, Technologies and Applications, Springer International Publishing.

Onwulata, C.I., Tunick, M.H., Thomas-Gahring, A.E., 2014. Pasting and extrusion properties of mixed carbohydrate and whey protein isolate matrices. J. Food Process. Preserv. 38, 1577–1591.

Ostrikov, A.N., Vasilenko, V.N., Frolova, L.N., Kochkin, I.B., Eremin, I.D., Mathematical model of the extrusion process of grain crops during non-isothermal flow of crop melting to the temperature at the beginning of the Maillard reaction. IOP Conference Series: Earth and Environmental Science. 1755–1307. **1052** (1) (2022), 012137, doi: 10.1088/1755-1315/1052/1/012137.

Oumeddour, D.Z., Lin, W., Lian, C., Zhao, L., Wang, X., Zhao, L., Guo, L., 2024. The Anti-Diabetic Effect of Non-Starch Polysaccharides Extracted from Wheat Beer on Diet/STZ-Induced Diabetic Mice. Foods. 23048158. 13 (17). Available from: www.mdpi.com/journal/foods. doi:10.3390/foods13172692.

Padmanabhan, M., Bhattacharya, M., 1989. Analysis of Pressure Drop in Extruder Dies. J. Food Sci. 0022-1147. 54 (3), 709–713. doi:10.1111/j.1365-2621.1989.tb04687.x.

Patil, S., Brennan, M.A., Mason, S., Brennan, C.S., 2017. Investigation of the combination of legumes and cereals in the development of extrudate snacks and its effect on physicochemical properties and in vitro starch digestion. J. Food Nutr. Res. 13368672. 56 (1), 32–41. Available from: http://www.vup.sk/en/.

Patil, S.S., Kaur, C., 2018. Current trends in extrusion: development of functional foods and novel ingredients. Food Sci. Technol. Res. 13446606. 24 (1), 23–34. Available from: http://www.jstage.jst.go.jp/browse/fstr/17/6/_contents. doi:10.3136/fstr.24.23.

Pedrosa, M.M., Guillamón, E., Arribas, C., 2021. Autoclaved and extruded legumes as a source of bioactive phytochemicals: a review. Foods 10, 379.

Person, T.J., Denn, M.M., 1997. The effect of die materials and pressure-dependent slip on the extrusion of linear low-density polyethylene. J. Rheol. 01486055. 41 (2), 249–265. Available from: http://scitation.aip.org/content/sor/journal/jor2. doi:10.1122/1.550877.

Philipp, C., Emin, M.A., Buckow, R., Silcock, P., Oey, I., 2018. Pea protein-fortified extruded snacks: linking melt viscosity and glass transition temperature with expansion behavior. J. Food Eng. 02608774. 217, 93–100. Available from: http://www.sciencedirect.com/science/journal/02608774. doi:10.1016/j.jfoodeng.2017.08.022.

Plattner, B., 2020. Extrusion techniques for meat analogues. Cereal Foods World 65 (4), 0043.

Qi, M., Jiang, L., Song, J., Han, F., Xu, M., Li, Y., Ma, C., Chen, S., Li, H., 2024. Investigating the impact of moisture levels on structural alterations and physicochemical properties of cassava flour through extrusion: a comprehensive study. Plant Foods Hum. Nutr. 0921-9668. 79 (4), 909–914. doi:10.1007/s11130-024-01237-z.

Rafiq, A., Sharma, S., Singh, B., 2018. Effect of pregelatination on rheology, cooking and antioxidant activity of pasta. J. Food Sci. Technol. 55, 1756–1766.

Rostamabadi, H., Nowacka, M., Kumar, Y., Xu, S., Colussi, R., Frasson, S.F., Singh, S.K., Falsafi, S.R., 2024. Green modification techniques: sustainable approaches to induce novel physicochemical and technofunctional attributes in legume starches. Trends Food Sci. Technol. 09242244. 146. Available from: https://www.sciencedirect.com/science/journal/09242244. doi:10.1016/j.tifs.2024.104389.

Rough, S.L., Wilson, D.I., 2005. Extrudate fracture and spheronisation of microcrystalline cellulose pastes. J. Mater. Sci. 0022-2461. 40 (16), 4199–4219. doi:10.1007/s10853-005-2840-8.

Saitkulov, F., Azimov, I., Ergasheva, M., Jo'Raqulov, H., 2022. Carbohydrates are the main source of energy in the body. Solution Social Problems Manag. Econ. 1, 68–71.

Sánchez-Aldana Villarruel, D., Ramos, M.Y.L., Galicia García, T., Meléndez, R.M., Talamás Abbud, R., 2021. Physicochemical Properties of Products Made from Mixtures of Corn and Legumes. Informa UK Limited, pp. 61–93. doi:10.1201/9781003020004-3.

Sasanam, S., Rungsardthong, V., Thumthanaruk, B., Wijuntamook, S., Rattananupap, V., Vatanyoopaisarn, S., Puttanlek, C., Uttapap, D., Mussatto, S.I., 2021. Properties and volatile profile of process flavorings prepared from d-xylose with glycine, alanine or valine by direct extrusion method. Food Biosci. 22124292. 44, 101371. doi:10.1016/j.fbio.2021.101371.

Sasanam, S., Thumthanaruk, B., Wijuntamook, S., Rattananupap, V., Vatanyoopaisarn, S., Puttanlek, C., Uttapap, D., Mussatto, S.I., Rungsardthong, V., Arshad, M.S., 2023. Extrusion of process flavorings from methionine and dextrose using modified starch as a carrier. Plos one. 1932-6203. 18 (2), e0269857. doi:10.1371/journal.pone.0269857.

Schirmer, M., Jekle, M., Becker, T., 2015. Starch gelatinization and its complexity for analysis. Starch-Stärke 67, 30–41.

Schmid, E.M., Farahnaky, A., Adhikari, B., Torley, P.J., 2022. High moisture extrusion cooking of meat analogs: a review of mechanisms of protein texturization. Compr. Rev. Food Sci. Food Saf. 15414337. 21 (6), 4573–4609. Available from: http://onlinelibrary.wiley.com/journal/10.1111/(ISSN)1541-4337. doi:10.1111/1541-4337.13030.

Schmid, V., Trabert, A., Schäfer, J., Bunzel, M., Karbstein, H.P., Emin, M.A., 2020. Modification of apple pomace by extrusion processing: studies on the composition, polymer structures, and functional properties. Foods. 23048158. 9 (10). Available from: https://www.mdpi.com/2304-8158/9/10/1385. doi:10.3390/foods9101385.

Selling, G.W., 2010. The effect of extrusion processing on Zein. Polym. Degrad. Stab. 01413910. 95 (12), 2241–2249. doi:10.1016/j.polymdegradstab.2010.09.013.

Serna Saldívar, S.O., Hernández, D.S., 2020. Dietary Fiber in Cereals, Legumes, Pseudocereals and Other Seeds Food Engineering Series. Springer, Mexico, pp. 87–122. Available from https://springer.com/series/5996. doi:10.1007/978-3-030-38654-2_5.

Sethy, K., Mishra, S., Mohanty, P.P., Agarawal, J., Meher, P., Satapathy, D., Sahoo, J.K., Panda, S., Nayak, S.M., 2015. An overview of non starch polysaccharide. J. Anim. Nutr. Physiol. 1, 17–22.

Shakoor, A., Zhang, C., Xie, J., Yang, X., 2022. Maillard reaction chemistry in formation of critical intermediates and flavour compounds and their antioxidant properties. Food Chem. 03088146. 393, 133416. doi:10.1016/j.foodchem.2022.133416.

Singh, N., Kaur, A., Shevkani, K., 2014. Maize: grain structure, composition, milling, and starch characteristics. In: Chaudhary, D., Kumar, S., Langyan, S. (Eds.), Maize: Nutrition Dynamics and Novel Uses. Springer, New Delhi, pp. 65–76. doi:10.1007/978-81-322-1623-0_5.

Singh, S., Gamlath, S., Wakeling, L., 2007. Nutritional aspects of food extrusion: a review. Int. J. Food Sci. Technol. 42, 916–929.

Sule, S., Okafor, G.I., Momoh, O.C., Gbaa, S.T., Amonyeze, A.O., 2024. Applications of food extrusion technology. MOJ Food Process. Technol. 2381182X. 12 (1), 74–84. doi:10.15406/mojfpt.2024.12.00301.

Tauferova, A., Pospiech, M., Javurkova, Z., Tremlova, B., Dordevic, D., Jancikova, S., Tesikova, K., Zdarsky, M., Vitez, T., Vitezova, M., 2021. Plant Byproducts as Part of Edible Coatings: a Case Study with Parsley, Grape and Blueberry Pomace. Polymers. 2073-4360. 13 (15), 2578. doi:10.3390/polym13152578.

Tedeschi, G., Benitez, J.J., Ceseracciu, L., Dastmalchi, K., Itin, B., Stark, R.E., Heredia, A., Athanassiou, A., Heredia-Guerrero, J.A., 2018. Sustainable Fabrication of Plant

Cuticle-Like Packaging Films from Tomato Pomace Agro-Waste, Beeswax, and Alginate. ACS Sustain. Chem. Eng. 21680485. 6 (11), 14955–14966. Available from: http://pubs.acs.org/journal/ascecg. doi:10.1021/acssuschemeng.8b03450.

Thymi, S., Krokida, M.K., Pappa, A., Maroulis, Z.B., 2005. Structural properties of extruded corn starch. J. Food Eng. 02608774. 68 (4), 519–526. doi:10.1016/j.jfoodeng.2004.07.002.

Trigui, K., De Loubens, C., Magnin, A., Putaux, J.-L., Boufi, S., 2020. Cellulose nanofibrils prepared by twin-screw extrusion: effect of the fiber pretreatment on the fibrillation efficiency. Carbohydr. Polym. 01448617. 240, 116342. doi:10.1016/j.carbpol.2020.116342.

Vamadevan, V., Bertoft, E., 2020. Observations on the impact of amylopectin and amylose structure on the swelling of starch granules. Food Hydrocoll. 0268005X. 103, 105663. doi:10.1016/j.foodhyd.2020.105663.

van der Sman, R.G.M., Vollebregt, H.M., Meinders, M.B.J., Beri, A., 2018. Effects of filler ingredients on the structure and texture of starchy, extruded snacks. Food Struct. 22133291. 18, 1–13. doi:10.1016/j.foostr.2018.10.001.

Wang, N., Dai, J., Miao, D., Li, C., Yang, X., Shu, Q., Zhang, Y., Dai, Y., Hou, H., Xu, S., 2023. Influence of enzymatic modification on the basis of improved extrusion cooking technology (IECT) on the structure and properties of corn starch. Int. J. Biol. Macromol. 01418130. 253 (Pt 5), 127274. doi:10.1016/j.ijbiomac.2023.127274.

Wibawa, C., Huang, Y., Patterson, D.H., FENG, Z., Serventi, L., 2023. Carbohydrates for Energy. Sustainable Food Innovation, Springer.

Wolf, B., 2010. Polysaccharide functionality through extrusion processing. Curr. Opin. Colloid Interface Sci. 13590294. 15 (1-2), 50–54. doi:10.1016/j.cocis.2009.11.011.

Xiao, Q., Ye, S., Wang, H., Xing, S., Zhu, W., Zhang, H., Zhu, J., Pu, C., Zhao, D., Zhou, Q., Wang, J., Lin, L., Liang, D., Lv, X., 2024. Soluble sugar, organic acid and phenolic composition and flavor evaluation of plum fruits. Food Chem. X. 25901575. (24), 101790. doi:10.1016/j.fochx.2024.101790.

Xu, E., Campanella, O.H., Ye, X., Jin, Z., Liu, D., BeMiller, J.N., 2020. Advances in conversion of natural biopolymers: a reactive extrusion (REX)–enzyme-combined strategy for starch/protein-based food processing. Trends Food Sci. Technol. 09242244. 99, 167–180. Available from: http://www.elsevier.com/wps/find/journaldescription.cws_home/601278/description#description. doi:10.1016/j.tifs.2020.02.018.

Yan, X., Ye, R., Chen, Y., 2015. Blasting extrusion processing: the increase of soluble dietary fiber content and extraction of soluble-fiber polysaccharides from wheat bran. Food Chem. 180, 106–115.

Yang, Y., Deng, Q., Jia, X., Shi, J., Wan, C., Zhou, Q., 2021. Qiang Wang, Characterization of key odorants in peeled and unpeeled flaxseed powders using solvent-assisted flavor evaporation and odor activity value calculation. LWT. 00236438. 138, 110724. doi:10.1016/j.lwt.2020.110724.

Yi, C., Qiang, N., Zhu, H., Xiao, Q., Li, Z., 2022. Extrusion processing: a strategy for improving the functional components, physicochemical properties, and health benefits of whole grains. Food Res. Int. 09639969. 160, 111681. doi:10.1016/j.foodres.2022.111681.

Yu, J., Xu, Y., Tu, X., Xia, S., Xue, Y., Xue, C., 2024. Effects of high temperature, high pressure, and high shear during extrusion on maize starch-fish protein extrudates: based on physical properties and multiscale structure. Food Chem. 142364.

Zhang, B., Qiao, D., Zhao, S., Lin, Q., Wang, J., Xie, F., 2021. Starch-based food matrices containing protein: recent understanding of morphology, structure, and properties. Trends Food Sci. Technol. 09242244. 114, 212–231. doi:10.1016/j.tifs.2021.05.033.

Zhang, D., Wang, L., Tan, B., Zhang, W., 2020. Dietary fibre extracted from different types of whole grains and beans: a comparative study. Int. J. Food Sci. Technol. 0950-5423. 55 (5), 2188–2196. doi:10.1111/ijfs.14472.

Zhang, J., Liu, P., Wu, A., Song, Y., Li, Q., Liao, X., Zhao, J., 2024. Towards understanding pectin-protein interaction and the role of pectin in plant-based meat analogs constructing. LWT. 00236438. 202, 116325. doi:10.1016/j.lwt.2024.116325.

Zhang, Z., Zhang, L., He, S., Li, X., Jin, R., Liu, Q., Chen, S., Sun, H., 2023. High-moisture Extrusion Technology Application in the Processing of Textured Plant Protein Meat Analogues: a Review. Food Rev. Int. 8755-9129. 39 (8), 4873–4908. doi:10.1080/87559129.2021.2024223.

Zhao, Y.H., Manthey, F.A., Chang, S.K.C., Hou, H.J., Yuan, S.H., 2005. Quality characteristics of spaghetti as affected by green and yellow pea, lentil, and chickpea flours. J. Food Sci. 00221147. 70 (6), s371. Available from: http://onlinelibrary.wiley.com/journal/10.1111/(ISSN)1750-3841. doi:10.1111/j.1365-2621.2005.tb11458.x.

Zhao, Y., Zhao, C., Tang, X., Zhou, J., Li, H., Zhang, H., Liu, J., 2021. Physicochemical properties and microstructure of corn flour–cellulose fiber extrudates. Food Sci. Nutr. 2048-7177. 9 (5), 2497–2507. doi:10.1002/fsn3.2195.

Zhihua, S., Li, S., Jian, Y., Jiang, S., Zhang, S., Yu, L., 2024. Research on cutting force of extrusion cutting process based on material flow characteristics. J. Manuf. Process. 122, 83–96.

Role of protein in extrusion

9

Srijita Banerjee[a], *Poonam Singha*[a], *Sushil Kumar Singh*[a], *and Kasivishwanathan Muthukumarappan*[b]
[a]Department of Food Process Engineering, National Institute of Technology Rourkela, Rourkela, Odisha, India
[b]Agricultural and Biosystems Engineering, South Dakota State University, Brookings, SD, United States

9.1 Introduction

"Extrusion" originates from the Latin word "extrude," meaning to "force or thrust out." Extrusion is a process that involves forcing a material to flow through a die or aperture with the appropriate cross section under high pressure and temperature. It combines a number of unit operations, all of which operate on the same basis. First, in the hopper, raw materials are fed into the extruder barrel, whereupon the food is conveyed by the screw. Inside the barrel, there are small flights that reduce the volume of food and make movement more difficult. The material is squeezed and kneaded by the screw as it moves forward in the barrel to form a plasticized, semisolid mass from a molten mass. Here, the heat from friction causes the temperature to rise quickly. It is then moved to the area where it is subjected to more pressure and shearing by the tiniest flights. Ultimately, under pressure, it is forced through dies at the barrel's discharge end. As the moisture is released as steam, the resultant product expands to take on its final shape and cools quickly. By adjusting the die, a range of shapes can be produced. The final product undergoes modifications as a result of kneading, mixing, shaping, shearing, heat and mass transfer, caramelization, cooking, and plasticizing, creating partially or fully cooked food with the least amount of nutritional loss. Extrusion cooking is a combination of mechanical and thermal processes by which starch-rich and/or protein ingredients are plasticized and cooked to obtain a predefined shape. It is a high-temperature, short-time approach that modifies size reduction, texture, and fluid flow. It is also known as "food extrusion" in the food processing industry (Prabha et al., 2021). Thus an extruder produces a new, distinct product when the organic macromolecules present, mainly fiber, starch, and protein, undergo physical and chemical transformation (Singha et al., 2019). A plasticized continuous mass is produced as a result of the thermomechanical action that occurs during extrusion, which also denatures protein, gelatinizes starch, and inactivates enzymes, microorganisms, and several anti-nutritional agents. All of these processes take place in a shear environment. The qualities of the raw materials and the

Extrusion Cooking for Food and Feed. https://doi.org/10.1016/B978-0-443-29987-2.00009-9

process parameters employed have a significant impact on the rate and degree of mixing, heating, compressing, and shearing of the materials inside the barrel and eventually the die (Singha and Muthukumarappan, 2017).

9.2 Interactions during extrusion

Depending on the exposed residues, proteins can either impede or enhance the retrogradation of starch (Fig. 9.1). Hydrophobic groups limit the release of amylose and its reassociation, charged residues enhance charge–dipole interactions between protein and starch-bound phosphate, and hydrophilic groups affect water and molecular mobility. Glycosidic connections produced during high-temperature processing between starch and protein may inhibit starch retrogradation, while covalent links (disulfide linkages) between proteins may accelerate starch retrogradation (Scott and Awika, 2023). In the mixing zone, protein structures were mostly retained by hydrophobic interactions and hydrogen bonds. It was observed that fatty acids improved hydrophobic interactions but decreased hydrogen bonds. The fatty acids exposed hydrophobic areas on proteins and altered the distribution of water around them. Furthermore, they may impair protein–protein interactions by forming hydrogen bonds between their hydroxyl groups and carbonyl groups. Fatty acids aided in the transition of β-sheet and α-helix structures into β-turn and random coil structures within the melting zone. Intermolecular aggregates that were formed were linked to an enhanced β-turn structural content, whereas more flexible protein and less compact structures were connected with random coils (Fig. 9.2) (Chen et al., 2023). The content of disulfide bonds rises, and surface hydrophobicity reduces during high-moisture extrusion. The disulfide bond concentration in the die is at its maximum throughout the entire extrusion process. This indicates that in the melting zone, protein aggregation takes place (Jiang et al., 2024).

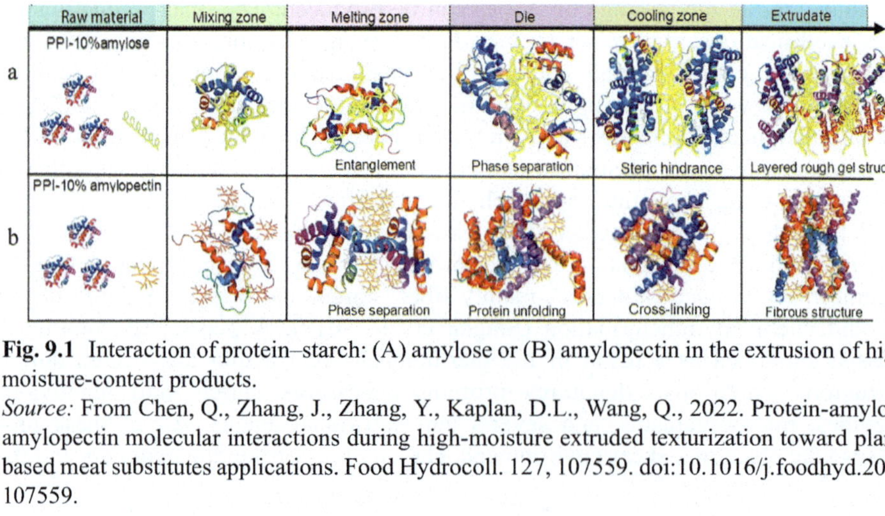

Fig. 9.1 Interaction of protein–starch: (A) amylose or (B) amylopectin in the extrusion of high-moisture-content products.
Source: From Chen, Q., Zhang, J., Zhang, Y., Kaplan, D.L., Wang, Q., 2022. Protein-amylose/amylopectin molecular interactions during high-moisture extruded texturization toward plant-based meat substitutes applications. Food Hydrocoll. 127, 107559. doi:10.1016/j.foodhyd.2022.107559.

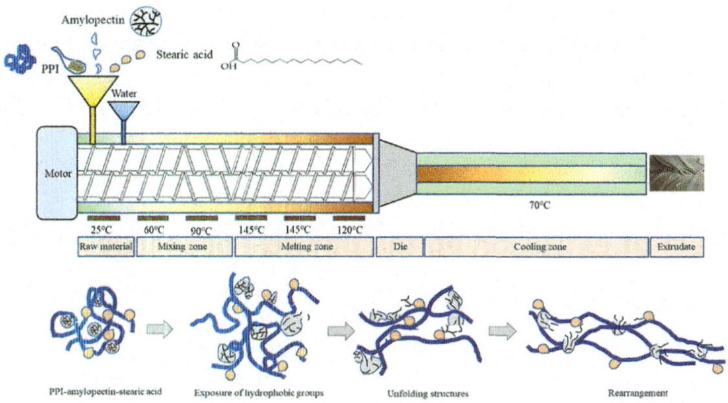

Fig. 9.2 Intermolecular interactions of protein (pea protein isolate), starch (amylopectin), and stearic acid in different zones of extrusion processing.
Source: From Chen Q, Zhang J, Liu H, Li T and Wang Q. 2023. Mechanism of high-moisture extruded protein fibrous structure formation based on the interactions among pea protein, amylopectin, and stearic acid. Food Hydrocoll 136, 108254. https://doi.org/10.1016/j.foodhyd.2022.108254.

9.3 Effect of extrusion on the chemical properties of protein

Protein solubility during and after processing is frequently considered a reliable predictor of the creation of cross-links during processing. Cross-linking reactions occur, and at high temperatures, it is possible that some covalent bonds form. Elevated temperature, increased screw speed, and increased moisture lead to an increase in insoluble extruded products, which is a sign of cross-linking. High temperatures during extrusion encourage the production of insoluble aggregates, similar to the fibrous structure of meat substitutes, through hydrophobic and electrostatic interactions. When proteins are denatured at high temperatures during the extrusion cooking process, antinutritional factors (lectins, antitrypsin factor, etc.) are inactivated, and digestibility is improved. While acidic conditions increase solubility but reduce the pasting characteristics of WPI, alkaline conditions greatly increase solubility and pasting properties. Essential amino acid availability, digestibility, and quantity all affect the nutritional value of protein. Excessively high drying temperatures have been found to reduce lysine bioavailability and protein digestibility. During extrusion, starch and protein constituents undergo slight molecular alterations. The protein fraction is mostly impacted by the drying stage. Drying at higher temperatures enables the protein network to be preserved even after an extended cooking time. At the molecular level, protein in-solubilization and aggregation increase with drying temperature. Furthermore, extremely high temperatures may encourage the creation of protein aggregates stabilized by irreversible protein connections, most likely through interpeptide cross-linking or Maillard-type reactions, depending on the drying parameters and moisture content employed. In cooked pasta, structural alterations, particularly those at high drying temperatures, may fortify the protein network and better protect starch from enzymatic destruction. The extrusion process is impacted by two factors that are primarily responsible for the changes in the alveogram characteristics: modification of starch and degradation of the gluten matrix. As the intensity of the treatment increases, the degradation of the gluten matrix also increases (Maurya, 2014).

9.4 Effect of extrusion on the physical properties of protein

The physical properties include texture, density, expansion ratio, and microstructure. It is important to optimize the parameters of extrusion for desired product characteristics by analyzing raw materials, parameters (such as the temperature of extrusion, moisture content of the feed, and speed of the screw), and the phenomena of expansion.

9.4.1 Expansion

By dividing the extrudates' radial diameter by the die nozzle diameter (Fig. 9.3), the extrudates' expansion ratio can be computed (Singha et al., 2018). Extruded foods should be extremely enlarged, light, and puffy. Two often used terms to characterize extrudate expansion in distinct directions are expansion indices/ratios in the radial and

Fig. 9.3 Die head with different types of dies.

longitudinal orientations. Depending on how it is computed, radial expansion can occasionally be expressed as cross-sectional expansion. Extruded items rich in protein typically have a lower expansion index compared to items with higher starch content. For instance, under the processing conditions, adding whey protein, soybean flour, and pea protein to ingredients containing starch negatively impacted the extrudates' expansion ratio. It is observed that the density and total expansion of extruded puffed foods are inversely related. Thus it is typical to anticipate that the addition of protein will result in an increase in the extrudate's piece/bulk density, as the bulk density of a product refers to the changes and expansion in cell structure (Brishti et al., 2021).

9.4.2 Texture

Puffed extruded snacks possess several significant characteristics, including crispiness, crunchiness, and hardness. These properties are determined by the cellular structure fractures and the amount of force needed to break the product. Both low Young's modulus and low fracture stress are characteristics of an ideal puffed product. Hard food items are defined as those whose fracture stress is comparable to that of human teeth. Puffed snacks exhibit tactile sensory qualities due to the sequential collapse of their interior cells during fracture, which is how consumers perceive the texture of the food. The texture of extruded products is routinely investigated using mechanical, acoustic, and sensory panel examinations, with mechanical analysis being the most commonly employed method. Mechanical analysis studies typically use shear, bending, and compression tests, with the latter imitating chewing. When the sample is crushed, the applied force gradually increases until it fractures. When a fracture ends, the force decreases but resumes when another fracture occurs. This simulates the fractures that happen during food chewing. Crispiness is determined by the number of fracture events during compression, while maximum force corresponds to food hardness. It has been frequently shown that extrudate hardness and protein content are positively correlated. Since thinner cell walls break more quickly, greater hardness is likely linked to thicker cell walls and less expansion. It is also proposed that extrudates with a high protein content have tougher cell walls. However, extrudate hardness is not necessarily directly impacted by protein content, as factors such as microstructure, density, and expansion also play a role (Luo and Koksel, 2023). As a marker for the production of fibrous structures, the degree of texturization is defined as the ratio of the longitudinal to crosswise shear forces. Shear force indicates the internal texture alignment of a material (Brishti et al., 2021).

9.4.3 Solubility

One of the most significant functional characteristics of proteins is the nitrogen solubility index (NSI), which is intimately associated with the gelation, foaming, and emulsification of proteins. Shear force, pressure, and temperature all work together to modify the protein and alter the NSI during the extrusion process. The extruder's shear strength increases as the screw speed rises. Large molecule breakdown is facilitated by the increased molecular space structure caused by shear force, which makes water molecules easier to penetrate. Water is equally distributed throughout the material at the same moment as a result of the screw's stirring action. On the contrary, if the speed is too high, the macromolecular material's incomplete degradation and fragmentation, as well as the material's shortened action time in the extruder, will cause the NSI to begin to drop. Water acts as a lubricant in the extruder. Thus as the material's moisture content increases, it leads to a synergistic effect between the disulfide and hydrogen bonds, the hydrophobic interaction in protein and disulfide bonds, and protein elongation (Gao et al., 2022).

9.5 Need for modification of protein

Proteins must be altered to increase their nutritional value and digestibility while lowering their allergic and antinutritional effects. Plant proteins that have not been altered typically have restricted functional qualities or cannot be used in food processing systems (Table 9.1). Furthermore, protein modification makes it possible to efficiently isolate and use novel proteinaceous materials with unconventional origins. It also creates products with proteins that possess functional properties that are less susceptible to changes in temperature, pressure, pH, or ionic strength during the processing and digestion of food (Sun-Waterhouse et al., 2014).

9.5.1 Enzymatic modification of protein

Modification of protein is an important aspect to improve nutritional and functional properties. Enzymes are highly specific, and under mild conditions, they exhibit accelerated reaction rates. They are mainly used to alter the physicochemical characteristics of food. This is done primarily for plant proteins, as they have fewer functional properties compared to animal proteins. The plant protein solutions, mainly soy, become less viscous due to the action of proteolytic enzymes, which alter the proteins' surface activity and hydration characteristics. For example, enzymes are used to modify enzymatically extruded soy protein concentrates, through which a new product with intended functional characteristics can be obtained. Concentrates of extruded soy protein are a valuable starting material for the enzymatic synthesis of hydrolyzates. It appears that two concurrent stages of the process have a considerable influence on their composition and functional characteristics. First comes hydrolysis, which results in the passage of proteinaceous materials into the solution; then there is further breakdown of dissolved components through proteolysis. By limiting the extent of hydrolysis, one can obtain products with modified qualities that combine the characteristics of the protein hydrolyzate and the extruded material (Surówka et al., 2004).

Table 9.1 Effects of extrusion parameters on protein interactions and functional properties.

Type of extruder and configuration	**Protein type**	**Interaction**	**Findings**	**References**
Corotating ZSK 26 Mc twin-screw extruder, L/D ratio: 29	Soy protein isolate	Protein–protein	Formation of anisotropic structure	Wittek et al. (2021)
Co-rotating twin-screw extruder ZSK 26 Mc; screw diameter: 25.5 mm, L/D ratio: 29, die diameter: 3 mm	Whey protein concentrate	Carbonyl-amine compounds	Formation of whey protein–citrus pectin conjugate	Koch et al. (2017)
Co-rotating intermeshing twin-screw extruders; screw diameter: 3 cm, L/D ratio: 23, length: 69 cm	Isolated soy protein	Protein–protein, hydrophobic interactions, between hydrogen bonds, between noncovalent bonds, and covalent bonds	Effect of wheat gluten and type of extrusion on meat analog's physicochemical properties	Samard et al. (2019)
Corotating intermeshing twin screw extruder, L/D ratio: 40; screw diameter: 16 mm; length of barrel: 640 mm	Soy protein concentrate	Protein–starch	Effect of the dimension of die on fibrous plant protein	Guyony et al. (2022)
Corotating, intermeshing, twin-screw food extruder L/D ratio: 20	Soy protein isolate	Protein-protein	Effect of system parameters and the properties of the product at a wide range of moisture	Chen et al. (2010)
Lab-scale twin-screw extruder, L/D ratio: 44, barrel diameter: 11 mm	Soy protein concentrate	Electrostatic and hydrophobic interactions, interactions between hydrogen and disulfide bonds	Formation of fibrous blend of soy protein and wheat gluten	Zhang et al. (2022)

Continued

Table 9.1 Effects of extrusion parameters on protein interactions and functional properties—cont'd

Type of extruder and configuration	Protein type	Interaction	Findings	References
Laboratory-scale corotating twin-screw extruder L/D ratio: 25, length: 40 cm, and diameter: 1.6 cm	Protein concentrates or isolates of pulses from different sources (lentil, white lupin, faba bean, and blue lupin) and flours from pseudocereals, namely buckwheat, amaranth, and quinoa	Protein–starch	Study on the sensory, structural, and functional properties of extruded products	Martin et al. (2022)
Co-rotating intermeshing twin-screw extruder, L/D ratio: 40, barrel diameter: 11 mm, and length: 440 mm	Lentil seeds, mung beans, peas, and dehulled pea seeds, soy protein isolates	Protein–protein	Effect on the composition and structure of pulse globulin protein	Shrestha et al. (2024)
Corotating twin-screw extruder, along with supercritical carbon dioxide injection, L/D ratio of screw: 28.5 is adjusted for supercritical carbon dioxide injection into the barrel with an L/D ratio of 24	Milk protein concentrate	Protein–protein	Effect of supercritical fluid extrusion on the physicochemical properties of milk protein extrudates	Arora et al. (2021)

9.5.2 Heat treatment

It is a type of physical modification. Heat treatment causes the peptide chains to become more mobile over time. This can result in the breaking of electrostatic, hydrogen, and disulfide bonds, as well as inter- and intramolecular hydrophobic interactions. When this happens, protein denaturation is exceeded, causing an initial, reversible unfolding that eventually becomes permanent, especially at higher temperatures. The structure of the protein molecule, mainly secondary and tertiary, is lost along with this irreversible denaturation. When this occurs, the protein's hydrophobic core is typically exposed, and new, enhanced crosslinking through interactions, namely, hydrophobic, disulfide, or sulfhydryl and hydrogen bonding, occurs. The unfolding and permanent denaturation of the proteins, together with the corresponding new form of aggregations, occur with heat treatment, especially above the denaturation temperature. When proteins are heated, their solubility is usually decreased due to aggregation as opposed to their native state (Akharume et al., 2021).

9.5.3 Chemically modified protein

The creation of modified proteins has made chemical protein modification an indispensable tool. The concurrent application of chemical and genetic techniques has produced a vast toolkit that facilitates the creation of nearly infinite protein constructions with residues that are either.

Naturally occurring or artificially altered, a number of key cellular activities, such as transport, differentiation, migration, and signaling, depend on these modifications, which include methylation, acylation, phosphorylation, farnesylation, glycosylation, sulfation, and ubiquitination (Boutureira and Bernardes, 2015).

9.5.4 Glycosylation

Protein molecules' hydroxyl bonds with amino acids are browned nonenzymatically during the modification process known as glycosylation modification, which occurs under weakly alkaline conditions. Polysaccharides or reducing sugars contain reducing groups that covalently bind to the amino residue in the backbone of proteins during the process of glycation. Protein physicochemical characteristics, such as film-forming and emulsifying capabilities, can be enhanced via glycosylation modification (Zheng et al., 2024).

9.5.5 Acylation

The chemical alteration of proteins, known as acylation, occurs when acylation groups react with ε-amino groups (lysine and arginine), aliphatic hydroxyl groups (serine and threonine), and other nucleophilic groups, such as phenolic groups (tyrosine). The result is the conversion of an amino group into an amide (Abedi and Pourmohammadi, 2021).

9.5.6 Phosphorylation

Reversible phosphorylation modifies the folding and activity of proteins, controlling vital processes such as enzyme activity, signal transduction, and cell division. One of the most important and common posttranslational modifications (PTMs) is phosphorylation. Because of its low stoichiometry and extremely dynamic nature, protein phosphorylation analysis has proven to be one of the most difficult processes, despite its relevance (Paradela and Albar, 2008). Target proteins' regulatory domains contain certain amino acids that can undergo PTMs that affect the stability of the protein. PTMs, which function as signals to either accelerate protein degradation (PTM-activated degrons) or stabilize a protein by preventing degradation (PTM-inactivated degrons), frequently regulate these areas, also known as degrons (Lee et al., 2023).

9.5.7 Deamidation

An organic molecule undergoes a chemical process called deamidation to remove its amide functional group. As a posttranslational alteration, deamidation of proteins can take place in vivo or in vitro, resulting in the conversion of asparagine residues to aspartic acid and glutamic acid, respectively (Hao et al., 2017). The isoelectric point and charge of a protein are significantly impacted by deamidation, another type of chemical alteration (Mirmoghtadaie et al., 2016).

9.6 Sources of proteins

9.6.1 Plant protein

These days, research into novel plant proteins for use as meat substitutes is receiving more attention. The most common sources of protein in meat substitutes are pulses, legume seeds, wheat gluten, soy proteins, and pea proteins. They are favored due to their widespread availability, low cost, high nutritional value, and simplicity in processing. Moreover, to improve the textural, structural, and sensory qualities of meat analog formulations, rapeseed, hemp, oat, rice, and lupin proteins can be incorporated. Beyond conventional protein sources, recent research has focused on developing innovative meat substitutes derived from mushrooms and mycoproteins. These alternatives have been evaluated for functionality and advancements in nutritional, structural, and sensory aspects in an effort to increase consumer acceptability (Dinali et al., 2024). For the majority of people, legumes provide a convenient, abundant, and inexpensive source of carbohydrates, protein, vitamins, fiber, and minerals, making them a great choice for human nutrition. The protein that has been processed and researched the most is soybean protein. Common beans are the main available source of vegetable protein in underdeveloped nations. In addition, peas are a highly nutritious meal that can be incorporated into a variety of products to increase protein intake and, thus, consumption. Other important sources of protein include chickpeas and pigeon peas (Sá et al., 2020). In addition to bioactive substances, proteins from plant sources have been utilized in a variety of culinary applications, most notably in the creation of RTE snacks because of their increased protein content, greater nutritional value, reduced cost, and sustainability. They are extracted from a variety of foods, such as root vegetables, grains, legumes, oilseeds, pulses, and microalgae (Pavani et al., 2024).

9.6.2 Animal protein

Proteins, vitamins, minerals, fatty acids, and fiber are abundant in the majority of insect species. Thus insects serve as an excellent alternative source of high-quality proteins and nutrients. When it comes to protein content, certain insects like grasshoppers, termites, caterpillars, houseflies, and weevils surpass beef and chicken by weight. All around the world, protein-rich edible insects, particularly those belonging to *Orthoptera* (which includes locusts, crickets, and grasshoppers), are considered a viable substitute for meat (Akhtar and Isman, 2018).

9.6.3 Other sources

Using fishmeal as the primary protein source in feed is no longer sustainable, both ecologically and economically. Soy white flakes and high-protein distillers' dried grains are a high-quality alternative source of protein that can be used in aquaculture feed (Singh and Muthukumarappan, 2016). Microalgae, filamentous fungi, bacteria, and nonpathogenic yeasts can all be used to create microbial protein products. They are also excellent natural sources of vitamin B. Protein products derived from microbial biomass have a higher protein content than cereals and legumes, and they are of a quality similar to legumes when consumed by humans (Kuhad et al., 1997).

9.7 Supercritical fluid extrusion

Due to the inactivation of antinutritional agents, extrusion can also boost the bioavailability of specific nutrients, such as protein and iron. However, typical extrusion's high heat, low moisture, and high shear can destroy a significant quantity of nutrients that are heat-labile.

Water performs two functions in traditional steam-based extrusion processes. When it flashes into steam at the die exit, it functions as a blowing agent, while also acting as a plasticizer by forming a melt in the extruder barrel. While steam-based extrusion is commonly utilized in the processing of cereals, it typically requires harsh processing conditions, including high temperatures (over 130°C), strong shear (above 150 rpm screw speed), and low moisture levels (as low as about 10%), especially when expansion is needed. Certain proteins, vitamins, flavors, and bioactive compounds are among the heat-sensitive components that are

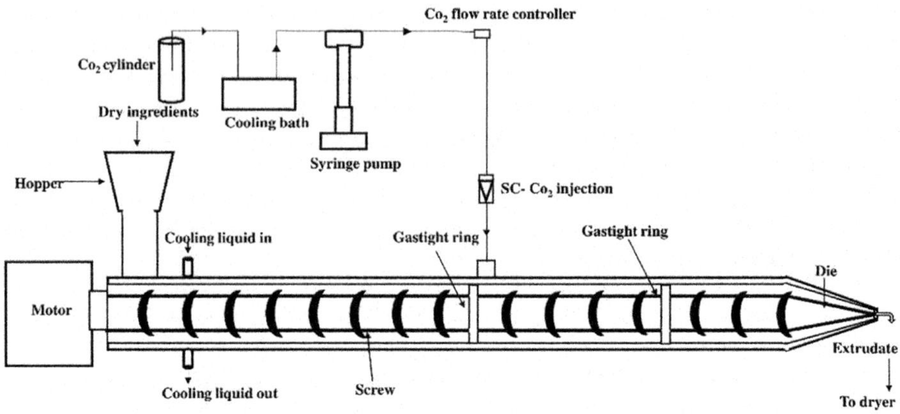

Fig. 9.4 Schematic diagram of supercritical fluid extrusion.

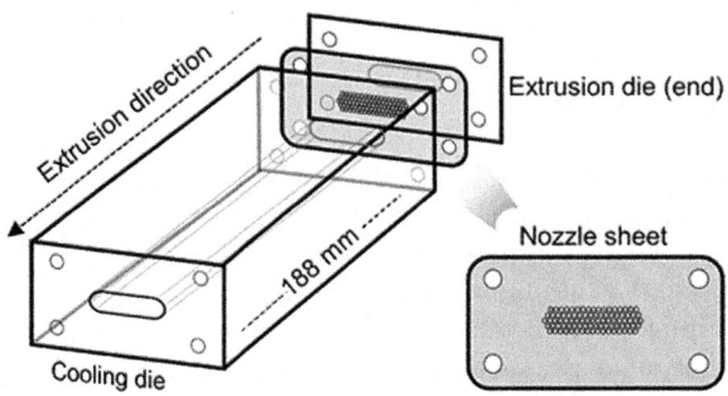

Fig. 9.5 Elements of a cooling die set: cooling die, extrusion die, and nozzle sheet.
Source: From Zheng, Y., Xu, J., Sun, C., Zhao, Y., Cao, Y., Lu, W., Zhang, Y., and Fang, Y. 2024. Multihole nozzle-mediated high-moisture extrusion of soy proteins into fiber-rich structures. Food Hydrocoll. 151, 109819. doi:10.1016/j.foodhyd.2024.109819.

frequently severely harmed by such extreme working conditions. In supercritical fluid extrusion (Fig. 9.4), supercritical carbon dioxide (SC-CO_2) is used in a blend process that combines extrusion with supercritical fluid technology. SC-CO2 acts as a plasticizer that lowers viscosity and as an agent that promotes expansion and foaming. SCFX uses low temperatures and shear, which minimize processing losses for heat-sensitive materials (Fig. 9.5). Moreover, SC-CO_2-assisted extrusion yields products with less explosive puffing than steam-based extrusion, resulting in a more uniform internal cell structure and a smoother surface. Better internal and external structures improve the end product's chemical stability and its sensory qualities, including reduced moisture absorption, less oxidative loss, better flavor, and improved nutrient retention. An example of this is the manufacturing of expanded rice products enhanced with proteins and heat-labile micronutrients (Paraman et al., 2012).

9.8 Conversion energy

Within the extrudate, conservative energy can be absorbed in two different ways: first, through the shear stress produced in pressurization chambers; and second, through the shear rate produced by the screw revolution rate and chambers where the screw and screw, or screw and barrel, are in close proximity to one another (Guy, 2001). They both produce thermal energy. Shear stress is preferable, as when a system is under pressure, it helps to forcefully hydrate the granules, whereas shear rate is detrimental to the starch and protein particles. The product absorbs the conversion energy in three different ways:

1. The product becomes pressurized as a result of the screws' conveying or pumping action. Higher pressures of 1500 psi are typically needed for highly proteinaceous products in order for moisture and thermal energy to completely penetrate the protein globules and for the protein matrix to form and extend at the die, forming a fibrous structure. While very low pressures are necessary for the expansion of starch granules, reaching a particular pressurization point for plasticized products is crucial for allowing the bubbles of steam produced at high pressure in the continuous medium to completely penetrate and impact all of the starch molecules in the matrix. In contrast, the inner matrix of the pellet may be under-sheared and under-cooked, resulting in the formation of a harder inner core and a skin on the extrudate during its exit from the die. Alternatively, variations in the extent of cooking may occur where the shear added at the die wall allows the pellet's exterior to be well-sheared, generating a continuous texture.
2. The energy utilized to raise the product's temperature inside the extruder is known as sensible heat. Placing a thermocouple to measure the product temperature away from the barrel and the surrounding environment makes it simple to ascertain this reaction. The composition of the ingredients and the liquid plasticizer's boiling point within the ingredient mix can affect this energy source.
3. The energy needed to propel a chemical reaction is known as the heat of reaction. Cooking refers to the quantity of energy absorbed by the final product's elements in order for the reaction to continue, as it is an endothermic process. The heat absorption equation can be as follows:

$$\frac{E_t}{m} = \int_{T_1}^{T_2} C_p \ dt + \int_{P_1}^{P_2} \frac{dp}{\sigma} + \Delta H^0 \ + \Delta H_{st}$$

where the total energy (E_t) absorbed per unit mass (m) is expressed as E_t/m; T is the temperature of the product; C_p is the product's specific heat; p is the product's density under pressure p; ΔH^0 is the heat that cooking reaction absorbs; H_{st} is the lipid fusion heat; and subscripts 1 and 2 denote the initial and final conditions.

Further generalized,

$$\frac{E_t}{m} = \int C_p \ dt + \Delta H^0$$

$$H = 14\ X_s + 95\ X_p$$

where X_s and X_p are the starch and protein fractions, respectively. A product's specific heat depends on its composition and temperature. This becomes important in an extrusion cooker that runs continuously because high temperatures are frequently obtained without much necessity. Based on a few experimental data, the formula can be more precise as:

$$C_p \approx 1.424\ X_c + 1.549\ X_p + 1.675\ X_f + 0.837\ X_a + 4.187\ X_m$$

where X is the weight of formula parts (X_c = carbohydrate c; X_p = protein p; X_f = fat f; X_a = ash a; X_m = moisture m) and C_p is the specific heat of the product in kJ/(kgK), which are further simplified to produce the following:

$$C_p \approx 1.71[X_m(4.187\ C_d)] + 0.00304T - 0.292$$

where T is the temperature in degrees Celsius, C_d is the dry matter's specific heat (kJ/kg·K), and X is the proportion of components by weight.

9.9 Extrusion and nutrition

Protein digestibility is increased during extrusion by denaturation, which makes the enzyme binding sites visible. Denaturation causes proteins, including enzymes and their inhibitors, to lose their active properties. A change in a protein's solubility in aqueous solutions or water is commonly used to measure the degree of denaturation. High-shear extrusion conditions make these changes more noticeable; however, mass moisture and temperature also have a significant impact. For instance, even at the comparatively moderate process temperatures required to make pasta, the solubility of wheat protein is decreased. Weaning foods, other specialty products, and nutritional assessments of extruded feeds have received significant attention because the majority of extruded foods are low in protein. Maillard reactions are facilitated by low moisture content and high barrel temperatures during extrusion. Reducing sugars, such as those produced when starch and sucrose are sheared, may react with lysine, reducing the nutritional value of protein. The limiting necessary amino acid found in grains is lysine, and further nutrient loss can stunt the growth of young animals and children. Combinations of soy isolates or concentrates, cornmeal, or full-fat soy flakes result in nutrient-dense components that can be reconstituted as gruel or porridge while maintaining a high lysine content. It was observed that Maillard reactions were enhanced by an acidic pH in a system containing lysine, glucose, and wheat starch. In addition, it has been determined that soy protein lowers cholesterol when added to food (Camire, 2001).

9.10 Applications of extrusion technology in protein based food products

9.10.1 *Meat analogs*

Meat substitutes or meat analogs, which are made entirely of proteins from plants, are all referred to as "plant-based meat alternatives". These products closely mimic the physicochemical characteristics of proteins from animals, even though they are derived from plants. Their main goal is to give customers a vegan product, that is, a meat-like sensory experience with similar fibrous muscle structure, equivalent moisture content, texture, and chewiness. The continuous need for plant-derived substitutes that meet the expectations of meat-eating customers for taste, texture, and appearance has made it difficult to achieve the desired fibers in meat analogs. Meat substitutes are typically made by combining largely plant-based whole materials (such as legumes, grains, and mushrooms) or processed ingredients (such as lipids, carbohydrates, vitamins, proteins, or minerals) (Dinali et al., 2024).

9.10.2 *Effects of extrusion on meat analogs*

Cutting force: It is an indirect indicator of hardness and fibrous structures, as it depends on the degree of hardness or texturization. The cutting force dramatically drops as the concentration of protein increases, suggesting a reduction in the formation of fibrous structures. Only at higher temperatures does the cutting force rise, as the proteins become more denatured at higher temperatures. The cutting force of meat analogues significantly increased with an increase in the speed of the screw (Fig. 9.6).

Yield of cooking: During cooking, water absorption impacts sensory qualities like juiciness. Hard structures can better hold water without dripping compared to soft structures. Increasing temperature, screw speed, and reducing water flow may lead to more insoluble protein clumps during extrusion. Improved water entrapment across protein networks leads to increased water-holding capacity at higher protein concentrations, potentially explaining the improved cooking output in extrudates (Singh and Sit, 2022).

9.10.3 *Fortification of extruded products*

One could argue that the extrusion procedure underutilizes legume flours and grains in snack food products. Chickpeas, lentils, and soybeans are examples of legumes that provide a good source of protein for humans, especially in regions of the world where access to meat and dairy products is limited due to allergies, availability, or moral concerns. The two main proteins present in legume seeds are albumin and globulin; globulins generate around 70% of the protein in legume seeds. In addition, legumes are a good source of minerals and vitamins. It is well recognized that the content of legume materials is important in reducing metabolic disorders, including diabetes and

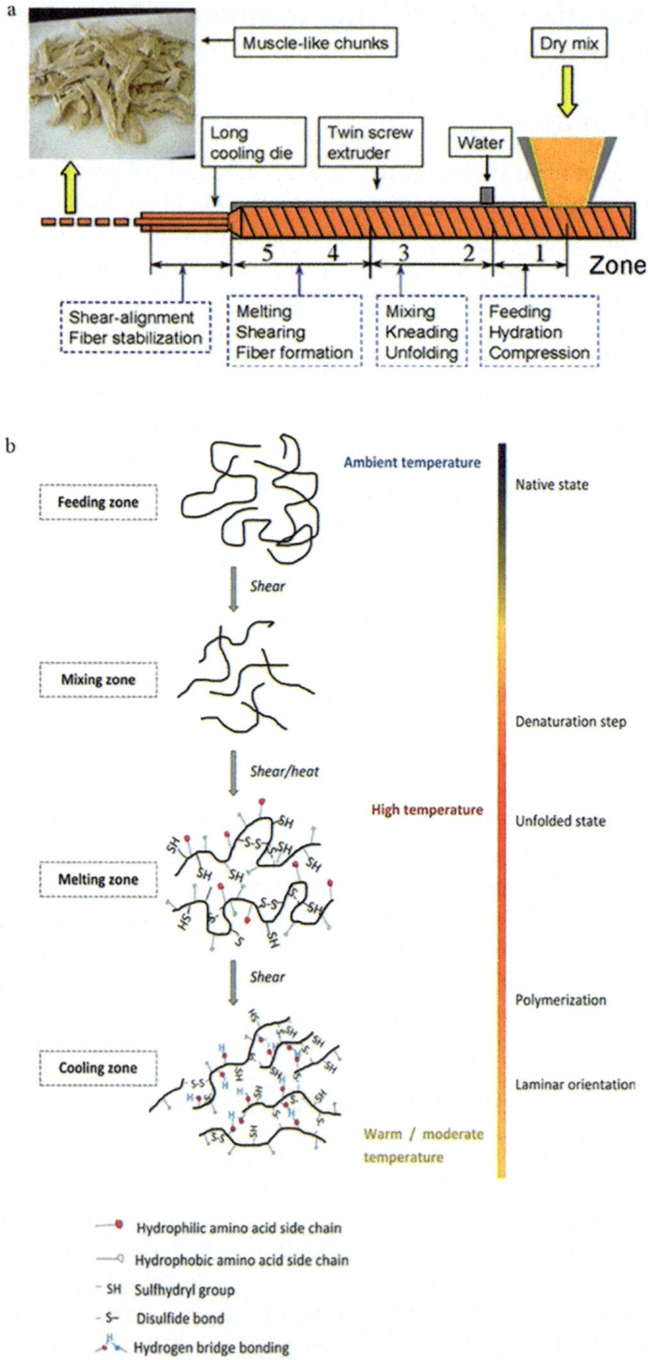

Fig. 9.6 (A) twin screw extruder for the production of meat analogues with high moisture content, and (B) alterations in native protein structures during mixing, heating, shearing, and cooling.

heart disease. Therefore it might be conceivable to create extruded snack products with higher protein levels and possibly even improved protein digestibility by mixing legume and wheat grain, respectively. Snack products can have more nutritional value when starch sources like maize, oat, barley, and wheat are mixed with protein sources like peas or beans. Cereals and legumes together can yield nutrient-dense goods. The impact of adding legume flour to conventional cereal-based flours and the addition of pigeon pea to biscuits made with wheat showed higher content of protein and possible nutritional benefits, such as an increase in protein digestibility. The addition of black gram flour to pasta increased its nutritional content, and processing food material lengthens its shelf life (Patil et al., 2016).

9.10.4 Protein functionality in extruded snack foods

9.10.4.1 Browning and toasting of extruded cereals

Caramelization and/or the Maillard reaction, which is a reaction between proteins or amines and carbohydrates, are involved in nonenzymatic browning. Condensation of a reducing sugar with an amino group (nonionized) is the first step in the Maillard reaction. These interactions yield brown pigments, which give toasted goods their color, as well as volatile ingredients with pleasant bakery flavors. When processing recipes for breakfast cereal, adding sugars, milk powder, and whey promotes Maillard browning. Furthermore, heat treatments at high temperatures significantly increase Maillard reactions since they have a large activation energy. It should be noted that the highest Maillard reaction rates occur at middle water activity levels (between 0.3 and 0.5). Cereal is toasted at a temperature of roughly 180°C during the initial drying phases, when the product's moisture level allows for high Maillard browning rates (Bouvier, 2001).

9.10.4.2 Snack foods

The main starch-rich components used in the recipes include potatoes, rice, wheat, and maize. These materials differ from one another in terms of composition, and tiny variations in their physical properties in the extruder, as well as notable variations in flavor and color, are caused by minor elements, including protein, lipids, carbohydrates, and fibers. Any of the components can be combined to create a starch spectrum, resulting in an inflated foam structure. Similar enlarged snack structures can be produced by varying the processing inputs as long as the starch amount is kept constant. Nonetheless, there are noticeable variations in extrudate color due to the substantial color disparities between the primary raw materials. By adding colors to the mix, these can be modified to uniform levels within specific color ranges. The primary raw materials, like rice, wheat, potatoes, and maize, also produce distinct flavors in snacks, each with

Source: From Dinali, Malsha, Liyanage, Rumesh, Silva, Mayumi, Newman, Lisa, Adhikari, Benu, Wijesekara, Isuru & Chandrapala, Jayani. (2024). Fibrous structure in plant-based meat: high-moisture extrusion factors and sensory attributes in production and storage. Food Rev. Int. 40 (9), 2940–2968. https://doi.org/10.1080/87559129.2024.2309593.

a unique character and intensity. In addition, processing affects them in terms of temperature and duration of exposure to the extruder's hot zones. Snack flavor can be changed by adding more flavoring to their coating during postextrusion procedures. Nonetheless, the snack's background flavor makes a substantial contribution to the total flavor and influences consumer preferences. Typically, the addition of flavor precursors to the raw material is mainly a mix of substances that are active, like peptides, amino acids, and reducing sugars. To create new flavors and colors, they go through the Maillard reaction and thermochemical processes in the extruder's high temperatures. They might include trace amounts of substances that are involved in these chemical interactions, which could alter their composition. For instance, the addition of milk solids will result in the introduction of a combination of proteins and peptides, as well as the reducing sugar lactose. These will go through Maillard reactions, giving the extrudates flavor and a reddish-brown coloration (Guy, 2001).

9.10.4.3 Effects of different physical methods on the functional properties of plant-based proteins

Because the underlying principles of different physical approaches change, so too do their effects on the functional qualities of proteins. Protein isolates have been successfully altered to increase their functionality, mostly by the use of cold plasma, ultrasound (US), high-pressure processing (HPP), pulsed electric field (PEF), and microwave. Protein oil or water holding capacity, emulsification, foaming qualities, solubility, and dispersion are all positively impacted by ultrasonication. The creation of bubbles during cavitation was thought to be the cause of the increase in solubility. This led to a rapid increase in localized pressure and temperature, which in turn caused protein unfolding, structural changes, enhanced protein–water interaction, and a decrease in the hydrodynamic diameter of globular proteins. US is known to primarily impact the foaming characteristics, solubility, and in vitro protein digestibility (IVPD) of legumes. It was observed that after ultrasonication, the foaming qualities and solubility of pea protein isolate and fava bean protein increased. Reduced surface tension, improved hydrophobicity, decreased apparent viscosity, and enhanced adsorption at the interface of oil and water were all suggested as causes of the improvement in solubility and foaming characteristics. Similar effects of the US on the functional characteristics of sunflower protein isolates have also been observed. It has been shown that protein solubility can occasionally be reduced by the addition of NaCl. It has been observed that applying PEF to protein isolates, such as sunflower protein isolate, alters the structures of proteins, both secondary and tertiary, leading to a reduced content of α-helix and an increase in random coils and β-sheets. The functional characteristics of the protein are directly correlated with these structural alterations. Generally speaking, higher voltage, residence duration, and frequency of pulses have a beneficial impact on the functional protein characteristics. HPP did not show a clear trend in emulsion characteristics or protein solubility. More research is required to demonstrate a definitive pattern for changes in emulsion and solubility qualities with the application of HPP. Solubility of protein, water holding capacity (WHC)/oil holding capacity (OHC), emulsifying, and foaming characteristics can all be increased by cold plasma. The negative impact of cold plasma on allergenicity and zeta-potential of proteins, which increases the range of applications for proteins, is one of its main advantages. Protein aggregation is the main effect of irradiation, such as radio and gamma frequencies, which reduces the functionality of the affected proteins (Goyal et al., 2024).

9.11 Advantages and disadvantages of extrusion

One of the most important methods for processing food is extrusion cooking, as it increases the shelf life of the product. It is carried out at high temperatures for a short duration of time. It destroys antinutrient enzymes and harmful microorganisms in a continuous process, which yields high productivity. In addition, it can also be used to produce nutrient-dense foods that address social requirements related to food and nutrition insecurity, as well as malnutrition. However, extrusion food processing has drawbacks, such as a high installation cost and the optimization of process variables like the configuration of the screw, moisture content, particle size of feed, feed rate, speed of the screw, temperature, and shape of the die to prevent the formation of reactive species. There should be trained personnel for handling, and proper training should be provided to them. Therefore the benefits of extrusion food processing greatly exceed the drawbacks (Egal and Oldewage-Theron, 2020).

9.12 Challenges and future perspectives

To obtain extrudates with the desired properties, it is important to consider factors beyond technological ones, such as the nature of the protein sources, the molecular structure, the initial composition of the material in terms of the percentage of lipids and proteins, and the physical characteristics of the material in terms of particle size, granulometry, and other factors. Utilizing online sensors, such as Raman, fluorescence, ultrasonic, and IR spectroscopy, mounted at the die's exit or barrel, can be an effective way to track the extrusion process and gather precise molecular data about the chemical composition, depending on the quality of the protein sources. This is a crucial factor to remember, particularly when working with proteins derived from insects. Currently, there is very little knowledge about the molecular properties of these proteins, and it is impossible to draw any conclusions about how their behavior is related to the extrusion conditions. This is an untapped market that must be further explored (Mosibo et al., 2022). The key benefits of extrusion include an increase in solubility, foaming capacity (FC), WHC, and IVPD. One of the main issues with the utilization of this technology is its limited control over parameters, which restricts its applicability (Goyal et al., 2024).

Three-dimensional (3D) food printing opens up new possibilities for creating visually appealing food structures without loss of nutritional value or sensory elements (taste, color, flavor, etc.) with the introduction of the digital culinary market. Even though 3D printing technology, which is based on extrusion, is thriving from a commercial and research standpoint, there are still issues with fine-tuning printing parameters, such as material compositions, printing accuracy, and postdeposition cooking to improve the goods' consumer appeal. The food industry is looking for new, high-value products with beneficial properties that can be made from plant-based byproducts such as bagasse, oilseed hulls, pomace, bran, fruit, and vegetable peels. The creation of functional foods can benefit from extrusion technology's ability to combine various basic ingredients into innovative foods. Food companies mainly utilize high-moisture extrusion technologies to create meat substitutes in response to the growing demand for healthy, cholesterol-free food options and the growing vegan movement. It is anticipated that further studies will improve the meaty flavor and sensory acceptability of plant-based meat substitute products. This is now seen as a challenge to the global marketing of such products. Lastly, it is important to identify sustainable methods in the food packaging industry, where extrusion technology excels in creating biodegradable packaging materials from biopolymers derived from food (Prabha et al., 2021).

9.13 Conclusion

Protein is an essential ingredient in extruded products, influencing both their nutritional content and textural qualities. As a **nutritional powerhouse**, protein is a crucial nutrient for building and maintaining body tissues. Extruded products enriched with protein offer a convenient way to boost dietary protein intake. While protein offers nutritional benefits, it also has a noticeable effect on the final product's physical characteristics or texture, as follows:

1. Higher protein content can result in a denser and firmer texture.
2. It can hinder expansion during extrusion, leading to a less airy and puffed product.
3. Extruded products with high protein content may also be more prone to cracking or breaking.

Therefore it is crucial to strike a balance between the amount of protein and other ingredients, such as starches, to achieve the desired textural properties and functionality in the final extruded product. This involves adding plant-based proteins to already-existing starch-based snacks, as well as fortifying them with bioactive substances and byproducts to improve their nutritional content and reduce health risks. By combining efficiency, versatility, and the ability to improve food safety and nutrition, extrusion has become an indispensable technology within the modern food processing industry.

References

Abedi, E., Pourmohammadi, K., 2021. Chemical modifications and their effects on gluten protein: an extensive review. Food Chem. 0308-8146. 343, 128398. doi:10.1016/j.foodchem.2020.128398.

Akharume, F.U., Aluko, R.E., Adedeji, A.A., 2021. Modification of plant proteins for improved functionality: a review. Compr. Rev. Food Sci. Food Saf. 1541-4337. 20 (1), 198–224.

Available from: http://onlinelibrary.wiley.com/journal/10.1111/(ISSN)1541-4337. doi:10.1111/1541-4337.12688.

Akhtar, Y., Isman, M.B., 2018. Insects as an Alternative Protein Source. Elsevier BV, pp. 263–288. doi:10.1016/b978-0-08-100722-8.00011-5.

Arora, B., Singha, P., Rizvi, S.S.H., 2021. Supercritical fluid extrusion: die design and physicochemical properties of milk protein extrudates. Innov. Food Sci. Emerg. Technol. 1466-8564. 68. Available from: http://www.elsevier.com/locate/ifset. doi:10.1016/j.ifset.2021.102637.

Boutureira, O., Bernardes, G.J.L., 2015. Advances in chemical protein modification. Chem. Rev. 1520-6890. 115 (5), 2174–2195. Available from: http://pubs.acs.org/journal/chreay. doi:10.1021/cr500399p.

Bouvier, J.-M., 2001. Breakfast Cereals. Elsevier BV, pp. 133–160. doi:10.1533/9781855736313.2.133.

Brishti, F.H., Chay, S.Y., Muhammad, K., Ismail-Fitry, M.R., Zarei, M., Saari, N., 2021. Texturized mung bean protein as a sustainable food source: effects of extrusion on its physical, textural and protein quality. Innov. Food Sci. Emerg. Technol. 1466-8564. 67. Available from: http://www.elsevier.com/locate/ifset. doi:10.1016/j.ifset.2020.102591.

Camire, M.E., 2001. Extrusion and Nutritional Quality. Elsevier BV, pp. 108–129. doi:10.1533/9781855736313.1.108.

Chen, F.L., Wei, Y.M., Zhang, B., Ojokoh, A.O., 2010. System parameters and product properties response of soybean protein extruded at wide moisture range. J. Food Eng. 0260-8774. 96 (2), 208–213. doi:10.1016/j.jfoodeng.2009.07.014.

Chen, Q., Zhang, J., Zhang, Y., Kaplan, D.L., and Wang, Q., 2022. Protein-amylose/amylopectin molecular interactions during high-moisture extruded texturization toward plant-based meat substitutes applications. Food Hydrocoll. 127, 107559. doi:10.1016/j.foodhyd.2022.107559.

Chen, Q., Zhang, J., Zhang, Y., Liu, H., Li, T., Wang, Q., Kaplan, D.L., 2023. Microscopic insight into the interactions between pea protein and fatty acids during high-moisture extrusion processing. Food Chem. 1873-7072. 404. Available from: https://www.elsevier.com/locate/foodchem. doi:10.1016/j.foodchem.2022.134176.

Dinali, M., Liyanage, R., Silva, M., Newman, L., Adhikari, B., Wijesekara, I., Chandrapala, J., 2024. Fibrous structure in plant-based meat: high-moisture extrusion factors and sensory attributes in production and storage. Food Rev. Int. 8755-9129. 40 (9), 2940–2968. doi:10.1080/87559129.2024.2309593.

Egal, A., Oldewage-Theron, W., 2020. Extruded food products and their potential impact on food and nutrition security. S. Afr. J. Clin. Nutr. 1607-0658. 33 (4), 142–143. Available from: https://www.tandfonline.com/loi/ojcn20. doi:10.1080/16070658.2019.1583043.

Gao, Y., Sun, Y., Zhang, Y., Sun, Y., Jin, T., 2022. Extrusion modification: effect of extrusion on the functional properties and structure of rice protein. Processes. 2227-9717. 10 (9), 1871. doi:10.3390/pr10091871.

Goyal, N., Thakur, R., Yadav, B.K., 2024. Physical approaches for modification of vegan protein sources: a review. Food Bioprocess Technol. 1935-5149. 17 (12), 4405–4428. Available from: https://www.springer.com/journal/11947. doi:10.1007/s11947-024-03368-2.

Guy, R., 2001. Extrusion Cooking: Technologies and Applications, Vol. 61. Woodhead publishing.

Guyony, V., Fayolle, F., Jury, V., 2022. Die dimensions impact on fibrous plant protein formation during high moisture extrusion. Appl. Food Res. 2772-5022. 2 (2), 100228. doi:10.1016/j.afres.2022.100228.

Hao, P., Adav, S.S., Gallart-Palau, X., Sze, S.K., 2017. Recent advances in mass spectrometric analysis of protein deamidation. Mass Spectrom. Rev. 1098-2787. 36 (6), 677–692.

Available from: http://onlinelibrary.wiley.com/journal/10.1002/(ISSN)1098-2787. doi:10.1002/mas.21491.

Jiang, W., Feng, J., Yang, X., Li, L., 2024. Structure of pea protein-based complexes on high-moisture extrusion: raw materials and extrusion zones. LWT. 0023-6438. 194, 115823. doi:10.1016/j.lwt.2024.115823.

Koch, L., Emin, M.A., Schuchmann, H.P., 2017. Influence of processing conditions on the formation of whey protein-citrus pectin conjugates in extrusion. J. Food Eng. 0260-8774. 193, 1–9. Available from: http://www.sciencedirect.com/science/journal/02608774. doi:10.1016/j.jfoodeng.2016.08.012.

Kuhad, R.C., Singh, A., Tripathi, K.K., Saxena, R.K., Eriksson, K.E.L., 1997. Microorganisms as an alternative source of protein. Nutr. Rev. 0029-6643. 55 (3), 65–75. Available from: http://nutritionreviews.oxfordjournals.org/. doi:10.1111/j.1753-4887.1997.tb01599.x.

Lee, J.M., Hammarén, H.M., Savitski, M.M., Baek, S.H., 2023. Control of protein stability by post-translational modifications. Nat. Commun. 2041-1723. 14 (1). Available from: https://www.nature.com/ncomms/. doi:10.1038/s41467-023-35795-8.

Luo, S., Koksel, F., 2023. Application of physical blowing agents in extrusion cooking of protein enriched snacks: effects on product expansion, microstructure, and texture. Trends Food Sci. Technol. 0924-2244. 133, 49–64. doi:10.1016/j.tifs.2023.01.012.

Martin, A., Schmidt, V., Osen, R., Bez, J., Ortner, E., Mittermaier, S., 2022. Texture, sensory properties and functionality of extruded snacks from pulses and pseudocereal proteins. J. Sci. Food Agric. 1097-0010. 102 (12), 5011–5021. Available from: http://onlinelibrary.wiley.com/journal/10.1002/(ISSN)1097-0010. doi:10.1002/jsfa.11041.

Maurya, A., 2014. Extrusion processing on physical and chemical properties of protein rich products-an overview. J. Bioresour. Eng. Technol. 2 (4), 61–67.

Mirmoghtadaie, L., Aliabadi, S.S., Hosseini, S.M., 2016. Recent approaches in physical modification of protein functionality. Food Chem. 1873-7072. 199, 619–627. Available from: https://www.elsevier.com/locate/foodchem. doi:10.1016/j.foodchem.2015.12.067.

Mosibo, O.K., Ferrentino, G., Alam, M.R., Morozova, K., Scampicchio, M., 2022. Extrusion cooking of protein-based products: potentials and challenges. Crit. Rev. Food Sci. Nutr. 1549-7852. 62 (9), 2526–2547. Available from: https://www.tandfonline.com/loi/bfsn20. doi:10.1080/10408398.2020.1854674.

Paradela, A., Albar, J.P., 2008. Advances in the analysis of protein phosphorylation. J. Proteome Res. 1535-3893. 7 (5), 1809–1818. doi:10.1021/pr7006544.

Paraman, I., Wagner, M.E., Rizvi, S.S.H., 2012. Micronutrient and protein-fortified whole grain puffed rice made by supercritical fluid extrusion. J. Agric. Food Chem. 1520-5118. 60 (44), 11188–11194. doi:10.1021/jf3034804.

Patil, S., Brennan, M., Mason, S., Brennan, C., 2016. The effects of fortification of legumes and extrusion on the protein digestibility of wheat based snack. Foods. 2304-8158. 5 (2), 26. doi:10.3390/foods5020026.

Pavani, M., Singha, P., Rajamanickam, D.T., Singh, S.K., 2024. Impact of extrusion processing on bioactive compound enriched plant-based extrudates: a comprehensive study and optimization using RSM and ANN-GA. Future Foods. 2666-8335. 9. Available from: https://www.sciencedirect.com/science/journal/26668335. doi:10.1016/j.fufo.2023.100286.

Prabha, K., Ghosh, P., S, A., Joseph, R.M., Krishnan, R., Rana, S.S., Pradhan, R.C., 2021. Recent development, challenges, and prospects of extrusion technology. Future Foods. 2666-8335. 3. Available from: https://www.journals.elsevier.com/future-foods. doi:10.1016/j.fufo.2021.100019.

Sá, A.G.A., Moreno, Y.M.F., Carciofi, B.A.M., 2020. Plant proteins as high-quality nutritional source for human diet. Trends Food Sci. Technol. 0924-2244. 97, 170–184. Available from: http://www.elsevier.com/wps/find/journaldescription.cws_home/601278/description#description. doi:10.1016/j.tifs.2020.01.011.

Samard, S., Gu, B.Y., Ryu, G.H., 2019. Effects of extrusion types, screw speed and addition of wheat gluten on physicochemical characteristics and cooking stability of meat analogues. J. Sci. Food Agric. 1097-0010. 99 (11), 4922–4931. Available from: http://onlinelibrary.wiley.com/journal/10.1002/(ISSN)1097-0010. doi:10.1002/jsfa.9722.

Scott, G., Awika, J.M., 2023. Effect of protein–starch interactions on starch retrogradation and implications for food product quality. Compr. Rev. Food Sci. Food Saf. 1541-4337. 22 (3), 2081–2111. Available from: http://onlinelibrary.wiley.com/journal/10.1111/(ISSN)1541-4337. doi:10.1111/1541-4337.13141.

Shrestha, S., Van't Hag, L., Haritos, V., Dhital, S., 2024. Fate of pulse globulin proteins molecular structure and composition on high moisture extrusion. Food Hydrocoll. 0268-005X. 149, 109512. doi:10.1016/j.foodhyd.2023.109512.

Singh, A., Sit, N., 2022. Meat analogues: types, methods of production and their effect on attributes of developed meat analogues. Food Bioprocess Technol. 1935-5130. 15 (12), 2664–2682. doi:10.1007/s11947-022-02859-4.

Singh, S.K., Muthukumarappan, K., 2016. Effect of feed moisture, extrusion temperature and screw speed on properties of soy white flakes based aquafeed: a response surface analysis. J. Sci. Food Agric. 1097-0010. 96 (6), 2220–2229. Available from: http://onlinelibrary.wiley.com/journal/10.1002/(ISSN)1097-0010. doi:10.1002/jsfa.7339.

Singha, P., Muthukumarappan, K., 2017. Effects of processing conditions on the system parameters during single screw extrusion of blend containing apple pomace. J. Food Process Eng. 1745-4530. 40 (4). Available from: http://www.blackwellpublishers.co.uk/journal_new.asp?site=1. doi:10.1111/jfpe.12513.

Singha, P., Singh, S.K., Muthukumarappan, K., 2019. Textural and structural characterization of extrudates from apple pomace, defatted soy flour and corn grits. J. Food Process Eng. 1745-4530. 42 (4). Available from: http://onlinelibrary.wiley.com/journal/10.1111/(ISSN)1745-4530. doi:10.1111/jfpe.13046.

Singha, P., Singh, S.K., Muthukumarappan, K., Krishnan, P., 2018. Physicochemical and nutritional properties of extrudates from food grade distiller's dried grains, garbanzo flour, and corn grits. Food Sci. Nutr. 2048-7177. 6 (7), 1914–1926. doi:10.1002/fsn3.769.

Sun-Waterhouse, D., Zhao, M., Waterhouse, G.I.N., 2014. Protein modification during ingredient preparation and food processing: approaches to improve food processability and nutrition. Food Bioprocess Technol. 1935-5149. 7 (7), 1853–1893. Available from: http://www.springer.com/life+sci/food+science/journal/11947. doi:10.1007/s11947-014-1326-6.

Surówka, K., Żmudziński, D., Surówka, J., 2004. Enzymic modification of extruded soy protein concentrates as a method of obtaining new functional food components. Trends Food Sci. Technol. 0924-2244. 15 (3-4), 153–160. doi:10.1016/j.tifs.2003.09.013.

Wittek, P., Zeiler, N., Karbstein, H.P., Emin, M.A., 2021. High moisture extrusion of soy protein: investigations on the formation of anisotropic product structure. Foods. 2304-8158. 10 (1). Available from: https://www.mdpi.com/2304-8158/10/1/102/pdf. doi:10.3390/foods10010102.

Zhang, X., Zhao, Y., Zhang, T., Zhang, Y., Jiang, L., Sui, X., 2022. High moisture extrusion of soy protein and wheat gluten blend: an underlying mechanism for the formation of fibrous structures. LWT. 0023-6438. 163, 113561. doi:10.1016/j.lwt.2022.113561.

Zheng, L., San, Y., Xing, Y., Regenstein, J.M., 2024. Rice proteins: a review of their extraction, modification techniques and applications. Int. J. Biol. Macromol. 1879-0003. 268. Available from: https://www.sciencedirect.com/science/journal/01418130. doi:10.1016/j.ijbiomac.2024.131705.

Role of fat and lipid in extrusion

10

Shubham Chandrakar and *R. Mahendran*
National Institute of Food Technology, Entrepreneurship and Management (NIFTEM-T), Thanjavur, India

10.1 Introduction

Lipids and fats have been fundamental to human life for centuries, serving multiple purposes across various industries, including food, cosmetics, biofuels, and pharmaceuticals (Cerone and Smith, 2021). In food science, lipids contribute to the sensory qualities of products (Rosa et al., 2020), act as a medium for heat transfer during cooking, and facilitate the transport of fat-soluble vitamins. In cosmetics, they are key components due to their emollient, moisturizing, and bioactive characteristics (Bowen-Forbes and Goldson-Barnaby, 2024; De Luca et al., 2021). Christie W.W. describes lipids as a diverse group of natural substances, including fatty acids and their derivatives, as well as steroids, terpenes, bile acids, and carotenoids. What they all have in common is that they dissolve easily in organic solvents like diethyl ether, hexane, chloroform, methanol, or benzene (Christie and Han, 2012).

Among the various lipids found in foods, triglycerides, also known as triacylglycerols, are the most common. As the name implies, a triglyceride consists of three fatty acid molecules bonded to one glycerol molecule (Singh et al., 2017). Food extrusion is the process of making food materials undergo various actions, such as melting, kneading, and shearing, before pushing them through a specially designed die to shape or expand the final product (Joy et al., 2012; Offiah et al., 2019). Fats and oils play three main roles in recipes for extrusion cooking: they serve as a valuable source of essential fatty acids, deliver fat-soluble vitamins, and improve the texture and taste of the final extruded product. In addition, fats aid in the extrusion process by acting as lubricants in the melt. As the screw applies shear force, the oils are either broken into small droplets or spread onto the polymers. When starch mixtures are heated and stirred in the barrel, amylose–lipid complexes form. Once the mixture cools, these complexes break apart and reorganize (Lazou, 2024; Singh et al., 2017).

10.2 Classification of lipids

Lipids are classified into various types based on composition, chain length, unsaturation, and saponification, as shown in Table 10.1. Based on composition, lipids are categorized as simple lipids, compound lipids, and derived lipids, all of which can be

Extrusion Cooking for Food and Feed. https://doi.org/10.1016/B978-0-443-29987-2.00010-5

Table 10.1 Classification of lipids based on composition, chain length, and saponification value.

Classification		Description
Based on composition		
1.	Simple lipids	Esters resulting from the combination of fatty acids and different alcohols.
	i. Fats	Primarily consist of triglycerides and are stored in adipose tissue. Animal sources: butter, oils, lard, and fish. Plant source: corn or olive.
	ii. Waxes	Fatty acid esters that include monohydric alcohols with a higher molecular weight, excluding glycerol. Lanolin, carnauba oil, and beeswax.
2.	Compounds lipids	Fatty acid esters that include an additional group alongside an alcohol and a fatty acid.
	i. Phospholipids	Found in animal tissues.
	ii. Glycolipids	Contains sphingosine, carbohydrate, and fatty acid in its respective structure.
	iii. Lecithin	Found in egg yolks and organ meats.
	iv. Cephalin	Found in nervous tissue and **cell membranes**.
	v. Plasmalogen	Found in anaerobic bacteria, invertebrate, and vertebrate animal species.
	vi. Lipositol	Found in plants, such as soybeans, and in animals, particularly in the brain and spinal cord.
	vii. Sphingomyelin	**Found in chicken egg yolk.**
	viii. Cerebrosides	*Found in neural tissue.*
	ix. Ganglioside	Found in tissues and body fluids.
	x. Sulfolipids	Found in the thylakoid membranes of plants, algae, and cyanobacteria.
	xi. Proteolipids	Found in the cytosol, the cytosolic membrane.
3.	Precursor and derived lipids	These lipids are obtained through the hydrolysis of simple or conjugated lipids. This includes glycerol, steroids, hormones, ketone bodies, and fatty acids.
4.	Terpenes, terpenoids, and steroids	Found in essential oils, resin acids, rubber, as well as plant pigments like lycopene and carotenoids.
Based on chain length and unsaturation		
1.	Short-chain fatty acids	<8 carbon atoms in the chain.
2.	Medium-chain fatty acids	8–12 carbon atoms in the chain.
3.	Long-chain fatty acids	13 and above carbon atoms in the chain.
4.	Saturated	Fatty acids with single bonds.
5.	Unsaturated	Fatty acids with double bonds.
Basis of saponification		
1.	Saponifiable	Includes sphingolipids, triacylglycerides, and glycerophospholipids.
2.	Nonsaponifiable	Includes terpenes, steroids, and prostaglandins.

obtained from different plant and animal sources. Based on chain length, lipids are classified as short, medium, or long, depending on the number of carbon atoms present in the chain. Based on saponification, lipids are either saponifiable or nonsaponifiable.

10.3 Role of fat and lipids

Lipids, which include fats and oils, are important in extrusion cooking because they act as lubricants. They reduce friction between the particles in the mixture and between the screw and barrel surfaces. Once the temperature in the extruder exceeds 40°C, fats and oils melt, blend with the other ingredients, and break down into tiny oil droplets (Joy et al., 2012). Lipid levels below 3% have minimal effect on expansion properties, but when they rise above 5%, the expansion rate significantly decreases. To preserve the expansion index in second-generation products like directly expanded snacks, this increase in lipid content can be managed by reducing the conditioning moisture content (Yazar and Rosell, 2023). The type of starch and lipid in the raw materials affects how the amylose–lipid complex forms. Free fatty acids and monoglycerides are more likely to support the formation of this complex compared to triglycerides (Camire et al., 1990). In wet protein extrusion, lipids hinder the formation of protein fibers because their lubricating effect reduces shear forces and prevents proper alignment of particles (Joy et al., 2012). Lipids make up a small portion of most extruded products. Typically, low levels of lipids (<5%) help stabilize specific mechanical energy and reduce the glassy texture of extrudates (Singh et al., 2007). However, higher lipid content lowers specific mechanical energy by increasing lubrication within the extruder barrel and reducing product expansion (Gulati et al., 2020).

10.4 Influence of fats and lipids on extrusion processing

10.4.1 Lubrication effect

Lipids and fats as lubricants decrease the friction between the extruder surfaces and the food material. This reduces energy requirements and mechanical wear on the equipment, which allows for more efficient processing. The longevity of the extruder and the overall energy costs are significantly impacted by the amount and type of fat used. Emulsifiers, which are unique kinds of lipids, have higher melting points compared to triglycerides but still possess properties similar to oils, offering lubrication during extrusion cooking. Different emulsifiers, such as monoglycerides, fatty acids, and their esters, interact with the amylose part of starch, forming complexes that create a protective layer around the starch during the process of extrusion. This plays an important role in transforming the texture and reducing the extrudate stickiness (Guy, 1994; Ilo et al., 2000).

Lipids melt and transition into oils as the extrusion temperature rises. By the shearing action of the screw elements, these oils are blended into the extrudates, forming tiny droplets (<1 μm). However, oils and fats spread across the biopolymers if the concentration exceeds a certain limit (Guy, 1994). Oil separation can occur if lipids are not properly mixed before or during the extrusion process, which coats the barrel and screw surface, lowering friction (Jager et al., 1990; Ilo et al., 2000). In low-moisture extrusion of starches with low lipid content, such as potato starch, this lubrication property is particularly important. When in contact with metal surfaces, these starches tend to overheat due to low-moisture conditions, leading to potential blockages and flow issues caused by the excessive formation of starch degradation products (Alam et al., 2016; Lazou, 2024; Rao and Artz, 1989).

10.4.2 Melt viscosity

It refers to the molten material (such as dough, polymer, or other substances) resistance to flow under applied force. In extrusion, it is an important property, where the material is heated and forced through a die to form a desired shape. In this molten state, the viscosity of the material determines how easily it can flow or be shaped. Wang et al. (1993) explained how palm oil and other components affect the melt viscosity of wheat dough processed using a twin-screw extruder. They revealed a complex interaction between melt viscosity and oil concentration. As the concentration of oil increased from 0% to 1%, melt viscosity increases, but at concentrations higher than 1%, it decreases significantly. The lubrication attained by the oils helps reduce damage and shield the starch granules. At low moisture levels, starch dispersal is mainly driven by mechanical shearing, which results in higher melt viscosity and less granule degradation at lower oil concentrations (Guy, 1994). If the oil content is further increased to 2%–3%, it hinders starch dispersal, causing a sharp decrease in melt viscosity (Ilo et al., 2000).

10.4.3 Extrudate expansion

The addition of small amounts of lipids (<3%) at the die has little effect on the expansion of the extrudate. However, if it exceeds 5%, extrudate expansion decreases significantly (Ilo et al., 2000). Mechanical energy input decreases due to high lipid content, leading to a less dispersed starch in the continuous phase and undercooked dough, which ultimately weakens die expansion. At higher levels of oil, instead of the typical sponge structure, the extrudate exhibits a biscuit-like texture, with the surplus fat affecting the overall quality and texture of low-moisture extruded products (Guy, 1994; Ilo et al., 2000). In addition, with high oil levels and moisture content (30%–35%), foam-like structures may form during extrusion cooking, where starch conversion occurs through diffusion and swelling rather than granule dispersal (Delgado et al., 2019; Guy, 1994).

Various studies have explained the direct addition of oil throughout the extrusion cooking of starch-based materials. Mohamed (1990) found that expansion increased with increased oil content due to the incorporation of 4% corn oil into maize grits. Similarly, Faller and Heymann (1996) found that higher oil levels increased both expansion and the fracturability of the extrudate due to the addition of 4% vegetable oil to flakes or potato meal. During extrusion cooking, the fat content of starch-based materials greatly affects their behavior. In many cases, extrusion characteristics improve by removing fat, and overall expansion increases in directly expanded products (Ilo et al., 2000). This change is mainly noticeable in maize, which is used for making directly expanded snack items. Degerminated maize is generally favored over whole maize meal because it yields better expansion (Guy, 1994). In degerminated maize meal, the lower oil content enhances expansion, as it contains <1% oil compared to 4% oil in whole maize meal. Similarly, other starch-based ingredients with high fat levels, such as quinoa, oats, and amaranth, are important in extrusion cooking because they provide considerable fiber or protein (Offiah et al., 2019).

10.4.4 Extrudate texture and structure

Fats contribute to extruded items by providing a softer, more attractive structure that improves the texture of the product. Depending on the product preparation, the fat content and its interaction with other ingredients can lead to variations in hardness and crispiness. For determining the texture and structure of extruded products, starch–lipid complexes play an important role (De Pilli et al., 2016). In relation to starch, proteins and lipids negatively affect the mechanical properties of thermoplastic flour. However, in extruded products, these components do not exhibit a plasticizing effect on the physicochemical behavior of the matrix based on corn materials (Arêas and Lawrie, 1984; Ayadi et al., 2011). For extrudates made entirely of wheat flour, the lowest break strength values occurred at minimal water feed levels and for those including oleic acid (De Pilli et al., 2016). Wheat starch-based extrudates showed more break strength than those made from wheat flour. In both starch- and wheat flour-based samples, the addition of oleic acid led to a reduction in break strength. This reduction is due to the formation of protein–lipid and starch–lipid complexes, which create molecular reorganization and result in a highly brittle texture in extrudates with oleic acid (De Pilli et al., 2016). The texture properties of the extrudates are influenced by the lower glass transition temperatures because oleic acid exhibits a plasticizing effect on wheat flour and starch (De Pilli et al., 2016).

10.4.5 Lipid oxidation

Due to high temperatures, extrusion cooking leads to an enhanced peroxide value. Alvarez et al. (1990) found that as die temperature rose from 71°C to 115°C during twin-screw extrusion, lipid oxidation in a meat-based product decreased. Likewise, Rao and Artz (1989) utilized a twin-screw extruder with barrel temperatures between 115°C and 175°C for lipid oxidation in a corn-based mixture with soybean oil added. It was observed that with rising temperatures, the peroxide value decreased, as higher temperatures promoted the decay of peroxides. In addition, during storage at 37°C, they examined the oxidative stability of the extruded products and found that improved extrusion temperatures led to more peroxide formation. This is due to the presence of transition metals in the extrudate, which increase with higher extrusion temperatures and may catalyze lipid oxidation (Alam et al., 2016; Lazou, 2024; Rao and Artz, 1989).

10.4.6 Lipid binding

During extrusion, lipids tend to bind with protein and starch. In nonpolar solvents, lipids bind to starch, making them less soluble. In a series of experiments, De Pilli et al. (2012) extruded mixtures of rice starch and pistachio flour. The mixtures contained 6.5% lipid due to the natural lipids in pistachio flour. They discovered that when high moisture levels and high barrel temperatures were used, the starch–lipid complex formation increased. However, under these circumstances, the extrudates exhibited less desirable qualities, such as reduced hardness, increased oil loss, and higher density from the extruder barrel (Gulati et al., 2020). Most studies have focused on using natural lipids to form starch–fatty acid complexes. Thachil et al. (2014) showed that compared to unsaturated fatty acids, saturated fatty acids tend to interact more with starch. Moreover, starches containing more amylopectin form fewer complexes with fatty acids than those with higher amylose content. Lipid binding was greatly influenced by the extrusion temperature and the moisture level of the feed, while the screw speed had a negligible effect. When cornmeal reached the shear conditions in the extruder, two-thirds of the free lipids became bound. With the temperature of extrusion, the amount of bound lipids showed a slight correlation (De Pilli et al., 2011). In both bound and free lipids, extrusion cooking increases the peroxide value. However, due to the presence of a greater proportion of nonpolar lipids, bound lipids show a lower peroxide value compared to free fats. Subsequently, during extrusion, the binding of lipids could contribute to extending the shelf life of extruded products (Chen et al., 2022; Ilo et al., 2000).

10.5 Drug delivery systems based on lipids

In the 1930s, hot melt extrusion (HME) was first introduced for processing plastic and food materials (Akdogan, 1999; Tambe et al., 2021). Over the past few years, HME has been used for producing lipid-based formulations such as lipid implants, nanostructured lipid carriers, nanoemulsions, and solid lipid nanoparticles (Khinast et al., 2013). Compared to conventional methods for creating lipid-based drug delivery systems (LBDDS), using HME provides better product quality and greater efficiency (Crowley et al., 2007). Extruders, which may have either a single-screw or twin-screw configuration, are used for HME. For LBDDS, twin-screw extruders are preferred as the intermeshing screws apply additional shear force, encouraging a more uniform lipid mixture (Chauhan et al., 2020). In the feeding zone, where solid or semisolid material is transferred, the process starts in the extruder (Alshetaili et al., 2016).

Increasing the feed rate enhances the quality due to a higher compression ratio, although it sometimes results in insufficient mixing. To improve mixing, a solvent is added in the feeding zone. Subsequently, the temperature of the barrel is an important factor for inducing the properties of lipid-based formulations (Patil et al., 2024).

10.5.1 Extrusion of LBDDS

To meet specific formulation requirements, LBDDS are manufactured using extrusion methods. In either single or twin-screw extruders for lipid extrusion, a blend of drugs, excipients, and lipids is introduced. This process produces a uniform dispersion, which can result in lipid emulsions made using techniques like ultrasonication and homogenization. Some factors, such as barrel length, screw speed, and design, significantly influence material handling, residence time, and mixing within the extruder (Qi et al., 2017). Plasticizers are not required to lower the transition temperature, as lipids have lower viscosities and melting points (Patil et al., 2024). Lipids may become liquid, degrade, and lose structure if the temperature is too high. If the temperature is too low, it leads to an uneven mixture, as they may not melt completely (Shukla et al., 2019). LBDDS are broadly applied for a variety of applications, such as taste masking, gastroretentive drug delivery, oral-dispersible formulations, lipid implants, and modified release (Wang et al., 2010).

HME offers important benefits for LBDDS, as in the melting process it involves the use of thermally stable lipids, which facilitate continuous manufacturing with minimal oxygen exposure and without solvents. This approach requires fewer processing steps and provides better control over processing conditions, leading to improved uniformity compared to conventional methods, which require reheating solid lipids (Patil et al., 2014; Repka et al., 2018). Compared to other methods like prilling and direct compression, lipid extrusion results in faster drug release (Aleksovski et al., 2016). Subsequently, without undergoing chemical decomposition, extrudates maintain their structural integrity, resulting in tougher matrices that provide better sustained release for water-soluble drugs. However, the miscibility of drugs and lipids, as well as lipid excipients, is adversely affected due to exposure to elevated temperatures. Various analytical techniques are employed to assess the stability of extruded lipid formulations. Lipid matrices retain drug stability after extrusion, with no signs of aggregation or hydrolysis (Chauhan et al., 2020; Duque et al., 2018).

10.6 Conclusion

In extrusion processing, lipids and fats play various critical roles that directly impact the efficiency and stability of the final extrudates. As lubricants, they reduce friction between the food material and extruder components, decrease energy consumption, enhance process efficiency, and lower mechanical wear. In low-moisture conditions, the lubrication effect is particularly important, as overheating and flow issues can occur. Fats and lipids also influence melt viscosity, with small amounts increasing viscosity due to starch protection, while extreme fat and lipid content significantly reduce melt viscosity by hampering starch dispersal. Furthermore, fat and lipid concentrations impact texture and extrudate expansion. At low levels, lipids have a slight effect on expansion, but higher concentrations can decrease expansion and change the texture from a sponge-like to a denser, biscuit-like structure. The interaction between starch and lipids affects both the texture of extrudates and their structural properties, often resulting in softer and more brittle products. Moreover, fats and lipids contribute to the oxidative stability of extruded products. While extrusion cooking can increase peroxide values due to high temperatures, bound lipids in starch–lipid complexes help reduce oxidation, possibly increasing the shelf life of extruded goods. Overall, the role of fats and lipids in extrusion processing is multifaceted, influencing lubrication, product expansion, melt viscosity, texture, and oxidation stability depending on their concentration and interaction with other components during the process.

References

Akdogan, H., 1999. High moisture food extrusion. Int. J. Food Sci. Technol. 0950-5423. 34 (3), 195–207. doi:10.1046/j.1365-2621.1999.00256.x.

Alam, M.S., Kaur, J., Khaira, H., Gupta, K., 2016. Extrusion and extruded products: changes in quality attributes as affected by extrusion process parameters: a review. Crit. Rev. Food Sci. Nutr. 1549-7852. 56 (3), 445–473. Available from: www.tandf.co.uk/journals/titles/10408398.asp. doi:10.1080/10408398.2013.779568.

Aleksovski, A., Van Bockstal, P.J., Roškar, R., Sovány, T., Regdon, G., De Beer, T., Vervaet, C., Dreu, R., 2016. Comparison of metoprolol tartrate multiple-unit lipid matrix systems produced by different technologies. Eur. J. Pharm. Sci. 1879-0720. 88, 233–245. Available from: www.elsevier.com/locate/ejps. doi:10.1016/j.ejps.2016.03.011.

Alshetaili, A.S., Almutairy, B.K., Alshahrani, S.M., Ashour, E.A., Tiwari, R.V., Alshehri, S.M., Feng, X., Alsulays, B.B., Majumdar, S., Langley, N., Kolter, K., Gryczke, A., Martin, S.T., Repka, M.A., 2016. Optimization of hot melt extrusion parameters for sphericity and hardness of polymeric face-cut pellets. Drug Dev. Ind. Pharm. 1520-5762. 42 (11), 1833–1841. Available from: http://www.tandfonline.com/loi/iddi20#.Vmuk8Lfovcs. doi:10.1080/03639045.2016.1178769.

Alvarez, V.B., Smith, D.M., Morgan, R.G., Booren, A.M., 1990. Restructuring of mechanically deboned chicken and nonmeat binders in a twin-screw extruder. J. Food Sci. 55, 942–946. doi:10.1111/j.1365-2621.1990.tb01570.x.

Arêas, J.A.G., Lawrie, R.A., 1984. Effect of lipid-protein interactions on extrusion of offal protein isolates. Meat Sci. 0309-1740. 11 (4), 275–299. doi:10.1016/0309-1740(84)90058-5.

Ayadi, F., Bliard, C., Dole, P., 2011. Materials based on maize biopolymers: effect of flour components on mechanical and thermal behavior. Starch – Stärke 63, 604–615. doi:10.1002/star.201100010.

Bowen-Forbes, C.S., Goldson-Barnaby, A., 2024. Fats. In: Pharmacognosy. Elsevier BV, 471–489. doi:10.1016/b978-0-443-18657-8.00019-0.

Camire, M.E., Camire, A., Krumhar, K., 1990. Chemical and nutritional changes in foods during extrusion. Crit. Rev. Food Sci. Nutr. 29, 35–57. doi:10.1080/10408399009527513.

Cerone, M., Smith, T.K., 2021. A brief journey into the history of and future sources and uses of fatty acids. Front. Nutr. 2296-861X. 8. Available from: journal.frontiersin.org/journal/nutrition. doi:10.3389/fnut.2021.570401.

Chauhan, G., Shaik, A.A., Kulkarni, N.S., Gupta, V., 2020. The preparation of lipid-based drug delivery system using melt extrusion. Drug Discov. Today. 1878-5832. 25 (11), 1930–1943. Available from: www.elsevier.com/locate/drugdiscov. doi:10.1016/j.drudis.2020.07.025.

Chen, Y., Liang, Y., Tian, X., Wang, X., Liu, H., Jia, F., Zhang, X., Wang, J., 2022. Effect of oil oxidation on aggregation of wheat gluten–peanut oil complexes during extrusion. Int. J. Food Sci. Technol. 0950-5423. 57 (4), 2467–2478. doi:10.1111/ijfs.15611.

Christie, W.W., Han, X., 2012. Lipids: their structures and occurrence. In: Lipid Analysis, 3–19. doi:10.1533/9780857097866.3.

Crowley, M.M., Zhang, F., Repka, M.A., Thumma, S., Upadhye, S.B., Battu, S.K., McGinity, J.W., Martin, C., 2007. Pharmaceutical applications of hot-melt extrusion: part I. Drug Dev. Ind. Pharm. 1520-5762. 33 (9), 909–926. doi:10.1080/03639040701498759.

De Luca, M., Pappalardo, I., Limongi, A.R., Viviano, E., Radice, R.P., Todisco, S., Martelli, G., Infantino, V., Vassallo, A., 2021. Lipids from microalgae for cosmetic applications. Cosmetics. 2079-9284. 8 (2). Available from: www.mdpi.com/journal/cosmetics. doi:10.3390/cosmetics8020052.

De Pilli, T., Derossi, A., Talja, R.A., Jouppila, K., Severini, C., 2011. Study of starch-lipid complexes in model system and real food produced using extrusion-cooking technology. Innov. Food Sci. Emerg. Technol. 1466-8564. 12 (4), 610–616. doi:10.1016/j.ifset.2011.07.011.

De Pilli, T., Derossi, A., Talja, R.A., Jouppila, K., Severini, C., 2012. Starch–lipid complex formation during extrusion-cooking of model system (rice starch and oleic acid) and real food (rice starch and pistachio nut flour). Eur. Food Res. Technol. 1438-2377. 234 (3), 517–525. doi:10.1007/s00217-012-1662-6.

De Pilli, T., Giuliani, R., Buléon, A., Pontoire, B., Legrand, J., 2016. Effects of protein–lipid and starch–lipid complexes on textural characteristics of extrudates based on wheat flour with the addition of oleic acid. Int. J. Food Sci. Technol. 0950-5423. 51 (5), 1063–1074. doi:10.1111/ijfs.13070.

Delgado, E., Gamero-Barraza, M., Flores-Rosas, W., Valles-Rosales, D.J., Medrano-Rold án, H., Reyes-Jáquez, D., 2019. Effect of lipid contents and process parameters on the physicochemical, rheological, calorimetric and structural properties of an extruded canine food. Anim. Nutr. Feed Technol. 0972-2963. 19 (3), 455. doi:10.5958/0974-181x.2019.00042.8.

Duque, L., Körber, M., Bodmeier, R., 2018. Impact of change of matrix crystallinity and polymorphism on ovalbumin release from lipid-based implants. Eur. J. Pharm. Sci. 117, 128–137. doi:10.1016/j.ejps.2018.02.019.

Faller, J.Y., Heymann, H., 1996. Sensory and physical properties of extruded potato puffs1. J. Sens. Stud. 0887-8250. 11 (3), 227–245. doi:10.1111/j.1745-459x.1996.tb00043.x.

Gulati, P., Brahma, S., Rose, D.J., 2020. Impacts of extrusion processing on nutritional components in cereals and legumes: carbohydrates, proteins, lipids, vitamins, and minerals. In: Extrusion Cooking. Elsevier, Lincoin, pp. 415–443. doi:10.1016/B978-0-12-815360-4.00013-4.

Guy, R.C.E., 1994. Raw materials for extrusion cooking processes. In: The Technology of Extrusion Cooking. Springer Science and Business Media LLC, 52–72. doi:10.1007/978-1-4615-2135-8_2.

Ilo, S., Schoenlechner, R., Berghofe, E., 2000. Role of lipids in the extrusion cooking processes. Grasas y Aceites 51. doi:10.3989/gya.2000.v51.i1-2.410

Jager, T., van Zuilichem, D.J., Stolp, W., Van't Riet, K., 1990. Residence time distributions in extrusion-cooking. Part 6: viscous dissipation in a drossel element in the feed zone of a counter-rotating, twin-screw extruder. J. Food Eng. 0260-8774. 12 (1), 67–82. doi:10.1016/0260-8774(90)90020-9.

Joy, C., Leoro, M.G.V., Schmiele, M., Eduardo, R., Kil, Y., 2012. Thermoplastic extrusion in food processing. In: Thermoplastic Elastomers. InTech. doi:10.5772/36874.

Khinast, J., Baumgartner, R., Roblegg, E., 2013. Nano-extrusion: a one-step process for manufacturing of solid nanoparticle formulations directly from the liquid phase. AAPS PharmSciTech 14, 601–604. doi:10.1208/s12249-013-9946-0.

Lazou, A.E., 2024. Food extrusion: an advanced process for innovation and novel product development. Crit. Rev. Food Sci. Nutr. 1549-7852. 64 (14), 4532–4560. Available from: https://www.tandfonline.com/loi/bfsn20. doi:10.1080/10408398.2022.2143474.

Mohamed, S., 1990. Factors affecting extrusion characteristics of expanded starch-based products. J. Food Process. Preserv. 0145-8892. 14 (6), 437–452. doi:10.1111/j.1745-4549.1990.tb00145.x.

Offiah, V., Kontogiorgos, V., Falade, K.O., 2019. Extrusion processing of raw food materials and by-products: a review. Crit. Rev. Food Sci. Nutr. 59, 2979–2998. doi:10.1080/10408398.2018.1480007.

Patil, H., Kulkarni, V., Majumdar, S., Repka, M.A., 2014. Continuous manufacturing of solid lipid nanoparticles by hot melt extrusion. Int. J. Pharm. 1873-3476. 471 (1-2), 153–156. Available from: www.elsevier.com/locate/ijpharm. doi:10.1016/j.ijpharm.2014.05.024.

Patil, H., Vemula, S.K., Narala, S., Lakkala, P., Munnangi, S.R., Narala, N., Jara, M.O., Williams, R.O., Terefe, H., Repka, M.A., 2024. Hot-melt extrusion: from theory to application in pharmaceutical formulation—where are we now? AAPS PharmSciTech. 1530-9932. 25 (2). Available from: https://www.springer.com/journal/12249. doi:10.1208/s12249-024-02749-2.

Qi, F., Heindel, T.J., Wright, M.M., 2017. Numerical study of particle mixing in a lab-scale screw mixer using the discrete element method. Powder Technol. 308, 334–345. doi:10.1016/j.powtec.2016.12.043.

Rao, S.K., Artz, W.E., 1989. Effect of extrusion on lipid oxidation. J. Food Sci. 0022-1147. 54 (6), 1580–1583. doi:10.1111/j.1365-2621.1989.tb05164.x.

Repka, M.A., Bandari, S., Kallakunta, V.R., Vo, A.Q., McFall, H., Pimparade, M.B., Bhagurkar, A.M., 2018. Melt extrusion with poorly soluble drugs—an integrated review. Int. J. Pharm. 1873-3476. 535 (1-2), 68–85. Available from: www.elsevier.com/locate/ijpharm. doi:10.1016/j.ijpharm.2017.10.056.

Rosa, A., Isola, R., Nieddu, M., Masala, C., 2020. The role of lipid composition in the sensory attributes and acceptability of the salted and dried Mullet Roes (Bottarga): a study in human and animal models. Nutrients. 2072-6643. 12 (11), 3454. doi:10.3390/nu12113454.

Shukla, S.K., Kulkarni, N.S., Chan, A., Parvathaneni, V., Farrales, P., Muth, A., Gupta, V., 2019. Metformin-encapsulated liposome delivery system: an effective treatment approach against breast cancer. Pharmaceutics. 1999-4923. 11 (11). Available from: https://www.mdpi.com/1999-4923/11/11/559/pdf. doi:10.3390/pharmaceutics11110559.

Singh, B., Sharma, C., Sharma, S., 2017. Fundamentals of extrusion processing. In: Nanda, V., Sharma, S. (Eds.), Novel Food Processing Technologies, New India Publishing Agency, pp. 1–45. doi:10.31219/osf.io/xqa5n.

Singh, S., Gamlath, S., Wakeling, L., 2007. Nutritional aspects of food extrusion: a review. Int. J. Food Sci. Technol. 42, 916–929. doi:10.1111/j.1365-2621.2006.01309.x.

Tambe, S., Jain, D., Agarwal, Y., Amin, P., 2021. Hot-melt extrusion: highlighting recent advances in pharmaceutical applications. J. Drug Deliv. Sci. Technol. 63, 102452. doi:10.1016/j.jddst.2021.102452.

Thachil, M.T., Chouksey, M.K., Gudipati, V., 2014. Amylose-lipid complex formation during extrusion cooking: effect of added lipid type and amylose level on corn-based puffed snacks. Int. J. Food Sci. Technol. 49, 309–316. doi:10.1111/ijfs.12333.

Wang, L., Wang, J., Lin, X., Tang, X., 2010. Preparation and in vitro evaluation of gliclazide sustained-release matrix pellets: formulation and storage stability. Drug Dev. Ind. Pharm. 0363-9045. 36 (7), 814–822. doi:10.3109/03639040903520967.

Wang, S., Casulli, J., Bouvier, J.M., 1993. Effect of dough ingredients on apparent viscosity and properties of extrudates in twin-screw extrusion-cooking. Int. J. Food Sci. Technol. 28, 465–479. doi:10.1111/j.1365-2621.1993.tb01296.x.

Yazar, G., Rosell, C.M., 2023. Fat replacers in baked products: their impact on rheological properties and final product quality. Crit. Rev. Food Sci. Nutr. 1549-7852. 63 (25), 7653–7676. Available from: https://www.tandfonline.com/loi/bfsn20. doi:10.1080/10408398.2022.2048353.

Effect on phytochemical properties

11

Abhishek Dasore[a,b] *and Norhashila Hashim*[a]
[a]Department of Biological and Agricultural Engineering, Universiti Putra Malaysia, Serdang, Malaysia
[b]School of Computer Science and Artificial Intelligence, SR University, Warangal, Telangana, India

11.1 Introduction

Extrusion stands out as a flexible and extensively utilized processing method in both the food and feed sectors, enabling the transformation of raw ingredients into a wide range of products. This method involves the deliberate utilization of heat, pressure, and mechanical shear on a combination of ingredients, typically in fluid forms such as dough or paste, resulting in a continuous and uniform product with specific texture, shape, and desirable attributes. The process includes multiple phases, such as conveying, mixing, cutting, separating, heating or cooling, shaping, discharging gases or moisture, adding flavors, enclosing, and sterilizing (Dalbhagat et al., 2019). It operates across a range of temperatures, from lower temperatures for pasta production to very high temperatures for creating bread and snacks. Extruders apply pressures ranging from nearly 20 to 250 bars to control shaping, maintain water in a superheated liquid state, and enhance cutting forces in specific screw configurations (Della Valle et al., 1987).

The extrusion process functions within a dynamic equilibrium, balancing input and output variables. To achieve desired characteristics in the extrudate, precise setting of multivariate inputs is essential, determining dependent physical conditions and chemical process changes within the extrusion machine. These variables are reflected in the product, necessitating a narrow processing window for optimal maintenance. Advancements in equipment and processing units have improved extrusion processes, leading to efficient production lines for specific products. Raw materials play an active role, influencing key variables such as pressure, load, and temperature. Extrusion technology's popularity has grown over the last two decades due to its versatility, lower processing costs, higher productivity, superior product quality, and environmentally friendly attributes (Patil et al., 2016; Bouvier and Campanella, 2014).

Phytochemicals, particularly secondary metabolites, are nonnutritional bioactive compounds synthesized by plants for self-defense. This diverse group encompasses flavonoids, carotenoids, allicin, anthocyanidins, polyphenols, tannins, lignans, and

Extrusion Cooking for Food and Feed. https://doi.org/10.1016/B978-0-443-29987-2.00011-7

phytosterols. Phytochemicals provide a level of protection to humans similar to that observed in plants. The significance of phytochemicals in human health, along with their presence in various plant extracts and food products, has been extensively explored in numerous studies. These substances display diverse pharmaceutical characteristics, including antioxidant effects, hormonal activity, stimulation of enzymes, disruption of DNA replication, and antimicrobial features. However, a comprehensive understanding of these phytochemicals is essential for the development of novel disease-preventive properties (Mendoza and Silva, 2018).

Given the widespread use of extrusion cooking in food processing today, it significantly impacts the quality attributes of food products. This technology can alter the nutritional and sensory properties of foods, making it a critical area of study for food scientists and technologists (Emin et al., 2012). Therefore, the primary objective of this chapter is to understand the impact on phytochemical properties in food and feed arising from the process of extrusion cooking. To delve deeper into this objective, the chapter begins by exploring the fundamental principles governing the extrusion cooking process, essential process parameters, and various types of extruders. It then delves into phytochemicals and their classification, emphasizing the importance of phytochemicals in human and animal nutrition and common sources of phytochemicals. The chapter then proceeds to explain changes in phytochemical content during extrusion, effects of process and feed parameters on phytochemical stability, and methods for analyzing phytochemical changes in extruded products and finally summarizes the entire chapter while discussing future trends in extrusion technology and phytochemical research.

11.2 Extrusion cooking

Extrusion cooking stands out as a versatile thermal processing method extensively utilized in the food and feed industries. It encompasses the concurrent application of heat, mechanical energy, and shear forces to convert raw ingredients into a diverse range of final products. This versatile process not only offers the standard benefits of thermal processing but also allows for the modification of functional properties in added ingredients (Koch et al., 2018). The extrusion technique was originally developed for the production of lead tubes and was later adapted for polymer processing in the plastic industry, as well as for the manufacturing of molded metals and synthetic materials. Subsequently, this technique gained significant recognition for its remarkable adaptability within the food and feed industries. Food items produced using extrusion commonly exhibit elevated starch levels, including pasta, various bread varieties such as croutons, breadsticks, and flatbreads, breakfast cereals, ready-to-eat snacks, confectionery, premade cookie dough, specific baby foods, full-fat soy, textured vegetable protein, selected beverages, dry or semimoist pet foods, and aquafeed (Riaz et al., 2009).

The extrusion process involves pushing a blend of various ingredients through an opening in a perforated plate or die, specifically designed for the food industry.

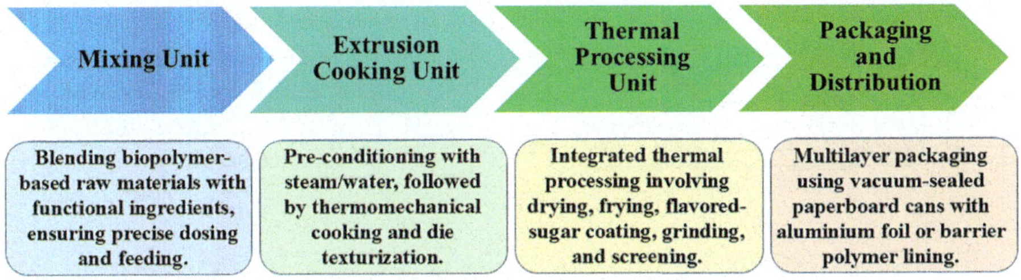

Fig. 11.1 Key steps and components in industrial extrusion processing. An overview of the central processes and key components in an industrial extrusion cooking processing line.

Following this, blades are employed to cut the mixture into a predetermined size. The extruder, a specialized device, propels the mixture through the die. The final product of this process is referred to as the extrudate (Riaz, 2000). Within the extruder, a sizable rotating screw is securely enclosed in a stationary barrel, shaping the products through the orifice or die positioned at the end. Fig. 11.1 outlines the main steps in conventional industrial extrusion. Beginning with raw materials like flours and grits, the process involves powder processing, extrusion cooking, thermal processing, and postprocessing. The last step includes final conditioning through appropriate packaging for storage, ensuring the product's preservation and quality (Bouvier and Campanella, 2014).

Extrusion processes in the food industry are commonly categorized into two types: cold extrusion, or forming, and hot extrusion, or extrusion cooking. Cold extrusion, also known as forming, and hot extrusion, specifically known as extrusion cooking, serve distinct purposes. Cold extrusion is employed for shaping foods like biscuit dough and pasta without the need for cooking. Conversely, extrusion cooking is utilized in producing a diverse range of items, including crispy snacks, sugar confectionery, and soya-based weaning foods. Both processes involve the use of extruders. In the food industry, three main types of extruders are piston extruders, roller-type extruders, and screw extruders. Screw extruders, the predominant type in contemporary usage, can be thermally classified as cold extruders or extruder-cookers and geometrically as single or twin-screw extruders (Kleinebudde and Lindner, 1993).

Cold extruders operate without heating the material; instead, they force it through openings in a "die" at the discharge end of the barrel, shaping it into forms such as rods, tubes, strips, or shells. In contrast, hot extruders elevate the material's temperature to 70°C and above, achieved through friction or supplementary heaters. The material then emerges from the dies under pressure, and in certain snack food applications, rapid expansion occurs, resulting in a light and crisp texture due to the release of steam during the sudden pressure drop upon exiting the die (Brennan et al., 2013). Cold-extruded items are preserved through methods like chilling, baking, or drying, ensuring an extended shelf life. In contrast, hot-extruded products, being free of microorganisms, naturally experience prolonged freshness without the need for additional preservation methods. Packaging is employed for both types to prevent moisture absorption and oxidation during storage. Cold extruders prove adaptable for

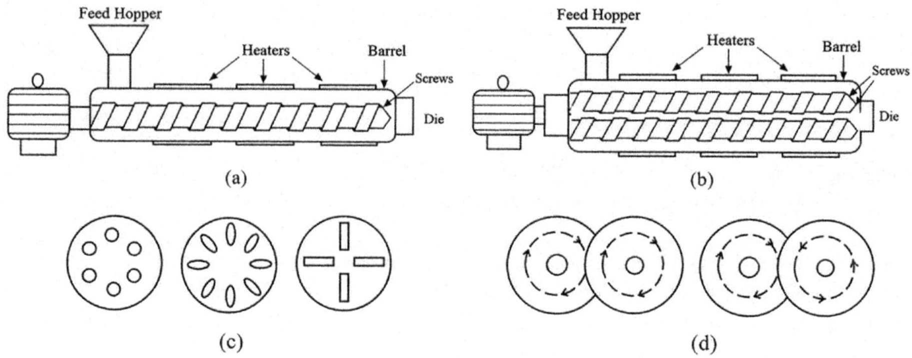

Fig. 11.2 Cross-sectional views and components of food extruders. Extrusion technology: (A) cross section of a single-screw food extruder; (B) cross section of a twin-screw food extruder; (C) various die opening types; and (D) co- and counterrotating screws.
Data from Dalbhagat, C.G., Mahato, D.K., Mishra, H.N., 2019. Effect of extrusion processing on physicochemical, functional, and nutritional characteristics of rice and rice-based products: a review. Trends Food Sci. Technol. 09242244. 85, 226–240. Available from: http://www.elsevier.com/wps/find/journaldescription.cws_home/601278/description#description. https://doi.org/10.1016/j.tifs.2019.01.001.

various scales, ranging from household to small-scale operations. Conversely, hot extruders, with higher costs, are generally more suitable for larger-scale producers.

In Fig. 11.2, a cross-sectional view reveals a single-screw extruder equipped with a segmented worm screw and barrel. This configuration involves a single rotating screw that continuously revolves within the barrel. While different types of these extruders exist, many commonly used ones have a constant pitch. However, their mixing capabilities are often limited, necessitating the premixing of materials or the use of a preconditioner for proper blending of ingredients (Bruin et al., 1978). Typically, a single-screw extruder is divided into three distinct sections: the feed section, the compression section, and the metering section. Granular raw materials are introduced through the hopper in the feed section, where the screw's rotational motion transports the material into the compression section. Here, the screw channel becomes shallower, resulting in material compaction. This process dissipates a significant amount of mechanical energy, leading to a temperature increase, that causes starch gelatinization and enhances material cohesion.

Subsequently, the material is conveyed through the metering section and extruded through the die opening (Arora et al., 2020). For efficient material conveyance and mixing, the inner surfaces of barrels in single-screw extruders typically incorporate helical or axial grooves. Characterization of these extruders relies on parameters such as the length-to-diameter (L/D) ratio and compression ratio (defined as the ratio of maximum to minimum channel depth), with a commonly employed compression ratio being 3:1. The mass flow rate capacity of a single-screw extruder depends on factors such as screw speed, screw geometry, and material properties. The extrusion process's impact on product properties is influenced by the degree of shear applied. Elevated shear levels result in the breakdown of starch and protein molecules, influencing characteristics such as solubility, viscosity, and water-holding capacity. In the contemporary food industry, single-screw extruders dominate, constituting approximately 70% of the total extruders (Berk, 2009).

On the contrary, as Fig. 11.2B illustrates, a twin-screw extruder consists of two equal-length parallel screws that rotate within its barrel. Twin-screw machines can be classified into three types: nonintermeshing counterrotating screws, as shown in Fig. 11.2D; intermeshing corotating screws (where both screws rotate in the same direction, with one screw's flight moving inside the channel of the other); and intermeshing counterrotating screws (where the screws rotate in opposite directions). A variety of die apertures are shown in Fig. 11.2C.

Nonintermeshing twin screws, regardless of their relative rotation, function similarly to two distinct single screws, with the exception of a larger central gap allowing increased backflow when the exit pressure is high (Pinho et al., 2021). Despite the increased intricacy and higher acquisition costs, twin-screw extruders offer superior control and flexibility. The positive pumping action of the screw flights ensures a consistent product flow through the barrel. This feature has contributed to the growing popularity of twin-screw food extruders for producing high-quality products, showcasing high versatility, lower energy consumption, and an expanded range of products in production (Patil et al., 2016).

In twin-screw extruders, the motion of one screw's flight inside the channel of the other propels the material forward, which is similar to the very low-friction operation of a positive displacement gear pump. Unlike single-screw extruders, twin-screw extruders have separate heat control from an external source and remain unaffected by screw speed. The self-wiping and intermeshing flights of the screws effectively move the material forward, preventing any material from becoming trapped in the gap between the barrel surface and the screw. Consequently, twin-screw extrusion cookers are frequently called self-cleaning devices (Crowley et al., 2007).

Moreover, twin-screw extruders ensure a smooth flow of mixed materials, sustaining equilibrium without interruptions or distinctive angular waves on the product surface. The contemporary design of these extruders is modular, incorporating screws with diverse elements mounted on the screw shaft. These elements serve specific functions such as handling transport, mixing, compressing, and melting. The precise configuration of these elements allows operators to control the material's behavior within the extruder, influencing the scope of structural and chemical changes during extrusion cooking. In modern twin-screw food extruders, raw materials and fluid components can be independently introduced through multiple feeders positioned at different locations along the barrel (Rodríguez-Miranda et al., 2011).

The selection and arrangement of elements in these machines require expertise and knowledge of production goals. The extruder manufacturer plays a crucial role in customizing screw, barrel, and supporting equipment configurations to align with specific processing requirements for targeted products. Selecting the appropriate extruder entails evaluating factors like the physical and sensory attributes of the end product and the characteristics of formula ingredients (e.g., moisture content and availability), product type (food grade or intended for feed/pet food), and desired attributes such as color, shape, target bulk density, production rate, and energy source. Once the ideal extruder is identified, precise assembly and thorough maintenance become imperative. Operator training is significant, and the extrusion equipment supplier should be well-equipped to provide this essential service (Prabha et al., 2021).

Extrusion technology offers numerous advantages over traditional methods in food and feed processing, showcasing flexibility in processing a diverse range of products with varying shapes, textures, colors, and appearances. Notably, it is recognized for its energy-efficient processing, proving to be a cost-effective alternative. The modern availability of automated extruders further enhances productivity, especially when handling heat-sensitive ingredients. Furthermore, extrusion processes seamlessly scale up from pilot plant to commercial production.

While food extrusion appears deceptively simple, it is considered a complex process influenced by a myriad of variables, encompassing both process and feed parameters. Processing conditions in extrusion are affected by independent and dependent variables. Independent variables, such as feed composition, moisture content, rate of feed, screw speed, and barrel temperature, can be controlled. In contrast, dependent variables, encompassing extrudate properties such as viscosity, flow rate, system pressure, power, specific energy, residence time, and product characteristics (texture, gelatinization, and chemical composition), assume values based on the magnitude of the independent variables (Cubeddu et al., 2014). Even minor adjustments in these variables significantly impact the final product's quality and characteristics. Temperature, screw speed, and system pressure emerge as the most influential variables in extrusion processes. Notably, among the feed parameters, moisture content plays a crucial role, influencing properties such as viscosity, residence time, and shear stress applied to the food and ultimately affecting the features of extrudates or energy consumption (Elsey et al., 1997).

To achieve products with desired characteristics, effective management of extrusion process operations is essential, along with a comprehensive understanding of the impacts of variables like temperature, screw speed, and feed moisture content. Beyond these operational factors, intrinsic characteristics of extrusion equipment, including the screw profile, die size and shape, and barrel length and diameter, must align with the desired attributes of the food product. Research interest is increasingly focusing on comprehending the physicochemical, functional, and nutritionally relevant effects of extrusion processing. Significantly, preventing or minimizing nutrient destruction, alongside enhancing starch or protein digestibility, holds crucial importance across various extrusion applications. Extrusion cooking further offers an advantage in eliminating antinutritional factors, particularly trypsin inhibitors, hemagglutinins, tannins, and phytates, all of which hinder protein digestibility. This underscores the importance of understanding the influence of phytochemical properties in extrusion cooking for both food and feed applications. Consequently, this chapter is dedicated to exploring the effects of extrusion cooking on the phytochemical properties of food and feed.

11.3 Phytochemicals and their significance

Phytochemicals, commonly known as phytonutrients, are natural and bioactive compounds synthesized by plants as part of their defense mechanisms. These compounds are present in various plant-based foods, such as whole grains, fruits, vegetables, nuts, and herbs. Despite the identification of nearly 4000 phytochemicals, only a fraction have undergone extensive study. The term "phytochemical" originates from the Greek word "phyto," which means "plant." These compounds belong to the secondary plant metabolites, going beyond the primary needs of the cell and contributing to the complete survival of the plant. They play a crucial role in the plant's immune system, protecting it from microorganisms, and contribute to specific characteristics such as color, flavor, aroma, and texture. Some act as signaling molecules within the plant or in interactions with other organisms, while others are believed to serve as natural pesticides. It is important to note that certain phytochemicals exhibit pharmacological activity, while others may be unpalatable or toxic (Tiwari et al., 2013).

Phytochemicals offer humans a level of protection similar to that observed in plants. However, their analysis presents challenges due to considerable variation within the same group of compounds, potential metabolic degradation during plant processing, and the need for mass spectrometry, which is a time-consuming and expensive process. In addition, the binding of certain phytochemicals to macromolecules complicates quantitative extraction. Numerous phytochemicals present in plant foods are poorly absorbed by humans, undergo rapid metabolism, and necessitate doses significantly higher than those naturally occurring in edible plant foods for effective action. Extracting or synthesizing active ingredients becomes crucial for imparting prophylactic or therapeutic value in humans. Phytochemicals exhibit diverse biochemical and physiological effects, with optimal effects often requiring varying plasma or tissue concentrations.

11.3.1 Classification of phytochemicals

The categorization of phytochemicals is intricate, but they can be classified into four major classes according to their chemical structure and properties, as depicted in Fig. 11.3.

11.3.1.1 Nitrogen-containing compounds

Nitrogenous compounds play a vital role in plant defense against herbivores and pathogens and have applications in the fields of pharmaceuticals, stimulants, narcotics, and poisons. They are categorized based on their chemical structures, including alkaloids, glucosinolates, amino acids, and isothiocyanates (ITCs).

- **Alkaloids:** Exhibiting diverse chemical forms, alkaloids often feature the prevalent presence of heterocyclic tertiary nitrogen in their structures. Abundant in nature, alkaloids concentrate in plant seeds and roots, frequently coexisting with vegetable acids (Heinrich et al., 2021).
- **Glucosinolates:** These sulfur-and nitrogen-containing phytochemicals are present in cruciferous or brassica vegetables like mustard, horseradish, broccoli, etc. Comprising a

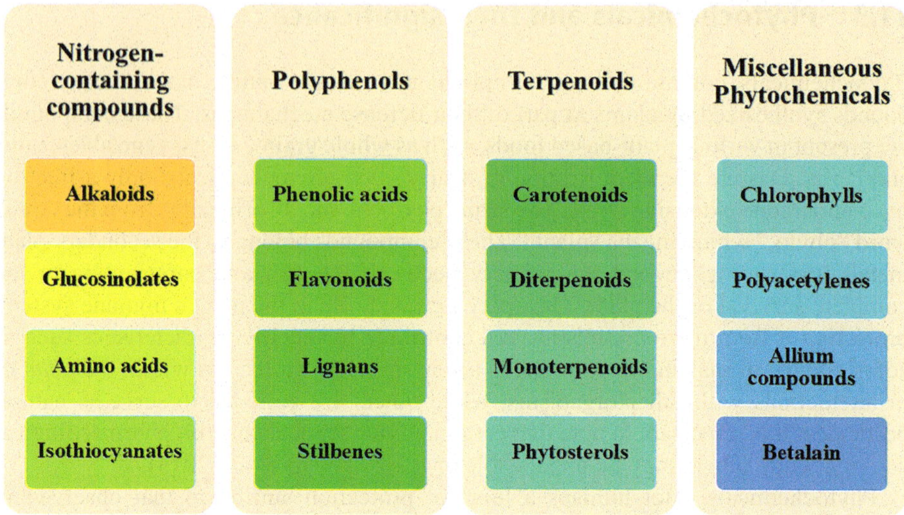

Fig. 11.3 Classification of phytochemicals and the phytochemicals in each class.

β-d-glycopyranose residue connected to a hydroximinosulfate ester through a sulfur bridge, they feature an R-group derived from one of eight amino acids (Connolly et al., 2021).

- **Amino acids:** Nonprotein amino acids are predominantly found in legumes, playing a crucial role in plant defense against stress and acting as essential mediators in response to abiotic factors (Odukoya et al., 2021).
- **ITCs:** They constitute the volatile phytochemicals resulting from the hydrolysis of glucosinolates, catalyzed by the myrosinase enzyme. ITCs are identified as the key compounds responsible for the characteristic pungent aroma and bitter taste found in cruciferous vegetables (Yadav et al., 2022).

11.3.1.2 Polyphenols

Significant secondary metabolites in plants, polyphenols stand out as the predominant antioxidants in human diets. They are categorized into subgroups based on chemical structures, including phenolic acids, flavonoids, lignans, and stilbenes.

- **Phenolic acids:** They are found in nearly all fruits and play various roles in plants. These include derivatives of benzoic acid (C6–C1) and derivatives of cinnamic acid (C6–C3) (Kumar and Goel, 2019).
- **Flavonoids:** Originating from parent compounds known as flavans, these compounds structurally consist of multiple benzene rings (C15 aromatic compounds) and are recognized for their varied antioxidant and free radical scavenging properties (Ullah et al., 2020).
- **Lignans:** They are a varied group of plant secondary metabolites that originate in fruits, vegetables, and cereals and are composed of two phenylpropanoid (C6C3) units (Touré and Xueming, 2010).
- **Stilbenes:** They originate from various plant families, featuring a 1,2-diphenylethylene nucleus with hydroxyl substitutions on aromatic rings. They are biosynthesized in response to biotic and abiotic stresses (Teka et al., 2022).

11.3.1.3 Terpenoids

The most extensive class of secondary metabolites, terpenoids, is widespread among all categories of living organisms. They are categorized into subgroups based on chemical structures, including carotenoids, diterpenoids, monoterpenoids, and phytosterols.

- **Carotenoids:** They belong to the isoprenoid family, including lycopenes, carotenes, and xanthophylls. They are precursors to vitamin A and are essential in photosynthetic organisms, fungi, nonphotosynthetic bacteria, and animals (Yoon et al., 2012).
- **Diterpenoids:** They originate from (E,E,E)-geranylgeranyl diphosphate, constituting a chemically diverse group of compounds with a C20 carbon skeleton formed by four isoprene units (Zhao et al., 2022).
- **Monoterpenoids:** They are categorized into acyclic, monocyclic, and bicyclic groups and consist of a C10 carbon backbone with two isoprene units. They are primarily identified in arthropod species, playing roles as defensive agents and pheromones (Kabir et al., 2020).
- **Phytosterols:** They are unique phytochemicals with a structural resemblance to cholesterol, found in all plant-based foods, with the highest concentrations in unrefined plant oils, nuts, and olive oil (Bradford and Awad, 2007).

11.3.1.4 Miscellaneous phytochemicals

This category encompasses chlorophyll, polyacetylenes, *Allium* compounds, and betalains.

- **Chlorophyll:** It is vital for photosynthesis and is found in almost all green plant parts, primarily in chloroplasts. In plants, it exists in two forms: chlorophyll-a ($C_{55}H_{72}MgO_5N_4$) with a methyl ($-CH_3$) group at the carbon-7 position and chlorophyll-b ($C_{55}H_{70}MgN_4O_6$) with an aldehyde (–CHO) group at the same position (Ebrahimi et al., 2023).
- **Polyacetylene:** These phytochemicals are gaining recognition for their potential contribution to the chemoprotective benefits in apiaceous vegetables. They are characterized by having two or more carbon–carbon triple bonds within their carbonic framework (Lai et al., 2023).
- ***Allium* compounds:** They are secondary metabolites rich in flavonoids, saponins, phytosterols, and organosulfur compounds. They play a vital role in the distinctive characteristics of the Allium genus, acting as a defense mechanism against various abiotic stresses (Deka et al., 2022).
- **Betalain:** They are hydrophilic pigments that accumulate in the cell vacuoles of certain plants. Betalains, derived from the Latin name of the common beet, contribute to the vibrant red hues in various plants. These pigments are classified into betacyanins, which provide red–violet colors, and betaxanthins, which impart yellow–orange tones. Betalains replace anthocyanin pigments in plants of the Caryophyllales order and are also found in some higher-order fungi (Gengatharan et al., 2016).

11.3.2 Medicinal benefits of phytochemicals

The historical utilization of plants as medicinal remedies traces back to early civilizations. Written records from ancient Chinese, Indian, and North African cultures offer evidence of human ingenuity in harnessing plants for treating various diseases.

The diverse properties inherent in plants have facilitated their application in disease treatments, attributed to the presence of phytochemicals. Although phytochemicals are not essential nutrients vital for sustaining life, they play a crucial role in preventing and combating common diseases.

The incorporation of phytochemicals into medicine and pharmacy is closely linked to phytotherapy, which involves utilizing plant-derived medications for disease treatment and prevention. Various trials and pharmacological studies have explored specific phyto-therapeutic preparations. While in certain countries phytotherapy is viewed as a form of traditional medicine, the World Health Organization estimates that around 80% of the population in certain Asian and African nations depends on complementary and traditional medicine, predominantly featuring plant extracts, for disease prevention and treatment (Kumar et al., 2023).

Each category of these functional agents includes a diverse array of chemicals, each with different potency levels, and certain phytochemicals perform multiple functions. Phytochemicals play a substantial role in preventing chronic diseases like cancer, diabetes, and coronary heart disease (Yuan and Macquarrie, 2015). Phytochemicals in plants can offer a diverse array of medicinal benefits, including combating microorganisms, reducing inflammation, protecting the heart, preventing cancer, supporting neurological health, regulating metabolism, and providing other therapeutic effects. Fig. 11.4 outlines the medicinal benefits of different phytochemicals, emphasizing their

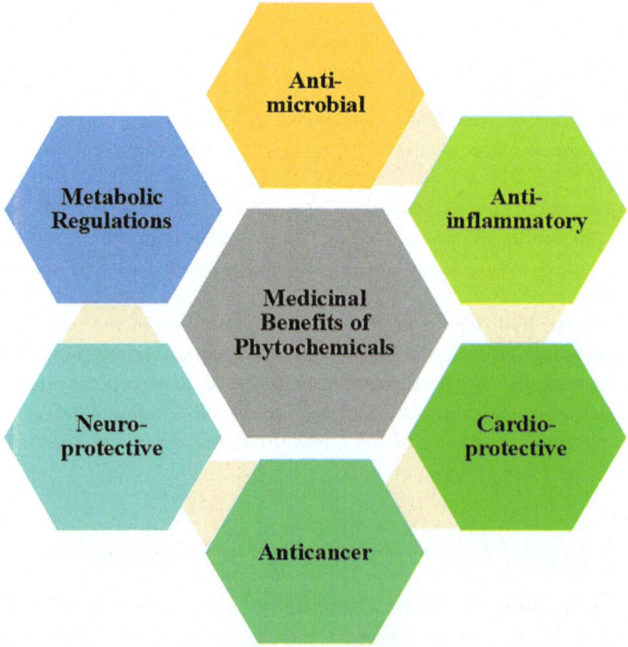

Fig. 11.4 Medicinal benefits of phytochemicals. Therapeutic attributes of phytochemicals.

diverse and valuable roles in promoting well-being. These encompass antimicrobial, antiinflammatory, cardioprotective, anticancer, neuroprotective, and metabolic regulatory properties (Jaeger and Cuny, 2016).

11.3.2.1 Antimicrobial properties

Phytochemicals exhibit promising antiviral properties by disrupting viral membranes, inhibiting viral replication stages, and displaying antiviral, antiinflammatory, and immunomodulatory effects. Essential oils, rich in active phytochemicals, penetrate viral membranes, disrupt their structure, and combat viral infections. Consequently, a chemoherbal approach emerges as a feasible and effective strategy in addressing viral challenges, including those posed by the SARS-CoV-2 virus (Reichling et al., 2009). Moreover, phytochemicals act as potent antifungal agents by disrupting fungal cell membranes, inhibiting key enzymes, and altering cellular balance. Certain herbal plants showcase antifungal and antimycotoxigenic activities, indicating their potential as natural biofungicides to combat food spoilage caused by oxidation (Gizaw et al., 2022).

11.3.2.2 Antiinflammatory

Phytochemicals, found in plant-derived compounds like flavonoids and tannins, show promise in treating inflammatory conditions. These bioactive compounds impact crucial signaling pathways, including toll-like receptor 4, and contribute to macrophage repolarization. Studies suggest that phytochemicals can downregulate proinflammatory genes, promote antiinflammatory genes, and modulate gene expression through epigenetic modifications. Notably, plant-derived compounds may regulate proinflammatory microRNAs, particularly those activated by nuclear factor-kappa B, highlighting their potential as effective antiinflammatory agents (Gonfa et al., 2023).

11.3.2.3 Cardioprotective

Phytochemicals, found in various plant extracts, offer significant medicinal benefits in cardiovascular diseases. While traditional pharmaceutical drugs like β-blockers and angiotensin receptor blockers are common for treating hypertension, plant extracts, such as *Helichrysum ceres* and *Ekebergia capensis*, demonstrate promising hypotensive effects by influencing vascular smooth muscles (Musabayane et al., 2008). Prominent plants like *Allium sativum*, *Sclerocarya birrea*, *Ficus thonningii,* and *Olea europaea* are employed in South Africa for their cardioprotective, vasodilatory, and bradycardic properties in traditional medicine and culinary uses. *O. europaea* extracts, abundant in phytochemicals, demonstrate antihypertensive, diuretic, and antiatherosclerotic effects, guarding against hypertension and atherosclerosis development (Al-Qattan et al., 2008; Musabayane et al., 2007).

The antihypertensive potential of phytochemicals becomes evident as they act on the renin–angiotensin–aldosterone system (RAAS), a pivotal pathway in blood pressure regulation. Medicinal plants, including *Tulbaghia violacea*, *Stangeria eriopus*, and *Justicia flava*, demonstrate inhibitory activity against angiotensin-converting enzyme, a critical step in the RAAS. This suggests a potential role for phytochemicals in developing antihypertensive interventions and managing cardiovascular health (Morgan et al., 2001).

11.3.2.4 Anticancer

Phytochemicals play a crucial role in preventing and treating cancer. Chemoprevention, employing natural, biological, or synthetic agents, strives to suppress, prevent, or reverse the process of carcinogenesis. Notable phytochemicals like capsaicin, catechins, lycopene, and isoflavones exhibit potent anticancer properties, effectively inhibiting tumor growth and inducing apoptosis. The main green tea component, epigallocatechin gallate, enhances the activity of anticancer drugs. These natural compounds show promise in various cancer models, providing alternative or synergistic approaches alongside conventional treatments. Moreover, they contribute to reducing the risk of precancerous lesions and preventing cancer recurrence. The efficacy of phytochemicals presents a valuable pathway for the development of effective and less harmful anticancer strategies (Pitot, 1993).

11.3.2.5 Neuroprotective

Phytochemicals show promise in alleviating neuropathic pain, which is linked to tissue injury and nerve damage. Traditional drugs have limitations due to adverse effects, leading to the exploration of alternative approaches. Medicinal plants such as *B. monosperma, A. calamus, N. latifolia, and P. granatum* demonstrate efficacy in managing neuropathic pain. *Panax ginseng, Centella asiatica*, and *Curcuma longa* hold potential for treating Alzheimer's disease. Phenolic compounds, flavonoids, glycosides, terpenes, and lignans demonstrate neuroprotective effects in various neurodegenerative disorders, providing avenues for addressing neuropathy (Ayaz et al., 2024).

11.3.2.6 Metabolic disorders

Phytochemicals play a crucial role in addressing metabolic disorders like obesity and diabetes. Specific species such as *Strychnos henningsii, Euclea undulata,* and *Schkuria pinnata*, as well as medicinal plants such as *Cucurbita pepo*, *Nuxia floribunda,* and *Cymbopogon citratus,* exhibit potential solutions by influencing glucose uptake, regulating postprandial hyperglycemia, and preserving pancreatic β-cells. Plants like *Hypoxis argentea* and *Carica papaya* possess antidiabetic properties, enhancing insulin release. These plant-derived compounds offer alternatives to conventional approaches in the management of metabolic disorders, addressing issues such as weight and alleviating symptoms of diabetes (Oyedemi et al., 2013).

11.3.3 Sources of phytochemicals

Phytochemicals, present in vegetables, fruits, cereal grains, and plant-based beverages such as tea and wine, are bioactive compounds possessing antioxidant properties. These compounds serve diverse functions, such as mimicking hormones, modifying

Fig. 11.5 Phytochemical sources based on food color. Color Code for phytochemicals in different fruits and vegetables.

cholesterol absorption, inhibiting inflammatory responses, and blocking specific enzymes, as detailed in earlier sections. Despite the knowledge of over 4000 phytochemicals, their health benefits remain largely unknown. In addition, their potential for toxicity is uncertain, especially when consumed in large amounts through supplements. A color-based method for selecting fruits and vegetables, aligned with phytochemical content, translates nutritional science into practical dietary guidelines (Barnes et al., 2013). Fig. 11.5 illustrates phytochemical sources based on food color, and Table 11.1 presents a diverse array of food sources containing phytochemicals and their therapeutic benefits.

11.4 Impact of extrusion cooking on phytochemicals

Extrusion cooking, despite enabling a wide range of food applications, introduces uncertainty regarding its effects on the quality characteristics of food products. Alterations in both primary and secondary metabolites can result in either favorable or detrimental outcomes. Detrimental effects include the compromise of heat-labile vitamins, leading to substantial degradation, particularly in products renowned for their sensory appeal (Singh et al., 2007). The Maillard reactions occurring between proteins and sugars during extrusion can diminish the nutritional value of the protein, a consequence influenced by raw material types, their composition, and specific process conditions (Alonso et al., 2000; Cheftel, 1986). In contrast, extrusion also brings about favorable effects, including the eradication of antinutritional factors, starch gelatinization, heightened soluble dietary fiber, and reduced lipid oxidation (Björck and Asp, 1983). Thus the impact of extrusion cooking on food quality is intricate, presenting both enhancements and challenges.

Extrusion parameters, such as residence time, moisture content, and temperature, must be precisely regulated to achieve the best balance between these effects (Alonso et al., 2001). For instance, moderate extrusion conditions, such as high moisture content, low residence time, and low temperature, have been shown to enhance the

Table 11.1 Sources of phytochemicals and their therapeutic benefits.

Food category	Source	Phytochemicals	Medicinal benefits	Reference
Vegetables	**Cruciferous** (broccoli, cabbage, Brussels sprouts, cauliflower)	Glucosinolates (upon ingestion, convert into biologically active substances such as isothiocyanates and indoles); polyphenols, alkaloids, tannins, chlorophyll, anthocyanins, phytosteroids	They provide protection against various cancers by inducing an oxidative stress response and activating multiple signaling pathways. This significantly contributes to both delaying cancer initiation and hindering its progression.	(Singh et al., 2006; Paul and Jed, 2001)
	Stem (asparagus, cardoon, celery, fennel, fiddlehead)	Polysaccharides, polyphenols, anthocyanins, saponins, flavones	Through in vitro and in vivo experiments, they exhibit anticancer, antitumor, antioxidant, immunomodulatory, hypoglycemic, antihypertensive, and antiepileptic effects.	(Guo et al., 2020)
	Leafy greens (kale, spinach, lettuce, Swiss chard, watercress)	Dietary fibers, carotenoids, lutein, polyphenols, chlorophyll	They offer potential health benefits, including cardioprotection, anticancer properties, antidiabetic effects, and antiaging benefits.	(Sarker and Oba, 2020)
	Root (beetroot, cassava, horseradish, carrot, radish, potato)	Phenolics, flavonoids, polyamines, carotenoids, alkaloids, glycosides, tannins, saponins, proteins, amino acids, terpenoids, and steroids	They play a role in vision, cell differentiation, glycoprotein synthesis, mucus secretion, bone growth, and overall development. In addition, these compounds help reduce the risk of cancer and cardiovascular diseases by offering antioxidant,	(Ahmad et al., 2019)

			antiinflammatory, plasma lipid modification, antitumor, and anxiety-reducing properties.	
	Allium (garlic, onions, chives, leeks, shallots, scallions)	Flavonoids, steroids, saponins, alliin, fructans, and organosulfur compounds	These constituents exhibit activities such as anticarcinogenic, antioxidant, antidiabetic, renoprotective, antiatherosclerotic, antibacterial, antifungal, and antihypertensive properties.	(Zhang et al., 2020; Chakraborty et al., 2022)
	Legumes (lentils, peas, chickpeas, soybeans, beans)	α-Galactosides, phytosterols, tocopherols, saponins, alkaloids, phytic acid, carotenoids, and (poly) phenolic compounds	Legumes regulate blood glucose, provide cardiovascular protection, and show potential in cancer prevention. Their antibacterial properties include suppressing biofilm formation and disrupting microbial metabolism.	(Tor-Roca et al., 2020; Sitohy et al., 2013)
	Gourd and squash (bottle gourd, pumpkin, apple gourd, ash gourd, bitter gourd)	Polysaccharides, saponins, phenolics, carotenoids (beta-carotene, zeaxanthin, lutein), flavonols	Gourd and squash exhibit antimicrobial and anthelmintic activities, antidiabetic effects, hypolipidemic and hypotensive activities, along with antidementia, anticancer, and antioxidative properties.	(Gayathry and John, 2022)
Fruits	**Berries** (blueberry, raspberry, strawberry, cranberry, blackberry, goji berry)	Carotenoids, sterols, terpenoids, phenolic compounds, flavonoids, ellagitannins, lignans, and anthocyanins	They showcase antioxidant, antiinflammatory, antidiabetic, and antimicrobial properties, conferring benefits against cardiovascular diseases, metabolic syndrome, and potential anticancer effects.	(Akin et al., 2016)

Continued

Table 11.1 Sources of phytochemicals and their therapeutic benefits—cont'd

Food category	Source	Phytochemicals	Medicinal benefits	Reference
	Citrus (lemon, orange, pomelo, lime, grapefruit)	Carotenoids, terpenes, flavonoids, phenolic acids, limonoids, and essential oils	They demonstrate antioxidant, antiinflammatory, antimicrobial, anticancer properties and a neuroprotective effect.	(Zahr et al., 2023)
	Drupes (cherry, plum, mango, peach, apricot)	Phenolics, flavonoids, carotenoids, and anthocyanins	They demonstrate antiparasitic, antimicrobial, antitumor, antidiabetic, antihyperlipidemic, analgesic, and antiinflammatory activities. In addition, they exhibit immunomodulatory effects, antipyretic and antiallergic activity, as well as hepatoprotective activity.	(Kumar et al., 2021)
	Tropical fruits (papaya, jackfruit, guava, pomegranate, passion fruit, sapota, pineapple, banana)	Carotenoids, flavonoids, volatile acids, sterols, polysaccharides, triterpenoid acids, alkaloids, essential oils, steroids, glycosides, tannins, and saponins	They offer diverse medicinal benefits, including antioxidant, antiviral, antiinflammatory, and potential anticancer properties, along with effects on myocardial inotropism and diabetes control.	(Naseer et al., 2018)
	Pommes (apples, pears, quinces, medlars, loquats)	Flavonoids, quercetin, catechin, and chlorogenic acid	They exhibit antiproliferative and antioxidant effects and are linked to reduced cancer and cardiovascular disease risk, asthma prevention, and potential cholesterol-lowering abilities.	(Boyer and Liu, 2004)

	Melons (watermelon, crane melon, cantaloupe, Ananas melon)	Carotenoids, lycopene, citrulline, anthocyanins, cucurbitacins, and other polyphenolic compounds	They exhibit hepatoprotective, antiinflammatory, antitumor, antimicrobial, and anthelmintic effects and are employed for addressing gastrointestinal issues and rheumatism, while also serving as an appetite stimulant and relieving constipation.	(Nadeem et al., 2022)
Miscellaneous	**Grains and cereals** (rice, barley, corn, millet, buckwheat, oats)	Polysaccharides, phenolics, carotenoids, vitamin E compounds, lignans, β-glucan, inulin, phenolic acids, flavones, phytic acid, flavonoids, coumarins, and terpenes	Cures immunodeficiency, cancer, inflammation, hypertension, hyperlipidemia; reduces diabetes; improves gut microbiota.	(Sidhu et al., 2007)
	Fats and oils (flaxseed oil, sesame oil, sunflower oil, canola oil, olive oil, soybean oil, peanut butter)	Phytosterols, lignans, fatty acids, flavonoids, phenolic acids, tocopherols, and phospholipids	They exhibit antioxidative, anticholesterolemic, antihypertensive, cardioprotective, hypoglycemic, and hypolipidemic effects. In addition, these oils have been traditionally used for treating burns.	(Li et al., 2024; Kamal-Eldin and Appelqvist, 1994)
	Herbs and Spices (basil, bay leaf, Lemon grass, rosemary, turmeric, ginger, saffron, chili peppers)	Saponins, tannins, phenolics, flavonoids, steroids, terpenoids, cardiac glycosides, and alkaloids	They offer medicinal benefits, such as enhancing gut health, supporting the immune system, and synergizing with antibiotics to combat infections, potentially mitigating chronic conditions like cancer and inflammatory diseases.	(Wong and Chow, 2024)

Continued

Table 11.1 Sources of phytochemicals and their therapeutic benefits—cont'd

Food category	Source	Phytochemicals	Medicinal benefits	Reference
	Nuts and seeds (almonds, peanuts, hazelnuts, cashew nuts, walnuts, chia seeds, flax seeds, pumpkin seeds, sesame seeds)	Tocols, phytosterols, carotenoids, chlorophylls, squalene, lignans, policosanols, thymoquinone, polyphenols, and stanols	They exhibit diverse medicinal benefits, including antioxidant, antiinflammatory, antidiabetic, antihypertensive, neuroprotective, and anticarcinogenic effects, supporting health and well-being.	(Alasalvar et al., 2021)
	Beverages (green tea, black tea, hibiscus tea, ceylon tea, chamomile tea, Arabica coffee, mocha coffee, robusta coffee)	Flavonoids, caffeine, cafestol, theobromine, epigallocatechin, gallic acid, hyperin, isoquercitrin, piceatannol, trifolin, rutin, theogallin, phenolic acids	They showcase a range of properties, including antioxidant, antiapoptotic, antiinflammatory, antiamyloidogenic, antimicrobial, anticancer, anticariogenic, antidegranulation, antilithogenic, antiviral, phytoestrogenic, memory-improving, nuclear factor erythroid 2-related factor 2 (Nrf2)-inhibiting, neuroprotective, and skin-protective effects. Moreover, these compounds contribute to stress and anxiety alleviation, support digestion, boost the immune system, aid in weight management, and exhibit potential antidiabetic effects.	(Cao et al., 2024; Pandit et al., 2023)

retention of amino acids, improve the digestibility of protein and starch, elevate soluble dietary fiber, reduce lipid oxidation, and increase the retention of vitamins and minerals. On the contrary, given the typical residence period of 0.5–1 min in the hot-screw segments, severe extrusion conditions, such as high extrusion temperature (≥200°C) and low moisture content (≤15%), as well as inappropriate formulation (e.g., the presence of highly reactive sugars), might result in nutritional degradation (Lobato et al., 2011; Plahar et al., 2003). Therefore meticulous optimization of extrusion parameters is crucial in achieving the highest product quality, encompassing both sensory attributes and phytochemical properties. This section aims to elucidate the modifications in phytochemical properties by highlighting the impact of process factors and feed properties.

11.4.1 Effect on phytochemical properties in extrusion cooking for food

Numerous studies highlight that extrusion cooking induces modifications in the microstructure, chemical attributes, and morphology of products (Ti et al., 2015; Wolfe and Liu, 2003). As food materials undergo extrusion, the elevated temperatures within the extruder barrel bring about transformations in phytochemical composition, indirectly influencing the medicinal properties of the product (Han and Baik, 2008; Xu and Chang, 2008). The elevated temperature of the food mass in the extruder barrel depends on numerous factors, categorized into primary and secondary factors, as depicted in Fig. 11.6.

Primary factors encompass the extruder model, feed composition, moisture content, feed particle size, feed rate, barrel temperature profile, screw configuration, screw speed, die geometry, and die size and number. Meanwhile, secondary factors comprise pressure, specific energy, and extrudate temperature (Camire, 2012). Understanding these compounds and their breakdown products in extruded foods is crucial for comprehending how the extrusion process affects these substances and their potential health implications. This section provides an overview of phytochemical changes during extrusion cooking.

11.4.1.1 Changes in nitrogen-containing compounds

Alkaloids, essential nitrogen-containing phytochemicals, have historically served medicinal and harmful roles across diverse cultures. *Papaver somniferum*, for instance, yields morphine and codeine for pain relief, while *Aconitum* species produce aconitine, historically employed in spear poison and as a murder weapon (Wink, 1998). In plants, four main groups of nitrogen-containing secondary metabolites include alkaloids, glucosinolates, nonprotein amino acids, and ITCs. During extrusion cooking, these compounds undergo alterations. For instance, Lupine, a protein- and fiber-rich legume, may contain quinolizidine alkaloids that can accumulate to harmful levels, necessitating efforts to reduce total alkaloid content through extrusion processes. However, attempts, including the use of 2% citric acid, have proven ineffective in achieving safe concentrations (Chamone et al., 2023).

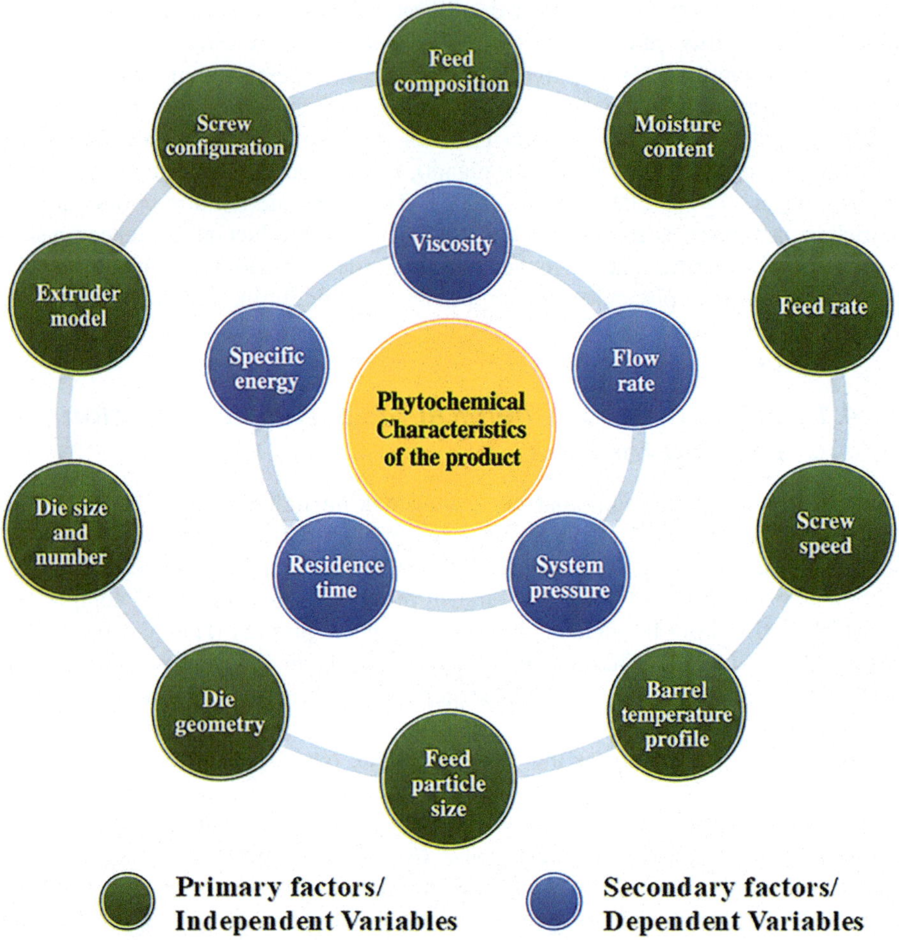

Fig. 11.6 Effect of extrusion cooking on phytochemical properties. Influential factors on the phytochemical composition of the product.

In French fry production, potato peelings, which are rich in dietary fiber, face limitations due to natural glycoalkaloids. Twin-screw extrusion under different conditions did not significantly reduce glycoalkaloids in potato peels (Zhao et al., 1994). Similarly, various extrusion conditions showed mixed results in reducing glycoalkaloid levels in potato flakes (Camire et al., 1990). In another extrusion study, commercial high-glucosinolate rapeseed meal underwent extrusion with varying screw speeds (200–400 rpm), moisture contents (14–20%), and ammonia concentrations (0–3%). Product temperature exhibited the strongest correlation with glucosinolate removal. Increasing screw speed lowered product temperature and decreased glucosinolate removal, while higher moisture content reduced removal in nonammonia treatments but improved detoxification when ammonia was present (Liang et al., 2002). A separate study investigated the impact of extrusion conditions on radish leaves and stems to reduce off-flavors and enhance antioxidant activity. Under optimized conditions (130°C, 250 rpm, and 20% moisture), extruded radish showed a significant decrease in glucosinolate levels, which are associated with off-flavors during processing, compared to nonextruded radish (Sung et al., 2016).

11.4.1.2 Changes in polyphenols

Polyphenols, secondary metabolites in plants, serve as crucial defenses against environmental stressors like UV radiation and pathogenic attacks (Shahidi and Yeo, 2016). This class encompasses phenolic acids, flavonoids, lignans, and stilbenes, which are widely distributed in plant-based foods and play a pivotal role in influencing the Maillard reaction (Marta et al., 2012). Phenolic compounds exist in two forms—free and bound—with the latter often associated with cell wall components (Rocchetti et al., 2022). During extrusion cooking, phenolic acids undergo intricate changes. The extractable phenolic content decreases as heat-labile phenolic compounds break down and some phenolics polymerize. However, there is a trade-off, as reducing the extrusion temperature could affect the finished product's porosity and perceived flavor (Wang et al., 2014). Extrusion-induced changes in phytochemical and functional properties, including antioxidant activity, depend on factors such as raw material and extrusion variables. Advanced laboratory techniques, like mass spectrometry and high-performance liquid chromatography (HPLC), have enabled the identification of phenolic acids separately during extrusion.

Continuing with the exploration of specific alterations in phenolic compounds observed during the extrusion process, recent research studies provide valuable insights. For instance, a study involving three flour mixes—corn, Bengal gram, and tree bean—in the ratios 70:30:00, 70:25:05, and 70:20:10, respectively, was extruded into ready-to-eat snacks using a corotating twin-screw extruder (barrel diameter: 37 mm, L/D ratio: 27:1). The 10% tree bean variant showed decreased proline and total carotenoids but higher levels of total phenolics (2.34 mg g^{-1} FW), ascorbate content (2.23 mg g^{-1} FW), total flavonoids (0.16 mg g^{-1} FW), and reduced glutathione (8.53 μM g^{-1} FW) compared to other formulations (Sahoo et al., 2022). Another study explored enriched extruded snacks, combining by-products like defatted soybean meal, germinated brown rice meal, and mango peel fiber with corn grit at 20% (w/w). This formulation showed a remarkable increase in phenolics (22.33–33.53 times) and antioxidant activity (5.30–11.53 times). Snacks extruded with 13% (wb) moisture content at 180 rpm and 160°C displayed superior expansion, total phenolic content (TPC), and antioxidant capacity. These fortified snacks have potential as a healthier alternative to regular extruded snacks (Korkerd et al., 2016).

The impact of extrusion on soy isoflavones, investigated in products like textured soy and soy protein concentrate blends, showcased changes in genistein, daidzein, and their derivatives (Mahungu et al., 1999). Other studies have explored the effects of screw speed, feed moisture content, and feed rate on antioxidant activity and TPC in extruded snacks, providing valuable insights into the overall process. Total phenolics were measured as millimolar gallic acid per gram of dry sample by UV spectroscopy with the Folin–Ciocalteu reagent, and total antioxidant capacity was assessed through the decolorization of the 2,2′-azino-bis(3-ethylbenzothiazoline-6-sulfonic acid) diammonium salt (ABTS) radical cation. Total antioxidant activity declined with higher screw speed and lower moisture content, while total phenolic values exhibited negligible alterations postextrusion (within a 95% confidence interval) (O″zer et al., 2006). Corn and rice flour extrudates were fortified with varying levels of beta-glucan, lignans, and gamma oryzanol isolated from barley flour, flaxseed, and rice bran oil, respectively. Postextrusion, the fortified and optimized samples retained significant proportions of beta-glucan (82.67% to 90.83%), lignans (86.31% to 66.66%), and gamma oryzanol (71.33% to 51.67%) (Kour et al., 2019).

In addition, metabolic analysis of white quinoa and extruded white quinoa revealed changes in phenolic composition postextrusion. Among the identified 50 phenolics, key compounds like ferulic acid, rutin, oleanolic acid, and coumarin decreased, while others, including madecassoside, botulin, jasmone, and curcumol, increased postextrusion. Some phenolics, like mangiferin and epigallocatechin, were entirely removed, signifying alterations in quinoa flour's phenolic composition after extrusion (Song and Tang, 2023). Lastly, the extrusion processing of rice-legume blends, incorporating carob fruit and peas, resulted in a notable decrease in inositol phosphates and lectin contents, accompanied by the complete elimination of protease inhibitors. Formulations with different pea and carob fruit proportions exhibited alterations in total phenol and total flavonoid levels. The impact of extrusion on phenolic groups varied; for instance, it led to a decrease in anthocyanins (4–50%) but an increase in flavonols (approximately threefold) (Arribas et al., 2019). This diverse range of studies provides valuable insights into the intricate alterations in phenolic compounds during the extrusion process.

11.4.1.3 Changes in terpenoids

The inclusion of lipophilic phytochemicals like carotenoids and phytosterols in food products holds promise for improving health outcomes, including preventing certain cancers and cardiovascular illnesses (Awad and Fink, 2000). Carotenoids find application in foods not just for their bioactive potential but also for their coloring capabilities, enhancing the visual appeal of products to consumers (Emin et al., 2012). Significant instability in carotenoids can occur during the extrusion cooking process. Depending on extrusion parameters like residence time and temperature, heat experienced during processing can cause carotenoids to partially isomerize or completely degrade. Recent investigations have shed light on specific alterations in carotenoids observed during extrusion cooking.

For instance, in a study involving Hokkaido pumpkin powder (PP) and ascorbic acid (AA), these were added to corn grits at different levels and subjected to extrusion at varying temperatures. Compared to carotenes, xanthophylls (lutein and zeaxanthin) were less impacted by extrusion. Despite its own breakdown, AA preserved the color of the extrudate and shielded carotenoids due to its increased sensitivity to heat. Antioxidant activity and polyphenol content increased with the additions. The levels of crude fiber, polyphenols, and antioxidant activity were all positively impacted by higher extrusion temperatures (Obradović et al., 2015). According to MARTY and BERSET (1986), the degradation of β-carotene during extrusion at 180°C was more noticeable than the loss that occurred with prolonged heating at the same temperature. This demonstrates how shear pressures have a major effect on carotenoid degradation.

Goji berries' carotenoid stability was assessed postextrusion at temperatures of 110, 130, and 150°C. There were no statistically significant variations observed in the approximately 60% loss of zeaxanthin dipalmitate, regardless of the process temperature. This implies that carotenoid degradation may have been caused by additional process factors, such as shear (Kosińska-Cagnazzo et al., 2017). By dissolving the compound in oil and adding it during the last stages of the extrusion process, about 40% of β-carotene can be maintained (Emin et al., 2012). A 90% β-carotene retention increase during extrusion was achieved by incorporating the compound into a heated protein–carbohydrate matrix before introducing it into the process. Carotenoids can be added to organic solutions in an efficient manner to increase their bioavailability in the human body since they are highly soluble in nonpolar solvents such as vegetable oils. Thus including carotenoids in saturated oils increases the stability of lipid oxidation by preventing the production of radical species that could react with carotenoids (Ambati et al., 2014). Ascorbic acid was added to an extrusion formulation to improve the retention of carotenoids included in PP, which in turn affected the stability of β-carotene isomers. The ensuing extrudates' retention of their crimson color was indicative of this effect (Obradović et al., 2015).

11.4.1.4 Changes in other phytochemical compounds

Other significant phytochemical compounds, including chlorophylls, polyacetylenes, *Allium* compounds, and betalains, exhibit sensitivity to extrusion cooking. In a research study by (Bekele et al., 2021), chickpea–sorghum blends with varying ratios (50:50, 60:40, and 70:30 chickpea/sorghum, w/w) were extruded under different moisture content and barrel temperature conditions. The 70:30 chickpea–sorghum blend, extruded at maximum expansion, showed the highest protein quality, indicating its suitability for snack production. Extrusion altered the limiting amino acids in certain blends. Overall, protein quality improved for chickpea–sorghum extrudates but not for sorghum extrudates.

In addition, Pitalua et al. (2010) observed changes in the vibrant red color of beetroot, attributed to betanin pigments, during extrusion. The retention of red beetroot color increased with higher moisture content and screw speed but decreased with an increase in die temperature. The addition of 1% citric acid proved effective, reducing the color degradation by almost 40% when used for preextrusion coloring. An observed inverse relationship between betalain stability and light intensity suggests that higher light intensity stimulates electrons in the pigment chromophore to a more energetic state, leading to greater reactivity and lower activation energy, which results in pigment loss (Herbach et al., 2004). Table 11.2 provides additional insights into phytochemical alterations in diverse food materials pre- and postextrusion cooking.

Table 11.2 Phytochemical changes in various food materials before and after the extrusion cooking process.

Raw material	Phytochemical	Process condition	Analysis methods	Before extrusion	After extrusion	Reference
Rice bran, polished rice, brown rice	Total phenolic content	Twin-screw extruder; L/D ratio: 27:1; screw diameter: 37 mm; die opening: 6 mm; feed rate: 25 kg/h; screw speed: 200 rpm	Colorimetric method	RB: 2945.0±114.0 (*mg of GA equiv/100 g DW*); PR: 80.1±1.2; BR: 545.8±3.0	RB: 3449.0±17.2; PR: 37.3±0.9; BR: 111.9±2.5	(Ti et al., 2015)
Cold-pressed oilseed (soybean, canola, sunflower) cakes	Phenolic compounds (extractable polyphenols)	Twin-screw extruder; screw diameter: 27 mm; L/D ratio: 24:1; moisture content: 17%; total throughput: 13.2 kg/h; screw speed: 500 rpm	Colorimetric method	Soybean: 2.5±0.1 (mg *GAE/g cake; d.b.*); canola: 35.7±2.8; sunflower: 25.4±1.3	Soybean: 2.6±0.4; canola: 43.1±3.2; sunflower: 8.2±1.3	(Vidal et al., 2022)
Rapeseed meal	High glucosinolate	Twin-screw extruder; speed of screw (200–400 rpm); moisture content (14–20%); NH_3 concentration (0–3%)	Thymol method	116.2 μ *mol/g*$^{-1}$	38.3 (Maximum reduction of 67% occurred at 200 rpm and 2% NH_3 concentration)	(Huang et al., 1995)
Banana flour (BF)	Total phenolic content (extractable polyphenols)	Twin-screw extruder with corotating screws; screw diameter: 25 mm; L/D ratio: 25:1; feed input flow rate: 10 kg/h; specific mechanical energy (SME): 140–160 kJ/kg	Chromatography and mass spectrometry	BF oven-dried: 337.79 ±9.79 *mg/100 g of gallic acid equivalents*; BF freeze-dried: 335.66±9.79	BF oven-dried: 1195.97 ±14.28; BF freeze-dried: 515.82b±2.65	(Pico et al., 2019)
Debittering lupin (*L. mutabilis*) seeds	Carotenoids	Flour hydrated to 35.0%, at a pressure of 20.0 MPa, with different sections in the extruder set at temperatures of 95°C, 120°C, 140°C, and 130°C	Normal-phase high-performance liquid chromatography (NP-HPLC)	Total carotenoid content for the Altagracia location: 1.15 *mg/kg DM*	Total carotenoid content for the Altagracia location: 0.86 *mg/kg DM*	(Brandolini et al., 2022)
White maize (70%) +common bean kernels (30%)	Free and bound phenolic compunds	Single-screw extruder; screw diameter: 19 mm; L/D ratio: 20:1; compression ratio: 3:1; die opening size: 3 mm; barrel temperature: 164°C; screw speed: 187 rpm	Ultra-performance liquid chromatography coupled to diode array detection and mass spectrometry (UPLC-DAD-ESI-MS^n)	Total phenolic content: free: 36.64±0.94 *mg/100 g, dw*; bound: 260.83±3.60	Total phenolic content: free: 12.21±0.30; bound: 256.69±3.89	(Félix-Medina et al., 2021)

Taro flour and mango pulp	β-carotene	Single-screw extruder; L/D ratio: 22.5:1; compression ratio: 3:1; temperature at 4 zones of the extruder: 40°C, 60°C, 80°C, 80–150°C; feed moisture content: 23 g/100 g; screw speed: 100 rpm; feed rate: 20 g/min; extrusion temperature: 115°C	Spectrophotometer	49.91 *μg/g* in the mango pulp portion: *0 g/100 g*	142.50 *μg/g* in mango pulp portion: *10 g/100 g*	(Pensamiento-Niño et al., 2018)
White quinoa	Phenolics	corotating and twin-screw (L/D ratio: 40:1), moisture content: 25%	Ultra-High-Performance Liquid Chromatography with Tandem Mass Spectrometry (UHPLC-MS/MS)	Ferulic acid: 5.88 *mg/g*; rutin: 7.92 mg/g; oleanolic acid: 12.56 mg/g; coumarin: 5.19 mg/g; madecassoside: 0.19 mg/g; botulin: 1.76 mg/g; jasmone: 0.14 mg/g	Ferulic acid: 2.15 *mg/g*; rutin: 2.99 mg/g; oleanolic acid: 2.26 mg/g; coumarin: 0.79 mg/g; madecassoside: 1.22 mg/g; botulin: 2.27 mg/g; jasmone: 0.40 mg/g	(Song and Tang, 2023)
Pecan nuts	Total Phenolic content	Corotating and twin-screw extruder (L/D ratio: 26.6:1), feed rate of 7.79 kg/h dry matter, screw speed: 88.75–211.75 rpm, barrel temperature of 33.25–106.75°C	Folin–Ciocalteu method	p-Hydroxybenzoic acid: 1.46±0.17 *μg/g DS*; gallic acid: 0.90±0.08 μg/g DS; ellagic acid: 1.56±0.23 μg/g DS	p-Hydroxybenzoic acid: 0.88±0.10 *μg/g DS*; gallic acid: 1.55 ±0.09 μg/g DS; ellagic acid: 1.74 ±0.17 μg/g DS	(Villasante et al., 2019)
Lupin seed coat	Polyphenols	Extruder type: Corotating intermeshing twin-screw, feed rate: 4 kg/h, barrel temperatures: varies between 120°C to 150°C, screw speed: 400 rpm, moisture content: 40%	Folin–Ciocalteu assay	54.26±0.69 *mg GAE/100 g db*	At 120°C: 47.55±1.35 At 135°C: 50.32±1.49 At 150°C: 46.65±0.56	(Zhong et al., 2021)
Corn grits+cocoa shell	Total polyphenols	Single-screw extruder, 4:1 screw compression ratio, die opening: 4 mm, temperature profile: 135°C, 170°C, 170°C	Folin-Ciocalteau method	Corn grits: 55.17 *mg GAE/100 g d.m.* Corn grits: cocoa shell (95:5): 84.37 Corn grits: cocoa shell (90:10): 105.14 Corn grits: cocoa shell (85:15): 109.91	Corn grits: 48.63 *mg GAE/100 g d.m.* Corn grits: cocoa shell (95:5): 72.25 Corn grits: cocoa shell (90:10): 83.99 Corn grits: cocoa shell (85:15): 105.68	(Jozinović et al., 2019)

Continued

Table 11.2 Phytochemical changes in various food materials before and after the extrusion cooking process—cont'd

Raw material	Phytochemical	Process condition	Analysis methods	Before extrusion	After extrusion	Reference
Milled quinoa	Extractable and hydrolyzable phenolic compounds (EPP and HPP); carotenoids (C)	Moisture content: 18%; twin-screw extruder with three zones of barrel temperature set at 75°C, 105°C, and 135°C; screw rotation speed of 251–253 rpm; and three nozzles with a diameter of 2.6 mm	Folin–Ciocalteu reaction and spectrophotometric method	EPP: 1.79±0.06 *mg AG/g* HPP: 4.58±0.10 C: 11.33±0.82 *μg β-Caroteno/g*	EPP: 0.75±0.03 *mg AG/g* HPP: 7.23±0.78 C: 8.39±0.69 *μg β-Caroteno/g*	(Muñoz-Pabon et al., 2022)
Corn grits (70%) +tomato powder (30%)	Total lycopene content (TLC) and total phenolics content (TPC)	single screw polytropic extruder, L: D=12:1, screw compression ratio of 3:1; barrel three zones were set at 125°C, 145°C, and 135°C temperatures; die opening: 3 mm; screw speed: 120 rpm	UV–Vis spectrophotometer (TLC) and Folin–Ciocalteu's method (TPC)	TLC=31.2±0.3 *mg/100 g*; TPC=109.7 ±6.9 *mg GAE/100 g*	TLC=22.3±0.1; TPC = 239.6±4.3	(Wójtowicz et al., 2018)
Corn grits (CG), brewer's spent grain (BSG), sugar beet pulp (SBP), apple pomace (AP)	Total phenolic content	Single-screw extruder, temperature steps of 135/170/170°C, compression ratio of 4:1; 4 mm die opening; blend's moisture content−15%	Folin–Ciocalteu colorimetric method	CG: 61.38±1.64 *mg GAE/100 g d.m.* CG+10% BSG: 69.74±0.79 CG+10% SBP: 53.30±2.88 CG+5% AP: 337.86±2.21	CG: 48.39±1.06 *mg GAE/100 g d.m.* CG+10% BSG: 60.10±1.71 CG+10% SBP: 51.04±2.88 CG+5% AP: 285.36±3.47	(Jozinović et al., 2021)
Brassica juncea (mustard) meal concentrate (MMC)	Total phenolic content	Twin-screw extruder, barrel temperature of 100–150°C, screw speed of 250–350±5 rpm, and moisture content of 12–18%	Folin–Ciocalteu method	5% MMC: 76.47 *mg of GAE/100 g db* 10% MMC: 84.49 15% MMC: 110.76±0.01	5% MMC: 75.11 *mg of GAE/100 g db* 10% MMC: 82.63 15% MMC: 108.54±0.02	(Chakraborty et al., 2021)
Rice (RF) and corn (CF) flour, +fenugreek seed powder (FSP), oat flour (OF), dried green pea flour (DGPF), and fenugreek leaf powder (FLP)	Total flavonoid content	Twin-screw extruder, feed moisture content: 12%, barrel temperatures: 110°C, screw speed: 200 rpm	Dowd technique	22% OF+9% DGPF+2% FSP+0.70% FLP +53.04% RF+13.26% CF: 1003±13 *μg (CE)/g*	802±9 *μg(CE)/g*	(Wani et al., 2021)

maize/common bean (70/30%) mixture	Phenolic acids and flavonoids	Single-screw extruder, screw diameter: 19 mm, die opening: 3 mm, barrel temperature: 164°C, screw speed: 187 rpm	Ultra-performance liquid chromatography coupled to diode array detection and mass spectrometry (UPLC-DAD-ESI-MS^n)	260.83±3.60 *mg/100 g, dw* 87.37 *mg/100 g, dw*	256.69±3.89 *mg/100 g, dw*; 86.23 *mg/100 g, dw*	(Félix-Medina et al., 2021)
Corn grits (CG) +dried carrot strips (CP) +ascorbic acid (AA)	Carotenoid (C)	Single-screw extruder: barrel temperature at three zones: 135°C, 170°C, 170°C; compression ratio: 4:1; screw speed: 100 rpm; feed rate: 15 rpm	HPLC-UV/VIS/PDA method (C)	• *CG*: 32.8 ±2.5 $mg\ kg^{-1}$ • *4% CP+1% AA*: 21.7±1.9 • *6% CP+1% AA*: 21.2±2.08% *CP* +*1% AA*: 16.7±3.0	• *CG*: 84.5 ±1.7 $mg\ kg^{-1}$ • *4% CP+1% AA*: 53.9±2.1 • *6% CP+1% AA*: 90.0±1.88% *CP* +*1% AA*: 75.0±2.4	(Obradović et al., 2021)
	Total polyphenol content (TPC)		Folin–Ciocalteu reagent (TPC)	• *CG*: 0.729±0.016 $gGAE\ kg^{-1}$ • *4% CP+1% AA*: 0.752±0.042 • *6% CP+1% AA*: 0.822±0.041 • *8% CP+1% AA*: 1.01±0.032	• *CG*: 2.393±0.014 $gGAE\ kg^{-1}$ • *4% CP+1% AA*: 1.629±0.029 • *6% CP+1% AA*: 1.970±0.022 • *8% CP+1% AA*: 1.937±0.002	
Carrot pulp+corn grits (1:4)	phenolic content (TPC), β-carotene, and lutein contents	Twin-screw corotating intermeshing; L/D ratio: 25:1; die opening: 3 mm; feed flow rate: 36±1 g/min; screw speed: 225 rpm; barrel zone temperatures: 80, 105, 130, 160°C	Folin–Ciocalteu reagent & HPLC analysis	TPC: 18.15±2.83 *mg/g* β-carotene: 11.13±2.43 *μg/g d.w.* lutein content: 2.66±0.47 *μg/g d.w.*	TPC: 10.01±0.36 *mg/g* β-carotene: 9.26 ±2.02 *μg/g d.w.* lutein content: 1.38 ±0.30 *μg/g d.w.*	(Ortak et al., 2017)
Cassava-soy composite with grape pomace (GP) [at three levels: 0%, 10%, and 20% (w/w)]	Total phenolic content	Twin-screw, corotation extruder: Barrel temperature at 5 zones: 60°C, 80°C, 100°C, 140°C, 140°C; moisture dosing rate-3Lh^{-1}, L/h; die opening: 3 mm; feed rate: 25 kg/h; screw speed: 200 rpm	Folin Ciocalteu procedure	0% GP: 2.1±0.2 *mg CE/g* 10% GP: 5.5±0.2 20% GP: 7.7±0.1	0% GP: 0.9±0.0 *mg CE/g* 10% GP: 3.5±0.2 20% GP: 5.9±0.1	(Oladiran and Emmambux, 2018)

Continued

Table 11.2 Phytochemical changes in various food materials before and after the extrusion cooking process—cont'd

Raw material	Phytochemical	Process condition	Analysis methods	Before extrusion	After extrusion	Reference
Corn with pea and rosehip	Total phenols (TP)	Single-screw extruder, compression ratio: 3:1; barrel diameter: 19 mm; feed rate: 3.51 kg/h; rotation speed: 150 rpm; barrel temperatures: 25°C, 70°C, 170°C, 175°C; nozzle diameter: 3 mm	Folin–Ciocalteu reagent	Control (corn flour): 23.68 ± 1.15 *mg/100 g d.w.* 10% rosehip: 334.00 ± 9 10% rosehip + pea protein: 272.00 ± 3	Control (corn flour): 14.40 ± 0.3 10% rosehip: 169.20 ± 0.7 10% rosehip + pea protein: 169.30 ± 0.8	(Igual et al., 2022)
	Total carotenoids (TC)		UV-3100PC spectrophotometer	Control (corn flour): 3.75 ± 0.09 10% rosehip: 43.74 ± 0.14 10% rosehip + pea protein: 22.73 ± 0.03	Control (corn flour): 2.11 ± 0.06 10% rosehip: 7.85 ± 0.04 10% rosehip + pea protein: 5.76 ± 0.02	
Sweet potato flours	Total carotenoid content (TCC), $\beta-$carotene	Twin-screw extruder, corotating, moisture (30%, 35%, and 40%), and screw speed (150, 220, 300 rpm); feed rate: 1.5 kg/h; maximum barrel temperature: 120°C	Spectrophotometric method	TCC- 1152 μg/g of solids $\beta-$carotene: 317 μg/g of solids	TCC: 517 μg/g of solids $\beta-$carotene: 194 μg/g of solids	(Waramboi et al., 2013)
Five formulations: maize semolina (M); maize + corn extract (AO); maize + chia (C). Maize + quinoa (Q); Maize + chia + corn extract (CAO);	Lutein, zeaxanthin	Counterrotating twin-screw extruder; Max. L/D ratio: 22:1; die opening: 4.5 mm; feed rate: 45 kg/h; screw speed: 82 rpm; temperatures of barrel sections: 50°C, 80°C, 100°C, 150°C	HPLC	Lutein (*in ppm*) M: 0.46 AO: 0.6 Q: 0.5 C: 0.5 CAO: 0.47 Zeaxanthin (*in ppm*) M: 0.26 AO: 0.37 Q: 0.3 C: 0.3 CAO: 0.27	Lutein (in ppm) M: 0.32 AO: 0.3 Q: 0.24 C: 0.15 CAO: 0.1 Zeaxanthin (in ppm) M: 0.17 AO: 0.15 Q: 0.13 C: 0.9 CAO: 0.65	(Cueto et al., 2017)

6 types of mixtures • Control corn grits (CCG) • CCG+2% spirulina powder (SP) • CCG+4%SP • CCG+6%SP • CCG+8%SP • CCG+8% SP +2% baking powder (BP)	Phycocyanin Chlorophyll [a] β−carotene	Single-screw extruder; screw speed: 125 rpm; screw L/D ratio: 12:1; compression ratio: 3:1; die opening: 4.5 mm; feed rate: 25 kg/h; temperatures of barrel sections: 105°C, 130°C, 110°C	Spectrophotometric method	**In CCG***Phycocyanin*: 0.0 *(mg/100 g d.b.)* Chlorophyll [a]: 0.0 β−carotene: 0.6 **In Spirulina powder, phycocyanin**: 2446.0 *(mg/100 g d.b.)* *Chlorophyll [a]*- 703.5 β−carotene: 156.5	*Phycocyanin* CCG: 0.0 2%SP−0.0 4%SP: 0.0 6%SP-0.0 8%SP-traces (tr) 8%SP+2%BP- tr *Chlorophyll [a]* CCG: 0.0 2%SP: 3.2 4%SP: 6.2 6%SP: 12.6 8%SP: 20.8 8%SP+2%BP: 27.5 β−carotene CCG: 0.5 2%SP: 0.8 4%SP: 1.1 6%SP: 1.5 8%SP: 6.8 8%SP+2%BP: 8.8	(Tańska et al., 2017)

11.4.2 Effect on phytochemical properties in extrusion cooking for feed

Feed technology plays a critical role in animal production systems, where livestock are raised in controlled environments with tailored diets to enhance growth rates and feed conversion efficiencies, producing high-quality, nutritious animal food. The ingredients in animal feeds are sourced from diverse materials, including plants, animals, pharmaceuticals, and industrial sources. The future of livestock production, vital for providing animal-derived food products and quality proteins for human well-being, relies on the judicious use of raw ingredients in feed production. Specific feed ingredients vary based on the animal type, such as pets, goats, poultry, rabbits, swine, and cattle (Tran et al., 2011).

Approximately 90% of animal and aqua food is extruded, highlighting the widespread use of extrusion cooking in feed production (Julie and Jr, 2004). This technique offers versatility to mix diets and enhance various food commodities, providing functions such as detoxification, sterilization, and texturization. However, a comprehensive understanding of the structures and functional properties of feed components, their changes during extrusion cooking, and their specific conditions is essential to accurately meet animal nutrient requirements.

Pet food extrusion involves steam-conditioning and compressing a mixture of ingredients, forcing it through a heated die to create kibbles cut by rotating knives. The process improves digestibility but adversely affects vitamins due to heat, pressure, and shear conditions. Vitamin loss is a significant concern, as commercially prepared pet food is a primary nutrient source for dogs and cats, and deficiencies can lead to severe illnesses or death (Morin et al., 2021). A $3 \times 2 \times 2$ factorial trial investigated the impact of various extrusion conditions on canine diet kibble. Parameters included mass temperature (110, 130, or 150°C), moisture content (200 or 300 g/kg), and number of extrusions (once or twice). Reactive lysine increased with temperature and extrusion but decreased after a second extrusion. Initial moisture content influenced lysine reactivity. Protein digestibility remained unaffected. Glucose digestibility and starch gelatinization showed an increasing trend with parameter increments. Kibble durability decreased with higher temperature and second extrusion, while increased moisture content improved durability (Lankhorst et al., 2007).

Fibrous ingredients, commonly used for weight management and gut health in pet diets, were evaluated. Diets composed of 90% basal ration and 10% beet pulp (BP), cellulose (CE), or miscanthus grass (MG) were assessed during extrusion. Total dietary fiber remained similar across processing conditions, suggesting that the extrusion parameters minimally affected any fiber source (Donadelli et al., 2021). Examining lipid oxidation in extruded dry pet food, a study evaluated the impact of fat type (beef tallow and poultry fat), fat content (0–0.75 g kg^{-1}), and processing conditions (initial moisture: 1.6–2 g kg^{-1}; screw speed: 200–400 rpm). The results revealed that the lipid oxidation rate depended on fat types, added fat content, and feed moisture. Control samples (no fat) exhibited the highest oxidation rate, while poultry fat had a higher rate than beef tallow. Higher fat content and moisture resulted in lower oxidation rates during storage. Extrudates at 300 rpm showed significantly higher lipid oxidation rates compared to those at 200 and 400 rpm. The degree of extrudate expansion, affecting cell size and wall thickness, appeared to be the main factor influencing lipid oxidation (Boakye et al., 2023).

Moving beyond pet food, Yang et al. (2020) found that microencapsulated vitamin retention in fortified swine feeds, including vitamin A, was up to 11% higher than the nonmicroencapsulated form. In another study, β-carotene retention was reported to be 95% in an emulsion stabilized by a heated protein–carbohydrate matrix, 70% in an emulsion stabilized by Tween 80 (emulsifier), and 65% when dissolved in medium-chain triglycerides (MCT) oil during corn-based product extrusion at 170°C and a screw speed of 200 rpm (Lu et al., 2020).

In the context of aquaculture, an experiment evaluated the impact of extrusion processing on digestibility coefficients in rainbow trout diets containing soybean meal, barley, corn gluten meal, and whole wheat. Extrusion significantly improved dry matter (DM), crude fat, and gross energy apparent digestibility coefficients. The study recommended supplementing trace minerals like copper, iron, and zinc by 10–20% in extruded rainbow trout diets due to reduced bioavailability in plant-based feed ingredients (Cheng and Hardy, 2003).

Shifting to poultry, a study investigated the impact of feed processing methods on production parameters. Identical broiler feeds were administered in three forms, namely, nonheat-treated mash, pellets, or extruded pellets. The results revealed distinct effects on broiler weight based on the form of the feed. Broilers consuming extruded crumbled starter rations exhibited higher weights compared to those on pelleted diets. Conversely, uncrumbled extruded diets led to lower bird weights than pelleted diets. The extruded feeds, having a lower weight per unit volume, may contribute to the observed variations. Notably, traditional ether extraction indicated lower fat content in extruded feeds than acid hydrolysis. The extrusion process, however, resulted in some protein and amino acid degradation, emphasizing the importance of adjusting nutrient density in formulated extruded rations to compensate for potential nutrient loss (Huang et al., 1995).

Another study evaluated the impact of extrusion on the nutritional value of pea seeds (*Pisum sativum* L.) from various fields. One hundred and fifty male *Ross 308* chickens were randomly assigned to three dietary treatments and kept individually. From days 1 to 16, all birds received basal diets. From days 17 to 21, the control group continued with the basal diet, while the other two groups had 20% of the basal diet replaced by raw or extruded peas. Extrusion of pea seeds resulted in reduced levels of crude fiber, acid and neutral detergent fiber, trypsin inhibitor activity, phytic P, and resistant starch, while increasing apparent metabolizable energy (AMEN) by approximately 2.25 MJ/kg DM. Extrusion significantly improved DM and crude protein digestibility, along with enhanced digestibility of all analyzed amino acids. In summary, extrusion had a positive impact on the chemical composition of peas, reducing antinutritional factors and improving nutrient digestibility and AMEN values (Hejdysz et al., 2016).

11.5 Conclusions

Extrusion stands as a practical method for crafting innovative snacks infused with functional elements derived from plant sources. Cereals, vegetables, fruits, legumes, seeds, and their by-products are common ingredients used to make extruded snacks. The degree to which the primary and secondary metabolites vary depends critically on the chemical composition and structure of the food matrix. Screw speed, barrel temperature profile, and pressure are some of the crucial processing parameters that have a major impact on the bioactive composition of the finished products. Thorough exploration of this aspect, coupled with a deeper understanding of the interactions among bioactive components during extrusion, is imperative. Furthermore, it is advisable to conduct studies that specifically explore the industrial application and impacts of extrusion on a larger scale, as a significant portion of the current literature predominantly focuses on laboratory-scale experiments.

Acknowledgment

The authors acknowledge the financial support of the Ministry of Higher Education, Malaysia, through the Transdisciplinary Research Grant Scheme Project (TRGS/1/2020/UPM/02/7).

References

Ahmad, C., Iqbal, A., Batool, T., Azam, A., 2019. Phytochemicals in Daucus carota and Their Health Benefits—Review Article. Foods. 2304-8158. 8 (9), 424. doi:10.3390/foods8090424.

Akin, M., Eyduran, S.P., Ercisli, S., Kapchina-Toteva, V., Eyduran, E., 2016. Phytochemical profiles of wild blackberries, black and white mulberries from southern Bulgaria. Biotechnol. Biotechnol. Equip. 13102818. 30 (5), 899–906. Available from: http://www.tandfonline.com/loi/tbeq20#.Vx8TFE1f3cs. doi:10.1080/13102818.2016.1204943.

Al-Qattan, K., Thomson, M., Ali, M., 2008. Garlic (Allium sativum) and ginger (Zingiber officinale) attenuate structural nephropathy progression in streptozotocin-induced diabetic rats. e-SPEN, the European e-Journal of Clinical Nutrition and Metabolism. 17514991. 3 (2), e62. doi:10.1016/j.eclnm.2007.12.001.

Alasalvar, C., Chang, S.K., Bolling, B., Oh, W.Y., Shahidi, F., 2021. Specialty seeds: nutrients, bioactives, bioavailability, and health benefits: a comprehensive review. Compr. Rev. Food Sci. Food Saf. 15414337. 20 (3), 2382–2427. Available from: http://onlinelibrary.wiley.com/journal/10.1111/(ISSN)1541-4337. doi:10.1111/1541-4337.12730.

Alonso, R., Aguirre, A., Marzo, F., 2000. Effects of extrusion and traditional processing methods on antinutrients and in vitro digestibility of protein and starch in faba and kidney beans. Food Chem. 03088146. 68 (2), 159–165. doi:10.1016/s0308-8146(99)00169-7.

Alonso, R., Rubio, L.A., Muzquiz, M., Marzo, F., 2001. The effect of extrusion cooking on mineral bioavailability in pea and kidney bean seed meals. Anim. Feed Sci. Technol. 03778401. 94 (1-2), 1–13. doi:10.1016/s0377-8401(01)00302-9.

Ambati, R.R., Moi, P.S., Ravi, S., Aswathanarayana, R.G., 2014. Astaxanthin: sources, extraction, stability, biological activities and its commercial applications - A review. Mar. Drugs. 16603397. 12 (1), 128–152. Available from: http://www.mdpi.com/1660-3397/12/1/128/pdf. doi:10.3390/md12010128.

Arora, B., Yoon, A., Sriram, M., Singha, P., Rizvi, S.S.H., 2020. Reactive extrusion: a review of the physicochemical changes in food systems. Innov. Food Sci. Emerg. Technol. 14668564. 64, 102429. doi:10.1016/j.ifset.2020.102429.

Arribas, C., Cabellos, B., Cuadrado, C., Guillamón, E., Pedrosa, M.M., 2019. The effect of extrusion on the bioactive compounds and antioxidant capacity of novel gluten-free expanded products based on carob fruit, pea and rice blends. Innov. Food Sci. Emerg. Technol. 14668564. 52, 100–107. Available from: http://www.elsevier.com/locate/ifset. doi:10.1016/j.ifset.2018.12.003.

Awad, A.B., Fink, C.S., 2000. Phytosterols as anticancer dietary components: evidence and mechanism of action. J. Nutr. 00223166. 130 (9), 2127–2130. Available from: http://jn.nutrition.org. doi:10.1093/jn/130.9.2127.

Ayaz, M., Mosa, O.F., Nawaz, A., Hamdoon, A.A.E., Elkhalifa, M.E.M., Sadiq, A., Ullah, F., Ahmed, A., Kabra, A., Khan, H., Murthy, H.C.A., 2024. Neuroprotective potentials of Lead phytochemicals against Alzheimer's disease with focus on oxidative stress-mediated signaling pathways: pharmacokinetic challenges, target specificity, clinical trials and future perspectives. Phytomedicine. 1618095X (124) Available from: https://www.sciencedirect.com/science/journal/09447113. doi:10.1016/j.phymed.2023.155272.

Barnes, S., Prasain, J., Kim, H., 2013. In Nutrition, Can We "See" What Is Good for Us? Adv. Nutr. 21618313. 4 (3), 327S. doi:10.3945/an.112.003558.

Bekele, E.K., Tyler, R.T., Henry, C.J., House, J.D., Nosworthy, M.G., 2021. In vitro protein digestibility of direct-expanded chickpea–sorghum snacks. Legume Sci. 26396181. 3 (4). Available from: https://onlinelibrary.wiley.com/journal/26396181. doi:10.1002/leg3.87.

Berk, Z., 2009. Extrusion. Elsevier BV, pp. 333–350. doi:10.1016/b978-0-12-373660-4.00015-6.

Björck, I., Asp, N.-G., 1983. The effects of extrusion cooking on nutritional value — a literature review. J. Food Eng. 02608774. 2 (4), 281–308. doi:10.1016/0260-8774(83)90016-x.

Boakye, P.G., Okyere, A.Y., Bharathi, R., Murai, T., Annor, G.A., 2023. Physicochemical and nutritional properties of extruded products from cereals of the Triticeae tribe – a review. Food Chem. Adv. 3. 10.1016/j.focha.2023.100379

Bouvier, J.M., Campanella, O.H., 2014. Extrusion Processing Technology: Food and Non-Food Biomaterials (John Wiley & Sons, Ltd.). 9781444338119, pp. 1–518, Available from: https://onlinelibrary.wiley.com/doi/book/10.1002/9781118541685. doi:10.1002/9781118541685.

Boyer, J., Liu, R.H., 2004. Apple phytochemicals and their health benefits. Nutr. J. 14752891. 3, 1–45. Available from: http://www.nutritionj.com/content/3/1/5. doi:10.1186/1475-2891-3-1.

Bradford, P.G., Awad, A.B., 2007. Phytosterols as anticancer compounds. Mol. Nutr. Food Res. 16134133. 51 (2), 161–170. doi:10.1002/mnfr.200600164.

Brandolini, A., Glorio-Paulet, P., Estivi, L., Locatelli, N., Cordova-Ramos, J.S., Hidalgo, A., 2022. Tocopherols, carotenoids and phenolics changes during Andean lupin (Lupinus mutabilis Sweet) seeds processing. J. Food Compos. Anal. 08891575. 106, 104335. doi:10.1016/j.jfca.2021.104335.

Brennan, M.A., Derbyshire, E., Tiwari, B.K., Brennan, C.S., 2013. Ready-to-eat snack products: the role of extrusion technology in developing consumer acceptable and nutritious snacks. Int. J. Food Sci. Technol. 13652621. 48 (5), 893–902. doi:10.1111/ijfs.12055.

Bruin, S., Van Zuilichem, D.J., Stolp, W., 1978. A review of fundamental and engineering aspects of extrusion of biopolymers in a single-screw extruder. J. Food Process Eng. 0145-8876. 2 (1), 1–37. doi:10.1111/j.1745-4530.1978.tb00193.x.

Camire, M., 2012. Nutritional Changes during Extrusion Cooking. Informa UK Limited, pp. 87–102. doi:10.1201/b11286-5.

Camire, M.E., Camire, A., Krumhar, K., 1990. Chemical and nutritional changes in foods during extrusion. Crit. Rev. Food Sci. Nutr. 1040-8398. 29 (1), 35–57. doi:10.1080/10408399009527513.

Cao, Q., Mei, S., Mehmood, A., Sun, Y., Chen, X., 2024. Inhibition of pancreatic lipase by coffee leaves-derived polyphenols: a mechanistic study. Food Chem. 03088146. 444, 138514. doi:10.1016/j.foodchem.2024.138514.

Chakraborty, A.J., Uddin, T.M., Zidan, B.M.R.M., Mitra, S., Das, R., Nainu, F., Dhama, K., Roy, A., Hossain, M.J., Khusro, A., Emran, T.B., 2022. Allium cepa: a Treasure of Bioactive Phytochemicals with Prospective Health Benefits. Evid. Based Complement. Alternat. Med. 17414288. 2022. Available from: http://www.hindawi.com/journals/ecam/contents.html. doi:10.1155/2022/4586318.

Chakraborty, P., Bhattacharyya, D.K., Ghosh, M., 2021. Extrusion treated meal concentrates of Brassica juncea as functionally improved ingredient in protein and fiber rich breadstick preparation. LWT. 00236438. 142, 111039. doi:10.1016/j.lwt.2021.111039.

Chamone, M.E.R., Ascheri, J.L.R., Vargas-Solórzano, J.W., Stephan, M.P., Carvalho, C.W.P., 2023. Chemical Characterization of White Lupin (Lupinus albus) Flour Treated by Extrusion Cooking and Aqueous Debittering Processes. Plant Foods Hum. Nutr. 15739104. 78 (2), 292–298. Available from: https://www.springer.com/journal/11130. doi:10.1007/s11130-023-01050-0.

Cheftel, J.C., 1986. Nutritional effects of extrusion-cooking. Food Chem. 03088146. 20 (4), 263–283. doi:10.1016/0308-8146(86)90096-8.

Cheng, Z.J., Hardy, R.W., 2003. Effects of extrusion processing of feed ingredients on apparent digestibility coefficients of nutrients for rainbow trout (Oncorhynchus mykiss). Aquac. Nutr. 13535773. 9 (2), 77–83. doi:10.1046/j.1365-2095.2003.00226.x.

Connolly, E.L., Sim, M., Travica, N., Marx, W., Beasy, G., Lynch, G.S., Bondonno, C.P., Lewis, J.R., Hodgson, J.M., Blekkenhorst, L.C., 2021. Glucosinolates From Cruciferous Vegetables and Their Potential Role in Chronic Disease: investigating the Preclinical and Clinical Evidence. Front. Pharmacol. 16639812. 12. Available from: http://www.frontiersin.org/Pharmacology. doi:10.3389/fphar.2021.767975.

Crowley, M.M., Zhang, F., Repka, M.A., Thumma, S., Upadhye, S.B., Battu, S.K., McGinity, J.W., Martin, C., 2007. Pharmaceutical applications of hot-melt extrusion: part I. Drug Dev. Ind. Pharm. 15205762. 33 (9), 909–926. doi:10.1080/03639040701498759.

Cubeddu, A., Rauh, C., Delgado, A., 2014. Hybrid artificial neural network for prediction and control of process variables in food extrusion. Innov. Food Sci. Emerg. Technol. 14668564. 21, 142–150. doi:10.1016/j.ifset.2013.10.010.

Cueto, M., Farroni, A., Schoenlechner, R., Schleining, G., Buera, P., 2017. Carotenoid and color changes in traditionally flaked and extruded products. Food Chem. 03088146. 229, 640–645. doi:10.1016/j.foodchem.2017.02.138.

Dalbhagat, C.G., Mahato, D.K., Mishra, H.N., 2019. Effect of extrusion processing on physicochemical, functional and nutritional characteristics of rice and rice-based products: a review. Trends Food Sci. Technol. 09242244. 85, 226–240. Available from: http://www.elsevier.com/wps/find/journaldescription.cws_home/601278/description#description. doi:10.1016/j.tifs.2019.01.001.

Deka, B., Manna, P., Borah, J.C., Talukdar, N.C., 2022. A review on phytochemical, pharmacological attributes and therapeutic uses of Allium hookeri. Phytomedicine Plus. 26670313. 2 (2). Available from: https://www.journals.elsevier.com/phytomedicine-plus. doi:10.1016/j.phyplu.2022.100262.

Della Valle, G., Tayeb, J., Melcion, J.P., 1987. Relationship of extrusion variables with pressure and temperature during twin screw extrusion cooking of starch. J. Food Eng. 02608774. 6 (6), 423–444. doi:10.1016/0260-8774(87)90003-3.

Donadelli, R.A., Dogan, H., Aldrich, G., 2021. The effects of fiber source on extrusion parameters and kibble structure of dry dog foods. Anim. Feed Sci. Technol. 03778401. 274, 114884. doi:10.1016/j.anifeedsci.2021.114884.

Ebrahimi, P., Shokramraji, Z., Tavakkoli, S., Mihaylova, D., 2023. Anna Lante, Chlorophylls as Natural Bioactive Compounds Existing in Food By-Products: a Critical Review. Plants. 2223-7747. 12 (7), 1533. doi:10.3390/plants12071533.

Elsey, J., Riepenhausen, J., McKay, B., Barton, G.W., Willis, M., 1997. Modeling and control of a food extrusion process. Comput. Chem. Eng. 00981354. 21 (1), S361. doi:10.1016/S0098-1354(97)87528-X.

Emin, M.A., Mayer-Miebach, E., Schuchmann, H.P., 2012. Retention of β-carotene as a model substance for lipophilic phytochemicals during extrusion cooking. LWT. 00236438. 48 (2), 302–307. doi:10.1016/j.lwt.2012.04.004.

Félix-Medina, J.V., Gutiérrez-Dorado, R., López-Valenzuela, J.A., López-Ángulo, G., Quintero-Soto, M.F., Perales-Sánchez, J.X.K., Montes-Ávila, J., 2021. Nutritional, antioxidant and phytochemical characterization of healthy ready-to-eat expanded snack produced from maize/common bean mixture by extrusion. LWT. 00236438. 142. Available from: https://www.journals.elsevier.com/lwt. doi:10.1016/j.lwt.2021.111053.

Gayathry, K.S., John, J.A., 2022. A comprehensive review on bitter gourd (Momordica charantia L.) as a gold mine of functional bioactive components for therapeutic foods. Food Production, Processing and Nutrition. 2661-8974. 4 (1). doi:10.1186/s43014-022-00089-x.

Gengatharan, A., Dykes, G.A., Choo, W.S., 2016. Stability of betacyanin from red pitahaya (Hylocereus polyrhizus) and its potential application as a natural colourant in milk. Int. J. Food Sci. Technol. 13652621. 51 (2), 427–434. doi:10.1111/ijfs.12999.

Gizaw, A., Marami, L.M., Teshome, I., Sarba, E.J., Admasu, P., Babele, D.A., Dilba, G.M., Bune, W.M., Bayu, M.D., Tadesse, M., Abdisa, K., 2022. Phytochemical Screening and in Vitro Antifungal Activity of Selected Medicinal Plants against Candida albicans and Aspergillus Niger in West Shewa Zone, Ethiopia. Adv. Pharmacol. Pharm. Sci. 26334690. 2022. Available from: https://www.hindawi.com/journals/aps/about/. doi:10.1155/2022/3299146.

Gonfa, Y.H., Tessema, F.B., Bachheti, A., Rai, N., Tadesse, M.G., Singab, A.N., Chaubey, K.K., Bachheti, R.K., 2023. Anti-inflammatory activity of phytochemicals from medicinal plants and their nanoparticles: a review. Curr. Res. Biotechnol. 25902628. 6. Available from: https://www.journals.elsevier.com/current-research-in-biotechnology/. doi:10.1016/j.crbiot.2023.100152.

Guo, Q., Wang, N., Liu, H., Zhenjing, L., Laifeng, L., Wang, C., 2020. The bioactive compounds and biological functions of Asparagus officinalis L. – a review. J. Funct. Foods. 17564646. 65, 103727. doi:10.1016/j.jff.2019.103727.

Han, H., Baik, B.K., 2008. Antioxidant activity and phenolic content of lentils (Lens culinaris), chickpeas (Cicer arietinum L.), peas (Pisum sativum L.) and soybeans (Glycine max), and their quantitative changes during processing. Int. J. Food Sci. Technol. 13652621. 43 (11), 1971–1978. doi:10.1111/j.1365-2621.2008.01800.x.

Heinrich, M., Mah, J., Amirkia, V., 2021. Alkaloids Used as Medicines: structural Phytochemistry Meets Biodiversity—An Update and Forward Look. Molecules. 1420-3049. 26 (7), 1836. doi:10.3390/molecules26071836.

Hejdysz, M., Kaczmarek, S.A., Rutkowski, A., 2016. Effect of extrusion on the nutritional value of peas for broiler chickens. Arch. Anim. Nutr. 14772817. 70 (5), 364–377. Available from: http://www.tandf.co.uk/journals/titles/1745039X.asp. doi:10.1080/1745039X.2016.1206736.

Herbach, K.M., Stintzing, F.C., Carle, R., 2004. Impact of thermal treatment on color and pigment pattern of red beet (Beta vulgaris L.) preparations. Institute of Food Technologists, Germany Journal of Food Science. 00221147. 69 (6), C491. Available from: http://onlinelibrary.wiley.com/journal/10.1111/(ISSN)1750-3841. doi:10.1111/j.1365-2621.2004.tb10994.x.

Huang, S., Liang, M., Lardy, G., Huff, H.E., Kerley, M.S., Hsieh, F., 1995. Extrusion processing of rapeseed meal for reducing glucosinolates. Anim. Feed Sci. Technol. 03778401. 56 (1-2), 1–9. doi:10.1016/0377-8401(95)00826-9.

Igual, M., Păucean, A., Vodnar, D.C., García-Segovia, P., Martínez-Monzó, J., Chiş, M.S., 2022. In Vitro Bioaccessibility of Bioactive Compounds from Rosehip-Enriched Corn Extrudates. Molecules. 1420-3049. 27 (6), 1972. doi:10.3390/molecules27061972.

Jaeger, R., Cuny, E., 2016. Terpenoids with special pharmacological significance: a review. Nat. Prod. Commun. 15559475. 11 (9), 1373–1390. Available from: http://members.naturalproduct.us/Secure/Login.aspx?ReturnUrl=%2fSecure%2fViewDoc.aspx%3fdocId%3d9297&docId=9297. doi:10.1177/1934578x1601100946.

Jozinović, A., Balentić, J.P., Ačkar, Đ., Babić, J., Pajin, B., Miličević, B., Guberac, S., Vrdoljak, A., Šubarić, D., 2019. Cocoa husk application in the enrichment of extruded snack products. J. Food Process. Preserv. 17454549. 43 (2). Available from: http://onlinelibrary.wiley.com/journal/10.1111/(ISSN)1745-4549. doi:10.1111/jfpp.13866.

Jozinović, A., Šubarić, D., Ačkar, Đ., Babić, J., Orkić, V., Guberac, S., Miličević, B., 2021. Food Industry By-Products as Raw Materials in the Production of Value-Added Corn Snack Products. Foods. 2304-8158. 10 (5), 946. doi:10.3390/foods10050946.

Julie, S.K., Jr, 2004. Resistant starch as related to companion animal nutrition. J. AOAC Int. 87 (3), 2004.

Kabir, A., Cacciagrano, F., Tartaglia, A., Lipsi, M., Ulusoy, H.I., Locatelli, M., 2020. Analysis of monoterpenes and monoterpenoids. In: Sanches Silva, A., Saeedi Nabavi, S.F., Nabavi, M., Seyed Mohammad (Eds.), Recent Advances in Natural Products Analysis. Elsevier, United States, pp. 274–286. Available from: https://www.elsevier.com/books/recent-advances-in-natural-products-analysis/nabavi/978-0-12-816455-6. doi:10.1016/B978-0-12-816455-6.00007-X.

Kamal-Eldin, A., Appelqvist, L.Å., 1994. Variation in fatty acid composition of the different acyl lipids in seed oils from four Sesamum species. J. Am. Oil Chem. Soc. 0003-021X. 71 (2), 135–139. doi:10.1007/bf02541547.

Kleinebudde, P., Lindner, H., 1993. Experiments with an instrumented twin-screw extruder using a single-step granulation/extrusion process. Int. J. Pharm. 03785173. 94 (1-3), 49–58. doi:10.1016/0378-5173(93)90008-4.

Koch, L., Hummel, L., Schuchmann, H.P., Emin, M.A., 2018. Improving the emulsifying properties of whey protein isolate-citrus pectin blends by a novel reactive extrusion approach. J. Food Eng. 02608774. 223, 175–188. doi:10.1016/j.jfoodeng.2017.10.027.

Korkerd, S., Wanlapa, S., Puttanlek, C., Uttapap, D., Rungsardthong, V., 2016. Expansion and functional properties of extruded snacks enriched with nutrition sources from food processing by-products. J. Food Sci. Technol. 0022-1155. 53 (1), 561–570. doi:10.1007/s13197-015-2039-1.

Kosińska-Cagnazzo, A., Bocquel, D., Marmillod, I., Andlauer, W., 2017. Stability of goji bioactives during extrusion cooking process. Food Chem. 03088146. 230, 250–256. doi:10.1016/j.foodchem.2017.03.035.

Kour, J., Singh, S., Saxena, D.C., 2019. Effect of nutraceuticals (beta-glucan concentrate, flaxseed lignan concentrate and gamma oryzanol concentrate) on nutritional, textural, pasting, thermal, structural and morphological properties of corn and rice flour blend based RTE extrudates. J. Food Meas. Charact. 21934134. 13 (2), 988–1003. Available from: http://rd.springer.com/journal/11694. doi:10.1007/s11694-018-0013-0.

Kumar, A., Nirmal, P., Kumar, M., Jose, A., Tomer, V., Emel, O., Proestos, C., Zeng, M., Elobeid, T., Sneha, K., Fatih, O., 2023. Major Phytochemicals: recent Advances in Health Benefits and Extraction Method. Molecules. 1420-3049. 28 (2), 887. doi:10.3390/molecules28020887.

Kumar, M., Saurabh, V., Tomar, M., Hasan, M., Changan, S., Sasi, M., Maheshwari, C., Prajapati, U., Singh, S., Prajapat, R.K., Dhumal, S., Punia, S., Amarowicz, R., Mekhemar, M., 2021. Mango (Mangifera indica l.) leaves: nutritional composition, phytochemical profile, and health-promoting bioactivities. Antioxidants. 20763921. 10 (2), 1–23. Available from: https://www.mdpi.com/2076-3921/10/2/299. doi:10.3390/antiox10020299.

Kumar, N., Goel, N., 2019. Phenolic acids: natural versatile molecules with promising therapeutic applications. Biotechnol. Rep. 2215017X. 24, e00370. doi:10.1016/j.btre.2019.e00370.

Lai, J.X., Dai, S.F., Xue, B.X., Zhang, L.H., Chang, Y., Yang, W., Wu, H.H., 2023. Plant polyacetylenoids: phytochemical, analytical and pharmacological updates. Arab. J. Chem. 18785352. 16 (10). Available from: http://colleges.ksu.edu.sa/Arabic%20Colleges/CollegeOfScience/ChemicalDept/AJC/default.aspx. (ScienceDirect) http://www.sciencedirect.com/science/journal/18785352. doi:10.1016/j.arabjc.2023.105137.

Lankhorst, C., Tran, Q.D., Havenaar, R., Hendriks, W.H., van der Poel, A.F.B., 2007. The effect of extrusion on the nutritional value of canine diets as assessed by in vitro indicators. Anim. Feed Sci. Technol. 03778401. 138 (3-4), 285–297. doi:10.1016/j.anifeedsci.2006.11.015.

Li, M., Luo, J., Nawaz, M.A., Stockmann, R., Buckow, R., Barrow, C., Dunshea, F., Suleria, H.A.R., 2024. Phytochemistry, Bioaccessibility, and Bioactivities of Sesame Seeds: an Overview. Food Rev. Int. 15256103. 40 (1), 309–335. Available from: www.tandf.co.uk/journals/titles/87559129.asp. doi:10.1080/87559129.2023.2168280.

Liang, M., Huang, S.C., Huff, H.E., Kerley, M.S., Hsieh, F.H., 2002. Extrusion cooking of rapeseed meal for feeding value improvement. Appl. Eng. Agric. 18 (3), 325–330.

Lobato, L.P., Anibal, D., Lazaretti, M.M., Grossmann, M.V.E., 2011. Extruded puffed functional ingredient with oat bran and soy flour. LWT. 00236438. 44 (4), 933–939. doi:10.1016/j.lwt.2010.11.013.

Lu, Y., Mao, L., Zheng, H., Chen, H., Gao, Y., 2020. Characterization of β-carotene loaded emulsion gels containing denatured and native whey protein. Food Hydrocoll. 102. 10.1016/j.foodhyd.2019.105600.

Mahungu, S.M., Diaz-Mercado, S., Li, J., Schwenk, M., Singletary, K., Faller, J., 1999. Stability of Isoflavones during Extrusion Processing of Corn/Soy Mixture. J. Agric. Food Chem. 0021-8561. 47 (1), 279–284. doi:10.1021/jf980441q.

Marta, C.-M., Corzo, N., Villamiel, M., Dolores Del Castillo, M., 2012. Browning Reactions. In: Simpson, B.K. Food Biochemistry and Food Processing. Wiley. doi:10.1002/9781118308035.ch4.

MARTY, C., BERSET, C., 1986. Degradation of trans -β-Carotene during Heating in Sealed Glass Tubes and Extrusion Cooking. J. Food Sci. 0022-1147. 51 (3), 698–702. doi:10.1111/j.1365-2621.1986.tb13915.x.

Mendoza, N., Silva, E.M.E., 2018. Introduction to Phytochemicals: secondary Metabolites from Plants with Active Principles for Pharmacological Importance. IntechOpen. doi:10.5772/intechopen.78226.

Morgan, T.O., Anderson, A.I.E., MacInnis, R.J., 2001. ACE inhibitors, beta-blockers, calcium blockers, and diuretics for the control of systolic hypertension. Am. J. Hypertens. 08957061. 14 (3), 241–247. doi:10.1016/S0895-7061(00)01266-8.

Morin, P., Gorman, A., Lambrakis, L., 2021. A literature review on vitamin retention during the extrusion of dry pet food. Anim. Feed Sci. Technol. 03778401. 277, 114975. doi:10.1016/j.anifeedsci.2021.114975.

Muñoz-Pabon, K.S., Roa-Acosta, D.F., Luis Hoyos-Concha, J., Eduardo Bravo-Gómez, J., Ortiz-Gómez, V., 2022. Quinoa Snack Production at an Industrial Level: effect of Extrusion and Baking on Digestibility, Bioactive, Rheological, and Physical Properties. Foods. 2304-8158. 11 (21), 3383. doi:10.3390/foods11213383.

Musabayane, C.T., Gondwe, M., Kamadyaapa, D.R., Chuturgoon, A.A., Ojewole, J.A.O., 2007. Effects of Ficus thonningii (Blume) [Morarceae] stem-bark ethanolic extract on blood glucose, cardiovascular and kidney functions of rats, and on kidney cell lines of the proximal (LLC-PK1) and distal tubules (MDBK). Ren. Fail. 15256049. 29 (4), 389–397. doi:10.1080/08860220701260735.

Musabayane, C.T., Kamadyaapa, D.R., Gondwe, M., Moodley, K., Ojewole, J.A.O., 2008. Cardiovascular effects of Helichrysum ceres S Moore [Asteraceae] ethanolic leaf extract in some experimental animal paradigms. Cardiovasc. J. Afr. 19951892. 19 (5), 246–253. Available from: http://search.sabinet.co.za/WebZ/images/ejour/cardio1/cardio1_v19_n5_a3.pdf?sessionid=01-42557-1069727030&format=F.

Nadeem, M., Navida, M., Ameer, K., Iqbal, A., Malik, F., Nadeem, M.A., Fatima, H., Ahmed, A., Din, A., 2022. A comprehensive review on the watermelon phytochemical profile and their bioactive and therapeutic effects. unchanged. 1738-7248. 29 (4), 546–576. doi:10.11002/kjfp.2022.29.4.546.

Naseer, S., Hussain, S., Naeem, N., Pervaiz, M., Rahman, M., 2018. The phytochemistry and medicinal value of Psidium guajava (guava). Clin. Phytosci. 2199-1197. 4 (1). doi:10.1186/s40816-018-0093-8.

O″zer, E.A., Emine, N.H., Salih, G., Paul, A., İbanoğlu, Ş., 2006. Effect of extrusion process on the antioxidant activity and total phenolics in a nutritious snack food. Int. J. Food Sci. Technol. 0950-5423. 41 (3), 289–293. doi:10.1111/j.1365-2621.2005.01062.x.

Obradović, V., Babić, J., Dragović-Uzelac, V., Jozinović, A., Ačkar, Đ., Šubarić, D., 2021. Properties of Extruded Snacks Prepared from Corn and Carrot Powder with Ascorbic Acid Addition. Processes. 2227-9717. 9 (8), 1367. doi:10.3390/pr9081367.

Obradović, V., Babić, J., Šubarić, D., Jozinović, A., Ačkar, Đ., Klarić, I., 2015. Influence of dried Hokkaido pumpkin and ascorbic acid addition on chemical properties and colour of corn extrudates. Food Chem. 03088146. 183, 136–143. doi:10.1016/j.foodchem.2015.03.045.

Odukoya, J.O., Odukoya, J.O., Mmutlane, E.M., Ndinteh, D.T., 2021. Phytochemicals and amino acids profiles of selected sub-saharan african medicinal plants' parts used for cardiovascular diseases' treatment. Pharmaceutics. 19994923. 13 (9). Available from: https://www.mdpi.com/1999-4923/13/9/1367/pdf. doi:10.3390/pharmaceutics13091367.

Oladiran, D.A., Emmambux, N.M., 2018. Nutritional and Functional Properties of Extruded Cassava-Soy Composite with Grape Pomace. Starch/Staerke. 1521379X. 70, 7–8.

Available from: http://onlinelibrary.wiley.com/journal/10.1002/(ISSN)1521-379X. doi:10.1002/star.201700298.

Ortak, M., Caltinoglu, C., Sensoy, I., Karakaya, S., 2017. Behic Mert, Changes in functional properties and in vitro bioaccessibilities of β-carotene and lutein after extrusion processing. J. Food Sci. Technol. 0022-1155. 54 (11), 3543–3551. doi:10.1007/s13197-017-2812-4.

Oyedemi, S., Koekemoer, T., Bradley, G., De Venter, M.V., Afolayan, A., 2013. In vitro anti-hyperglycemia properties of the aqueous stem bark extract from Strychnos henningsii (Gilg). Int. J. Diabetes Dev. Ctries. 19983832. 33 (2), 120–127. doi:10.1007/s13410-013-0120-8.

Pandit, N., Kulkarni, S., Singhvi, G., 2023. Effect of Green Tea on Human Brain Health Nutraceutical Fruits and Foods for Neurodegenerative Disorders. Elsevier, India, pp. 301–331. Available from: https://www.sciencedirect.com/book/9780443189517. doi:10.1016/B978-0-443-18951-7.00018-9.

Patil, H., Tiwari, R.V., Repka, M.A., 2016. Hot-Melt Extrusion: from Theory to Application in Pharmaceutical Formulation. AAPS PharmSciTech. 1530-9932. 17 (1), 20–42. doi:10.1208/s12249-015-0360-7.

Paul, T., Jed, 2001. Phytochemicals from Cruciferous Plants Protect against Cancer by Modulating Carcinogen Metabolism. J. Nutr. 131. 10.1093/jn/131.11.3027S2001

Pensamiento-Niño, C.A., Gómez-Aldapa, C.A., Hernández-Santos, B., Juárez-Barrientos, J.M., Herman-Lara, E., Martínez-Sánchez, C.E., Torruco-Uco, J.G., Rodríguez-Miranda, J., 2018. Optimization and characterization of an extruded snack based on taro flour (Colocasia esculenta L.) enriched with Mango pulp (Mangifera indica L.). J. Food Sci. Technol. 0022-1155. 55 (10), 4244–4255. doi:10.1007/s13197-018-3363-z.

Pico, J., Xu, K., Guo, M., Mohamedshah, Z., Ferruzzi, M.G., Martinez, M.M., 2019. Manufacturing the ultimate green banana flour: impact of drying and extrusion on phenolic profile and starch bioaccessibility. Food Chem. 18737072. 297. Available from: www.elsevier.com/locate/foodchem. doi:10.1016/j.foodchem.2019.124990.

Pinho, L.S., Rodrigues, C.E.C., Favaro-Trindade, C.S., Campanella, O.H., 2021. The Incorporation of Carotenoids on Ready to Eat Foods Studied Through Their Stability During Extrusion Processing. Food Eng. Rev. 18667929. 13 (4), 902–915. Available from: http://www.springer.com/life+sci/food+science/journal/12393. doi:10.1007/s12393-021-09285-w.

Pitalua, E., Jimenez, M., Vernon-Carter, E.J., Beristain, C.I., 2010. Antioxidative activity of microcapsules with beetroot juice using gum Arabic as wall material. Food Bioprod. Process. 09603085. 88 (2-3), 253–258. doi:10.1016/j.fbp.2010.01.002.

Pitot, H.C., 1993. The molecular biology of carcinogenesis. Cancer. 10970142. 72 (3 S), 962–970. doi:10.1002/1097-0142(19930801)72:3+<962::AID-CNCR2820721303>3.0.CO;2-H.

Plahar, W.A., Okezie, B.O., Gyato, C.K., 2003. Development of a high protein weaning food by extrusion cooking using peanuts, maize and soybeans. Plant Foods Hum. Nutr. 0921-9668. 58 (3), 1–12. doi:10.1023/B:QUAL.0000041157.35549.b3.

Prabha, K., Ghosh, P., S, A., Joseph, R.M., Krishnan, R., Rana, S.S., Pradhan, R.C., 2021. Recent development, challenges, and prospects of extrusion technology. Future Foods. 26668335. 3. Available from: www.journals.elsevier.com/future-foods. doi:10.1016/j.fufo.2021.100019.

Reichling, J., Schnitzler, P., Suschke, U., Saller, R., 2009. Essential oils of aromatic plants with antibacterial, antifungal, antiviral, and cytotoxic properties - An overview. unchanged. 16614127. 16 (2), 79–90. doi:10.1159/000207196.

Riaz, M.N., 2000. Extruders in Food Applications Extruders in Food Applications. CRC Press, United States, pp. 1–228. Available from: https://www.taylorfrancis.com/books/e/9781482278859.

Riaz, M.N., Asif, M., Ali, R., 2009. Stability of vitamins during extrusion. Crit. Rev. Food Sci. Nutr. 15497852. 49 (4), 361–368. doi:10.1080/10408390802067290.

Rocchetti, G., Gregorio, R.P., Lorenzo, J.M., Barba, F.J., García Oliveira, P., Prieto, M.A., Simal-Gandara, J., Mosele, J.I., Motilva, M.-J., Tomas, M., Patrone, V., Capanoglu, E., Lucini, L., 2022. Functional implications of bound phenolic compounds and phenolics–food interaction: a review. Compr. Rev. Food Sci. Food Saf. 21 (2), 811–842. doi:10.1111/1541-4337.12921.

Rodríguez-Miranda, J., Ruiz-López, I.I., Herman-Lara, E., Martínez-Sánchez, C.E., Delgado-Licon, E., Vivar-Vera, M.A., 2011. Development of extruded snacks using taro (Colocasia esculenta) and nixtamalized maize (Zea mays) flour blends. LWT. 00236438. 44 (3), 673–680. doi:10.1016/j.lwt.2010.06.036.

Sahoo, M.R., Kuna, A., Devi, M.P., Sowmya, M., Dasgupta, M., 2022. Fortification of ready–to–eat extruded snacks with tree bean powder: nutritional, antioxidant, essential amino acids, and sensory properties. J. Food Sci. Technol. 09758402. 59 (6), 2351–2360. Available from: https://www.springer.com/journal/13197. doi:10.1007/s13197-021-05251-w.

Sarker, U., Oba, S., 2020. Nutrients, minerals, pigments, phytochemicals, and radical scavenging activity in Amaranthus blitum leafy vegetables. Sci. Rep. 2045-2322. 10 (1). doi:10.1038/s41598-020-59848-w.

Shahidi, F., Yeo, J., 2016. Insoluble-Bound Phenolics in Food. Molecules. 1420-3049. 21 (9), 1216. doi:10.3390/molecules21091216.

Sidhu, J.S., Kabir, Y., Huffman, F.G., 2007. Functional foods from cereal grains. Int. J. Food Prop. 15322386. 10 (2), 231–244. doi:10.1080/10942910601045289.

Singh, J., Upadhyay, A.K., Bahadur, A., Singh, B., Singh, K.P., Rai, M., 2006. Antioxidant phytochemicals in cabbage (Brassica oleracea L. var. capitata). Sci. Hortic. 03044238. 108 (3), 233–237. doi:10.1016/j.scienta.2006.01.017.

Singh, S., Gamlath, S., Wakeling, L., 2007. Nutritional aspects of food extrusion: a review. Int. J. Food Sci. Technol. 0950-5423. 42 (8), 916–929. doi:10.1111/j.1365-2621.2006.01309.x.

Sitohy, M., Mahgoub, S., Osman, A., El-Masry, R., Al-Gaby, A., 2013. Extent and Mode of Action of Cationic Legume Proteins against Listeria monocytogenes and Salmonella Enteritidis. Probiotics Antimicrob. Proteins. 1867-1306. 5 (3), 195–205. doi:10.1007/s12602-013-9134-2.

Song, J., Tang, Y., 2023. Effect of extrusion temperature on characteristic amino acids, fatty acids, organic acids, and phenolics of white quinoa based on metabolomics. Food Res. Int. 09639969. 169, 112761. doi:10.1016/j.foodres.2023.112761.

Sung, N.Y., Park, W.Y., Kim, Y.E., Cho, E.J., Song, H., Jun, H.K., Park, J.N., Kim, M.H., Ryu, G.H., Byun, E.H., 2016. Increase in anti-oxidant components and reduction of off-flavors on radish leaf extracts by extrusion process. unchanged. 12263311. 45 (12), 1769–1775. Available from: http://jkfn.kfn.or.kr/archive/paper_file_download.kin?no=8050. doi:10.3746/jkfn.2016.45.12.1769.

Tańska, M., Konopka, I., Ruszkowska, M., 2017. Sensory, Physico-Chemical and Water Sorption Properties of Corn Extrudates Enriched with Spirulina. Plant Foods Hum. Nutr. 0921-9668. 72 (3), 250–257. doi:10.1007/s11130-017-0628-z.

Teka, T., Zhang, L., Xiaoyan, G., Yanjie, L., Han, L., Yan, X., 2022. Stilbenes: source plants, chemistry, biosynthesis, pharmacology, application and problems related to their clinical

Application-A comprehensive review. Phytochemistry. 00319422. 197, 113128. doi:10.1016/j.phytochem.2022.113128.

Ti, H., Zhang, R., Zhang, M., Wei, Z., Chi, J., Deng, Y., Zhang, Y., 2015. Effect of extrusion on phytochemical profiles in milled fractions of black rice. Food Chem. 03088146. 178, 186–194. doi:10.1016/j.foodchem.2015.01.087.

Tiwari, B.K., Brunton, N.P., Brennan, C.S., 2013. Plant Food Phytochemicals Handbook of Plant Food Phytochemicals: Sources, Stability and Extraction. Wiley, pp. 1–4. doi:10.1002/9781118464717.ch1.

Tor-Roca, A., Garcia-Aloy, M., Mattivi, F., Llorach, R., Andres-Lacueva, C., Urpi-Sarda, M., 2020. Phytochemicals in Legumes: a Qualitative Reviewed Analysis. J. Agric. Food Chem. 15205118. 68 (47), 13486–13496. Available from: http://pubs.acs.org/journal/jafcau. doi:10.1021/acs.jafc.0c04387.

Touré, A., Xueming, X., 2010. Flaxseed Lignans: source, Biosynthesis, Metabolism, Antioxidant Activity, Bio-Active Components, and Health Benefits. Compr. Rev. Food Sci. Food Saf. 1541-4337. 9 (3), 261–269. doi:10.1111/j.1541-4337.2009.00105.x.

Tran, Q.D., Hendriks, W.H., van der Poel, A.F.B., 2011. Effects of drying temperature and time of a canine diet extruded with a 4 or 8mm die on physical and nutritional quality indicators. Anim. Feed Sci. Technol. 03778401. 165 (3-4), 258–264. doi:10.1016/j.anifeedsci.2011.03.009.

Ullah, A., Munir, S., Badshah, S.L., Khan, N., Ghani, L., Poulson, B.G., Emwas, A.-H., Jaremko, M., 2020. Important Flavonoids and Their Role as a Therapeutic Agent. Molecules. 1420-3049. 25 (22), 5243. doi:10.3390/molecules25225243.

Vidal, N.P., Roman, L., Swaraj, V.J.S., Ragavan, K.V., Simsek, S., Rahimi, J., Kroetsch, B., Martinez, M.M., 2022. Enhancing the nutritional value of cold-pressed oilseed cakes through extrusion cooking. Innov. Food Sci. Emerg. Technol. 14668564. 77. Available from: http://www.elsevier.com/locate/ifset. doi:10.1016/j.ifset.2022.102956.

Villasante, J., Pérez-Carrillo, E., Heredia-Olea, E., Metón, I., Pilar Almajano, M., 2019. In Vitro Antioxidant Activity Optimization of Nut Shell (Carya illinoinensis) by Extrusion Using Response Surface Methods. Biomolecules. 2218-273X. 9 (12), 883. doi:10.3390/biom9120883.

Wang, T., He, F., Chen, G., 2014. Improving bioaccessibility and bioavailability of phenolic compounds in cereal grains through processing technologies: a concise review. J. Funct. Foods. 17564646. 7 (1), 101–111. doi:10.1016/j.jff.2014.01.033.

Wani, S.A., Ganie, N.A., Kumar, P., 2021. Quality characteristics, fatty acid profile and glycemic index of extrusion processed snacks enriched with the multicomponent mixture of cereals and legumes. Legume Sci. 26396181. 3 (2). Available from: https://onlinelibrary.wiley.com/journal/26396181. doi:10.1002/leg3.76.

Waramboi, J.G., Gidley, M.J., Sopade, P.A., 2013. Carotenoid contents of extruded and non-extruded sweetpotato flours from Papua New Guinea and Australia. Food Chem. 18737072. 141 (3), 1740–1746. Available from: www.elsevier.com/locate/foodchem. doi:10.1016/j.foodchem.2013.04.070.

Wink, M., 1998. A Short History of Alkaloids. unchanged LLC, 11–44. doi:10.1007/978-1-4757-2905-4_2.

Wójtowicz, A., Zalewska-Korona, M., Jabłońska-Rya, E., Skalicka-Woźniak, K., Oniszczuk, A., 2018. Chemical Characteristics and Physical Properties of Functional Snacks Enriched with Powdered Tomato. unchanged. 20836007. 68 (3), 251–261. Available from: http://www.degruyter.com/view/j/pjfns.2012.62.issue-4/issue-files/pjfns.2012.62.issue-4.xml. doi:10.1515/pjfns-2017-0028.

Wolfe, K.L., Liu, R.H., 2003. Apple peels as a value-added food ingredient. J. Agric. Food Chem. 00218561. 51 (6), 1676–1683. doi:10.1021/jf025916z.

Wong, Y.Y., Chow, Y.L., 2024. Exploring the potential of spice-derived phytochemicals as alternative antimicrobial agents. eFood. 26663066. 5 (1). Available from: https://onlinelibrary.wiley.com/journal/26663066. doi:10.1002/efd2.126.

Xu, B., Chang, S.K.C., 2008. Effect of soaking, boiling, and steaming on total phenolic contentand antioxidant activities of cool season food legumes. Food Chem. 03088146. 110 (1), 1–13. doi:10.1016/j.foodchem.2008.01.045.

Yadav, K., Dhankhar, J., Kundu, P., 2022. Isothiocyanates – a Review of their Health Benefits and Potential Food Applications. Curr. Res. Nutr. Food Sci. 23220007. 10 (2), 476–502. doi:10.12944/crnfsj.10.2.6.

Yang, P., Zhao, J., Wang, H., Longxian, L., Yongxi, M., 2020. Effects of vitamin forms and levels on vitamin bioavailability and growth performance in piglets. Appl. Sci. 10 (14). doi:10.3390/app10144903.

Yoon, G.A., Yeum, K.J., Cho, Y.S., Chen, C.Y.O., Tang, G., Blumberg, J.B., Russell, R.M., Yoon, S., Lee-Kim, Y.C., 2012. Carotenoids and total phenolic contents in plant foods commonly consumed in Korea. Nutr. Res. Pract. 20056168. 6 (6), 481–490. Available from: http://www.ncbi.nlm.nih.gov/pmc/articles/PMC3542437/pdf/nrp-6-481.pdf. doi:10.4162/nrp.2012.6.6.481

Yuan, Y., Macquarrie, D., 2015. Microwave assisted extraction of sulfated polysaccharides (fucoidan) from Ascophyllum nodosum and its antioxidant activity. Carbohydr. Polym. 01448617. 129, 101–107. Available from: http://www.elsevier.com/wps/find/journaldescription.cws_home/405871/description#description. doi:10.1016/j.carbpol.2015.04.057.

Zahr, S., Zahr, R., Hajj, R.E., Khalil, M., 2023. Phytochemistry and biological activities of Citrus sinensis and Citrus limon: an update. J. Herb. Med. 22108041. 41. Available from: http://www.elsevier.com/wps/find/journaldescription.cws_home/725778/description#description. doi:10.1016/j.hermed.2023.100737.

Zhang, Y., Liu, X., Ruan, J., Zhuang, X., Zhang, X., Zhiming, L., 2020. Phytochemicals of garlic: promising candidates for cancer therapy. Biomed. Pharmacother. 07533322. 123, 109730. doi:10.1016/j.biopha.2019.109730.

Zhao, H., Sun, L., Kong, C., Mei, W., Dai, H., FengQing, X., Huang, S., 2022. Phytochemical and pharmacological review of diterpenoids from the genus Euphorbia Linn (2012–2021). J. Ethnopharmacol. 03788741. 298, 115574. doi:10.1016/j.jep.2022.115574.

Zhao, J., Camire, M.E., Bushway, R.J., Bushway, A.A., 1994. Glycoalkaloid Content and in Vitro Glycoalkaloid Solubility of Extruded Potato Peels. J. Agric. Food Chem. 15205118. 42 (11), 2570–2573. doi:10.1021/jf00047a037.

Zhong, L., Fang, Z., Wahlqvist, M.L., Hodgson, J.M., Johnson, S.K., 2021. Multi-response surface optimisation of extrusion cooking to increase soluble dietary fibre and polyphenols in lupin seed coat. LWT. 00236438. 140. Available from: https://www.journals.elsevier.com/lwt. doi:10.1016/j.lwt.2020.110767.

Effect on antinutritional factors 12

Vaibav[a], *Shubhajit Sarkhel*[a], *Ankansha Kumari*[a], *and Anupam Roy*[a,b]
[a]Laboratory of Applied Food Chemistry, Microbiology, and Process Engineering, Centre for Food Engineering and Technology, Department of Chemical Engineering, Birla Institute of Technology Mesra, Ranchi, Jharkhand, India
[b]Centre for Rural Development and Technology, Indian Institute of Technology Delhi, New Delhi, Delhi, India

12.1 Introduction

Extrusion is a high-temperature, short-time (HTST) process involving the combination of moisture, pressure, temperature, and mechanical shear (Saha et al., 2021) to develop distinct products with unique physical and chemical properties (Maurya and Said, 2015). It is a commonly used process due to its versatility, efficiency, and ability to produce a wide range of products while maintaining quality and improving shelf life. Extrusion is particularly beneficial for the high-volume production of products like breakfast cereals, pasta, and snacks with a wide variety of shapes, textures, and sizes. During the extrusion process, food undergoes a combination of mechanical and thermal treatment, resulting in a range of changes, including cooking, changes in texture, and potential chemical reactions. The process typically involves high temperature, pressure, and shear, which can lead to the denaturation of proteins, gelatinization of starch, lipid oxidation, and the degradation of certain vitamins. The combination of heat and pressure can cause the food to swell and expand, leading to changes in texture, such as puffing or texturizing. High temperatures and shear can cause proteins to unfold and change their structure, potentially affecting their functionality and digestibility. Furthermore, during extrusion, the starches present in the food undergo gelatinization, which subsequently changes their structure and digestibility. Therefore extrusion stands out as an essential food processing technology that efficiently combines thermal and mechanical forces to produce diverse, shelf-stable products with great textures and nutritional profiles, making it an indispensable method in modern food manufacturing (Alonso et al., 2001; Yi et al., 2022).

The antinutrient factors (ANFs) are generally found in most cereals, oilseeds, pseudocereals, and pulses. Some of the common ANFs are tannins, oxalates, phytic acid, trypsin inhibitors, lectins, hemagglutinins, goitrogens, and saponins. These ANFs are recognized as food inhibitors that chelate macro- and micronutrients (such as protein, carbohydrates, iron, zinc, and calcium) and prevent them from being bioavailable. As a result, the bioavailability of indigenous micronutrients in pulses is retarded (Gupta et al., 2015). Moreover, the presence of high amounts of ANFs restricts micronutrient

Extrusion Cooking for Food and Feed. https://doi.org/10.1016/B978-0-443-29987-2.00012-9

fortification strategies, as they can impede fortified micronutrients, leading to ineffectiveness. A study reported that the consumption of an oxalate-rich diet can often cause kidney stones and other major health issues (Mitchell et al., 2019). On the contrary, the consumption of lectins, hemagglutinins, and goitrogens can lead to digestive distress. Therefore effective methods for minimizing these ANF compounds in various food matrices are necessary to enhance their nutritional benefits and ensure better dietary outcomes for consumers.

Most of the ANFs are heat-sensitive in nature. Extrusion operated at high temperatures and pressure (HTST) can improve mineral absorption by deactivating absorption inhibitors. A study reported that the high temperature and pressure in the extrusion process cause phytic acid to hydrolyze and increase mineral availability in extruded peas and kidney beans (Alonso et al., 2001). Therefore extrusion can be an effective process for eliminating the ANFs. The substantial reduction of ANFs by extrusion processing may be explained by the thermal destruction of ANF compounds and changes in their chemical reactivity or the creation of insoluble complexes. Furthermore, high moisture content during the soaking process prior to extrusion resulted in a significant reduction of ANFs. This phenomenon can be attributed to enhanced leaching, hydrolysis, and enzyme activity, as well as the initiation of partial fermentation, which promotes the production of enzymes that reduce ANFs (Rathod and Annapure, 2016; Nair and Duvnjak, 1990).

Several efforts have been undertaken to reduce ANFs both in pre- and postharvest stages (soaking, cooking, autoclaving, germination, fermentation, ultra-sonication, high-pressure processing, and extrusion) of grain processing for enhancing the bioavailability of micronutrients (Sarkhel and Roy, 2022). However, these processes fall short of completely eliminating the ANF content from food matrices. For example, the easiest way to lower the phytic acid in lentils is to cook them. Even so, it only lowers the phytic acid content by around 30%–45%, which is still sufficient to bind a significant amount of nutrients under intestinal conditions (Pal et al., 2017; Wang et al., 2009). On the other hand, extrusion technology has been reported to achieve the highest reduction of ANFs. Extrusion at 140°C–180°C, 150–250 rpm, and 14%–22% feed moisture reduced the phytic acid content in lentil splits by 99% (Rathod and Annapure, 2016; Nair and Duvnjak, 1990). Therefore it is posited that extrusion technology, with its integrated application of thermal and mechanical forces, holds considerable promise for the effective reduction of ANFs present in various food matrices, particularly in cereals and legumes, while concurrently enhancing the bioavailability of essential nutrients. Although numerous pre- and postharvest techniques have been explored for the reduction of ANFs, extrusion is distinguished as a superior, scalable, and industrially viable alternative. This chapter deals with a detailed discussion of the principles and fundamentals of extrusion technology, different sources of ANFs, the mechanism of ANF reduction, several extrusion-based low-ANF products, advantages, and limitations that illustrate the efficacy of extrusion in improving nutrient availability. The discussions herein are designed to provide an in-depth understanding of how extrusion can be strategically implemented to optimize both the safety and the nutritional quality of plant-based food products.

12.2 Effect of extrusion conditions on ANF reduction

Extrusion is a highly adaptable food processing technique that combines several unit operations (mixing, cooking, kneading, shearing, shaping, and forming) within a single continuous system. During extrusion, raw materials are first mixed to achieve a uniform composition. This is followed by cooking under controlled temperature and pressure conditions, which gelatinizes starches and denatures proteins. The material is then subjected to kneading and intense shearing forces through rotating screws, promoting molecular interactions and textural development. Finally, the processed mass is forced through a shaped die, where it is formed into specific shapes and structures before cooling and solidifying.

The primary extruder process parameters, like barrel temperature, feed moisture, feed rate, and screw speed, determine the secondary process parameters such as mass temperature, die pressure, torque, and specific mechanical energy (SME) (Dalbhagat et al., 2019). These processing variables not only influence the physical and structural transformation of food materials during extrusion but also significantly impact the reduction of antinutritional factors (ANFs) such as phytates, tannins, trypsin inhibitors, and lectins. ANFs are naturally occurring compounds that interfere with the bioavailability of nutrients and impair digestion. Optimizing extrusion conditions has, therefore, emerged as an effective strategy to degrade or inactivate these compounds, enhancing the nutritional quality and safety of the final product.

12.2.1 Primary extrusion parameters

12.2.1.1 Barrel temperature

Barrel temperature is one of the most critical parameters in the extrusion process, directly influencing the cooking, structural transformation, and final quality of the extruded products. It determines the thermal energy available to melt and cook the feed material, facilitating key changes such as starch gelatinization, protein denaturation, fiber solubilization, and Maillard browning reactions. In single-screw and twin-screw extruded rice-based products, the barrel temperature typically ranges from 60°C to 190°C, depending on the formulation and desired product characteristics. Snacks and fortified extrudates, such as those made from soybean flour and rice, are processed at higher temperatures (150°C–190°C) to ensure proper expansion and textural attributes, while more delicate products, like gluten-free pasta and vermicelli, are processed at lower ranges (70°C–120°C) to retain structural integrity and minimize degradation (Cabrera-Chávez et al., 2012; Charutigon et al., 2008). In twin-screw extrusion, higher temperatures (up to 184°C) are more common due to better mixing and heat transfer capabilities, allowing for more complex product development, such as soy-fortified millet snacks and pea-rice extrudates (Seth and Rajamanickam, 2012).

Extrusion temperature plays a crucial role in reducing ANFs in food materials. Heat-stable ANFs, such as trypsin inhibitors and lectins, can be effectively inactivated through thermal processing during extrusion. Extrusion at 140°C significantly reduces trypsin inhibitor activity in wheat and rice bran (Kaur et al., 2015). Similarly, Jatropha curcas extruded at 140°C exhibited complete removal of both lectins and trypsin inhibitors, while lower temperatures were less effective (Valdez-Flores et al., 2016). However, Kelkar et al. (2012) suggest that extremely high temperatures may not always be necessary to reduce antinutrients. Extrusion at 85°C was sufficient to reduce lectin content by over 90% in raw bean flour (Kelkar et al., 2012). These findings indicate that while higher temperatures can enhance ANF reduction, optimal conditions may vary depending on the specific ANF and raw material.

12.2.1.2 Feed moisture

Feed moisture is a primary extrusion parameter that significantly affects the rheology of the dough, energy consumption, and the structural and functional properties of the final extruded product. It refers to the amount of water present in the feed mixture before extrusion and plays a vital role in heat and mass transfer, plasticization, starch gelatinization, protein denaturation, and expansion behavior during extrusion. In rice-based extruded products, feed moisture generally ranges from 11% to 43%, depending on the formulation, processing type (single or twin-screw), and desired product attributes (Kamarudin et al., 2018; Barbiroli et al., 2013). Lower feed moisture levels (11%–18%) are often used for snack-type products, such as rice-grit snacks or soybean-rice extrudates, to promote higher expansion, porosity, and a crunchy texture (Pastor-Cavada et al., 2011). In contrast, higher moisture levels (above 28%) are employed in products like gluten-free pasta or rice vermicelli, where structural integrity, elasticity, and smooth texture are more critical. Feed moisture directly influences the dough's viscosity inside the barrel. High moisture reduces dough viscosity, easing flow and reducing shear stress and SME, while low moisture increases viscosity, enhancing shear and heat generation. Extruded flour using whey protein concentrate was processed between 16% and 23% moisture, balancing protein denaturation and texture (De Pilli et al., 2011; Dalbhagat and Mishra, 2019).

Moisture content during extrusion plays a significant role in the reduction of ANFs. Studies have shown that increasing feed moisture from 14% to 22% can enhance the reduction of phytates, tannins, and polyphenols in lentils (Rathod and Annapure, 2016). Similarly, Kaur et al. (2015) found that a higher moisture level (20%), combined with a lower temperature (115°C), was effective in reducing phytate content in wheat bran (Kaur et al., 2015). The likely mechanism is the thermal degradation and hydrolysis of inositol hexaphosphate (phytate) into lower molecular weight forms. However, findings are not universally consistent. Yağcı and Evci (2015) observed that increasing moisture content had little impact on phytate reduction in chickpea–wheat snacks (Yağci and Evci, 2015). Instead, they found that extrusion duration and pressure were more critical factors. These variations suggest that the effectiveness of moisture in reducing ANFs is dependent on the food matrix and its interaction with other process parameters.

12.2.1.3 Screw speed

Screw speed is a critical extrusion processing parameter that directly influences the mixing, shear rate, residence time, and energy input during the extrusion process. It refers to the rotational speed of the screws inside the extruder barrel, typically measured in revolutions per minute (rpm). In both single-screw and twin-screw extruders, adjusting screw speed allows for control over mechanical shear and heat generation, which in turn affects the physical and functional properties of the final product. In rice-based extruded products, screw speed has been reported to vary widely from 20 rpm (for rice vermicelli) to 600 rpm (for pea-fortified rice snacks or soybean-rice snacks). Lower screw speeds generally result in longer residence time and lower shear, which may be suitable for pasta-type or dense-textured products. Higher screw speeds increase shear and mechanical energy, which are favorable for snack-type products requiring puffed or expanded structure, crispiness, and lower density. Studies show that screw speeds around 150–300 rpm are commonly used in instant rice and rice-based snacks, which require significant expansion. Meanwhile, gluten-free brown rice pasta and chestnut-rice flour snacks are produced at lower speeds (80 rpm) to maintain structure and reduce breakage. Higher screw speeds also contribute to more homogeneous mixing of ingredients, improved dispersion of bioactive compounds, and better incorporation of fortificants like soy, whey, or pea protein. However, excessively high screw speeds can lead to over-shearing, reduced residence time, incomplete cooking, and nutrient degradation.

Screw speed is a critical extrusion parameter influencing the reduction of ANFs such as trypsin inhibitors, phytates, and tannins. In a study by Nwabueze (2007), the effect of varying screw speeds (100–180 rpm) on the ANF content of extruded African breadfruit-corn-soy mixtures was evaluated. The results showed that increasing the screw speed to a maximum of 180 rpm led to a substantial reduction in key ANFs (Nwabueze, 2007). Specifically, tannins were reduced by 89%–92%, phytates by 33%–44%, and trypsin inhibitor activity by 88%–91%. Higher screw speeds (150–250 rpm), in combination with elevated die temperatures (140°C–180°C), have been shown to significantly reduce ANFs in lentil extrudates, achieving reductions of up to 99% without affecting total protein content (Rathod and Annapure, 2016). The reduction is likely due to increased mechanical shear and frictional heat, leading to thermal degradation of these compounds and the formation of insoluble complexes. Similarly, Mukhopadhyay and Bandyopadhyay (2003) reported a 61.25% tannin reduction in sesame meal using a screw speed of 96.8 rpm at 80°C with 40% moisture (Mukhopadhyay and Bandyopadhyay, 2003). In linseed, optimal reduction of tannins (61.27%) was also achieved at a screw speed of 96.8 rpm, confirming the significance of higher mechanical shear (Mukhopadhyay et al., 2007). However, the effectiveness of screw speed is influenced by other parameters like temperature and moisture, suggesting that its impact on ANF reduction must be evaluated in conjunction with the overall extrusion setup.

12.2.2 Secondary extruder process parameters

12.2.2.1 Mass temperature

Mass temperature, also known as product temperature, is a critical secondary parameter in the extrusion process and is typically measured at the die head assembly. It serves as an indicator of the thermal and mechanical energy imparted to the material during extrusion. This temperature is generally higher than the barrel temperature due to internal heat generation from viscous dissipation and friction between the feed material and the barrel wall as it travels along the screw (Pansawat et al., 2008). One of the major contributors to increased mass temperature is mechanical shear, which occurs as the screw rotates and compresses the feed. In addition, higher feed rates tend to cause tighter packing of material in the barrel and at the die, which increases shear and internal pressure, resulting in more heat generation and a subsequent rise in product temperature (Vainionpää, 1991). For instance, melt temperature may increase by about 7°C–9°C due to shear alone, while a lower feed moisture content can further elevate mass temperature by 18°C–20°C, owing to reduced water acting as a heat absorber and lubricant (Bajaj and Singhal, 2019). Interestingly, in specific cases such as hulled rice extrusion, the inclusion of thermostable α-amylase has been found to reduce mass temperature. This is because part of the mechanical energy is diverted toward enzymatic starch breakdown, including gelatinization and dextrinization, thereby lowering the energy available for temperature rise (Xu et al., 2015). Overall, mass temperature is a vital parameter influencing product quality attributes such as texture, expansion, and nutritional stability, and it reflects the complex interplay between thermal settings, feed properties, and mechanical energy input.

12.2.2.2 Die pressure and torque

Die pressure and torque are important indicators in the extrusion process. Die pressure shows the resistance the material faces when exiting the die, while torque measures the mechanical energy needed to move and process the material through the extruder (Singh et al., 2007). Several factors affect both die pressure and torque. Higher feed moisture and barrel temperature make the material softer and less viscous, which lowers both pressure and torque. Faster screw speed improves mixing and heat transfer, also reducing viscosity and torque (Dalbhagat et al., 2019). The type of ingredients used matters too. Fat-rich materials, like amaranth or olive pomace, reduce friction and pressure. Low-amylose rice can become sticky and raise torque, while high-amylose rice may reduce torque at higher temperatures (Meng et al., 2010). Additives, like phosphate salts or enzymes such as thermostable α-amylase, can lower viscosity and reduce pressure. However, high-protein materials, like whey, may increase die pressure due to their thick, sticky nature.

12.2.2.3 Specific mechanical energy

SME is a key measure of the mechanical energy needed to process feed material during extrusion. It directly affects starch gelatinization, product expansion, and texture. Higher SME means more shear and energy, which leads to better starch breakdown and puffing (Dalbhagat et al., 2019). Several factors influence SME, such as feed moisture, screw speed, and barrel temperature. Increasing feed moisture reduces SME because water acts as a lubricant, lowering friction inside the barrel (Meng et al., 2010; Ilo et al., 1996). In contrast, low moisture increases resistance, raising SME and enhancing starch transformation. Screw speed also plays a major role; higher speeds boost shear and friction, increasing SME, though very high speeds can reduce residence time and limit processing. The composition of the feed matters too; high-protein ingredients like green gram raise SME due to their resistance to breakdown, while adding pea grit or amaranth can lower it by improving flow. Fat and fiber-rich ingredients, such as apple or olive pomace, affect SME depending on their concentration and interaction with starch. Finally, using enzymes like thermostable α-amylase can reduce SME by breaking down starch during extrusion, leading to easier processing and better product texture.

Secondary extrusion parameters are largely influenced by primary processing conditions like feed moisture, barrel temperature, and screw speed. Since primary parameters are well-documented to affect the reduction of ANFs, it can be inferred that these secondary parameters also play a significant role. For instance, higher SME and die pressure typically result from low feed moisture and high screw speed conditions, which are known to enhance the thermal and mechanical degradation of ANFs such as phytates, tannins, and trypsin inhibitors. Similarly, high mass temperature and torque reflect increased shear and thermal input, which can promote the breakdown or inactivation of heat- and shear-sensitive ANFs. However, direct studies examining the impact of these secondary parameters on ANFs are limited, and further research is needed to establish clear correlations.

12.3 Different ANFs present in different food matrices and their reduction process

Plant-based foods, such as cereals, pulses, legumes, and oilseeds, are consumed heavily due to their high nutritional content and easy accessibility. Nutrients available include fat, fiber, carbohydrates, proteins, etc. However, the bioavailability of these nutrients is hindered by the presence of certain ANFs in these food matrices. This section discusses the different ANFs present in these food matrices, their behavior within the food, their interactions with the nutrients, and their reduction.

12.3.1 Phytic acid

Phytic acid, or phytate, is present in the form of phosphorus in plant tissues, especially in rice, wheat, barley, rye, and beans. It is recognized as one of the significant antinutritional factors (ANFs). The phosphorus in this form is typically not bioavailable to humans and monogastric animals due to the absence of the enzyme phytase, which is crucial for detaching phosphorus atoms from the phytate molecule. Phytic acid has a strong attraction for metallic cations such as K, Mn, Mg, Ca, Fe, and Zn, forming mixed-salt molecules known as phytin, which, in turn, reduces the bioavailability of micronutrients for monogastric animals and humans. Phytin is known to accumulate within the protein bodies of seeds, either as dispersed or thick inclusions referred to as globoids. The stable complexes formed between these minerals and phytic acid contribute to micronutrient deficiencies (for instance, Fe and Zn), particularly in developing countries that rely on cereal-based diets. Given that monogastric animals, including poultry, pigs, and fish, cannot effectively digest phytic acid, phosphorus is added to their feed. Consequently, a significant quantity of undigested phytic acid is excreted by these animals, leading to pollution of water and soil. In wheat and rice, which are generally recognized as monocotyledon crops, phytates are present in the bran or aleurone layer, and in dicotyledons such as legumes, oilseeds, and nuts, phytates are found in close association with proteins (Dionisio et al., 2007).

High temperature, pressure, and moisture content during extrusion processing accelerate the reduction of total phytic acid content in lentil splits (Rathod and Annapure, 2016) by promoting phytic acid hydrolysis and initiating partial fermentation with phytase enzyme production (Nair and Duvnjak, 1990). Extrusion at 140°C–180°C, 150–250 rpm, and 14%–22% feed moisture reduced the phytic acid content in lentil splits by 99% (Rathod and Annapure, 2016). Similarly, extrusion performed at 100 rpm (twin screw) and an outlet temperature of 145°C–150°C significantly reduces phytic acid levels in peas, kidney beans, and faba beans in flour form (moisture content of 25%) by 13.5%, 20.8%, and 26.7%, respectively. Here, most of the molecules of inositol hexaphosphate (IP6), which were embedded in peas, kidney beans, and faba beans, were partially hydrolyzed to penta- (IP5) and tetraphosphates (IP4) during extrusion (Alonso et al., 1998, 2000). The moisture content before (soaking) and during extrusion has a major impact on the breakdown of phytic acid. Soaking of faba bean flour followed by extrusion (twin screw = 250 rpm; moisture content = 18% or 22%; temperature = 140°C or 180°C) resulted in a significant decrease in its phytate level (about 21%). Soaking before extrusion increases the moisture content and, in turn, activates natural phytase enzymes and reduces available phytic acid content, which is further degraded during extrusion (Abd El-Hady and Habiba, 2003). Therefore soaking followed by extrusion would be a promising technology to enhance mineral bioaccessibility by destroying mineral–ANF complexes.

12.3.2 Tannins

Tannin is a polyphenolic biomolecule that possesses an astringent and bitter taste, negatively influencing the nutritional value of food by chelating with proteins and forming a complex molecule that blocks the pathway for protein metabolism, consequently hindering nutritional availability. Tannins are found in a wide variety of plants, including legume seeds, cider, cereals, cacao, peas, some leafy and green vegetables, coffee, tea, and nuts (Suvanto et al., 2017; Fraga-Corral et al., 2020; Lochab et al., 2014). Tannins prevent the digestion of proteins, and the presence of phenolic compounds decreases the digestibility of both proteins and carbohydrates, as well as the bioavailability of vitamins. They also increase fecal nitrogen content in humans and animals. Tannins are found in a wide variety of plants, including legume seeds, cider, cereals, cacao, peas, some leafy and green vegetables, coffee, tea, and nuts (Suvanto et al., 2017; Fraga-Corral et al., 2020; Lochab et al., 2014). A barrel temperature of 82.5°C, screw speed of 90 rpm, and moisture content of 41% in a single-screw shrimp extruder successfully reduced tannin content by 61.2%

(Mukhopadhyay et al., 2007). Tannin content in sorghum also reduced significantly in a single-screw extruder at 150°C with varying feed moisture levels (2–3l/h) (Adarkwah-Yiadom and Duodu, 2017).

12.3.3 Trypsin inhibitors

Trypsin inhibitors are the ANFs found in plant-derived foods, including legumes, cereals, and certain vegetables. They garnered considerable interest in the 1970s and 1980s because of their impact on growth and digestion in animals (Sharma, 2021). These inhibitors are known for diminishing the biological functions of proteolytic enzymes like trypsin and chymotrypsin, which disrupt protein digestion and can lead to pancreatic issues (Li et al., 2017). The content of trypsin inhibitors in foods exhibits significant variation. Soybeans contain the highest levels, with concentrations ranging from 8.6 to 48.2 mg/g of sample or 20.3 to 122.6 mg/g of protein (Sharma, 2021). According to (Li et al., 2017), various foods demonstrate trypsin-inhibiting activity, with black beans showing the lowest activity and soy exhibiting the highest. Furthermore, it is evident that the inhibiting activity can differ among various food types. For instance, sweet lupines display greater inhibiting activity compared to bitter lupines (Avilés-Gaxiola et al., 2018). Trypsin inhibitors are heat-labile ANFs that impair protein digestibility by inhibiting the proteolytic enzyme trypsin. Mixtures of soy and sweet potato flour were extruded at a feed moisture content of 18%, utilizing various die diameters ranging from 6 to 10 mm and screw speeds between 80 and 140 rpm. The most significant inactivation of trypsin inhibitors occurred with 80% sweet potato inclusion, a die diameter of 9 mm, and a screw speed of 154 rpm, resulting in a residual TIA of 3.40 mg/g. Starting with an initial TIA of 58.8 mg/g in soy flour, this process accomplished a remarkable reduction of 94.2%. A lower soy content combined with decreased feed moisture levels contributed to enhanced inactivation rates, rendering the extruded product appropriate for ready-to-eat formulations intended for infants and mothers (Iwe and Ngoddy, 2000).

12.3.4 Oxalates

Oxalates are known as ANFs that occur in vegetables, including spinach, almonds, beets, and rhubarb. These compounds function as strong organic acids capable of forming water-soluble salts through their interaction with minerals like sodium or potassium, as well as creating water-insoluble salts by binding to calcium, iron, or zinc (Lo et al., 2018). Spinach contains approximately 1145 mg of total oxalate per 100 g of fresh weight, of which 803 mg is soluble oxalate. Almonds are notable as well, providing 469 mg per 100 g of product, with 153 mg being soluble oxalate (Petroski and Minich, 2020). Plant-derived oxalate exists in various forms; it can be found as either water-soluble oxalate (including oxalic acid, potassium, sodium, and ammonium oxalates) or as insoluble oxalate salts (mainly calcium oxalate)(SC Noonan and Savage, 1999). Soluble (unbound) oxalates have the ability to chelate minerals, which diminishes their absorption, or they may be absorbed through the intestines and colon. It is believed that absorbed dietary oxalates play a role in the formation of calcium oxalate kidney stones (Savage et al., 2000). Conversely, insoluble oxalates are eliminated through feces (Chai and Liebman, 2005). In the pigeon pea and unripe plantain flour blend, extrusion effectively reduced the oxalate content. The raw blend in a ratio of 75:25 contained 0.72% oxalates initially. It was treated in a single-screw extruder at 100°C and 120 rpm, which reduced the oxalate content to 0.04%, a 94.44% reduction. This dramatic reduction can be attributed to the high temperature and mechanical shear involved in the extrusion process, which facilitated the reduction of oxalic acid and disrupted the calcium oxalate complexes (Anuonye et al., 2012).

12.3.5 Lectins and hemagglutinins

Lectins represent a class of glycoproteins that possess noncatalytic carbohydrate-binding sites, organized according to their species of origin, encompassing animal, algal, bacterial, fungal, and plant lectins (Mishra et al., 2019) (Fig. 12.1). Specifically, plant lectins are present in nuts, cereals, and predominantly in the seeds of legumes (El-Araby et al., 2020). Lectins possess the ability to agglutinate red blood cells by reversibly binding to specific mono- and oligosaccharides found in glycoproteins and glycolipids (Sharon, 2007). Phytohemagglutinin is a glycoprotein with a tetrameric structure and a molecular weight of 120 kDa, located in kidney beans (Muramoto, 2017). Boiling and extrusion, prominent thermal processing methods, can effectively reduce lectins in leguminous seeds. Boiling (100°C for 15 min) significantly reduced lectins in soybean and pea seeds when subjected to boiling (100°C for 15 min). When extrusion was applied at 130°C–135°C under high pressure, it led to a more pronounced reduction in lectins compared to boiling. Nearly complete inactivation of lectins was observed, indicating a higher efficiency of extrusion in antinutrient reduction. This suggests that extrusion processing is more effective in diminishing heat-sensitive antinutrients due to its combination of high temperature, pressure, and shear (Leontowicz et al., 1999).

12.3.6 Goitrogens

Goitrogens are naturally occurring compounds that can interfere with the proper functioning of the thyroid gland (Fig. 12.1). Their name is derived from the term "goiter," which signifies an enlargement of the thyroid gland. When the thyroid gland faces challenges in producing adequate thyroid hormone, it may enlarge as a compensatory mechanism for insufficient hormone production. Like the conditions that result in goiter, goitrogens create obstacles for the thyroid in hormone synthesis. There are two main types of foods that have been associated with disrupted thyroid hormone production in humans: soybean-based foods and cruciferous vegetables (Khare et al., 2017). Goitrogens affect the absorption of iodine. They can inhibit the transporters that are crucial for iodine's cellular entry (e.g., the NIS), or they may disrupt the process of synthesizing thyroid hormones (e.g., T3 and T4). Pigeon pea

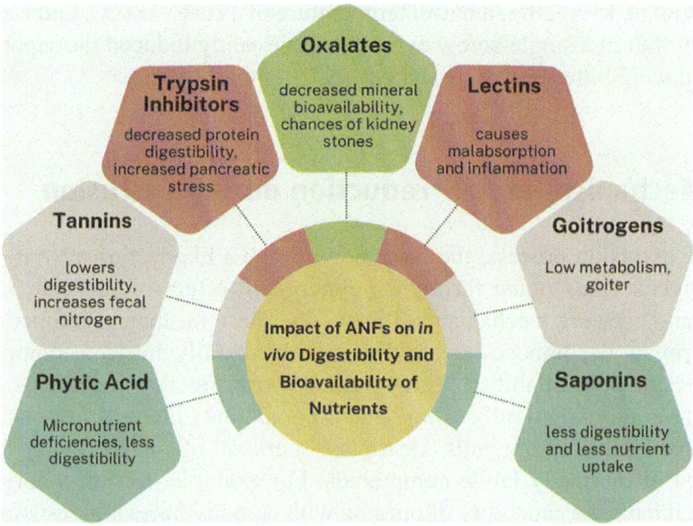

Fig. 12.1 ANFs and their impact on the bioavailability of micronutrients.

and unripe plantain blend has hydrogen cyanide content, which is a common goitrogen. The Brabender single-screw extruder (100°C, 120 rpm screw speed) was capable of significantly reducing the hydrogen cyanide content in the said blend. This could be attributed to the heat degradation of the compound (Anuonye et al., 2012).

12.3.7 Saponins

Saponins are classified as nonvolatile, surface-active secondary metabolites predominantly found in plants. They are composed of a sugar moiety that is attached to a steroid or triterpene group. These surface-active substances are present in legumes, as well as in various spices and herbs (Scott et al., 1985). Elevated levels of saponins can lead to astringency and a bitter taste in food plants. Traditionally, they have been viewed as ANFs because of their detrimental effects, such as inhibiting growth and their bitter flavor, along with their tendency to irritate the throat, which has resulted in decreased consumption and poses a significant barrier to their utilization (Shi et al., 2004). Studies have indicated that saponins can diminish the physiological availability of nutrients and enzymes, and they also obstruct the activity of certain metabolic catalysts like trypsin and chymotrypsin, thus affecting the digestibility of proteins. Extrusion carried out at 122°C in a twin-screw extruder with moisture levels of 20% and 10% significantly reduced tannin levels in mixed oilseed meals (flaxseed, hemp, rapeseed) with pea protein isolate in a 1:1 ratio, by 4.81%–21.31%, respectively. This showed that the low-moisture extrusion process is able to reduce the saponin content effectively in plant-based meat analogs (Han et al., 2025). Saponins in quinoa flour may be a cause of bitterness in the end product made from the flour. Hence, the reduction of saponin content helps reduce bitterness in the end product. A varying feed moisture of 15%–20%, a barrel temperature of 120°C–160°C, and a screw speed of 300–500 rpm in a single-screw extruder significantly reduced the saponin content, thereby reducing bitterness (Kowalski et al., 2016).

12.4 Mechanism of ANF reduction during extrusion

During the extrusion process, the reduction of ANFs like lectins, phytates, oxalates, and goitrogens takes place through a synergistic effect of thermal, mechanical, moisture, and pressure mechanisms (Fig. 12.2). These mechanisms function together to transform or decompose heat-labile and structurally intricate compounds that hinder nutrient bioavailability. These effects work together to change or degrade heat-sensitive and structurally complicated compounds that are responsible for the lower bioavailability of minerals. Heat plays a crucial role as the major factor in the inactivation of thermally labile compounds. For example, lectins, a form of glycoproteins that have the capability of binding with carbohydrates and resisting digestive enzymes, can be significantly destroyed at higher temperatures (Cuadrado et al., 2002; Shi et al., 2018). Similarly, phytic acid is broken down by heat treatments, which further inactivate it through enzymatic and hydrolytic processes (Shi et al., 2018). These effects are enhanced when these compounds undergo extrusion, a process in which heat is applied for brief periods. Due to the mechanical shear forces produced during extrusion, the integrity of plant cell walls is compromised, which in turn results in the degradation of intracellular ANFs. This disruption leads to the breakdown of enzyme inhibitors and complexes bound with proteins, such as phytate–protein

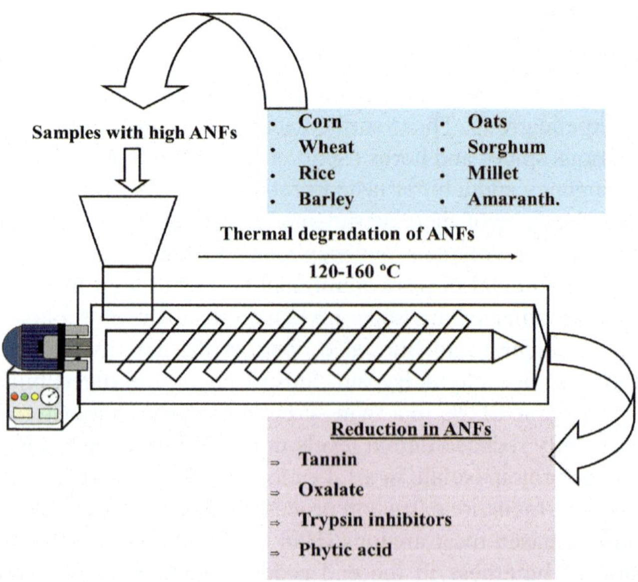

Fig. 12.2 Overall diagram of the mechanism of ANF reduction during extrusion.

complexes or phytate-mineral interactions, which were previously held intact by ionic and hydrogen bonds. This disruption is particularly essential for phytic acid, which creates insoluble complexes with divalent minerals such as calcium and iron, resulting in low bioavailability (Lesjak and Srai, 2019). Moisture content also plays a vital role during extrusion. Water acts as a plasticizer that softens the food matrix, enhancing thermal conductivity and mass transfer. These conditions aid the hydrolysis of ANFs like phytates and oxalates. For example, oxalates, which bind with minerals, are affected by this process. Therefore heat-moisture treatments lead to a reduction in oxalates. This demonstrates that moisture in the food matrix facilitates the mobilization of water-soluble forms of ANFs (Akhtar et al., 2011). Pressure and residence time are also two major parameters that play an essential role in the degradation of ANFs in the extrusion chamber (Hwang and Kim, 2013). When there is sufficient pressure and optimized residence time, it ensures that the food material receives adequate mechanical and thermal energy, which is essential for the maximum destruction of ANFs like phytates and goitrogens. For instance, glucosinolates, which are precursors to goitrogens, undergo enzymatic hydrolysis and diffusion from the food matrix due to cell lysis when subjected to heat and pressure. Due to the integration of heat, pressure, and mechanical forces within a controlled, short span of time, the extrusion process demonstrates the greatest efficiency. Therefore extrusion works on a complex mechanism that leads to thermal denaturation, shear-induced cell disruption, water-mediated diffusion, and pressure-assisted transformation to effectively destroy ANFs in plant-based foods. These modifications enhance the nutritional quality and safety of extruded products, making them more appropriate for human consumption. As shown in Table 12.1, various processing techniques such as boiling, fermentation, germination, and extrusion significantly reduce the levels of common ANFs like lectins, phytates, oxalates, and goitrogens.

12.5 Different extrusion-based products with reduced ANF factors

Extruded products offer enhanced nutritional benefits primarily due to the significant reduction of ANFs such as phytates, tannins, lectins, and protease inhibitors during the extrusion process. This reduction contributes to improved protein and starch digestibility, better mineral bioavailability, and overall enhanced nutritional quality. Mild extrusion conditions, such as moderate barrel temperatures, adequate feed moisture, and shorter residence times, are effective in degrading ANFs while preserving essential nutrients. Delgado et al. (2012) demonstrated that the extrusion of a bean–corn blend (30:70) at 150°C and 20% feed moisture completely eliminated lectin and trypsin inhibitor activity within 30 s (Delgado et al., 2012). Similarly, Marzo et al. (2011) found that the extrusion of kidney beans effectively reduced phytates and tannins and inactivated protease and amylase inhibitors. Rats fed with extruded beans showed improved growth and nutrient utilization compared to those fed raw beans, highlighting the negative impact of ANFs on protein digestibility and mineral

Table 12.1 Reducing ANFs through various processing methods.

Antinutrient	Processing method	Reduction range (%)	Food material	References
Lectins	Boiling	90%	Pulses	(López-Moreno et al., 2022)
	Autoclaving	100%	Lupin	(Embaby, 2010)
	Cooking and extrusion	80–90%	Indigenous bean	(Osuna-Gallardo et al., 2023)
Phytates	Autoclaving and soaking	34.8%	Lupin beans	(Embaby, 2010)
	Vacuum impregnation with simultaneous ZnO fortification	85.9%	Lentils	(Sarkhel and Roy, 2024)
	Soaking	26%	Faba Bean	(Patterson et al., 2017)
	Extrusion	55.83%	Cereal bran	(Kaur et al., 2015)
		1.1–1.3%	Kidney beans	(Li et al., 2024)
Oxalates	Soaking	42–46%	Soybean, sesame seeds	(López-Moreno et al., 2022)
		26.3%	Kidney beans	(Abera et al., 2023)
	Extrusion	36.84%	Cereal bran	(Kaur et al., 2015b)
Tripsin inhibitor	Instant controlled pressure drop	95–98%	Soybean	(Avilés-Gaxiola et al., 2018)
	Ultrafiltration	9.47%	Soybean	
	High hydrostatic pressure	76%	Soybean	
	Ultrasound	55%	Soybean	
	Soaking	35%	Soybean	
	Autoclaving	86.8%	Lupin	(Embaby, 2010)
	Extrusion	72.39%	Cereal bran	(Kaur et al., 2015)
		79–95%	Kidney beans	(Li et al., 2024)
Tannins	Autoclaving and soaking	71%	Lupin	(Embaby, 2010)
	Extrusion	46–89%	Kidney beans	(Li et al., 2024)
		68.82%	Indigenous bean	(Osuna-Gallardo et al., 2023)
Saponins	Soaking	20–23%	Faba bean	(Patterson et al., 2017)
	Extrusion	70.7%	Indigenous bean	(Osuna-Gallardo et al., 2023)
Goitrogens	Boiling	51%	Broccoli, cabbage	(López-Moreno et al., 2022)
	Microwave	17–24%	Brussels sprouts, broccoli	
	Extrusion	67%	Extruded rapeseed meal	(Huang et al., 1995)

absorption (Marzo et al., 2011). In a human study, Hurrell et al. (2002) compared iron absorption from extruded, roller-dried, and home-cooked cereals (Hurrell et al., 2002). Although absorption was slightly lower in extruded products due to residual phytates, the difference was minimal, suggesting that extrusion does not severely impair iron bioavailability if optimized (Hurrell et al., 2002; Nikmaram et al., 2017). Wani and Kumar (2016) developed an extruded multigrain snack using fenugreek, oats, peas, rice, and corn. They observed a significant reduction in tannins (0.38 mg/g) and phytates (5.02 mg/g) compared to raw flours, along with increased protein, fiber, carbohydrate content, and antioxidant activity, despite using high processing temperatures (110°C) (Wani and Kumar, 2016).

Extrusion enhances nutritional quality through ANF reduction; however, it may lead to some losses in heat-sensitive vitamins. Extrusion of *Digitaria exilis*–soybean blends resulted in a small loss of riboflavin (6%) and a significant loss of pyridoxine (86%) (Anuonye et al., 2010). Other studies have reported potential degradation of vitamin C during extrusion and storage. Moreover, the Maillard reaction, promoted by high sugar content and high extrusion temperatures, can reduce amino acid availability despite improving sensory attributes. This issue can be managed by optimizing barrel temperature, moisture, and sugar levels to minimize negative effects while maximizing ANF reduction. Overall, extruded products show strong potential as health-promoting foods due to effective ANF reduction and improved digestibility, making them suitable for both human consumption and animal feed formulations (Nikmaram et al., 2017).

12.6 Effect of the extrusion process on other nutritional compositions in food

The extrusion process significantly influences the physicochemical and nutritional composition of food, affecting components such as proteins, starch, vitamins, minerals, and the in vivo digestibility of macronutrients. The high temperature and pressure during the extrusion process cause lipid oxidation, protein denaturation and cross-linking, starch gelatinization and dextrinization, degradation of vitamins, denaturation of enzymes, browning, and flavor formation. Pulses are rich in protein, while cereals are high in starch. Therefore the extrusion of pulses at high temperature and pressure denatures the protein structure. Moreover, it loses its native structure and forms a viscoelastic mass. Denaturation of the protein decreases its solubility and improves digestibility (Verfaillie et al., 2024). The disulfide bonds present in the protein structure are broken, resulting in the formation of a lower molecular weight protein from a higher molecular weight one (Verfaillie et al., 2024). Also, during the extrusion process, realignment of the protein molecules may take place, which leads to cross-linking and restructuring of the protein molecules, resulting in a chewy, texturized product (Verfaillie et al., 2024). Simultaneously, the high shear force during the extrusion process causes gelatinization of the starch molecules in cereal grains, which affects their functionality, such as solubility, viscosity, and water holding capacity. The reorganization of the native structure of the starch is also achieved during the extrusion process. Several studies have reported that extrusion can result in the denaturation of protein and gelatinization of starch molecules (Offiah et al., 2019). However, this change in the physicochemical properties, mainly in textural properties, is sometimes beneficial in terms of new food product development. For example, extruded soybean protein can be used as a meat analog (Zahari et al., 2020). 3D food printing can also be achieved using this technology (Hussain et al., 2022). The gelatinization of starch during the extrusion process significantly helps in the production of iron-fortified rice.

The extrusion process also impacts the enzymes' functions. Some enzymes lose their biological activity during high pressure and temperature in the extrusion process. For example, the inactivation of lipoxygenase-1, 2, and 3 enzymes during extrusion cooking at different temperatures, ranging from 77°C to 121°C, was studied. The results showed that high temperatures inactivated all the enzymes. Furthermore, the protein digestibility index (PDI) for the extruded products was measured at approximately 22 when there was complete inactivation of all three lipoxygenases. The PDI serves as a useful indicator of the extent of heat treatment applied. Moreover, an increase in the moisture content of soybean feed from 9.2% to 16.3% resulted in a decrease in PDI from 68.4 to 24.2 (Offiah et al., 2019). The inactivation of enzymes by extrusion cooking lowers the risk of lipid oxidation and other enzyme-mediated adverse reactions. In addition, the high temperature and pressure during the extrusion process cause the formation of reducing sugars, which then react with the free amine groups of lysine and other amino acids. This reaction results in the formation of color, a phenomenon known as nonenzymatic browning (Luo et al., 2025). Moreover, mineral absorption may be improved by extrusion due to the destruction of inhibitory factors like ANFs. However, the loss of heat-sensitive vitamins is expected during extrusion, especially for water-soluble vitamins such as vitamin C (Moscicki et al., 2013). In conclusion, while the extrusion process induces significant structural and nutritional modifications such as protein denaturation, starch gelatinization, enzyme inactivation, and vitamin degradation, it also enhances digestibility and improves functional properties, making it a versatile technology for developing value-added, functional, innovative, and fortified food products.

12.6.1 Protein

Mild extrusion cooking improves the nutritional quality of vegetable proteins by enhancing their digestibility, mainly through protein denaturation and the deactivation of enzyme inhibitors. Higher temperatures (100°C–140°C) further increase protein digestibility by reducing protease inhibitors, especially in wheat flour, without damaging nutrients due to short residence times (Adeleye et al., 2020). Extrusion processing affects amino acid retention, especially lysine, which is often limited in cereals. High screw speeds (80–140 rpm) and smaller die sizes (6–10 mm) can improve lysine retention by reducing heat exposure through shorter residence times (Hoyos-Concha et al., 2022). However, maintaining low product temperatures

(<180°C) at low moisture levels (<15%) is challenging in industrial settings and may impact product texture. The Maillard reaction, occurring between amino acids and reducing sugars, reduces lysine availability and overall protein digestibility (Hoyos-Concha et al., 2022). Lysine, being highly reactive, is commonly used as a marker of protein damage. High barrel temperatures and low feed moisture during extrusion accelerate this reaction. Studies show up to 16% lysine loss in wheat bran and 9.1% in protein-rich blends. Increasing feed moisture (22%–26%) helps reduce lysine degradation by minimizing residence time and browning (Lund and Ray, 2017; Nandane et al., 2025).

12.6.2 Carbohydrates

Carbohydrates play a vital role in extrusion processing, influencing the texture and functional quality of the final product. The key carbohydrate components that transform under extrusion conditions are sugars, starch, and dietary fiber. Sugars such as sucrose, fructose, and lactose are important for providing quick energy and sweetness. During extrusion, they play a critical role in flavor and color development through Maillard reactions (chemical interactions between sugars and amino acids like lysine) (Pismag et al., 2024). These reactions enhance sensory appeal but can reduce nutritional quality, especially in infant or therapeutic foods. In adult-targeted foods, particularly for weight management, lower sugar levels are preferred. During extrusion, sucrose often breaks down into glucose and fructose, which then react with proteins, reducing both sugar content and perceived sweetness (Nandane et al., 2025; Pismag et al., 2024).

Starch, composed of linear amylose and branched amylopectin, plays an important role in developing texture during extrusion. Its ratio affects gelatinization and viscosity, which influence the final product's texture (Karakelle et al., 2020). Extrusion enables starch gelatinization at lower moisture levels (12%–22%) than traditional methods, improving digestibility for humans and animals (Kantrong et al., 2022). However, this also raises the glycemic index (GI) of the product, which can be concerning for diabetic or health-conscious consumers. Extrusion may increase the GI of millet-based products by 6%–9%. Incorporating dietary fibers from barley, oats, wheat, or rice bran can reduce the GI by 5%–8%, offering a way to tailor the health impact of starchy extruded foods (Geetha et al., 2020).

Dietary fiber undergoes significant changes during extrusion, mainly in its physicochemical properties rather than its total content. The process can reduce the molecular weight of insoluble fibers like pectin and hemicellulose, improving their water solubility and functional behavior. Extrusion has been shown to increase soluble fiber from 12.5 to 16.7 g per 100 g of dry matter (Nandane et al., 2025; Schmid et al., 2020). The modifications largely depend on processing conditions such as temperature (161°C–180°C), moisture level (~15%), and screw speed (150–200 rpm), which minimally change the dietary fiber (Korkerd et al., 2016). Although the overall fiber content in cereals like wheat remains mostly unchanged, extrusion enhances the viscosity and digestibility of the fiber.

12.6.3 Lipids

Lipids in extrusion undergo thermal and mechanical transformations. Cereal grains like wheat and corn are low in fat (~2%), while oilseeds like soybeans contain up to 50% oil. During milling, high-fat components (bran, germ) are often removed to improve shelf life (Nandane et al., 2025). In extrusion, high-fat materials (>5–6%) are less suitable for puffed or expanded products as they reduce internal pressure and hinder expansion. Small fat additions (<5%) can aid extrusion by acting as lubricants and improving texture (Leonard et al., 2020). However, fat loss occurs, and only about 40% of the original lipids may be recovered postextrusion (e.g., in maize), while some fats form complexes with amylose or proteins or leak out at the die (Kamau et al., 2020).

12.6.4 Vitamins

Extrusion can affect the bioavailability of vitamins, either increasing or decreasing their absorption depending on the type of vitamins and processing conditions. In particular, vitamin stability during HTST extrusion has been widely studied, especially for thiamine (B1), with some research also focusing on riboflavin (B2), vitamin C, and vitamin A. Water-soluble vitamins are more sensitive to heat, moisture, light, oxygen, and pH, making them more vulnerable during extrusion. The design of the extruder also influences vitamin retention. Short-barrel extruders can retain more B vitamins (44%–62%) compared to long-barrel types, indicating that processing design affects nutrient preservation (Mironeasa et al., 2023). Vitamin B6 (pyridoxine) shows better stability in corn-based extrudates than in those made from oats or corn/pea blends, though the reasons are not fully understood and need further study (Nandane et al., 2025; Mironeasa et al., 2023). Temperature changes during extrusion can reduce levels of B1 and B2, especially in fortified maize products. However, the extent of vitamin loss also depends on moisture levels, grain type, and any pretreatment applied. This highlights the need to carefully balance processing efficiency with nutrient retention in industrial food production.

12.6.5 Minerals

Minerals are generally stable during heat processing, like extrusion; however, extrusion can still influence their bioavailability. One major benefit is the breakdown of antinutritional factors, such as phytates and polyphenols, which normally bind with minerals and reduce their absorption in the body. By reducing these inhibitors, extrusion cooking can enhance the availability of minerals like iron and zinc. Research has shown that extrusion does not lead to mineral loss through steam at the die, but it can improve mineral absorption (Mironeasa et al., 2023). Extrusion of multigrain flour made from sorghum, pearl millet, and kidney beans increased the levels of iron, zinc, and sodium, although calcium content significantly decreased in all blends. This reduction in calcium may be due to its binding with proteins, making it less available, or due to high extrusion temperatures (Nandane et al., 2025; Mironeasa et al., 2023). Other minerals, like magnesium and phosphorus, are usually not significantly affected. These findings suggest that while total mineral content remains mostly stable, processing conditions play a key role in determining how well the body can absorb them.

12.7 Advantages and limitations of extrusion for ANF reduction

The extrusion technology is a widely accepted food processing technology that offers high ANF reduction. The high temperature, moisture, and pressure during this process are effective in terms of denaturation or deactivation of ANFs. Reduction of the ANFs results in a significant improvement in the bioavailability of macro- and micronutrients. Moreover, high mechanical shear and pressure help in disrupting the cellular matrix, as well as the native structure of protein and starch in raw food products, thereby improving digestibility. In addition, extrusion is a continuous, scalable process that enhances the functional and physicochemical properties, such as texture, flavor, color, and shelf life of the product. It helps in creating new product developments, such as meat analogs and fortified foods. However, it has certain limitations. It is very difficult to completely remove all the ANFs in one cycle of the extrusion process. Therefore performing multiple cycles may create process-induced alterations, such as loss of flavor, aroma, and essential vitamins (e.g., vitamin C). The process also requires precise optimization of parameters like temperature, moisture content, screw speed, and feed rate to ensure effective ANF reduction without compromising nutrient retention. Extrusion equipment is capital-intensive and demands skilled operation and maintenance, which can be a barrier for small-scale processors. In addition, under certain conditions, extrusion may lead to the formation of undesirable compounds, such as acrylamide or advanced glycation end-products. In conclusion, while extrusion is a highly effective method for reducing ANFs and enhancing the nutritional quality of food products, its efficacy is contingent upon the control of processing parameters. This control is essential to minimize nutrient loss and to prevent the formation of potentially harmful by-products.

References

Abd El-Hady, E.A., Habiba, R.A., 2003. Effect of soaking and extrusion conditions on antinutrients and protein digestibility of legume seeds. LWT. 00236438. 36 (3), 285–293. doi:10.1016/s0023-6438(02)00217-7.

Abera, S., Yohannes, W., Chandravanshi, B.S., 2023. Effect of processing methods on antinutritional factors (Oxalate, Phytate, and Tannin) and their interaction with minerals (calcium, iron, and zinc) in red, white, and black kidney beans. Int. J. Anal. Chem. 16878779. 2023. Available from: https://www.hindawi.com/journals/ijac/. doi:10.1155/2023/6762027.

Adarkwah-Yiadom, M., Duodu, K.G., 2017. Effect of extrusion cooking and simulated in vitro gastrointestinal digestion on condensed tannins and radical scavenging activity of type II and type III whole grain sorghum. Int. J. Food Sci. Technol. 13652621. 52 (10), 2282–2294. doi:10.1111/ijfs.13510.

Adeleye, O.O., Awodiran, S.T., Ajayi, A.O., Ogunmoyela, T.F., 2020. Effect of high-temperature, short-time cooking conditions on in vitro protein digestibility, enzyme inhibitor activity and amino acid profile of selected legume grains. Heliyon. 24058440. 6 (11). Available from: http://www.journals.elsevier.com/heliyon/. doi:10.1016/j.heliyon.2020.e05419.

Akhtar, M.S., Israr, B., Bhatty, N., Ali, A., 2011. Effect of cooking on soluble and insoluble oxalate contents in selected Pakistani vegetables and beans. Int. J. Food Prop. 15322386. 14 (1), 241–249. doi:10.1080/10942910903326056.

Alonso, R., Aguirre, A., Marzo, F., 2000. Effects of extrusion and traditional processing methods on antinutrients and in vitro digestibility of protein and starch in faba and kidney beans. Food Chem. 03088146. 68 (2), 159–165. doi:10.1016/s0308-8146(99)00169-7.

Alonso, R., Orúe, E., Marzo, F., 1998. Effects of extrusion and conventional processing methods on protein and antinutritional factor contents in pea seeds. Food Chem. 03088146. 63 (4), 505–512. doi:10.1016/s0308-8146(98)00037-5.

Alonso, R., Rubio, L.A., Muzquiz, M., Marzo, F., 2001. The effect of extrusion cooking on mineral bioavailability in pea and kidney bean seed meals. Anim. Feed Sci. Technol. 03778401. 94 (1-2), 1–13. doi:10.1016/S0377-8401(01)00302-9.

Anuonye, J.C., Jigam, A.A., Ndaceko, G.M., 2012. Effects of extrusion-cooking on the nutrient and anti-nutrient composition of pigeon pea and unripe plantain blends. J. Appl. Pharm. Sci. 22313354. 2 (5), 158–162. Available from: http://www.japsonline.com/admin/php/uploads/479_pdf.pdf. doi:10.7324/JAPS.2012.2533Nigeria.

Anuonye, J.C., Onuh, J.O., Egwim, E., Adeyemo, S.O., 2010. Nutrient and antinutrient composition of extruded acha/Soybean blends. J. Food Process. Preserv. 17454549. 34 (2), 680–691. doi:10.1111/j.1745-4549.2009.00425.x.

Avilés-Gaxiola, S., Chuck-Hernández, C., Serna Saldívar, S.O., 2018. Inactivation methods of trypsin inhibitor in legumes: a review. Journal of Food Science. 17503841. 83 (1), 17–29. Available from: http://www3.interscience.wiley.com/journal/118509799/issueyear?year=2008. doi:10.1111/1750-3841.13985.

Bajaj, S.R., Singhal, R.S., 2019. Effect of extrusion processing and hydrocolloids on the stability of added vitamin B12 and physico-functional properties of the fortified puffed extrudates. LWT. 00236438. 101, 32–39. Available from: https://www.journals.elsevier.com/lwt. doi:10.1016/j.lwt.2018.11.011.

Barbiroli, A., Bonomi, F., Casiraghi, M.C., Iametti, S., Pagani, M.A., Marti, A., 2013. Process conditions affect starch structure and its interactions with proteins in rice pasta. Carbohydr. Polym. 01448617. 92 (2), 1865–1872. doi:10.1016/j.carbpol.2012.11.047.

Cabrera-Chávez, F., de la Barca, A.M.C., Islas-Rubio, A.R., Marti, A., Marengo, M., Pagani, M.A., Bonomi, F., Iametti, S., 2012. Molecular rearrangements in extrusion processes for the production of amaranth-enriched, gluten-free rice pasta. LWT. 00236438. 47 (2), 421–426. Available from: https://www.journals.elsevier.com/lwt. doi:10.1016/j.lwt.2012.01.040.

Chai, W., Liebman, M., 2005. Effect of different cooking methods on vegetable oxalate content. J. Agric. Food Chem. 0021-8561. 53 (8), 3027–3030. doi:10.1021/jf048128d.

Charutigon, C., Jitpupakdree, J., Namsree, P., Rungsardthong, V., 2008. Effects of processing conditions and the use of modified starch and monoglyceride on some properties of extruded rice vermicelli. LWT. 00236438. 41 (4), 642–651. doi:10.1016/j.lwt.2007.04.009.

Cuadrado, C., Hajos, G., Burbano, C., Pedrosa, M.M., Ayet, G., Muzquiz, M., Pusztai, A., Gelencser, E., 2002. Effect of natural fermentation on the lectin of lentils measured by immunological methods. Food Agric. Immunol. 14 (1), 41–49. doi:10.1080/09540100220137655.

Dalbhagat, C.G., Mahato, D.K., Mishra, H.N., 2019. Effect of extrusion processing on physicochemical, functional and nutritional characteristics of rice and rice-based products: a review. Trends Food Sci. Technol. 09242244. 85, 226–240. Available from: http://www.elsevier.com/wps/find/journaldescription.cws_home/601278/description#description. doi:10.1016/j.tifs.2019.01.001.

Dalbhagat, C.G., Mishra, H.N., 2019. Effects of extrusion process conditions on system parameters; physicochemical properties and cooking characteristics of extruded fortified rice kernels. J. Cereal Sci. 10959963. 89. Available from: http://www.elsevier.com/inca/publications/store/6/2/2/8/5/9/index.htt. doi:10.1016/j.jcs.2019.05.016.

De Pilli, T., Derossi, A., Talja, R.A., Jouppila, K., Severini, C., 2011. Study of starch-lipid complexes in model system and real food produced using extrusion-cooking technology. Innov. Food Sci. Emerg. Technol. 14668564. 12 (4), 610–616. doi:10.1016/j.ifset.2011.07.011.

Delgado, E., Vences-Montaño, M.I., Hernández Rodríguez, J.V., Rocha-Guzman, N., Rodriguez-Vidal, A., Herrera-Gonzalez, S.M., Medrano-Roldan, H., Solis-Soto, A., Ibarra-Perez, F., 2012. Inhibition of the growth of rats by extruded snacks from bean (*Phaseolus vulgaris*) and corn (*Zea mays*). Emir. J. Food Agric. 20790538. 24 (3), 255–263. Available from http://ejfa.info/index.php/ejfa/article/viewFile/12009/6143.

Dionisio, G., Holm, P.B., Brinch-Pedersen, H., 2007. Wheat (*Triticum aestivum* L.) and barley (*Hordeum vulgare* L.) multiple inositol polyphosphate phosphatases (MINPPs) are phytases expressed during grain filling and germination. Plant Biotechnol. J. 14677652. 5 (2), 325–338. doi:10.1111/j.1467-7652.2007.00244.x.

El-Araby, M.M., El-Shatoury, E.H., Soliman, M.M., Shaaban, H.F., 2020. Characterization and antimicrobial activity of lectins purified from three *Egyptian leguminous* seeds. AMB Express. 21910855. 10 (1). Available from: http://www.amb-express.com/. doi:10.1186/s13568-020-01024-4.

Embaby, H.E.S., 2010. Effect of soaking, dehulling, and cooking methods on certain antinutrients and in vitro protein digestibility of bitter and sweet lupin seeds. Food Sci. Biotechnol. 12267708. 19 (4), 1055–1062. doi:10.1007/s10068-010-0148-1.

Fraga-Corral, M., García-Oliveira, P., Pereira, A.G., Lourenço-Lopes, C., Jimenez-Lopez, C., Prieto, M.A., Simal-Gandara, J., 2020. Technological application of tannin-based extracts. Molecules. 14203049. 25 (3). Available from: https://www.mdpi.com/1420-3049/25/3/614/pdf. doi:10.3390/molecules25030614.

Geetha, K., Yankanchi, G.M., Hulamani, S., Hiremath, N., 2020. Glycemic index of millet based food mix and its effect on pre diabetic subjects. J. Food Sci. Technol. 09758402. 57 (7), 2732–2738. Available from: http://www.springerlink.com/content/121580/. doi:10.1007/s13197-020-04309-5.

Gupta, R.K., Gangoliya, S.S., Singh, N.K., 2015. Reduction of phytic acid and enhancement of bioavailable micronutrients in food grains. J. Food Sci. Technol. 09758402. 52 (2), 676–684. Available from: http://www.springerlink.com/content/121580/. doi:10.1007/s13197-013-0978-y.

Han, R., McDowell, R., Gaunt, S., Mondor, M., Hernández-Álvarez, A.J., 2025. Transforming oilseed blends: the impact of low-moisture extrusion on antinutritional factors, protein structure, and nutritional value. Food Chem. 18737072. 484. Available from: https://www.sciencedirect.com/science/journal/03088146. doi:10.1016/j.foodchem.2025.144417.

Hoyos-Concha, J.L., Villada-Castillo, H.S., Fernández-Quintero, A., Ortega-Toro, R., 2022. Effect of the addition of high-protein hydrolyzed flour from oncorhynchus mykiss byproducts on the properties of an extruded feed. ACS Omega. 24701343. 7 (3),

2554–2564. Available from: https://www.pubs.acs.org/journal/acsodf. doi:10.1021/acsomega.1c02650.

Huang, S., Liang, M., Lardy, G., Huff, H.E., Kerley, M.S., Hsieh, F., 1995. Extrusion processing of rapeseed meal for reducing glucosinolates. Anim. Feed Sci. Technol. 03778401. 56 (1-2), 1–9. doi:10.1016/0377-8401(95)00826-9.

Hurrell, R.F., Reddy, M.B., Burri, J., Cook, J.D., 2002. Phytate degradation determines the effect of industrial processing and home cooking on iron absorption from cereal-based foods. Br. J. Nutr. 00071145. 88 (2), 117–123. doi:10.1079/BJNBJN2002594.

Hussain, S., Malakar, S., Arora, V.K., 2022. Extrusion-based 3D food printing: technological approaches, material characteristics, printing stability, and post-processing. Food Eng. Rev. 18667929. 14 (1), 100–119. Available from: http://www.springer.com/life+sci/food+science/journal/12393. doi:10.1007/s12393-021-09293-w.

Hwang, E.S., Kim, G.H., 2013. Effects of various heating methods on glucosinolate, carotenoid and tocopherol concentrations in broccoli. Int. J. Food Sci. Nutr. 14653478. 64 (1), 103–111. doi:10.3109/09637486.2012.704904.

Ilo, S., Tomschik, U., Berghofer, E., Mundigler, N., 1996. The effect of extrusion operating conditions on the apparent viscosity and the properties of extrudates in twin-screw extrusion cooking of maize grits. LWT. 00236438. 29 (7), 593–598. doi:10.1006/fstl.1996.0092.

Iwe, M.O., Ngoddy, P.O., 2000. Effect of extrusion on trypsin inhibitor contents of soy-sweet potato mixtures. J. Food Process. Preserv. 0145-8892. 24 (6), 453–463. doi:10.1111/j.1745-4549.2000.tb00434.x.

Kamarudin, M.S., de Cruz, C.R., Saad, C.R., Romano, N., Ramezani-Fard, E., 2018. Effects of extruder die head temperature and pre-gelatinized taro and broken rice flour level on physical properties of floating fish pellets. Anim. Feed Sci. Technol. 03778401. 236, 122–130. doi:10.1016/j.anifeedsci.2017.12.007.

Kamau, E.H., Nkhata, S.G., Ayua, E.O., 2020. Extrusion and nixtamalization conditions influence the magnitude of change in the nutrients and bioactive components of cereals and legumes. Food Sci. Nutr. 20487177. 8 (4), 1753–1765. Available from: onlinelibrary.wiley.com/journal/10.1002/%28ISSN%292048-7177 . doi:10.1002/fsn3.1473.

Kantrong, H., Klongdee, S., Jantapirak, S., Limsangouan, N., Pengpinit, W., 2022. Effects of extrusion temperature and puffing technique on physical and functional properties of purpled third-generation snack after heat treatment. J. Food Sci. Technol. 0022-1155. 59 (6), 2209–2219. doi:10.1007/s13197-021-05234-x.

Karakelle, B., Kian-Pour, N., Toker, O.S., Palabiyik, I., 2020. Effect of process conditions and amylose/amylopectin ratio on the pasting behavior of maize starch: a modeling approach. J. Cereal Sci. 07335210. 94 (102998). doi:10.1016/j.jcs.2020.102998.

Kaur, S., Sharma, S., Baljit Singh, B.N.D., 2015. Effect of extrusion variables (temperature, moisture) on the antinutrient components of cereal brans. J. Food Sci. Technol. 0022-1155. 52 (3), 1670–1676. doi:10.1007/s13197-013-1118-4.

Kelkar, S., Siddiq, M., Harte, J.B., Dolan, K.D., Nyombaire, G., Suniaga, H., 2012. Use of low-temperature extrusion for reducing phytohemagglutinin activity (PHA) and oligosaccharides in beans (*Phaseolus vulgaris* L.) cv. Navy and Pinto. Food Chem. 03088146. 133 (4), 1636–1639. doi:10.1016/j.foodchem.2012.02.044.

Khare, V., Sinha, K., Scientist, J., Bharti, L., 2017. Review on antinutritional factors in vegetable crop. Pharma Innov. J. 6 (12), 353–358.

Korkerd, S., Wanlapa, S., Puttanlek, C., Uttapap, D., Rungsardthong, V., 2016. Expansion and functional properties of extruded snacks enriched with nutrition sources from food processing by-products. J. Food Sci. Technol. 0022-1155. 53 (1), 561–570. doi:10.1007/s13197-015-2039-1.

Kowalski, R.J., Medina-Meza, I.G., Thapa, B.B., Murphy, K.M., Ganjyal, G.M., 2016. Extrusion processing characteristics of quinoa (*Chenopodium quinoa* Willd.) var. Cherry Vanilla. J. Cereal Sci. 10959963. 70, 91–98. Available from: http://www.elsevier.com/inca/publications/store/6/2/2/8/5/9/index.htt. doi:10.1016/j.jcs.2016.05.024.

Leonard, W., Zhang, P., Ying, D., Fang, Z., 2020. Application of extrusion technology in plant food processing byproducts: an overview. Compr. Rev. Food Sci. Food Saf. 1541-4337. 19 (1), 218–246. doi:10.1111/1541-4337.12514.

Leontowicz, H., Leontowicz, M., Kostyra, H., Gralak, M.A., Kulasek, G.W., 1999. The influence of extrusion or boiling on trypsin inhibitor and lectin activity in leguminous seeds and protein digestibility in rats. Pol. J. Food Nutr. Sci. 49 (4), 77–87.

Lesjak, M., Srai, S.K.S., 2019. Role of dietary flavonoids in iron homeostasis. Pharmaceuticals. 14248247. 12 (3). Available from: https://www.mdpi.com/1424-8247/12/3/119/pdf. doi:10.3390/ph12030119.

Li, J., Xiang, Q., Liu, X., Ding, T., Zhang, X., Zhai, Y., Bai, Y., 2017. Inactivation of soybean trypsin inhibitor by dielectric-barrier discharge (DBD) plasma. Food Chem. 03088146. 232, 515–522. doi:10.1016/j.foodchem.2017.03.167.

Li, X., Manickavasagan, A., Lim, L.T., 2024. Reduction of antinutrients and off-flavour in kidney bean flour by acidic and alkaline reactive extrusion. Food Res. Int. 18737145. 192. Available from: https://www.sciencedirect.com/science/journal/09639969. doi:10.1016/j.foodres.2024.114832.

Lo, D., Wang, H.I., Wu, W.J., Yang, R.Y., 2018. Anti-nutrient components and their concentrations in edible parts in vegetable families. CAB Rev. 17498848. 13. Available from: https://www.cabi.org/cabreviews/FullTextPDF/2018/20183230385.pdf. doi:10.1079/PAVSNNR201813015.

Lochab, B., Shukla, S., Varma, I.K., 2014. Naturally occurring phenolic sources: monomers and polymers. RSC Adv. 2046-2069. 4 (42), 21712–21752. doi:10.1039/c4ra00181h.

López-Moreno, M., Garcés-Rimón, M., Miguel, M., 2022. Antinutrients: lectins, goitrogens, phytates and oxalates, friends or foe? J. Funct. Foods. 17564646. 89 (104938). doi:10.1016/j.jff.2022.104938.

Lund, M.N., Ray, C.A., 2017. Control of maillard reactions in foods: strategies and chemical mechanisms. J. Agric. Food Chem. 15205118. 65 (23), 4537–4552. Available from: http://pubs.acs.org/journal/jafcau. doi:10.1021/acs.jafc.7b00882.

Luo, Y., Li, N., Lv, S., Li, D., Hu, Y., Liu, X., Zhou, D., 2025. Non-enzymatic browning in stored ready-to-eat oysters: regional differences in color changes, chemical reactions and mechanisms. LWT. 00236438. 224, 117861. doi:10.1016/j.lwt.2025.117861.

Marzo, F., Milagro, F.I., Urdaneta, E., Barrenetxe, J., Ibañez, F.C., 2011. Extrusion decreases the negative effects of kidney bean on enzyme and transport activities of the rat small intestine. J. Anim. Physiol. Anim. Nutr. 09312439. 95 (5), 591–598. doi:10.1111/j.1439-0396.2010.01088.x.

Maurya, A.K., Said, P.P., 2015. Extrusion processing on physical and chemical properties of protein rich products-an overview. J. Bioresour. Eng. Technol. 2 (4), 61–67 .

Meng, X., Threinen, D., Hansen, M., Driedger, D., 2010. Effects of extrusion conditions on system parameters and physical properties of a chickpea flour-based snack. Food Res. Int. 09639969. 43 (2), 650–658. doi:10.1016/j.foodres.2009.07.016.

Mironeasa, S., Coţovanu, I., Mironeasa, C., Ungureanu-Iuga, M., 2023. A review of the changes produced by extrusion cooking on the bioactive compounds from vegetal sources. Antioxidants. 2076-3921. 12 (7), 1453. doi:10.3390/antiox12071453.

Mishra, A., Behura, A., Mawatwal, S., Kumar, A., Naik, L., Mohanty, S.S., Manna, D., Dokania, P., Mishra, A., Patra, S.K., Dhiman, R., 2019. Structure-function and application of plant lectins in disease biology and immunity. Food Chem. Toxicol. 18736351. 134. Available from: https://www.elsevier.com/locate/foodchemtox. doi:10.1016/j.fct.2019.110827.

Mitchell, T., Kumar, P., Reddy, T., Wood, K.D., Knight, J., Assimos, D.G., Holmes, R.P., 2019. Dietary oxalate and kidney stone formation. J. Physiol. Renal Physiol. 15221466. 316 (3), F409. Available from: https://journals.physiology.org/doi/pdf/10.1152/ajprenal.00373.2018. doi:10.1152/ajprenal.00373.2018.

Moscicki, L., Mitrus, M., Wojtowicz, A., Oniszczuk, T., Rejak, A., 2013. Extrusion-cooking of starch. InTech. doi:10.5772/52323.

Mukhopadhyay, N., Bandyopadhyay, S., 2003. Extrusion cooking technology employed to reduce the anti-nutritional factor tannin in sesame (*Sesamum indicum*) meal. J. Food Eng. 02608774. 56 (2-3), 201–202. doi:10.1016/s0260-8774(02)00250-9.

Mukhopadhyay, N., Sarkar, S., Bandyopadhyay, S., 2007. Effect of extrusion cooking on anti-nutritional factor tannin in linseed (*Linum usitatissimum*) meal. Int. J. Food Sci. Nutr. 14653478. 58 (8), 588–594. doi:10.1080/09637480701343952.

Muramoto, K., 2017. Lectins as bioactive proteins in foods and feeds. Food Sci. Technol. Res. 1344-6606. 23 (4), 487–494. doi:10.3136/fstr.23.487.

Nair, V.C., Duvnjak, Z., 1990. Reduction of phytic acid content in canola meal by Aspergillus ficuum in solid state fermentation process. Appl. Microbiol. Biotechnol. 0175-7598. 34 (2), 183–188. doi:10.1007/bf00166777.

Nandane, A.S., Ganorkar, P.M., Ranveer, R.C., Patil, H., Al-Asmari, F., Sangsawad, P., Nirmal, N., Ozogul, F., 2025. Impact of extrusion process on the macro- and micro-nutrient in extruded food products: challenges and future trends. Food Bioprocess Technol. 19355149. 18 (8), 6939–6963. Available from: https://www.springer.com/journal/11947. doi:10.1007/s11947-025-03886-7.

Nikmaram, N., Leong, S.Y., Koubaa, M., Zhu, Z., Barba, F.J., Greiner, R., Oey, I., Roohinejad, S., 2017. Effect of extrusion on the anti-nutritional factors of food products: an overview. Food Control. 09567135. 79, 62–73. Available from: https://www.journals.elsevier.com/food-control. doi:10.1016/j.foodcont.2017.03.027.

Noonan, S.C., Savage, G.P., 1999. Oxalate content of foods and its effect on humans. Asia Pac. J. Clin. Nutr. 8, 64-74. doi:10.1046/j.1440-6047.1999.00038.x.

Nwabueze, T.U., 2007. Effect of process variables on trypsin inhibitor activity (TIA), phytic acid and tannin content of extruded African breadfruit-corn-soy mixtures: a response surface analysis. LWT. 00236438. 40 (1), 21–29. Available from: https://www.journals.elsevier.com/lwt. doi:10.1016/j.lwt.2005.10.004.

Offiah, V., Kontogiorgos, V., Falade, K.O., 2019. Extrusion processing of raw food materials and by-products: a review. Crit. Rev. Food Sci. Nutr. 15497852. 59 (18), 2979–2998. Available from: https://www.tandfonline.com/loi/bfsn20. doi:10.1080/10408398.2018.1480007.

Osuna-Gallardo, E.I., Cuevas-Rodríguez, E.O., Sepúlveda-García, C.I., Reyes-Moreno, C., León-López, L., Han, R., Hernández-Álvarez, A.J., 2023. Impact of cooking and extrusion processing on nutritional, antinutritional, and techno-functional characteristics of indigenous bean (*Phaseolus coccineus*). ACS Food Sci. Technol. 26921944. 3 (11), 1835–1853. Available from: https://pubs.acs.org/page/afsthl/about.html. doi:10.1021/acsfoodscitech.2c00416.

Pal, R.S., Bhartiya, A., Yadav, P., Kant, L., Mishra, K.K., Aditya, J.P., Pattanayak, A., 2017. Effect of dehulling, germination and cooking on nutrients, anti-nutrients, fatty acid composition and antioxidant properties in lentil (*Lens culinaris*). J. Food Sci. Technol. 0022-1155. 54 (4), 909–920. doi:10.1007/s13197-016-2351-4.

Pansawat, N., Jangchud, K., Jangchud, A., Wuttijumnong, P., Saalia, F.K., Eitenmiller, R.R., Phillips, R.D., 2008. Effects of extrusion conditions on secondary extrusion variables and physical properties of fish, rice-based snacks. LWT. 00236438. 41 (4), 632–641. doi:10.1016/j.lwt.2007.05.010.

Pastor-Cavada, E., Drago, S.R., González, R.J., Juan, R., Pastor, J.E., Alaiz, M., Vioque, J., 2011. Effects of the addition of wild legumes (*Lathyrus annuus* and *Lathyrus clymenum*) on the physical and nutritional properties of extruded products based on whole corn and brown rice. Food Chem. 18737072. 128 (4), 961–967. doi:10.1016/j.foodchem.2011.03.126.

Patterson, C.A., Curran, J., Der, T., 2017. Effect of processing on antinutrient compounds in pulses. Cereal Chem. 19433638. 94 (1), 2–10. Available from: http://aaccipublications.aaccnet.org/doi/pdf/10.1094/CCHEM-05-16-0144-FI. doi:10.1094/CCHEM-05-16-0144-FI.

Petroski, W., Minich, D.M., 2020. Is there such a thing as "anti-nutrients"? A narrative review of perceived problematic plant compounds. Nutrients. 20726643. 12 (10), 1–32. Available from: https://www.mdpi.com/2072-6643/12/10/2929/pdf. doi:10.3390/nu12102929.

Pismag, R.Y., Polo, M.P., Hoyos, J.L., Bravo, J.E., Roa, D.F., 2024. Effect of extrusion cooking on the chemical and nutritional properties of instant flours: a review. F1000Research. 20461402. 12. Available from: https://f1000research.com/. doi:10.12688/f1000research.140748.2.

Rathod, R.P., Annapure, U.S., 2016. Effect of extrusion process on antinutritional factors and protein and starch digestibility of lentil splits. LWT. 00236438. 66, 114–123. doi:10.1016/j.lwt.2015.10.028.

Saha, S., Jha, S., Tiwari, A., Jayapalan, S., Roy, A., 2021. Considerations for improvising fortified extruded rice products. Journal of Food Science. 0022-1147. 86 (4), 1180–1200. doi:10.1111/1750-3841.15656.

Sarkhel, S., Roy, A., 2022. Phytic acid and its reduction in pulse matrix: structure–function relationship owing to bioavailability enhancement of micronutrients. J. Food Process Eng. 0145-8876. 45 (5). doi:10.1111/jfpe.14030.

Sarkhel, S., Roy, A., 2024. Vacuum impregnation assisted simultaneous micronutrients fortification and phytic acid reduction in lentils. J. Food Eng. 02608774. 365, 111823. doi:10.1016/j.jfoodeng.2023.111823.

Savage, G.P., Vanhanen, L., Mason, S.M., Ross, A.B., 2000. Effect of cooking on the soluble and insoluble oxalate content of some New Zealand foods. J. Food Compos. Anal. 08891575. 13 (3), 201–206. doi:10.1006/jfca.2000.0879.

Schmid, V., Trabert, A., Schäfer, J., Bunzel, M., Karbstein, H.P., Emin, M.A., 2020. Modification of apple pomace by extrusion processing: studies on the composition, polymer structures, and functional properties. Foods. 23048158. 9 (10). Available from: https://www.mdpi.com/2304-8158/9/10/1385. doi:10.3390/foods9101385.

Scott, M.T., Goss-Sampson, M., Bomford, R., 1985. Adjuvant activity of saponin: antigen localization studies. Int. Arch. Allergy Appl. Immunol. 00205915. 77 (4), 409–412. doi:10.1159/000233817.

Seth, D., Rajamanickam, G., 2012. Development of extruded snacks using soy, sorghum, millet and rice blend – a response surface methodology approach. International Journal of Food Science & Technology. 0950-5423. 47 (7), 1526–1531. doi:10.1111/j.1365-2621.2012.03001.x.

Sharma, A., 2021. A review on traditional technology and safety challenges with regard to antinutrients in legume foods. J. Food Sci. Technol. 09758402. 58 (8), 2863–2883.

Available from: http://www.springerlink.com/content/121580/. doi:10.1007/s13197-020-04883-8.

Sharon, N., 2007. Lectins: carbohydrate-specific reagents and biological recognition molecules. J. Biol. Chem. 00219258. 282 (5), 2753–2764. doi:10.1074/JBC.X600004200.

Shi, J., Arunasalam, K., Yeung, D., Kakuda, Y., Mittal, G., Jiang, Y., 2004. Saponins from edible legumes: chemistry, processing, and health benefits. J. Med. Food. 1096-620X. 7 (1), 67–78. doi:10.1089/109662004322984734.

Shi, L., Arntfield, S.D., Nickerson, M., 2018. Changes in levels of phytic acid, lectins and oxalates during soaking and cooking of Canadian pulses. Food Res. Int. 18737145. 107, 660–668. Available from: https://www.elsevier.com/inca/publications/store/4/2/2/9/7/0. doi:10.1016/j.foodres.2018.02.056.

Singh, S., Gamlath, S., Wakeling, L., 2007. Nutritional aspects of food extrusion: a review. International Journal of Food Science & Technology. 0950-5423. 42 (8), 916–929. doi:10.1111/j.1365-2621.2006.01309.x.

Suvanto, J., Nohynek, L., Seppänen-Laakso, T., Rischer, H., Salminen, J.P., Puupponen-Pimiä, R., 2017. Variability in the production of tannins and other polyphenols in cell cultures of 12 Nordic plant species. Planta. 14322048. 246 (2), 227–241. Available from: https://www.link.springer.de/link/service/journals/00425/index.htm. doi:10.1007/s00425-017-2686-8.

Vainionpää, J., 1991. Modelling of extrusion cooking of cereals using response surface methodology. J. Food Eng. 02608774. 13 (1), 1–26. doi:10.1016/0260-8774(91)90035-q.

Valdez-Flores, M., Germán-Báez, L.J., Gutiérrez-Dorado, R., Medina-Godoy, S., Norzagaray-Valenzuela, C., Hernández-Verdugo, S., Reyes-Moreno, C., Valdez-Ortiz, A., 2016. Improving bioactivities of *Jatropha curcas* protein hydrolysates by optimizing with response surface methodology the extrusion cooking process. Ind. Crops Prod. 09266690. 85, 353–360. Available from: https://www.elsevier.com/inca/publications/store/5/2/2/8/2/5. doi:10.1016/j.indcrop.2015.12.084.

Verfaillie, D., Li, J., Janssen, F., Blontrock, E., Van Royen, G., Wouters, A.G.B., 2024. Relating the protein denaturation degree and solubility of soy protein isolates to the structure of high moisture extrudates. Food Hydrocoll. 0268005X. 155, 110211. doi:10.1016/j.foodhyd.2024.110211.

Wang, N., Hatcher, D.W., Toews, R., Gawalko, E.J., 2009. Influence of cooking and dehulling on nutritional composition of several varieties of lentils (*Lens culinaris*). LWT. 00236438. 42 (4), 842–848. Available from: https://www.journals.elsevier.com/lwt. doi:10.1016/j.lwt.2008.10.007.

Wani, S.A., Kumar, P., 2016. Effect of extrusion on the nutritional, antioxidant and microstructural characteristics of nutritionally enriched snacks. J. Food Process. Preserv. 01458892. 40 (2), 166–173. doi:10.1111/jfpp.12593.

Xu, E., Wu, Z., Long, J., Wang, F., Pan, X., Xu, X., Jin, Z., Jiao, A., 2015. Effect of thermostable α-amylase addition on the physicochemical properties, free/bound phenolics and antioxidant capacities of extruded hulled and whole rice. Food Bioprocess Technol. 1935-5130. 8 (9), 1958–1973. doi:10.1007/s11947-015-1552-6.

Yağci, S., Evci, T., 2015. Effect of instant controlled pressure drop process on some physicochemical and nutritional properties of snacks produced from chickpea and wheat. Int. J. Food Sci. Technol. 13652621. 50 (8), 1901–1910. Available from: http://onlinelibrary.wiley.com/journal/10.1111/(ISSN)1365-2621. doi:10.1111/ijfs.12843Turkey.

Yi, C., Qiang, N., Zhu, H., Xiao, Q., Li, Z., 2022. Extrusion processing: a strategy for improving the functional components, physicochemical properties, and health benefits of whole grains. Food Res. Int. 09639969. 160, 111681. doi:10.1016/j.foodres.2022.111681.

Zahari, I., Ferawati, F., Helstad, A., Ahlström, C., Östbring, K., Rayner, M., Purhagen, J.K., 2020. Development of high-moisture meat analogues with hemp and soy protein using extrusion cooking. Foods. 23048158. 9 (6). Available from: https://www.mdpi.com/2304-8158/9/6/772. doi:10.3390/foods9060772.

Development of texturized vegetable protein

13

Srutee Rout[a], *Soubhagya Tripathy*[a], *Vigneshwaran V.*[b], *and Prem Prakash Srivastav*[a]
[a]Department of Agricultural and Food Engineering, Indian Institute of Technology Kharagpur, Kharagpur, India
[b]Department of Biotechnology, Alagappa College of Technology, Anna University, Chennai, Tamil Nadu, India

13.1 Introduction

With the global population projected to reach approximately 9 billion by 2050, one of the pressing challenges for food security lies in providing sufficient protein in a manner that is cost-effective, nutritionally sound, and environmentally sustainable (Oluwole et al., 2023). Given the ecological strain associated with animal protein production and the increasing recognition of the health advantages linked to plant-based proteins, there has been growing interest in developing meat analogues (Rout et al., 2024). For over five decades, extrusion technology has been employed to produce texturized vegetable protein (TVP), which mimics the fibrous texture of animal muscle. TVP exhibits beneficial characteristics such as high-water absorption and retention capacities, absence of cholesterol, and a digestibility rate of up to 93%–97% in the human digestive system (Guerrero et al., 2025). Moreover, bioactive compounds present in plant proteins, such as isoflavones, saponins, and amino acids like arginine, are associated with health-promoting effects, including blood pressure regulation, antiaging benefits, and enhanced mineral uptake (Rout et al., 2024; Rao and Poonia, 2023). Consequently, consistent consumption of TVP-based products may help mitigate the risk of chronic diseases like cardiovascular disorders, hypertension, and obesity. Despite increasing interest, meat analogs still represent a small fraction of the overall meat market. For example, in the Netherlands, plant-based meat alternatives account for only about 1% of total meat sales, likely due to limitations in sensory attributes such as flavor and texture. Consumers are now more inclined toward sustainable, health-oriented, and novel dietary choices, yet the widespread adoption of meat alternatives hinges on their ability to closely replicate the sensory qualities of conventional meat (Baig et al., 2025).

The overall quality of TVP is influenced by several factors, including the composition and properties of raw materials (e.g., particle size, pretreatment, and additives), the design and configuration of the extrusion equipment (e.g., screw type, die shape, L/D

Extrusion Cooking for Food and Feed. https://doi.org/10.1016/B978-0-443-29987-2.00013-0

ratio), and operational conditions (e.g., temperature, moisture, feed rate, and screw speed) (Ribeiro et al., 2024). The extrusion process acts as a complex biochemical reactor, providing controlled heat, mechanical shear, and pressure. These forces drive a series of transformations, including protein unfolding, aggregation, carbohydrate breakdown, lipid oxidation, and molecular interactions that give rise to new structures and textures (Gao et al., 2022). However, accurately managing the extrusion process remains a significant challenge. The continuous physical and chemical changes occurring along the barrel make it difficult to pinpoint specific transformations (Patti, 2024). These dynamic reactions contribute to both visible changes, such as shape, texture, and color, and invisible alterations like molecular restructuring and bond formation. Understanding and optimizing these transformations requires segmenting the extrusion process into phases and studying how primary components behave in each stage. This chapter aims to present a comprehensive overview of advancements in food extrusion technology for TVP production, highlight conformational changes in key ingredients during processing, assess the influence of extrusion parameters on product quality, and explore methods for analyzing and monitoring the extrusion process.

13.2 Sources of proteins for texturization of vegetable protein

TVP offers various advantages over animal proteins, particularly in terms of enhanced well-being and more economical aspects (Rout and Srivastav, 2023; Roberts, 2013). The development of vegetable-based protein reduces dependency on animal meat protein (Le Bourgot et al., 2023; Rout and Srivastav, 2023). The major sources of vegetable proteins are listed in Fig. 13.1. The minor oilseed-based protein sources include sesame, linseed, safflower, and flaxseed (Riaz, 2011). The drivers for the texturization of vegetable proteins include population growth and consumers' shift toward sustainable protein intake (Peñaranda et al., 2025). In general, processing aids may be introduced into the texturization process, such as pH modifiers or reducing agents.

13.2.1 Legume proteins

Legumes are widely referred to as the poor man's meat and represent the second-largest cultivated crop after cereals. Legumes provide 33% of the human protein needs (Padha et al., 2024). Mung bean, also known as green pearl, is a rich source of protein but is limited in terms of its food applications. Mung bean protein could be utilized as a source for protein texturization. The mung bean protein isolate demonstrated similar outcomes to soy protein isolates (SPIs) in terms of absorption (water and oil) capacity, denaturation temperature, solubility, and even comparable values for its least gelation concentration (Rout and Srivastav, 2025). However, the foaming capacity of mung bean protein isolate is greater than that of SPI (Brisht et al., 2017). In addition, when mung bean is extruded at lower moisture content (49.33%), under specific screw speed (80.66 rpm) and barrel temperature (144.57°C), it exhibits improved properties. These texturized mung bean proteins' physical and structural properties, along with improved amino acid retention capabilities, make them a viable alternative to animal meat (Brishti et al., 2021). Similarly,

Fig. 13.1 Classification of vegetable proteins.

pea protein is another high-demand protein that is also used in the production of meat alternatives. The functional properties of pea proteins depend on their sources; however, the extrudate product showed similar textural properties when formed alone, while in patty formation, it exhibited different phenomena. This indicates that the texturization behavior could be better understood if studied in combination with other materials to achieve precise texturization results through similar trials. The texturized protein from one plant source (pea) is always different from that of other sources (Rout et al., 2024; Webb, 2021).

The incorporation of whole leguminous protein leads to the production of a leguminous flavor and makes it difficult to achieve the desired texture, color, and other meat-like properties. Several approaches have been taken, such as the incorporation of color enrichment or blending different proteins and texturizing. For instance, the enrichment of pea protein with *Chlorella* improved the color, minimized the legume flavor, and increased acceptability in terms of texture and flavor. This approach enhanced the usage of pea protein in vegetable-based hamburgers (Peñaranda et al., 2023). Texturized fava beans could be used as a complete or partial replacement for beef hamburgers. When texturized fava beans were used as-is, they produced a low cooking yield and retained less moisture compared to commercial products. However, the best results were observed with a 25% replacement in beef hamburger patties, which showed improved moisture, cooking yield, and reduced thickness compared to beef hamburger patties (Guerrero et al., 2025). texturization of fava bean and lentil proteins did not alter their protein content. However, as the moisture content increased, their ability to rehydrate improved. Notably, the process led to changes in color, which were attributed to the Maillard reaction. Specific changes were also observed, including alterations in moisture content, temperature, and screw speed. The water and oil absorption capacity of texturized fava bean protein remained unaltered, while it varied in lentil bean protein isolate. These texturized proteins, when used as a 10% replacer, increased cooking yield and decreased thickness and diameter changes in patties (Guerrero, 2023). It was found that a 20% incorporation of chickpea protein with SPI increased the extrudate bulk density but decreased the water-holding capacity. Without the introduction of chickpea protein, the texture showed high springiness; however, beyond 20% incorporation of chickpea protein, the texture was negatively affected. The substitution of chickpea protein between 10% and 20% is considered the optimal percentage for use in texturized pea protein in starch-based food systems to create meat-like textured products (Webb et al., 2020; Rout and Srivastav, 2024). Even existing commercial SPIs varied in terms of the degree of texturization, sensory scores, and the textural properties of the final products (Meng et al., 2023).

13.2.2 Cereal proteins

The introduction of texturized cereal protein into foods enhances the overall protein content, improves bioaccessibility, reduces the energy required for extrusion, alters the physicochemical properties, and improves product texture. Protein-rich textured food from wheat was developed using textured wheat protein combined with chickpea and quinoa (lysine) (Le Bourgot et al., 2023). Wheat is one of the most widely used cereals and is rich in gluten protein. The mechanism behind the texturization of wheat gluten (WG) (protein) during extrusion involves the subsequent reduction of the disulfide groups and the corresponding covalent bond formation of free thiol groups with the disulfide groups. For the texturization of WG to take place, heat, mechanical energy, and reducing agents are required (Roberts, 2013). Peñaranda et al. analyzed the texturization effect on the protein in vitro bioaccessibility of essential and nonessential amino acids with their corresponding protein isolates. Among the proteins from rice, pea, and an equal proportion of rice and pea, except for pea protein, the in vitro digestibility and bioaccessibility of nonessential amino acids were increased in the corresponding textured protein compared to the isolates. The process of texturization also reduced the in vitro bioaccessibility of specific amino acids, such as lysine. Rice protein showed high thermal stability at the extrusion temperature with a balanced amino acid profile (Peñaranda et al., 2025). When focusing on rice protein for texturization, its addition as a substitution to SPI, along with corn starch and WG, resulted in a textured product with improved nutritional content (methionine) and decreased porosity, mechanical energy, and water absorption capacity. The addition of rice protein isolate also demonstrated better rehydration properties and is suitable as an alternative to meat (Lee et al., 2022).

Even the true ileal digestibility (i.e., proportion of nutrient absorbed in the small intestine) values of wheat protein (95.3%) showed similar scores to beef meat (95.8%), increased after texturization (96.8%), and in combination with chickpea with texturized wheat protein (96.5%). Even the digestible indispensable amino acid scores for the texturized protein combined with chickpea and quinoa, chicken fillet, and textured soy protein are 91, 110, and 111, respectively. Thus by optimizing the sources, vegetable-based texturized meat could be an alternative to animal meat (Le Bourgot et al., 2023). The physicochemical properties of the texturized proteins are specific to the raw material and texturization conditions used. In the case of plant-based meat development in low-moisture extrusion, the incorporation of pea protein isolate (PPI) and soybean protein formulations requires less energy, more cross-linking between them, and produces a loosely (porous) packed product, while the structural characteristics are vice versa for formulations incorporated with texturized WG protein and soy protein. Understanding the physicochemical properties, such as water holding capacity and pasting behavior, will be necessary to understand the texturization behavior of protein (Webb et al., 2023). In addition, texturization (by extrusion) determines the microstructure, fractal dimensions, and dynamic mechanical properties. The introduction of wheat protein (gluten) improved the existing texture of soy protein. There exists an inverse relationship between elasticity and moisture content. However, a direct relationship exists between extrusion temperature (130°C–160°C) and both creep and nonrecoverable deformation. The fractal dimension is negatively influenced by both moisture content and WG introduction (Wu et al., 2019). In the development of meat analogs, rather than being fully made from pea protein, the incorporation of a barley fraction rich in protein (30%) during texturization (extrusion) improved the textural properties, such as hardness, gumminess, and chewiness values. With variations in the moisture content in the raw material and incorporation of varying percentages of barley fraction, a wide range of textured products could be made (Singh, 2024).

13.2.3 Algal proteins

The contribution of algae to the production of meat alternatives through extrusion is increasing. There are four different forms of structural and functional proteins obtained from various algae. These include lectin (agglutinin) or simple carbohydrate-binding protein, glycoprotein, the major cell wall protein, phycobiliproteins present in the stroma inside the chloroplast, and microsporine-like acids, which are secondary metabolite proteins (Espinosa-Ramírez et al., 2023). In general, during the process of texturization of products, algae are introduced into the food product as is in many cases. Spirulina can serve as a partial replacer of soy protein when subjected to high-moisture extrusion (Grahl et al., 2018). Spray-dried microalgae (*Auxenochlorella protothecoides*) were texturized and used as a meat analog. This meat analog was rich in vitamins (B and E), improved color, and reduced dependency on soy protein for meat alternatives (Caporgno et al., 2020). Algal-based protein demonstrated better performance in emulsification, solubility, gelation, and foaming properties, but its incorporation remains limited. The primary reasons include its negative effect on the sensory properties of the product. More research is needed to address this issue (Espinosa-Ramírez et al., 2023).

13.2.4 Dairy proteins

Whey and casein are the major protein components present in milk. Texturized whey protein has shown several changes from its native protein properties. For example, when whey proteins were extracted at acidic pH (3.0) in reactive supercritical fluid extrusion at three different temperatures, variations in their physical properties were observed. The texturized whey protein extruded at low temperatures (50°C and 70°C) was minimally denatured, with smaller aggregates and better solubility properties than whey protein produced at high temperature (90°C). During the texturization process, noncovalent bonds were formed, resulting in the generation of cold gel by the texturized protein-containing products. Unlike the thermally unstable low-temperature texturized whey proteins (pseudoplastic material), high-temperature texturized whey proteins were stable at higher temperatures (Nor Afizah and Rizvi, 2014). Texturized casein protein was made by subjecting renneted casein (gels) to sheet die extrusion under controlled shear and temperature, producing a similar textured protein to that obtained from soy–gluten mix and meat products. The texturized (fibrous) protein showed maximum shear stress values (1.9–4.1 kPa) that could be used in a variety of products as a new protein source (Kern et al. (2020). However, the major limitation associated with the introduction of dairy protein in snack products (made with corn flour) is that it forms a pasty texture upon interaction with (corn) starches. Texturization of dairy protein improved its textural properties for use in starchy food systems, thereby enhancing the nutrient content. When dairy proteins [whey protein isolate and whey protein concentrate (WPC)] were subjected to twin-screw extrusion at specific temperatures (50°C, 75°C, and 100°C), the water-binding ability of dairy proteins decreased without denaturation of the protein (Kern et al., 2020).

13.2.5 Miscellaneous proteins

Many improvements in TVP have been made so far. These include the introduction of green tea components into the texturization process of vegetable proteins, which increases antioxidant activity (such as phenolic and flavonoid content) and improves texture properties like hardness, integrity, and cutting strength. However, the incorporation negatively affects the nitrogen solubility index and the expansion of textured products by decreasing water holding capacity, cohesiveness, and springiness values (Ma and Ryu, 2019). Rather than adding one protein as an ingredient, the combination effect was studied. When chickpea flour was added to TVP, the proximate analysis of prepared chicken nuggets showed poor moisture and fat content, while the other proximate values were greater than those of control chicken nuggets. The imitation chicken nuggets made with chickpeas (10%) and TVP (30%) showed the best results. The texture could be further altered by incorporating specific percentages of carrageenan, mushroom, and seitan (Sharima-Abdullah et al., 2018). Such improvement includes the fermentation effect on TVP. When *Bacillus subtilis* was used to ferment the texturized protein at 37°C for 60 h, the best result was obtained when the TVP was made at a moisture content of 50% through extrusion (Maung et al., 2020).

Several approaches have been undertaken to introduce more than one TVP to enhance textural properties. A texturized mix of proteins, such as WG, PPI, and starch in a ratio of 4:5:1, was produced at 50% moisture content and 130°C through extrusion. This plant-based meat patty demonstrated improved yield (cooking), shape (reduced shrinkage and thickness), and textural properties (low hardness, cohesiveness, chewiness, and cutting strength values) compared to the meat-based commercial burger patty, except for lower springiness and moisture retention values (Samard et al., 2021). Even in this case, where more than one texturized protein product was used, the extrusion process parameters had a significant influence on the extrudate. For instance, in the production of textured vegetable protein from SPI, starch (corn), and wheat protein (gluten) in the ratio of 5:3:2, an increase in screw speed raised the energy required to form a textured protein. Similarly, an increase in die temperature enhanced the extension properties but reduced the bulk density of the textured protein (Lyu et al., 2023). TVP (pea, fava bean, or pea–fava bean–quinoa) was successfully incorporated into bread fortification. When wheat was substituted with 35% TVP, it improved the protein content without negatively affecting the bread's sensory properties (Laugesen et al., 2022). TVP also includes its own alternatives. Alternatives to the predominantly texturized protein from soybeans have been developed. Texturized protein made from PPI and gluten showed better results in terms of nutrition (protein, amino acid score) and taste. For partial substitution of soybean protein, a mixture of pea isolate and soybean concentrate could be used (Brugger et al., 2017). Soybean-based texturized protein, when used as a partial replacer (10%–40%) in beef patties, increased the fiber content, cohesiveness, gumminess, sourness, saltiness, umami, and astringency values of the patties. However, reductions were observed in parameters such as moisture, fat, water, bitterness, umami, cooking loss, hardness, thickness, and a slight reduction in the fatty acid composition of the patties. These results demonstrated that substitution is feasible within this range (Bakhsh et al., 2021).

Apart from the majorly utilized protein sources, the underutilized sources (like yeast and insects) were also reported for the texturization of their protein. When yeast protein was texturized along with soy protein, it increased the concentrations of amino acids and essential amino acids, improving the overall aroma of the meat analog. Though the in vitro digestibility remained unaltered, the surface hydrophobicity increased, and changes in protein (tryptophan) structure were observed (Xia et al., 2023). To satisfy the growing demand for protein sources, even insect-made texturized protein from high-moisture extrusion was developed. The best texturized meat analog was obtained with a mix of soy in the range of 5%–10% and *Alphitobius diaperinus* up to 40% as the maximum level (Smetana et al., 2018).

13.3 Techniques for processing vegetable protein

13.3.1 Extrusion

Extrusion is a thermomechanical process where food materials are moistened, expanded, and plasticized, followed by their extrusion. During the extrusion of protein, the change in protein from its folded (native) structure to an unfolded structure ultimately results in the creation of viscous and chewy extrudate. These texturized (extruded) proteins increase bioavailability but decrease protein solubility, antinutritional factors, and native

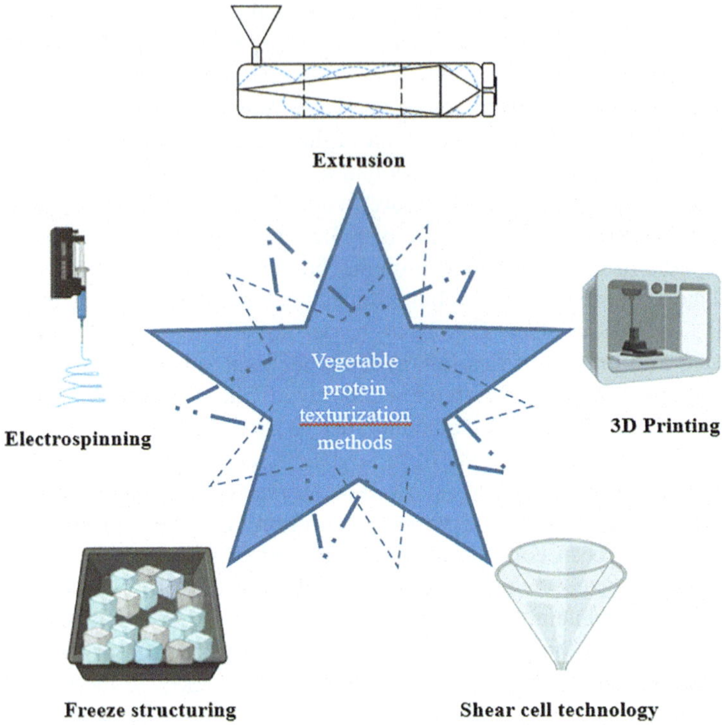

Fig. 13.2 Methods for producing texturized vegetable proteins.

off-flavors. All these processes are achieved inside the extruder configuration, which includes components such as the mixer, preconditioner, extrusion barrel (feeding, processing, and melting zones), final die, and knife (Vatansever et al., 2020). Moisture-based vegetable protein extrusion exists in two types, namely, low-moisture extrusion (15%–36%) and high-moisture extrusion (>50%) (Pennells et al., 2024; Vatansever et al., 2020). Various methods for the production of texturized proteins are shown in Fig. 13.2, and their impact on the final product is discussed in detail in the following sections.

13.3.1.1 High-moisture extrusion

High-moisture extrusion is an extensively used method for meat analog production, even at the industry level (Fig. 13.3). The fibrous texture of meat analogs is attributed to disulfide and hydrogen bonds formed irrespective of the moisture content, though more bonds are observed during high-moisture extrusion (Schmid et al., 2022). Wittek et al. (2021) proposed a possible mechanism for the formation of texture during high-moisture extrusion, which involves the creation of a multiphase system, i.e., a water-rich dispersion phase surrounded by a protein-rich phase. This result was observed by subjecting SPI to high-moisture extrusion (>40%) at elevated temperatures (124°C–135°C) during the extrusion (Witte et al., 2021). The introduction of polysaccharides with protein isolate

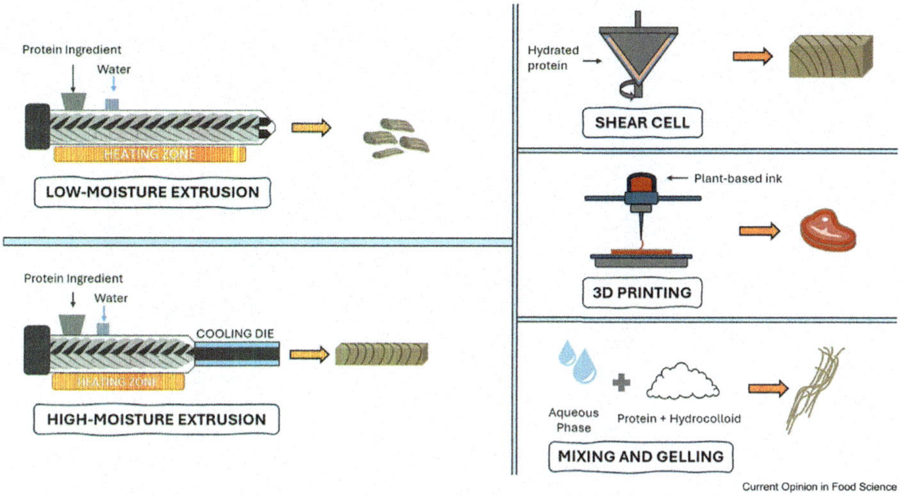

Fig. 13.3 Texturization technologies for the development of meat analogs and materials with a fibrous texture.
From Angelis, G., Pasqualone, A., Summo, C., 2024. Advancements in texturization processes for the development of plant-based meat analogs: a review. Curr. Opin. Food Sci. 58.

during extrusion is due to their ability to form a phase-separated mixture, which contributes to the meat-like texture in the extrudate. Various factors influencing the extrudate include moisture content, screw speed, and extrusion temperature (Schmid et al., 2022). For instance, in the case of peanut protein extrusion, the die temperature and pressure are influenced by the shear direction, extrusion temperature, and feed moisture content. Each of these factors affects the product in different ways. The shear direction influences fiber formation, extrusion temperature determines the product texture (springiness and tensile properties), and moisture content affects the hardness and color of the product (Zhang et al., 2020). Even the substitution of one plant-based protein with another can be achieved through high-moisture extrusion. When hemp protein concentrate was replaced with SPI and subjected to a twin-screw extruder, the extrudate exhibited satisfactory texture up to 60% substitution. Hemp protein concentrate's lower water absorption capacity allowed for a softer textured meat analog with an increased moisture content of 5%. The incorporation of hemp protein concentrate increased the denaturation temperature and the darkness of the pure SPI extrudates used to make meat analogs (Zahari et al., 2020). At high-moisture extrusion (55%), different protein combinations behave differently. For instance, blending pea protein powder with SPI increased the die temperature while decreasing specific mechanical energy, producing an extrudate with a dense and smooth texture. Conversely, the addition of WG to PPI resulted in a different phenomenon, except for no change in the die temperature (Zhang et al., 2018). A few more effects of high-moisture extrusion on the extrudate are listed in Table 13.1.

Table 13.1 Processing of TVP using high- and low-moisture extrusion.

Protein source	Type of extrusion	Process conditions	Impact on the final product	Reference
Pea protein (PP), soy protein isolate (SPI), chickpea protein (CP), and wheat gluten (WG)	High moisture	Barrel diameter: 11 mm Barrel length: 440 mm Screw speed: 150 rpm Temperature of extruder zones: 40°C–100°C, cooling zone: 20°C	• PP–CP–WG produced meat analogs with better chewiness and fiber structure	(Jiang et al., 2024)
Yeast protein (YP) and soy protein isolate	High moisture	Screw speed: 250 rpm Moisture content: 60% Cooling die: 100°C Heating temperature of zones: 30°C–140°C	• YP increased the protein realignment and networks to promote the formation of fiber structures	(Zhao et al., 2023)
Pea protein isolate (PPI) was mixed with L-cysteine	High moisture	Screw speed: 162 r/min Moisture content: 54% Feeding speed 10 kg/h Temperature: 40°C–140°C	• OHC reached the maximum value of 109.73% when 0.09% L-cysteine was added • Nitrogen solubility index increased from 3.28% to 5.14% with the increase of l-cysteine (0% to 0.06%) • l-cysteine significantly affected the textural properties by altering the cross-linking effect	(Peng et al., 2022)
Soy protein with insoluble dietary fiber (IDF)	High moisture	Length of screw: 1440 mm Diameter of screw: 36 mm Screw speed: 280 rpm Temperature of cooling die: 70°C Moisture content: 55%	• Textural properties decreased, and tensile properties increased with the increasing substitution of IDF • IDF increased the apparent G' and elasticity of the blends	(Deng et al., 2023)
Soy protein concentrate and wheat gluten	High moisture	Barrel diameter: 11 mm Length/diameter ratio of screw: 44:1 Screw speed: 150 rpm Temperature of cooling die: 20°C Moisture: 60%	• SPC-WG ratio of 50/50 exhibited highly fibrous structures	(Zhang et al., 2022)

Mung bean protein	High moisture	Extruder barrels: 19 mm Screw speed: 40 rpm Moisture content: 9%–10%	• The optimized extrusion process conditions are as follows: feed moisture 49.33%, screw speed 80.66 rpm, and barrel temperature 144.57°C	(Brishti et al., 2021)
Mung bean protein	High moisture	The screw speed and temperature of the extruder barrel were set at 80.66 rpm and 144.57°C. The feed moisture content was fixed at three levels: low (30.0%), intermediate (49.3%), and high (60.0%).	• Texturized MBP showed that 49.3% is the optimum feed moisture content	(Brishti et al., 2021)
Lima bean and African oil bean seed protein concentrate	High moisture	Barrel temperature: 92.45°C Screw speed: 101.48 rpm Feed moisture: 59.63%	• In terms of sensorial aspects, the best result was obtained when lima bean protein concentrate was extruded	(Arueya et al., 2017)
Faba bean protein concentrate	High-moisture extrusion	Temperature: 130°C–140°C Feed rate: 1.1 kg/h Water-to-feed ratio: 4	• Meat analog with better sensory scores and improved oligosaccharide content compared to the raw material	(Saldanha Do Carmo et al., 2021)
Soy protein isolate: whey protein concentrate (70:30)	High-moisture extrusion	Moisture content: 55% Material temperature: 115°C–133°C Screw speed: 600 rpm	• The addition of whey protein concentrate increased the anisotropy index but reduced the dispersed phase in the multiphase system	(Wittek et al., 2021)

13.3.1.2 Low-moisture extrusion

When the PPI was subjected to low-moisture extrusion (26%–35%) at temperatures above 130°C, at varying screw speeds, low water-soluble protein extrudate was obtained. Furthermore, a few protein fraction structures, like vicilin and convicilin, were unaltered, while legumin was altered (aggregated or proteolyzed) (Beck et al., 2017). Certain pretreatments and blending with carbohydrates can improve the functionalities of extruded protein isolate. In terms of preheating by the dry heat method at 120°C for 2 h, followed by low-moisture (36%) extrusion, the water holding capacity values of the soy, pea, and chick PPIs remained unaltered, while their pasting properties showed increased values. However, at elevated dry heat treatment at 160°C, the density and darkness values of the extrudate increased (Pennells et al., 2024). When subjecting the blend of β-glucan (5%) with SPI to low-moisture extrusion (36%), improved cooking yield and oil holding capacity of the extrudates were observed. The blended extrudate also became hydrophobic with improved nutrient retention (Xiao et al., 2023).

13.3.2 Shear cell technology

The shear cell was initially used to study the extruder-like effect on products until fibrils formed from calcium caseinate. Shear cells follow three basic steps: hydrating, thermo-mechanical treatment, and cooling, as in extrusion, but shear flow is more uniform in shear cells than in high-moisture extrusion. There are two different types of shear cells. In the cone-in-cone shear cell, the inner (top) cone is stationary and surrounded by a rotating inner (bottom) cone, while in another type, the Couette device, it operates in reverse to the shear cell. Though the extrusion effects were extensively explored for texturization, the shear cell has been less explored as a texturization technique (Cornet et al., 2022). By applying simple shear flow, a fibrous and layered texture was obtained from soy protein concentrate (SPC) (35%) at 140°C. Unlike SPI, the polysaccharides in SPC also contributed to the formation of the anisotropic (varied properties) material (Grabowska et al., 2016). Using a shear cell device, an alternative to fish meal was created from different protein concentrates (lupine, soy, and rapeseed). The texturized lupine protein concentrate showed similar properties to fish meal in terms of solubility, low water holding capacity, and low paste viscosity. Moreover, the incorporated texturized protein concentrate required less energy, low moisture (22%), and was easily dried after extrusion (Draganovic et al., 2014). Protein texturization from vegetable proteins depends on the simultaneous introduction of shear and heat. When the Couette cell device was used for the production of a meat analog from SPI (31%) in WG dispersion, fibers were obtained between 90°C and 100°C, while time and rotation had no significant impact on the produced fibers. This Couette cell method can be scalable and applicable for commercial continuous production (Krintiras et al., 2015).

13.3.3 Electrospinning

Electrospinning produces high-quality nanoscale fibers (≤500 nm) from linear or coil-shaped proteins. Most vegetable proteins have a globular texture, which limits their nanofiber formation (Mathew and Vigneshwaran, 2024) (Fig. 13.4). One such vegetable protein widely studied for its electrospinnability is zein. Zein is a protein compound present in corn or maize, also known as a biopolymer, and is a digestible material. These electrospun zein proteins have found applications in active substance encapsulation (lycopene, folic acid, saffron extract, etc.), either alone or in combination with other materials, and are also utilized in active packaging (Jiang et al., 2023). Another protein isolate, rapeseed protein, was studied for its electrospinnability properties after modification, as discussed by Zhao et al. A better spinnable fiber was obtained from rapeseed protein isolate at alkaline pH following subsequent ultrasonic treatment. This treatment improved the physicochemical properties, such as solubility, viscosity, and conductivity, of the rapeseed protein isolate compared to the untreated one (Zhao et al., 2023). The spinnability and texture of materials could also be improved by the addition of small quantities of polyethylene oxide. Trindade et al. reported that the addition of pea protein to a zinc–polyethylene oxide mix in the ratio of 20:1 improved the vegan meat texture, making it closely resemble that of animal meat (Trindade et al., 2024a). Another approach using zein protein (23%), carrageenan (1%), and polyethylene oxide (1%) improved the texture of meat analogs (da Trindade et al., 2024). Electrospinnability of a material also depends on its morphology and its interaction with the solvent used. When PPI and bean protein concentrate were used with various solvents (hexafluoroisopropanol, water, trifluoroethanol, formic acid, and trifluoroacetic acid), most of them formed bead-like structures since protein denaturation was insufficient to form fibers. All the protein isolate–solvent mixtures exhibited pseudoplastic behavior. Only the bean protein concentrate dissolved in hexafluor-

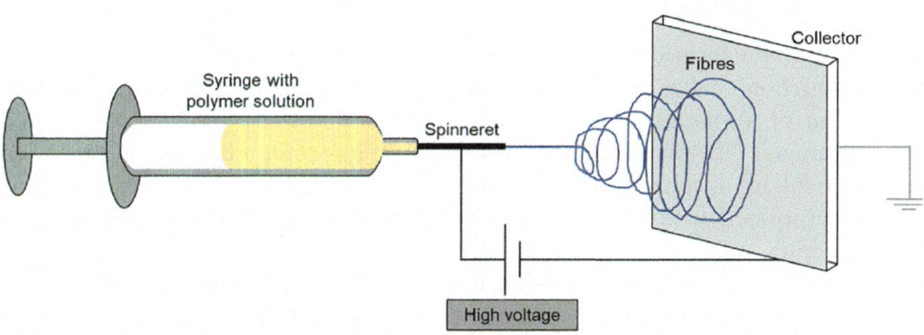

Fig. 13.4 Schematic structure of the electrospinning process of TVP.
From Baune, M.C., Terjung, N., Tülbek, M.Ç., Boukid, F., 2022. Textured vegetable proteins (TVP): future foods standing on their merits as meat alternatives. Future Foods 6. Available from: https://doi.org/10.1016/j.fufo.2022.100181.

oisopropanol showed a fiber-like structure, and overall, the bean protein isolate demonstrated better properties than the SPI (Aguilar-Vázquez et al., 2020).

13.3.4 Freeze-structuring

The thermomechanical extrusion process produces a low-quality meat analog with less juiciness and a varied microscopic structural arrangement compared to animal meat. To produce a product more similar to meat, freeze structuring creates a microscopic texture similar to meat at low temperatures. Different protein isolates result in different texture formations, and after freezing, the removal of ice crystals leads to the texturized product. Freeze structuring of protein isolates in forming meat analogs depends on their concentration and morphology. When soybean flour and SPIs were studied for freeze structuring, soybean flour (10%) produced a better meat analog with a dense and porous structure compared to its protein isolate, due to the latter's poor gel-forming ability (Chantanuson et al., 2022). Freeze structuring was also investigated in mixed protein isolates of pea (40%) and soy (60%). This combination produced an intermediate moisture (0.553 a_W) meat substitute with satisfactory sensory scores and improved nutritional content (16%) compared to the control meat product (Venkatachalam et al., 2024). Mixed protein isolates of wheat and pea (3:1) subjected to freeze structuring produced a product with a similar texture to commercially available chicken nuggets, without changes in protein and moisture content. The incorporation of wheat protein was identified as the principal contributor to the meat analog's texture. Unlike pure PPI, wheat dough exhibited low gel strength (Yuliarti et al., 2021). Dry heating also influences crystal formation during freezing. With an increase in the solid content and binding capacity of SPI, the formation of a porous structure decreases. A better meat-mimicking structure was obtained from dry-heated (70°C) SPI (10% solid content) at 70% relative humidity (Chantanuson et al., 2024).

13.3.5 3D printing

3D printing is a form of texturization primarily used for applications in personalized nutrition, enhanced bioavailability, meat-mimicking structures, and the production of a vast array of structured food products. In terms of customized food products, 3D texturized protein isolates have found applications in the development of dysphagia-friendly foods. One example is 3D-printed red cabbage incorporated with SPI as a transitional food (levels 5 and 6) with different infilling rates (12.5%, 25%, and 50%) made according to the International Dysphagia Diet Standardization Initiative (IDDSI) for dysphagia patients. The texture and printability were optimal when 25% SPI was used with 20% red cabbage (Carranza et al., 2023). Another level 4 food was developed for dysphagia patients by combining agar (0.2 g) with soy protein (6 g) enriched potato puree through 3D printing. This texturized potato puree was easy for elderly people to consume (Mirazimi et al., 2022). 3D printing has also demonstrated its application in food fortification by improving bioaccessibility in combination with nanostructured protein isolates. Riboflavin (vitamin B2) was encapsulated with whey protein isolate nanostructures (65 nm) with an efficiency of 60%, and vitamin B2 could be incorporated at a range of 0.1 mg/mL under safety guidelines. When incorporated into food matrices involving carrots in formulations, it not only improved the bioaccessibility of riboflavin (23%) but also enhanced the printability of the food matrix (Araújo et al., 2025).

3D printing could also be employed for the production of food that mimics meat patties or meat products. 3D printed texturized soy protein, when combined with guar gum (1.25%), corn starch, and water, showed good stability, printability, layer height, and lower syneresis value. The vegan burger made from this mixture, with added ingredients, showed comparable sensorial properties to those of meat burgers (Bertucci et al., 2016). In terms of meat products, 3D-printed protein-based food serves as an alternative texture-forming method to fried chicken. The best printability and frying characteristics were shown by the ink made with texturized soy protein and xanthan gum, outperforming other cellulose-based (carboxymethyl and hydroxy methyl), gum-based hydrocolloids (guar, konjac), and sodium alginate. This texturized steak-like product exhibited values of hardness, gumminess, and chewiness close to those of control fried chicken samples (Chen et al., 2021). The whole-cooked meat texture could be mimicked by incorporating polysaccharide (carrageenan) or enzyme (transglutaminase) into 3D-printed PPI and alginate, thereby improving textural properties. The addition of 0.9% transglutaminase improved the hardness of uncooked meat and, after cooking, produced a salmon-like texture for the meat mimic with reduced cooking losses. Meanwhile, the addition of 0.9% carrageenan increased the hardness and produced a salmon-like cooked texture, but more cooking losses were observed (Leelapunnawut et al., 2022). This 3D printing technology also finds applications in food ink formulation. One such promising source for printability and sensory acceptability was achieved with peanut protein isolate among the other tested protein isolates. Moreover, peanut protein isolate could easily be combined with various fruit and vegetable composites, with certain pretreatments, thereby expanding its versatility (Chen et al., 2021).

13.4 Main ingredients for texturization of vegetable proteins

Alongside primary plant protein sources, various functional additives are incorporated to replicate the appearance, texture, taste, aroma, and color of traditional meat products, enhancing consumer appeal (Devi et al., 2024). One of the key hurdles in developing plant-based meat is maintaining structural integrity after cooking, as these products often crumble due to their textural limitations. When created solely from plant particles, such as those found in burger patties, the absence of binding components can cause the structure to disintegrate (Rout and Srivastav, 2025). Therefore including ingredients that offer elasticity, cohesiveness, and strong binding properties is critical for improving the quality and performance of plant-based meat alternatives.

13.4.1 Protein

TVP primarily consists of protein, which plays a crucial role in its structure and functionality. Common protein sources used in TVP production include soy, peanut, and WG (Bakhsh et al., 2022). For the successful development of a fibrous texture through extrusion, the protein content typically needs to range between 50% and 70%. During the extrusion process, proteins undergo a series of structural transformations that include molecular chain unfolding, intermolecular association, aggregation, and cross-linking. These changes may also involve partial degradation or oxidative modifications, depending on the processing conditions.

13.4.1.1 Wheat gluten

Gluten is primarily made up of two types of proteins: monomeric gliadins (prolamins) and polymeric glutenins (glutelins), the latter of which include both low and high-molecular-weight (HMW) subunits. HMW glutenins are particularly important for imparting viscoelastic properties to gluten networks. Regions rich in α-helices within these subunits are abundant in cysteine residues, which serve as key sites for forming inter-molecular disulfide linkages. Analytical techniques such as NMR and FTIR spectroscopy have revealed that, in a hydrated dough state, the proportion of β-sheet structures in HMW glutenins increases (Gong et al., 2024). It is hypothesized that these subunits, initially arranged in loop conformations, become extended during mechanical stretching, promoting the alignment and formation of β-sheet-rich networks at the cost of β-turns. Historically, WG has been recognized as the only cereal-derived protein capable of forming highly functional viscoelastic matrices.

Due to its inherent viscoelasticity, several researchers have proposed the addition of WG, even in modest quantities, to plant protein systems to enhance fibrous texture development similar to that of meat analogues (Peng et al., 2023). In a model system based on soy protein, increasing the gluten concentration from 0% to 30% led to more defined fibrous structures and improved texture. These outcomes were linked to greater hydrogen bonding and disulfide bridge formation throughout the protein matrix (Fan et al., 2024). The extent of disulfide-mediated polymerization was identified as a key factor in creating anisotropic and viscoelastic protein networks, even when WG was the sole protein component. Furthermore, Peng et al. (2023) demonstrated that shear-induced alignment of soy and gluten phases could result in either fibrous or layered structures, depending on the orientation of the aligned regions. Collectively, these findings underscore the critical role of protein-protein interactions, particularly those that enhance cohesion and elasticity, in achieving meat-like textures in plant-based formulations.

13.4.1.2 Zein

Zein, a prolamin protein, is the predominant protein found in the endosperm of corn kernels, making up approximately 70% of the total protein content (Devi et al., 2024). It is characterized by a high concentration of hydrophobic amino acids such as glutamine, leucine, proline, alanine, and phenylalanine, which contribute to its limited water solubility and functionality in food systems. Its application in the food industry has traditionally been limited due to poor solubility, a lack of functional versatility, and the limited availability of odor- and color-neutral food-grade formulations. However, more recent research has highlighted its potential, especially in gluten-free bread formulations, once appropriately modified (Zhang et al., 2023). Zein can form a viscoelastic network when heated above its glass transition temperature, around 28°C, at 20% moisture content. This property is attributed to the formation of β-sheet structures, which can be further stabilized by introducing coproteins or lipids. In the context of "clean label" food innovation, where there is a push to eliminate chemically modified additives like methylcellulose, zein is gaining interest as a natural alternative that can impart desired texture in plant-based analogues (Rout et al., 2025). Studies have explored the application of zein in plant-based meat and cheese products. In meat substitutes, zein contributes to fibrous textural properties, while in cheese alternatives, it enables partial melting at elevated temperatures (60°C–70°C) (Ozturk and Hamaker, 2023).

A recent investigation compared several approaches to generating fibrous textures from zein: mechanical stretching, electrospinning, and antisolvent precipitation (Rahman et al., 2023). Among these, mechanical elongation proved most effective in producing WG-like viscoelastic fibers. When this mechanically treated zein was combined with soy protein in a model meat analog (at 1.5% [w/w]), the resulting product had an aligned fibrous matrix and demonstrated textural characteristics comparable to precooked chicken strips. Additional work by our group showed that incorporating zein into a pea protein matrix resulted in a dispersed viscoelastic network, especially when alkaline compounds like calcium hydroxide and arginine were included. These additives promoted structural rearrangement, increased β-sheet content, and enhanced elasticity at ambient temperatures (Ozturk and Hamaker, 2023). Furthermore, adjusting the calcium hydroxide levels in zein–pea protein blends allowed for the tailoring of texture to match that of commercial products such as fish fillets, chicken nuggets, cheeses (e.g., Colby, mozzarella), and plant-based burgers. These findings suggest that zein holds considerable promise as a natural structuring agent in next-generation plant-based food products.

13.4.2 Carbohydrates

In TVP, carbohydrates are generally classified into two types based on their hydrolysis level: low-molecular-weight sugars and high-molecular-weight polymers (Li and Li, 2024). Simple sugars, such as glucose, are highly reactive and can undergo Maillard reactions with free amino groups, influencing the color and flavor of TVP through browning. Research by Guerrero and de la Caba (2017) indicated that protein–sugar complexes resulting from this reaction lead to the formation of dark-colored, water-insoluble polymeric structures, contributing to a more organized texture in extruded soy protein products. In addition, Nielsen et al. (2022) noted that lactose exhibited a greater Maillard reaction potential than sucrose, likely due to the reactive hydroxyl group on its anomeric carbon. On the other hand, polysaccharides like starch and dietary fibers are essential for creating a fibrous texture in TVP, though their inclusion should generally not exceed 10% of the total mix (Rout and Srivastav, 2023). Among these, starch is frequently used in extrusion recipes due to its functionality during gelatinization and degradation (Mohamed, 2023). During processing, starch is thoroughly blended with proteins and other ingredients in the initial mixing stage, where it begins to hydrate and swell. In the melting zone, starch gelatinizes as its intermolecular hydrogen bonds break, altering the viscosity of the mixture. This process, alongside interactions between starch and protein, influences protein unfolding and aggregation, which are critical for fiber formation in the final product (Zhang et al., 2025).

At the die exit during low-moisture extrusion, rapid depressurization leads to water evaporation and starch puffing, forming air pockets in the structure. In high-moisture extrusion, starch becomes dispersed within the protein matrix due to phase separation, limiting protein-protein interactions and promoting the stability of hydrophobic bonds. As the product cools, these dispersed phases align with the extrusion flow, forming a layered structure that enhances the fibrous nature of TVP when torn.

13.4.2.1 Methylcellulose and its derivatives

Methylcellulose and its related compounds, such as hydroxypropyl methylcellulose and carboxymethylcellulose, are widely utilized in the development of plant-based food products. Although the exact gelation process is not fully understood, it is suggested that the gelling behavior between methylcellulose and proteins occurs when the structured water network surrounding methylcellulose molecules is disrupted at elevated temperatures (Solé-Martí et al., 2022). Wei et al. (2024) investigated how varying concentrations of methylcellulose (ranging from 0% to 4%) affected the performance of plant-based patties formulated with texturized soy protein (Webb et al., 2023). Their results indicated that higher methylcellulose levels were linked to reduced cooking loss and water release, likely due to improved ingredient cohesion and moisture retention. Notably, the formulation containing 3% methylcellulose achieved texture characteristics such as cohesiveness and springiness comparable to those of traditional beef patties.

13.4.2.2 K-carrageenan

A recent study utilizing coaxial nozzle-assisted 3D food printing explored the use of carrageenan and glucomannan in varying proportions within an SPI-based matrix to replicate the texture of beef (Inguglia et al., 2023). While sodium alginate, included in the formulation, contributed to gelation via ionic interactions, the enhancement of gel strength was primarily attributed to carrageenan and glucomannan, which promoted hydrogen bonding. When combined at concentrations of 2.5% carrageenan and 1.5% glucomannan, these hydrocolloids not only achieved mechanical characteristics closely resembling those of real beef but also led to improved cooking performance, including reduced moisture loss and less shrinkage during processing.

13.4.3 Combination of ingredients

Biopolymer interactions generally fall into two categories: associative, where different polymers attract one another, and segregative, where they repel, resulting in the formation of distinct phases. During extrusion, the high temperature and mechanical shear often lead to protein unfolding and polysaccharide degradation, promoting phase separation (Wang et al., 2023). When these separated macromolecular phases are compressed and forced through a small die, they can align into fibrous structures. Legumes are a rich source of protein and lysine, while cereals provide sulfur-containing amino acids, making their combination nutritionally beneficial. Cereal-legume blends are often used to enhance both the nutritional and structural attributes of extruded foods. Among these, mixtures involving soy protein, WG, and starch are frequently investigated due to their affordability and functional versatility (Peng et al., 2023). Moreover, replacing a portion of soy or wheat protein isolates with their corresponding flours has shown promising results in product development.

13.4.3.1 Carbohydrate level

The inclusion of polysaccharides in extrusion feed formulations can considerably impact the texture of the final product. Studies have found that formulations using SPI with high protein content ($\geq 90\%$) tend to produce uniform, cohesive structures. In contrast, those incorporating SPC, which contains about 67% protein and a higher proportion of polysaccharides, result in anisotropic structures characterized by layered or fibrous textures aligned along the extrusion flow direction (Wang and Smith, 2021). Polysaccharides such as soy arabinogalactans or maltodextrins in SPC encourage the development of a dispersed phase during thermal processing, likely leading to phase separation between proteins and carbohydrates as the material aligns within the die. In addition, combining dehulled soy flour with protein isolates has demonstrated promise in extrusion applications (Politiek et al., 2023). Dehulled soy flour has a lower protein content (under 40%) and fiber content (approximately 5%), but a significantly higher oil concentration (around 20%) compared to SPC. As soy flour content increases, so does the lipid content in the mixture. Lipids act as both plasticizers and lubricants during extrusion. They integrate with proteins and starches, forming protective coatings around protein molecules. When lipid content surpasses 15%, the extrudate becomes notably more slippery due to a lowered friction coefficient (Mikriukova et al., 2022). Regardless of the quantity, lipids tend to create complexes with proteins and starches that hinder protein aggregation, reducing the ability to form fibrous textures. The potential role of starch in enhancing protein aggregation and structure formation has also been examined, particularly in blends with SPI and WG. Starch can contribute to phase separation in the melt due to its interaction with proteins. In high-moisture extrusion cooking (HMEC), the combination of water and elevated temperature induces starch gelatinization, where intermolecular bonds are broken, allowing the granules to absorb more water, swell, and transition into a rubbery state. Gelatinized starch acts similarly to a molten polymer and restricts protein mobility (Sethi et al., 2022). Notably, starches with high amylose content, such as high-amylose maize starch, resist gelatinization under typical extrusion conditions, allowing them to remain intact and form distinct phases within the protein matrix. These starches may function as structural fillers, mimicking some of the behavior of proteins in textured formulations (Schmid et al., 2022).

13.4.3.2 Combination of legume protein and WG

The extrusion of soy protein combined with WG has shown promising results in creating fibrous, meat-like textures (Tian et al., 2023). Specifically, incorporating up to 20% WG into SPC formulations has been found to enhance the solubility of polysaccharides within the protein matrix, resulting in a more compact and cohesive texture after cooling. Increasing the WG content from 10% to 40% in SPC/WG blends led to the development of a well-structured network composed of interconnected, fine fibers (Sun et al., 2022). This structural enhancement is attributed to the role of gluten in promoting the cross-linking of protein molecules, primarily through disulfide bonds. The elastic gluten subunits, particularly those with high molecular weight, are prone to aggregation under extrusion conditions and may behave similarly to polysaccharides in promoting phase separation within the matrix. However, both soy and wheat proteins are known allergens, which restrict the use of these ingredients for individuals with sensitivities to either. Food allergies, characterized by immune reactions to specific proteins, are a significant public health issue (Schmid et al., 2022). Research into the effects of thermal processing on allergenic proteins, such as those from lupine, has shown that treatments like microwave heating, boiling, or extrusion do not significantly alter the recognition of allergens by the immune system. As a result, the most effective approach to managing such allergies is the complete avoidance of the problematic ingredients in food formulations. PPI has gained popularity as a soy-free alternative in recent years (Ma et al., 2022). Nonetheless, when blended with gluten and subjected to extrusion, PPI-based products tend to form weaker fibrous structures compared to those made with SPI and gluten. Despite this, PPI offers a desirable soft texture similar to that of cooked poultry, making it the key component in commercial products like Sunfed plant-based meat alternatives.

13.4.3.3 Animal proteins

Numerous dairy protein powders, such as whole milk powder, casein, and whey, are commercially available and have been explored as complementary ingredients in formulations with plant-based materials (El-Aidie and Khalifa, 2024). In nonextrusion applications, casein-rich blends have been used in fiber formation through technologies like melt spinning. A notable example of dairy-based innovation in meat alternatives comes from FrieslandCampina, a Dutch company that developed and marketed products such as schnitzels, nuggets, and burger patties made primarily from processed milk. These are sold under the Valess brand and are characterized by their incorporation of plant-based fibers into dairy matrices to mimic a meat-like texture.

In terms of extrusion technology, WPC, which typically contains up to 25% protein, has been utilized in the production of low-moisture starch-based snacks (moisture content < 10%), where it contributes to enhanced nutritional value (Wu et al., 2024). However, the integration of dairy proteins in HMEC systems, either for functional or nutritional benefits, remains largely unexplored in the current literature. Similarly, there is a lack of studies involving egg proteins in extrusion processes, although egg white remains a key component in binding mycoproteins, fungal-based proteins produced from *Fusarium venenatum* in products like Quorn.

When considering alternative animal proteins, such as those derived from fish or insects, their inclusion in meat analog formulations challenges traditional vegetarian definitions. Nevertheless, they may appeal to flexitarian consumers who seek to reduce meat consumption without strictly avoiding animal-derived ingredients. Proteins from fish and insects are believed to offer enhanced mimicry of meat's complex textural properties (Vallikkadan et al., 2023). Research into insect-based HMEC remains limited, but available studies suggest it is a promising direction. For instance, a blend comprising 10% soy fiber and 40% protein concentrate from *A. diaperinus* (lesser mealworm) has shown potential for developing a fibrous structure.

13.5 Future scope and limitations

The surge in demand for plant-based foods is largely driven by the growing number of flexitarian, vegetarian, and vegan consumers around the globe. This trend is significantly contributing to the expansion of the TVP market. To better replicate the texture and mouthfeel of traditional meat, TVP production methods continue to evolve. Currently, plant-based products that imitate processed meats have gained more traction due to their similarity in structure and flavor, making them easier to replicate and more widely accepted by consumers. However, alternatives that mimic whole muscle cuts, like steaks, are still limited in availability. Such products are difficult to replicate accurately, especially in terms of flavor and chewiness, though softer meats like chicken have shown more promise for replication in formats such as strips. Still, only a small portion of meat-eating consumers actively seek out these alternatives. In this context, hybrid meat products that combine plant and animal proteins are emerging as a viable option, although the differences between the two protein types can affect processing and end-product characteristics.

In the TVP market landscape, low-moisture texturized vegetable protein (LM-TVP) currently dominates. More recently, high-moisture texturized vegetable protein (HM-TVP) options have entered the market, offering improved sensory qualities such as a meat-like bite, enhanced taste, and better nutritional profiles. Achieving the complex taste and texture of real meat remains a significant challenge. One major issue during high-temperature extrusion is the degradation of heat-sensitive ingredients like flavors and natural colorants, which must be added postprocessing. However, this approach leads to uneven distribution, particularly in the product's core. As a result, innovative lower-temperature processing techniques, such as shear cell technology or freeze-structuring, are being explored to overcome these limitations.

From a health and nutrition standpoint, many consumers mistakenly view plant-based meat analogs as inherently healthier substitutes without considering the essential nutrients found in animal products. Nutritional gaps, especially in micronutrients such as B vitamins, must be addressed in vegan diets to avoid deficiencies. Furthermore, studies have shown that many commercial meat alternatives have lower protein content and inferior amino acid profiles compared to animal-based meats. This highlights the importance of blending multiple plant protein sources to create a complete amino acid profile and calls for more research into the digestibility and bioavailability of proteins in TVP-based products. Producers are actively innovating to diversify their product offerings beyond conventional formats like burgers, nuggets, and sausages, with new efforts focusing on replicating more complex items such as bacon. Despite these advancements, achieving the desired texture remains challenging. There is a continued need to identify effective binders that replicate the structural role of meat components.

13.6 Conclusions

Extruders function as high-temperature, short-duration biochemical reactors that convert raw ingredients into processed intermediates or final products. The production of LM-TVP and HM-TVP requires distinct extruder configurations and processing parameters, both of which greatly influence the end-product characteristics. Throughout extrusion, the structural configuration of ingredients is disrupted, enabling molecular rearrangement and the formation of new conformations. Due to thermodynamic incompatibilities among the components, phase separation naturally occurs, with proteins typically forming the continuous matrix, a crucial factor in establishing TVP's fibrous texture. As the proteins stretch under heat and shear, they transition from their native globular structures to elongated chains, stabilized by disulfide bridges, hydrophobic interactions, and hydrogen bonds. Simultaneously, secondary structures such as α-helices and β-sheets shift toward more flexible forms like β-turns and random coils. Polysaccharides, such as starch or dietary fibers, serve as the dispersed phase within the protein network. These large carbohydrates may help maintain protein structure by limiting excessive unfolding or aggregation. In addition, low-molecular-weight sugars, including glucose, play a significant role in the browning and flavor development of TVP through the Maillard reaction. Lipids, when present, can form complexes with proteins or starch, coating protein surfaces and helping to prevent aggregation, thereby supporting a more stable fibrous network. Nonetheless, these hypotheses on molecular transitions during extrusion remain largely qualitative due to the variability of raw material purity and the intricate nature of the extrusion process.

The structural development of TVP is primarily influenced by how extrusion parameters impact the molecular reconfiguration of core ingredients. Achieving a fibrous structure depends on the full thermal "melting" of the mixture inside the barrel and controlled laminar flow through the die and cooling section. As such, process conditions such as temperature profile, moisture level, feed rate, and screw speed must be carefully optimized to regulate the input of heat and mechanical energy. Furthermore, factors, including raw material structure, component distribution, and extruder and die design, all merit future exploration. Despite advancements, precise control over extrusion remains complex due to the interplay of thermal and mechanical energy and the multifaceted transformations of the food matrix. To better interpret the changes occurring within the extruder, there is a pressing need for advanced computational modeling that considers ingredient characteristics, process variables, molecular transitions, and the final quality metrics of the extruded products.

References

Aguilar-Vázquez, G., Ortiz-Frade, L., Figueroa-Cárdenas, J.D., López-Rubio, A., Mendoza, S., 2020. Electrospinnability study of pea (*Pisum sativum*) and common bean (*Phaseolus vulgaris* L.) using the conformational and rheological behavior of their protein isolates. Polym. Test. 01429418. 81, 106217. doi:10.1016/j.polymertesting.2019.106217.

Araújo, J.F., Fernandes, J.M., Madalena, D., Gonçalves, R.F.S., Vieira, J.M., Martins, J.T., Vicente, A.A., Pinheiro, A.C., 2025. Development of 3D-printed foods incorporating riboflavin-loaded whey protein isolate nanostructures: characterization and in vitro digestion. Food Funct. 2042650X. 16 (5), 2124–2135. Available from: http://pubs.rsc.org/en/journals/journal/fo. doi:10.1039/d4fo05102e.

Arueya, G.L., Owosen, B.S., Olatoye, K.K., 2017. Development of texturized vegetable protein from lima bean (Phaseolus lunatus) and african oil bean seed [Pentaclethrama crophylla (benth)]: optimization approach. Acta Univ. Cibiniensis Ser. E Food Technol. 21 (1), 61–68.

Baig, M.A., Ajayi, F.F., Hamdi, M., Baba, W., Brishti, F.H., Khalid, N., Zhou, W., Maqsood, S., 2025. Recent research advances in meat analogues: a comprehensive review on production, protein sources, quality attributes, analytical techniques used, and consumer perception. Food Rev. Int. 15256103. 41 (1), 236–267. Available from: https://www.tandf.co.uk/journals/titles/87559129.asp. doi:10.1080/87559129.2024.2396855.

Bakhsh, A., Lee, E.Y., Ncho, C.M., Kim, C.J., Son, Y.M., Hwang, Y.H., Joo, S.T., 2022. Quality characteristics of meat analogs through the incorporation of textured vegetable protein: a systematic review. Foods. 23048158. 11 (9). Available from: https://www.mdpi.com/2304-8158/11/9/1242/pdf. doi:10.3390/foods11091242.

Bakhsh, A., Lee, S.J., Lee, E.Y., Hwang, Y.H., Joo, S.T., 2021. Characteristics of beef patties substituted by different levels of textured vegetable protein and taste traits assessed by electronic tongue system. Foods. 23048158. 10 (11). Available from: https://www.mdpi.com/2304-8158/10/11/2811/pdf. doi:10.3390/foods10112811.

Beck, S.M., Knoerzer, K., Arcot, J., 2017. Effect of low moisture extrusion on a pea protein isolate's expansion, solubility, molecular weight distribution and secondary structure as determined by fourier transform Infrared spectroscopy (FTIR). J. Food Eng. 214, 166–174.

Bertucci, B.L., Matuda, T.G., Andreola, K., Tribess, T.B., 2016. Development of formulations based on vegetable proteins and polysaccharides for 3D food printing. In IMT GCSP Seminar: presented Dec (vol. **5**, p. 2024).

Brisht, F.H., Zarei, M., Muhammad, S.K.S., Ismail-Fitry, M.R., Shukri, R., Saari, N., 2017. Evaluation of the functional properties of mung bean protein isolate for development of textured vegetable protein. Int. Food Res. J. 24 (4), 1595–1605.

Brishti, F.H., Chay, S.Y., Muhammad, K., Ismail-Fitry, M.R., Zarei, M., Saari, N., 2021. Texturized mung bean protein as a sustainable food source: effects of extrusion on its physical, textural and protein quality. Innov. Food Sci. Emerg. Technol. 67, 102591.

Brugger, C., Dellemann, J.P., Petry, C., Laporte, M., Müller, N., Windhab, E.J., 2017. Next generation texturized vegetable proteins. Food Mark. Technol. 1, 20–25.

Caporgno, M.P., Böcker, L., Müssner, C., Stirnemann, E., Haberkorn, I., Adelmann, H., Handschin, S., Windhab, E.J., Mathys, A., 2020. Extruded meat analogues based on yellow, heterotrophically cultivated *Auxenochlorella protothecoides* microalgae. Innov. Food Sci. Emerg. Technol. 14668564. 59. Available from: http://www.elsevier.com/locate/ifset. doi:10.1016/j.ifset.2019.102275.

Carranza, T., Guerrero, P., de la Caba, K., Etxabide, A., 2023. Texture-modified soy protein foods: 3D printing design and red cabbage effect. Food Hydrocoll. 0268005X. 145, 109141. doi:10.1016/j.foodhyd.2023.109141.

Chantanuson, R., Nagamine, S., Kobayashi, T., Nakagawa, K., 2022. Preparation of soy protein-based food gels and control of fibrous structure and rheological property by freezing. Food Struct. 22133291. 32 (100258). doi:10.1016/j.foostr.2022.100258.

Chantanuson, R., Nagamine, S., Kobayashi, T., Nakagawa, K., 2024. Effect of dry heat treatment of soy protein powder on aligned structure formation in soy protein-based food gels during freezing. J. Food Eng. 02608774. 363, 111779. doi:10.1016/j.jfoodeng.2023.111779.

Chen, G., Zhang, X., Yang, M., Yu, H., Cao, Q., 2021. Performance of tropical cyclone forecasts in the western North Pacific in 2017. Trop. Cyclone Res. Rev. 22256032. 10 (1), 1–15. Available from: https://www.sciencedirect.com/journal/tropical-cyclone-research-and-review. doi:10.1016/j.tcrr.2021.03.002.

Cornet, S.H.V., Snel, S.J.E., Schreuders, F.K.G., van der Sman, R.G.M., Beyrer, M., van der Goot, A.J., 2022. Thermo-mechanical processing of plant proteins using shear cell and high-moisture extrusion cooking. Crit. Rev. Food Sci. Nutr. 15497852. 62 (12), 3264–3280. Available from: https://www.tandfonline.com/loi/bfsn20. doi:10.1080/10408398.2020.1864618.

da Trindade, L.G., Zanchet, L., Bonsanto, F.P., Braga, A.R.C., 2024. Spinning a sustainable future: electrospun polysaccharide–protein fibers for plant-based meat innovation. Foods 13 (18), 2962.

De Angelis, D., Pasqualone, A., Summo, C., Van Der Goot, A.J., 2024. Advancements in texturization processes for the development of plant-based meat analogs: a review. Curr. Opin. Food Sci. 58, 101192.

Deng, Q., Wang, Z., Liwei, F., Zhiyong, H., Zeng, M., Qin, F., Chen, J., 2023. High-moisture extrusion of soy protein: effects of insoluble dietary fiber on anisotropic extrudates. Food Hydrocoll. 0268005X. 141, 108688. doi:10.1016/j.foodhyd.2023.108688.

Devi, T.A., Rahul, R., Joshua, H.M., Naveen, N., Karthik, P., 2024. Formulation of plant-based meat alternatives and its optimization by experimental design using response surface methodology. Sustain. Food Technol. 27538095. 2 (4), 1139–1151. Available from: https://www.rsc.org/journals-books-databases/about-journals/sustainable-food-technology/. doi:10.1039/d3fb00210a.

Draganovic, V., Boom, R.M., Jonkers, J., Goot, A.J.V.D., 2014. Lupine and rapeseed protein concentrate in fish feed: a comparative assessment of the techno-functional properties using a shear cell device and an extruder. J. Food Eng. 02608774. 126, 178–189. doi:10.1016/j.jfoodeng.2013.11.013.

El-Aidie, S.A.M., Khalifa, G.S.A., 2024. Innovative applications of whey protein for sustainable dairy industry: environmental and technological perspectives—a comprehensive review. Compr. Rev. Food Sci. Food Saf. 1541-4337. 23 (2). doi:10.1111/1541-4337.13319.

Espinosa-Ramírez, J., Mondragón-Portocarrero, A.C., Rodríguez, J.A., Lorenzo, J.M., Santos, E.M., 2023. Algae as a potential source of protein meat alternatives. Front. Nutr. 2296861X. 10. Available from: journal.frontiersin.org/journal/nutrition. doi:10.3389/fnut.2023.1254300.

Fan, L., Wang, H., Li, M., Lei, M., Li, L., Ma, S., Huang, J., 2024. Impact of wheat bran dietary fiber on gluten aggregation behavior in dough during noodle processing. Int. J. Biol. Macromol. 18790003. 257. Available from: https://www.sciencedirect.com/science/journal/01418130. doi:10.1016/j.ijbiomac.2023.128765.

Gao, Y., Sun, Y., Zhang, Y., Sun, Y., Jin, T., 2022. Extrusion modification: effect of extrusion on the functional properties and structure of rice protein. Processes. 2227-9717. 10 (9), 1871. doi:10.3390/pr10091871.

Gong, Y., Sui, W., Wang, H., Wang, Y., Li, S., Cui, J., Xie, R., Liu, R., Wu, T., Zhang, M., 2024. In-depth understanding of the effects of different molecular weight pullulan interacting with protein and starch on dough structure and application properties. Int. J. Biol. Macromol. 01418130. 268 (Pt 1), 131556. doi:10.1016/j.ijbiomac.2024.131556.

Grabowska, K.J., Zhu, S., Dekkers, B.L., De Ruijter, N.C.A., Gieteling, J., Goot, A.J.V.D., 2016. Shear-induced structuring as a tool to make anisotropic materials using soy protein concentrate. J. Food Eng. 02608774. 188, 77–86. Available from: http://www.sciencedirect.com/science/journal/02608774. doi:10.1016/j.jfoodeng.2016.05.010.

Grahl, S., Palanisamy, M., Strack, M., Meier-Dinkel, L., Toepfl, S., Mörlein, D., 2018. Towards more sustainable meat alternatives: how technical parameters affect the sensory properties of extrusion products derived from soy and algae. J. Clean. Prod. 09596526. 198, 962–971. doi:10.1016/j.jclepro.2018.07.041.

Guerrero, M.P., 2023. Production of texturized vegetable proteins (TVPs) from pulses and their application in meat products.

Guerrero, M., Stone, A.K., Singh, R., Lui, Y.C., Koksel, F., Nickerson, M.T., 2025. Effect of extrusion conditions on the characteristics of texturized vegetable protein from a faba bean protein mix and its application in vegan and hybrid burgers. Foods 14 (4), 547.

Guerrero, P., de la Caba, K., 2017. Protein-based films and coatings. Edible Food Packag. Mater. Process. Technol. 81120.

Inguglia, E.S., Song, Z., Kerry, J.P., O'Sullivan, M.G., Hamill, R.M., 2023. Addressing clean label trends in commercial meat processing: strategies, challenges and insights from consumer perspectives. Foods. 23048158. 12 (10). Available from: https://www.mdpi.com/journal/foods. doi:10.3390/foods12102062.

Jiang, W., Du, Y., Huang, C., Ji, Y., Yu, D.-G., 2023. Electrospun zein nanofibers: from food to food. ES Food Agrofor. 26877295. 14 (3). doi:10.30919/esfaf863.

Jiang, W., Feng, J., Yang, X., Li, L., 2024. Structure of pea protein-based complexes on high-moisture extrusion: raw materials and extrusion zones. LWT. 00236438. 194, 115823. doi:10.1016/j.lwt.2024.115823.

Kern, C., Scharfe, M., Hinrichs, J., 2020. Texturization of renneted casein-based gel particles by sheet die extrusion: mechanical properties and numerical analysis of flow characteristics. J. Food Eng. 278.

Krintiras, G.A., Göbel, J., Goot, A.J.V.D., Stefanidis, G.D., 2015. Production of structured soy-based meat analogues using simple shear and heat in a Couette Cell. J. Food Eng. 02608774. 160, 34–41. Available from: http://www.sciencedirect.com/science/journal/02608774. doi:10.1016/j.jfoodeng.2015.02.015.

Laugesen, S.B., Dethlefsen, S.L., Petersen, I.L., Aaslyng, M.D., 2022. Texturized vegetable protein as a source of protein fortification of wheat buns. Foods. 23048158. 11 (22). Available from: https://www.mdpi.com/journal/foods. doi:10.3390/foods11223647.

Le Bourgot, C., Liu, X., Buffière, C., Hafanaoui, N., Salis, L., Pouyet, C., Dardevet, D., Rémond, D., 2023. Development of a protein food based on texturized wheat proteins, with high protein digestibility and improved lysine content. Food Res. Int. 09639969. 170, 112978. doi:10.1016/j.foodres.2023.112978.

Lee, J.S., Oh, H., Choi, I., Yoon, C.S., Han, J., 2022. Physico-chemical characteristics of rice protein-based novel textured vegetable proteins as meat analogues produced by low-moisture extrusion cooking technology. LWT. 00236438. 157. Available from: https://www.journals.elsevier.com/lwt. doi:10.1016/j.lwt.2021.113056.

Leelapunnawut, S., Ngamwonglumlert, L., Devahastin, S., Derossi, A., Caporizzi, R., Chiewchan, N., 2022. Effects of texture modifiers on physicochemical properties of 3D-printed meat mimics from pea protein isolate-alginate gel mixture. Foods. 2304-8158. 11 (24), 3947. doi:10.3390/foods11243947.

Li, J., Li, L., 2024. Physical modification of vegetable protein by extrusion and regulation mechanism of polysaccharide on the unique functional properties of extruded vegetable protein: a review. Crit. Rev. Food Sci. Nutr. 1040-8398. 64 (31), 11454–11467. doi:10.1080/10408398.2023.2239337.

Lyu, J.S., Lee, J.S., Chae, T.Y., Yoon, C.S., Han, J., 2023. Effect of screw speed and die temperature on physicochemical, textural, and morphological properties of soy protein isolate-based textured vegetable protein produced via a low-moisture extrusion. Food Sci. Biotechnol. 20926456. 32 (5), 659–669. Available from: https://www.springer.com/journal/10068. doi:10.1007/s10068-022-01207-8.

Ma, K.K., Grossmann, L., Nolden, A.A., McClements, D.J., Kinchla, A.J., 2022. Functional and physical properties of commercial pulse proteins compared to soy derived protein. Future Foods. 26668335. 6. Available from: https://www.journals.elsevier.com/future-foods. doi:10.1016/j.fufo.2022.100155.

Mathew, R., Vigneshwaran, V., 2024. Micronization and texturization. In: Structured Foods. CRC Press, India, pp. 154–170. Available from: https://www.routledge.com/Structured-Foods/EswaranU-Srivastav-Srivastava/p/book/9781032399485. doi:10.1201/9781003355441-7.

Maung, T.T., Gu, B.Y., Kim, M.H., Ryu, G.H., 2020. Fermentation of texturized vegetable proteins extruded at different moisture contents: effect on physicochemical, structural, and microbial properties. Food Sci. Biotechnol. 20926456. 29 (7), 897–907. Available from: http://www.springerlink.com/content/n3031612t262/. doi:10.1007/s10068-020-00737-3.

Meng, X.Y., Zhu, X.Q., An, H.Z., Yang, J.F., Dai, H.H., 2023. Study on the relationship between raw material characteristics of soybean protein concentrate and textured vegetable protein quality. Food Sci. Technol. (Brazil). 1678457X. 43. Available from: https://www.scielo.br/j/cta/i/2023.v43/. doi:10.1590/fst.121822.

Mikriukova, M., Lahti, J., Kuusipalo, J., 2022. Permeation of vegetable oils and slippery properties of extrusion coated paperboard. Packag. Technol. Sci. 35 (5), 445–458.

Mirazimi, F., Saldo, J., Sepulcre, F., Gràcia, A., Pujola, M., 2022. Enriched puree potato with soy protein for dysphagia patients by using 3D printing. Food Front. 2643-8429. 3 (4), 706–715. doi:10.1002/fft2.149.

Mohamed, I.O., 2023. Interaction of starch with some food macromolecules during the extrusion process and its effect on modulating physicochemical and digestible properties. A review. Carbohydr. Polym. Technol. Appl. 26668939. 5, 100294. doi:10.1016/j.carpta.2023.100294.

Nielsen, S.D., Knudsen, L.J., Bækgaard, L.T., Rauh, V., Larsen, L.B., 2022. Influence of lactose on the maillard reaction and dehydroalanine-mediated protein cross-linking in casein and whey. Foods. 23048158. 11 (7). Available from: https://www.mdpi.com/2304-8158/11/7/897/pdf. doi:10.3390/foods11070897.

Nor Afizah, M., Rizvi, S.S.H., 2014. Functional properties of whey protein concentrate texturized at acidic pH: effect of extrusion temperature. LWT. 00236438. 57 (1), 290–298. Available from: https://www.journals.elsevier.com/lwt. doi:10.1016/j.lwt.2014.01.019.

Oluwole, O., Ibidapo, O., Arowosola, T., Raji, F., Zandonadi, R.P., Alasqah, I., Lho, L.H., Han, H., Raposo, A., 2023. Sustainable transformation agenda for enhanced global food and nutrition security: a narrative review. Front. Nutr. 2296861X. 10. Available from: journal.frontiersin.org/journal/nutrition. doi:10.3389/fnut.2023.1226538.

Ozturk, O.K., Hamaker, B.R., 2023. Texturization of plant protein-based meat alternatives: processing, base proteins, and other constructional ingredients. Future Foods. 26668335. 8. Available from: https://www.journals.elsevier.com/future-foods. doi:10.1016/j.fufo.2023.100248.

Ozturk, O.K., Hamaker, B.R., 2023. Texturization of plant protein-based meat alternatives: processing, base proteins, and other constructional ingredients. Future Foods 8, 100248.

Padha, N., Salve, R.V., D, G., 2024. Legume-Based Meat Analog Products: a comprehensive analysis of nutritional and technological aspects. Bioscene 2422 (June), 1116–1132.

Patti, A., 2024. Challenges to improve extrusion-based additive manufacturing process of thermoplastics toward sustainable development. Macromol. Rapid Commun. 15213927. 45 (17). Available from: http://onlinelibrary.wiley.com/journal/10.1002/(ISSN)1521-3927. doi:10.1002/marc.202400249.

Peñaranda, I., Garrido, M.D., García-Segovia, P., Martínez-Monzó, J., Igual, M., 2023. Enriched Pea protein texturing: physicochemical characteristics and application as a substitute for meat in hamburgers. Foods. 23048158. 12 (6). Available from: https://www.mdpi.com/journal/foods. doi:10.3390/foods12061303.

Peñaranda, I., Garrido, M.D., García-Segovia, P., Martínez-Monzó, J., Igual, M., 2025. Effect of protein texturization on amino acids and protein in vitro bioaccessibility of pea and rice protein. Food Funct. 2042650X. 16 (5), 1820–1829. Available from: http://pubs.rsc.org/en/journals/journal/fo. doi:10.1039/d4fo04829f.

Peng, H., Zhang, J., Wang, S., Qi, M., Yue, M., Zhang, S., Song, J., Wang, C., Zhang, D., Wang, X., Chengye, M., 2022. High moisture extrusion of pea protein: effect of l-cysteine on product properties and the process forming a fibrous structure. Food Hydrocoll. 0268005X. 129, 107633. doi:10.1016/j.foodhyd.2022.107633.

Peng, Y., Zhao, D., Li, M., Wen, X., Ni, Y., 2023. The interactions of soy protein and wheat gluten for the development of meat-like fibrous structure. Molecules 28 (21), 7431.

Pennells, J., Trigona, L., Patel, H., Ying, D., 2024. Ingredient functionality of soy, chickpea, and pea protein before and after dry heat pretreatment and low moisture extrusion. Foods. 2304-8158. 13 (14), 2168. doi:10.3390/foods13142168.

Politiek, R.G.A., Pegiou, E., Balfoort, L.L., Bruins, M.E., Schutyser, M.A.I., Keppler, J.K., 2023. Functional and flavour properties of de-oiled flours and dry-enriched protein concentrates of lupin and soy. Future Foods. 26668335. 8. Available from: https://www.journals.elsevier.com/future-foods. doi:10.1016/j.fufo.2023.100274.

Rahman, M., Dip, T.M., Haase, T., Truong, Y.B., Le, T.C., Houshyar, S., 2023. Fabrication of zein-based fibrous scaffolds for biomedical applications—a review. Macromol. Mater. Eng. 14392054. Available from: http://onlinelibrary.wiley.com/journal/10.1002/(ISSN)1439-2054. doi:10.1002/mame.202300175.

Rao, V., Poonia, A., 2023. Protein characteristics, amino acid profile, health benefits and methods of extraction and isolation of proteins from some pseudocereals—a review. Food Prod. Process. Nutr. 5 (1), 37.

Riaz, M.N., 2011. Texturized Vegetable Proteins. In: Handbook of Food Proteins. Elsevier Inc., United States, pp. 395–418. Available from: http://www.sciencedirect.com/science/book/9781845697587. doi:10.1533/9780857093639.395.

Ribeiro, G., Piñero, M.-Y., Parle, F., Blanco, B., Roman, L., 2024. Optimizing screw speed and barrel temperature for textural and nutritional improvement of soy-based high-moisture extrudates. Foods. 2304-8158. 13 (11), 1748. doi:10.3390/foods13111748.

Roberts, R., 2013. Understanding the Mechanism of Texturization and the Relationship Between Properties of Wheat Gluten and Texturized Vegetable Protein (Doctoral dissertation). Kansas State University.

Rout, S., Dash, P., Panda, P.K., Yang, P.C., Srivastav, P.P., 2024. Interaction of dairy and plant proteins for improving the emulsifying and gelation properties in food matrices: a review. Food Sci. Biotechnol. 20926456. 33 (14), 3199–3212. Available from: https://www.springer.com/journal/10068. doi:10.1007/s10068-024-01671-4.

Rout, S., Panda, P.K., Dash, P., Srivastav, P.P., Hsieh, C.T., 2025. Cold plasma-induced modulation of protein and lipid macromolecules: a review. Int. J. Mol. Sci. 14220067. 26 (4). Available from: http://www.mdpi.com/journal/ijms. doi:10.3390/ijms26041564.

Rout, S., Srivastav, P.P., 2023. Cold atmospheric plasma processing of plant-based proteins. In: Novel Plant Protein Processing: Developing the Foods of the Future. CRC Press, India, pp. 172–188. Available from: http://www.tandfebooks.com/doi/book/9781003812678. doi:10.1201/9781003369790-9.

Rout, S., Srivastav, P.P., 2024. Modification of soy protein isolate and pea protein isolate by high voltage dielectric barrier discharge (DBD) atmospheric cold plasma: comparative study on structural, rheological and techno-functional characteristics. Food Chem. 447, 138914.

Rout, S., Srivastav, P.P., 2025. Dielectric barrier discharge cold plasma-modified pea protein nanoparticles: enhancing functional and thermal properties for food and biopolymer applications. Sustain. Food Technol. 27538095. Available from: https://www.rsc.org/journals-books-databases/about-journals/sustainable-food-technology/. doi:10.1039/d5fb00123d.

Saldanha Do Carmo, C., Knutsen, S.H., Malizia, G., Dessev, T., Geny, A., Zobel, H., Myhrer, K.S., Varela, P., Sahlstrøm, S., 2021. Meat analogues from a faba bean concentrate can be generated by high moisture extrusion. Future Foods. 26668335. 3. Available from: https://www.journals.elsevier.com/future-foods. doi:10.1016/j.fufo.2021.100014.

Samard, S., Maung, T.T., Gu, B.Y., Kim, M.H., Ryu, G.H., 2021. Influences of extrusion parameters on physicochemical properties of textured vegetable proteins and its meatless burger patty. Food Sci. Biotechnol. 20926456. 30 (3), 395–403. Available from: http://www.springerlink.com/content/n3031612t262/. doi:10.1007/s10068-021-00879-y.

Schmid, E.M., Farahnaky, A., Adhikari, B., Torley, P.J., 2022. High moisture extrusion cooking of meat analogs: a review of mechanisms of protein texturization. Compr. Rev. Food Sci. Food Saf. 21 (6), 4573–4609.

Sethi, S., Choudhary, P., Nath, P., Chauhan, O.P., 2022. Starch gelatinization and modification. In: Advances in Food Chemistry: Food Components, Processing and Preservation. Springer Nature, Singapore, pp. 65–116. Available from: https://link.springer.com/book/10.1007/978-981-19-4796-4. doi:10.1007/978-981-19-4796-4_3.

Sharima-Abdullah, N., Hassan, C.Z., Arifin, N., Huda-Faujan, N., 2018. Physicochemical properties and consumer preference of imitation chicken nuggets produced from chickpea flour and textured vegetable protein. Malaysia Int. Food Res. J. 22317546. 25 (3), 1016–1025. Available from http://www.ifrj.upm.edu.my/25%20(03)%202018/(18).pdf.

Singh, A., 2024. Extrusion Texturization of Air-Classified Barley Protein: A Sustainable Plant-Based Meat Alternative (Doctoral dissertation). University of Manitoba.

Smetana, S., Larki, N.A., Pernutz, C., Franke, K., Bindrich, U., Toepfl, S., Heinz, V., 2018. Structure design of insect-based meat analogs with high-moisture extrusion. J. Food Eng. 02608774. 229, 83–85. doi:10.1016/j.jfoodeng.2017.06.035.

Solé-Martí, X., Vilella, T., Labay, C., Tampieri, F., Ginebra, M.P., Canal, C., 2022. Thermosensitive hydrogels to deliver reactive species generated by cold atmospheric plasma: a case study with methylcellulose. Biomater. Sci. 20474849. 10 (14), 3845–3855. Available from: http://pubs.rsc.org/en/journals/journal/bm. doi:10.1039/d2bm00308b.

Sun, C., Fu, J., Chang, Y., Li, S., Fang, Y., 2022. Structure design for improving the characteristic attributes of extruded plant-based meat analogues. Food Biophys. 1557-1858. 17 (2), 137–149. doi:10.1007/s11483-021-09692-w.

Tian, T., Ren, K., Cao, X., Peng, X., Zheng, L., Dai, S., Tong, X., Zeng, Q., Qiu, S., Wang, H., Jiang, L., 2023. High moisture extrusion of soybean-wheat co-precipitation protein: mechanism of fibrosis based on different extrusion energy regulation. Food Hydrocoll. 0268005X. 144. Available from: https://www.journals.elsevier.com/food-hydrocolloids. doi:10.1016/j.foodhyd.2023.108950.

Vallikkadan, M.S., Dhanapal, L., Dutta, S., Sivakamasundari, S.K., Moses, J.A., Anandharamakrishnan, C., 2023. meat alternatives: evolution, structuring techniques, trends, and challenges. Food Eng. Rev. 18667929. 15 (2), 329–359. Available from: https://www.springer.com/journal/12393. doi:10.1007/s12393-023-09332-8.

Vatansever, S., Tulbek, M.C., Riaz, M.N., 2020. Low-and high-moisture extrusion of pulse proteins as plant-based meat ingredients: a review. Cereal Foods World 65 (4), 1214.

Venkatachalam, C.D., Rajan, D., Arunachalam, S., 2024. Optimization of legume based meat alternative using freeze structuring and evaluation of Its quality indices. Suranaree J. Sci. Technol. 31 (1).

Wang, F., Lian, W., Gu, X., Zhang, H., Gao, Y., Lü, M., Zhu, Y., Huang, Y., Sun, Y., Zhu, X., 2023. Effect of polysaccharides on the rheological behaviour of soy–wheat protein aggregation and conformational changes during high-moisture extrusion. J. Sci. Food Agric. 0022-5142. 103 (12), 5992–6004. doi:10.1002/jsfa.12669.

Wang, Z., Smith, D.E., 2021. A fully coupled simulation of planar deposition flow and fiber orientation in polymer composites additive manufacturing. Materials. 1996-1944. 14 (10), 2596. doi:10.3390/ma14102596.

Webb, D., Dogan, H., Yonghui, L., Alavi, S., 2023. Physico-chemical properties and texturization of pea, wheat and soy proteins using extrusion and their application in plant-based meat. Foods. 2304-8158. 12 (8), 1586. doi:10.3390/foods12081586.

Webb, D.M., 2021. Physicochemical Properties of Pea Proteins, Texturization Using Extrusion, and Application in Plant-Based Meats (Doctoral dissertation). Kansas State University.

Webb, D., Plattner, B.J., Donald, E., Funk, D., Plattner, B.S., Alavi, S., 2020. Role of chickpea flour in texturization of extruded pea protein. J. Food Sci. 17503841. 85 (12), 4180–4187.

Available from: http://onlinelibrary.wiley.com/journal/10.1111/(ISSN)1750-3841. doi:10.1111/1750-3841.15531.

Wei, A.S., Brishti, F.H., Sani, M.S.A., Ishamri, I., Sarbon, N.M., Ismail-Fitry, M.R., 2024. Methylcellulose replacement with different enzymatically treated plant fibres as a binder in the production of plant-based meat patties. LWT 201, 116231.

Wittek, P., Zeiler, N., Karbstein, H.P., Emin, M.A., 2021. High moisture extrusion of soy protein: investigations on the formation of anisotropic product structure. Foods. 23048158. 10 (1). Available from: https://www.mdpi.com/2304-8158/10/1/102/pdf. doi:10.3390/foods10010102.

Wu, M., Huang, X., Gao, F., Sun, Y., Duan, H., Li, D., 2019. Dynamic mechanical properties and fractal analysis of texturized soybean protein/wheat gluten composite produced by high moisture extrusion. Int. J. Food Sci. Technol. 0950-5423. 54 (2), 499–508. doi:10.1111/ijfs.13963.

Wu, Y., Wu, H., Hu, L., 2024. Recent advances of proteins, polysaccharides and lipids-based edible films/coatings for food packaging applications: a review. Food Biophys. 19 (1), 29–45.

Xia, S., Shen, S., Song, J., Li, K., Qin, X., Jiang, X., Xue, C., Xue, Y., 2023. Structure and flavor properties of meat analogues from yeast and soy protein prepared via high-moisture extrusion. Food Chem. 402 (October), 117013.

Xiao, T., Su, X., Jiang, R., zhou, H., Xie, T., 2023. Low moisture extrusion of soybean protein isolate: effect of β-glucan on the physicochemical properties of the product. LWT. 00236438. 179, 114660. doi:10.1016/j.lwt.2023.114660.

Xuelian, M., Ryu, G., 2019. Effects of green tea contents on the quality and antioxidant properties of textured vegetable protein by extrusion-cooking. Food Sci. Biotechnol. 1226-7708. 28 (1), 67–74. doi:10.1007/s10068-018-0437-7.

Yuliarti, O., Kovis, T.J.K., Yi, N.J., 2021. Structuring the meat analogue by using plant-based derived composites. J. Food Eng. 288, 110138.

Zahari, I., Ferawati, F., Helstad, A., Ahlström, C., Östbring, K., Rayner, M., Purhagen, J.K., 2020. Development of high-moisture meat analogues with hemp and soy protein using extrusion cooking. Foods. 23048158. 9 (6). Available from: https://www.mdpi.com/2304-8158/9/6/772. doi:10.3390/foods9060772.

Zhang, J., Liu, L., Jiang, Y., Faisal, S., Wang, Q., 2020. A new insight into the high-moisture extrusion process of peanut protein: from the aspect of the orders and amount of energy input. J. Food Eng. 02608774. 264, 109668. doi:10.1016/j.jfoodeng.2019.07.015.

Zhang, J., Liu, L., Zhu, S., Wang, Q., 2018. Texturisation behaviour of peanut–soy bean/wheat protein mixtures during high moisture extrusion cooking. Int. J. Food Sci. Technol. 0950-5423. 53 (11), 2535–2541. doi:10.1111/ijfs.13847.

Zhang, M., Jia, R., Ma, M., Yang, T., Sun, Q., Li, M., 2023. Versatile wheat gluten: functional properties and application in the food-related industry. Crit. Rev. Food Sci. Nutr. 1040-8398. 63 (30), 10444–10460. doi:10.1080/10408398.2022.2078785.

Zhang, T., Huang, Y., Liu, L., Lv, M., Zhu, Y., Sun, B., Zhu, X., 2025. Modified starch on quality of low moisture extruded soybean protein-wheat gluten. J. Sci. Food Agric. 0022-5142. 105 (9), 4949–4963. doi:10.1002/jsfa.14253.

Zhang, X., Zhao, Y., Zhang, T., Zhang, Y., Jiang, L., Sui, X., 2022. High moisture extrusion of soy protein and wheat gluten blend: an underlying mechanism for the formation of fibrous structures. LWT. 00236438. 163, 113561. doi:10.1016/j.lwt.2022.113561.

Zhao, Y., Zhang, X., Li, K., Shen, S., Li, J., Liu, X., 2023. Formation mechanism of yeast-soy protein extrudates during high-moisture extrusion and their digestive properties. Food Hydrocoll. 145, 109093.

Development of aquafood by extrusion processing

14

Kasiviswanathan Muthukumarappan
Agricultural and Biosystems Engineering, South Dakota State University, Brookings, SD, United States

14.1 Aquaculture feed production—introduction

In response to the burgeoning rate of world population growth, researchers continue to search for a secure, sustainable food supply. Seafood provides an excellent source of protein that can be used to fulfill this need. However, its sustainability amidst the overharvesting of edible, wild marine fish has become a valid concern. Researchers believe that aquaculture, one of the fastest-growing sectors of food production, will remain a major contributor to feeding the world's population (Riaz, 2008). Considering the techno-economic feasibility of industrial fish farming, feed cost is a primary concern (Pritchard, 1976). Protein is the single most important yet expensive dietary nutrient because it comes largely from ground, ocean-caught fishmeal (FM), which has ideal nutritional qualities and has thus far been readily available. The expanding global demand for aquaculture products, coupled with the depletion of wild fish stocks, has elevated the cost of FM and, consequently, fish feed prices (Lim and Lee, 2009). Production of a quality aquafeed, which enhances the feed conversion and growth performance of fish, is related to the physical and nutritional properties of the final product and the technological treatment that is applied to make the feed. Extrusion cooking technology has become one of the more effective methods used in the aquafeed industry.

14.2 Extrusion cooking technology

Extruder technology has been used in aquafeed production for half a century. It offers better flexibility of operation for a large-scale aquafeed production system compared with conventional technologies such as pellet milling and expanders (Riaz, 2012). An extrusion system consists of four basic units: preconditioner, feed delivery system, extruder, and die assemblies (Fig. 14.1). The main components of the extruder are the screw and barrel, which denote the type of extruder. Two main types of extruders—single-screw and twin-screw extruders—are used in aquafeed industries. They can be further classified by their hard structure design, operational

Extrusion Cooking for Food and Feed. https://doi.org/10.1016/B978-0-443-29987-2.00014-2

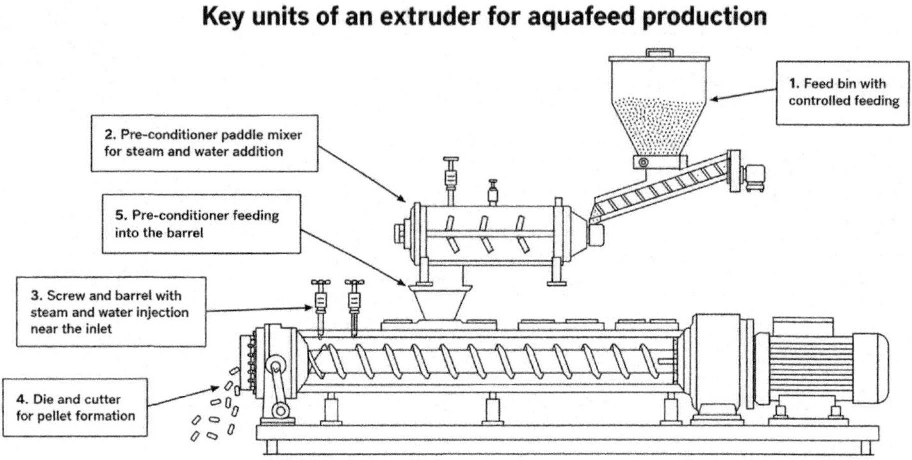

Fig. 14.1 Main units of an extruder for aquafeed production.

parameters (Fallahi et al., 2013a), and external aiding parts. Extrusion is a unique process for mixing the ingredients, cooking the mixture at a high temperature inside a barrel chamber, and compressing and texturizing the cooked dough by forcing it through a narrow die restriction within a short period (Fallahi et al., 2013a). Twin-screw extruders are more flexible and versatile in processing raw ingredients into the final products and can handle a wide range of ingredient blends from very low to high moisture contents. According to Riaz (2000), characteristics of the raw ingredients and the final product are the two essential factors in the selection of each of these components. It is recommended that the twin-screw extruder is more applicable for the production of aquafeeds with high-fat content, pellets with high moisture content, and small-sized extrudates with uniform shape and size (Riaz, 2003). He also suggested that at a commercial scale, when frequent product changeover is required due to small batch size and/or different product formulations, a twin-screw extruder is more cost-effective than a single-screw extruder. The major fish feed components consist of protein, starch, crude fat, raw fiber, crude ash, and micronutrients (vitamins and minerals, amino acids, fatty acids, etc.). This process involves independent variables such as operating conditions (i.e., temperature, screw speed, and moisture content), blend composition (quantity and type of protein, starch, lipids, and total acidity), die size, screw configuration, type of extruder, and dependent variables such as specific mechanical energy, residence time (Riaz, 2003), torque, and pressure. Residence time of the ingredients inside the barrel is a critical variable affected by feeding rate, screw speed, screw and die configurations, barrel length, barrel temperature, and water content of the feed. Residence time ultimately influences material throughput and the required total energy inputs to the system. In fact, the interaction effects of these variables induce chemical and physical changes to the processing ingredients and thus transform the state of the materials at both macroscopic and microscopic scales, mainly due to starch gelatinization and protein denaturation, changing the rheological properties of the molten dough inside the barrel, leading to the production of extrudates with a wide range of properties (Hurber, 2000; Fallahi et al., 2013a).

14.3 Advantages of extrusion processing in aquafeed production

Extrusion is a food-oriented process that provides a higher degree of control and consistency to achieve the desired aquafeed specifications on a large, commercial scale. Some of the advantages of extrusion processing include complete control of density and porosity to produce floating and sinking feeds; flexibility in using raw materials to develop feeds for a diverse range of species such as tilapia, trout, perch, and shrimp; particle size reduction and consistency, which are important for low-starch feeds; moisture equilibration; efficient cooking; pellet durability; and the physical appearance of the final product. In addition, extrusion offers the opportunity for cell structure determination, leading to an increase in the improved lipid retention characteristics of extrudates; versatility in producing diets with high protein and fat content; increased bulk density; enhanced production of microaquatic feeds; a high level of starch gelatinization to improve feed durability, water stability, and digestibility; increased nutrient density and palatability; microbial reduction and improved feed safety; inactivation of antinutritional factors; and a long shelf-life (Riaz, 2012).

14.4 Effect of extrusion processing on the quality of aquafeeds

In regard to product quality, researchers who examined the effect of extrusion processing on product quality reported that optimal process conditions (Tumuluru and Sokhansanj, 2008; Rolfe et al., 2001) coupled with an appropriate feed blend composition (Cavalcanti and Behnke, 2005a, b) must be applied to achieve a physically and nutritionally viable aquafeed. Not only does it improve fish feed consumption, but the feed is more digestible, and the nutrients are more accessible for the fish. Extrusion is a complex multivariate process (Singh et al., 2007), implying a strong interrelationship between the effect of extrusion processing on the physical and nutritional properties of the extrudates. Aquafeed quality can vary depending upon the fish species' requirements, therefore making it more complex to manufacture than other feeds.

The desired physical properties of aquafeed that can be pointed out as important are: high water stability and pellet durability, maximum density, floatability, expansion ratio, and cohesion (Wood, 1994). According to Riaz (2008), actual moisture, thermomechanical energy, and retention time in the extruder barrel are among the most critical parameters affecting the quality of extrudates. High water stability prevents the loss of nutrients from rapid disintegration in water and increases the time the feed is available for fish to consume (Fallahi et al., 2012). The extruder enhances the production of floating, sinking, and slow-sinking aquaculture feed (Opstvedt et al., 2003). Unit density and expansion ratio dictate the floatability of the aquafeed. The more expanded extrudate can improve feed digestibility because the process increases starch gelatinization (Chang and Wang, 1999). Expansion is governed by the energy released from the latent heat of moisture vaporization, the subsequent cooling step, and a drastic pressure drop at the die exit, which leads to a phase transition from a semi-liquid to a rubbery dough state. High temperature and screw speed normally result in a porous, open extrudate structure (Barrett, 2003) due to reduced dough apparent viscosity and increased water evaporation at the die exit (Mercier et al., 1989). However, the expansion of the feed extrudates is also related to the degree of starch gelatinization and protein denaturation (Raw Materials and Their Impact on the Extrusion of Aqua Feeds, 2007). Trout feed diets produced with the highest temperature showed the highest specific growth rate. If feed protein components are denatured before extrusion processing, extrudate expansion and texturization will be reduced (Sørensen et al., 2009). In addition, high protein content results in less expanded aquafeed extrudates because of the lower contribution of protein to binding and expansion (Riaz, 2013). Increasing the die nozzle diameter results in lower pressure, less moisture vaporization, and thus less expansion at the die exit (Mercier et al., 1989; Liu and Rosentrater, 2011). Typically, mild extrusion processing can expose the active proteolytic enzyme sites to attack, inactivate the heat-labile antinutritional factors, and enhance the digestibility of fish feed extrudates (Riaz, 2013; Sorensen, 2007). For economic purposes, the specific mass content of a given volume of fish feed extrudates (i.e., bulk density) needs to be considered (Rosentrater, 2006); the higher the bulk density, the lower the logistic costs (i.e., packaging, transport, and storage). Because the degree of expansion influences the bulk density of the extrudates, variables such as feed chemical compositions, extruder temperature, pressure developed inside the barrel, screw speed, feed moisture content, extruder water, conditioner steam (Fallahi et al., 2013a), and die geometry (Chevanan et al., 2007a; Tumuluru and Sokhansanj, 2008) significantly affect the bulk density of the fish feed extrudates. Fish feed chemical compositions, moisture content, the extent of heat treatment, and the degree of starch transformation during the extrusion process affect the overall cohesion of fish feed extrudates in terms of water stability and pellet durability. Using starch with a higher amylose/amylopectin ratio in fish feed formulas can improve these properties (Kannadhason et al., 2011). A high content of noninteractive ingredients such as fiber or highly transformed starch and proteins in the fish feed formula also results in the production of noncohesive extrudates (1989); (Chevanan et al., 2008). To control the cohesiveness of the fish feed extrudates, modifications can be made to the amount of liquid water or steam added during the extrusion process. If water and/or steam do not enter the system at a constant rate, it is difficult to achieve consistent production. Indeed, the plasticizing effect of water influences the extent of macromolecular transition (i.e., starch gelatinization and protein denaturation) and the final product texture. Increasing moisture content in the feed ingredients can improve the extrudates' cohesion regardless of the nature of the ingredients (Bandyopadhyay and Rout, 2001; Rolfe et al., 2001; Shukla et al., 2005). Engineering the cohesion and durability of the feed extrudates reduces economic losses due to reduced feed utilization, nutrient loss, feed wastage, and pond pollution caused by low-quality extrudates with mechanically weak structures. On the contrary, unused feed in water causes environmental problems due to algae growth. In addition, this high-temperature, short-time reactor acts as a sterilizer for the finished product (Bhandari et al., 2001).

As stated earlier, a well-balanced nutritional aquafeed is a must for fish growth. Fish require protein, fat, carbohydrates, essential amino acids (Table 14.1), vitamins, and minerals; however, these requirements vary by species and life stage. The effect of the extrusion cooking process on the nutritional quality of food/feed products is still unclear (Singh et al., 2007). The interactive effect of high temperature, pressure, and

Table 14.1 Essential amino acids requirements of Channel catfish, Rainbow trout, Pacific salmon, Common carp, and Nile tilapia (Percentage of Digestible Protein).[a]

Essential amino acid	**Channel catfish *(Ictalurus punctatus)***	**Rainbow trout *(Oncorhynchus mykiss)***	**Pacific salmon *(Oncorhynchus spp.)***	**Common crap *(Cyprinus carpio)***	**Nile tilapia *(Oreochromis niloticus)***
Arginine	5.0	5.3	5.3	4.8	4.9
Histidine	1.8	2.0	2.1	2.0	2.0
Isoleucine	3.3	3.3	3.3	3.3	3.3
Leucine	4.9	4.9	4.9	4.9	4.9
Lysine	5.0	5.3	5.0	5.4	5.6
Methionine + Cystine	3.3	3.5	3.5	3.6	3.4
Phenylalanine + Tyrosine	6.0	6.2	6.2	6.4	6.0
Threonine	3.3	3.6	3.6	3.5	3.7
Tryptophan	0.8	0.9	0.9	0.9	1.0
Valine	3.6	3.8	3.8	3.8	3.8

[a]Values represent an "ideal protein" pattern for fish derived from a synthesis of published essential amino acid requirement studies, classical reviews, NRC recommendations, and species-specific studies for channel catfish, salmonids (including rainbow trout and Pacific salmon), common carp, and Nile tilapia. Values were normalized to digestible protein (g/100 g digestible protein) and harmonized across species using lysine as the reference amino acid (Wilson, 1989; National Research Council (NRC), 2011; Rolland et al., 2023; Food and Agriculture Organization (FAO), 1985; Santiago and Lovell, 1988; Kaushik and Médale, 1994; Rumsey and Ketola, 1975; Tacon, 1992; U.S. Grains Council, 2018; Furuya and Furuya, 2007).

water during extrusion processing could eliminate antinutritional factors (such as trypsin inhibitors, hemagglutinins, tannins, and phytates), inactivate undesirable enzymes, increase the solubility of dietary fiber, reduce lectin activity (Alonso et al., 2001), and even reduce lipid oxidation (Singh et al., 2007). Increased extruder temperature could improve protein digestibility in fish–wheat flour blends due to the inactivation of protease inhibitors (Bhattacharya et al., 1988), while increased shearing forces due to high screw speed could denature the protein of corn-gluten more easily and improve protein digestibility (Bhattacharya and Hanna, 1985). The chemical composition of protein and carbohydrate components of feed ingredients, combined with the high temperature and relatively dry conditions of extrusion processing, favors the Maillard reaction, which can adversely affect the bioavailability of essential amino acids (such as lysine, arginine, cysteine, and tryptophan) and reduce the nutritional quality of protein (Singh et al., 2007). Iwe et al. (2001) observed that the bioavailability of cysteine in maize grits decreased during extrusion processing at moisture contents of <14.5%, extruder temperature of 181°C–187°C, and required torque of 35–79 Nm; bioavailability of arginine, histidine, aspartic acid, and serine was reduced by 21%, 15%, 14%, and 13%, at extrusion temperature of 135°C–160°C, and screw speed of 150 or 200 rpm. To avoid lysine loss during extrusion, the temperature must be less than 180°C, the moisture content must be >15%, and reduced sugars must be as low as possible (Singh et al., 2007). Increasing screw speed from 80 to 140 rpm and decreasing die diameter from 10 to 6 mm enhanced available lysine retention (Iwe et al.,

2004). Several researchers noted that increasing temperature and screw speed, while decreasing moisture content, feeding rate, and die size during extrusion processing, significantly reduced vitamin stability (Camire, 2000; Killeit, 1994; Camire et al., 1990; Björck and Asp, 1983). According to Riaz et al. (2009), vitamins A, E, C, and B1 are very sensitive to extrusion processing conditions; vitamin E alcohol, menadione sodium bisulfite complex, ascorbic acid, and coated ascorbic acid are the least stable vitamins; vitamins B2, B6, B12, niacin, Ca pantothenate, and biotin are highly stable during this process. Overall, these findings suggest additional steps may be needed to maintain the required vitamins. Perhaps extra vitamins can be added to compensate for losses during extrusion cooking or *via* a postprocessing step (i.e., coating, enrobing, spraying, dusting, etc.) (Moscicki, 2011; Singh et al., 2007).

14.5 Effect of preconditioning on extrusion process performance in aquafeed production

This initial step mixes and kneads dry ingredients with water, steam, and oils; hydrates by allowing moisture diffusion to feed particles; and initiates heat energy transfer from heating steam to feed ingredient particles (Fig. 14.2). Appropriate preconditioning initiates proper cooking through starch gelatinization and protein denaturation, lowers product viscosity, improves extrudate flow, increases product throughput, and enhances plasticity, cohesiveness, and water stability of the feed extrudate. Steam heating in the preconditioner provides additional energy to that generated from friction and mechanical energy dissipation, thereby aiding in reaching the targeted extrusion cooking

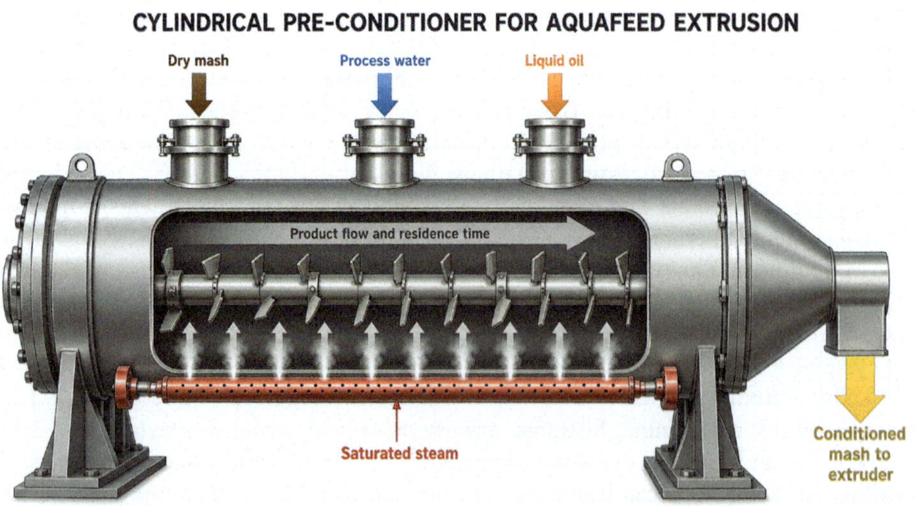

Fig. 14.2 Sequence of addition of ingredients during the preconditioning step for extrusion processing of aquafeed.

Table 14.2 Relation or indicative ranges for extruder die geometry and pellet/extrudate diameter for slow-sinking and floating aquafeeds.[a]

Pellet size	Slow sinking feed			Floating feed		
(mm)	Die diameter (mm)	Length of straight die (mm)	Depth of conical inlet conical (mm)	Die diameter (mm)	Length of straight die (mm)	Depth of conical inlet (mm)
2	1.5–1.7	1.5–2.5	1.0–2.0	1.1–1.3	1.0–1.5	1.0–1.5
3	2.2–2.5	2.5–3.5	2.0–3.0	1.7–1.9	1.3–1.8	1.5–2.5
5	3.3–3.7	4.0–5.5	4.0–5.0	2.7–2.9	2.5–3.5	3.0–4.0
7	4.8–5.2	6.0–7.5	6.0–7.0	3.8–4.2	3.5–4.5	4.0–5.0
9	5.8–6.2	8.0–9.5	8.0–9.0	4.8–5.2	4.5–5.5	4.0–6.0
Hole area	–	100%	–	–	≈70%	–
Minimum content of cereals (starch) in formula	–	≥14%	–	–	≥18%	–
Added water and steam	–	≈15%	–	–	≈20%	–

[a]Pellet size ranges and die geometries are indicative and maybe adjusted according to formulation and extruder design. Ranges were developed as indicative engineering guidelines by synthesizing information from technical and industrial literature on aquafeed and feed extrusion die design rather than a single published table; these sources describe typical fish-feed die diameters from about 0.5–12 mm, die land-length-to-diameter ratios L/D between about 1 and 3, and higher die land length and starch contents (around 18%–20%) for floating feeds compared with sinking feeds around (10%–15% starch) (Insta Pro International, 2016; Makine, 2022; Kearns, 2012; Anonymous, 2017 Bouvier and Brisset, 2006).

temperature (Mottaz and Bruyas, 2001), so the overall high-temperature processing time can be <5 min (Sorensen, 2007). The sequence in which liquids are added to dry ingredients is also very important (Table 14.2). The average temperature in the preconditioner can reach as high as 90°C–95°C, and the average moisture content of the ingredient mix is 20%–25% when it enters the extruder (Jauncey et al., 2008).

14.6 Important factors in the production of floatable aquafeeds

The floatability of feed can be crucial for some species. For instance, carp, salmonids, and sea bass require sinking feeds, while trout and tilapia need floating feed because they capture feed from the surface of the pond (Kazemzadeh, 1991). Even though fish can be trained to capture the feed wherever it is, using floatable feed reduces nutrient loss and water pollution and facilitates feed response evaluation for the producer. Controlling the density and thus buoyancy of the aquafeed can be achieved through appropriate feed ingredient compositions (the amount and type of starch [amylose/amylopectin], type of protein, protein/starch ratio, etc.), extrusion operating conditions (temperature, moisture content, and residence time), and extrusion hardware structure properties (geometry of screw, barrel, and die). The minimum starch required to produce a floating aquafeed is 18%–22%, while for sinking feed, it is 10% (Riaz, 1977). When using a low-fat fish feed formula (~3%) and a die nozzle size of ~1.5–2.0 mm, the maximum required starch content to produce floatable feed is 20%, while for a smaller die nozzle size of 1.0 mm, it must be at least 20% (Lan and Ma, 2012). Several extrusion studies were conducted to investigate the effect of different starches (corn, potato, cassava, sago, and tapioca starches) on the physical properties of extruded aquafeed (Rosentrater et al., 2009b; Rosentrater et al., 2009a; Kannadhason et al, 2011; Umar et al., 2013). In addition, the arrangement/sequencing of essential amino acids impacts protein functionality, binding capacity, density, and strength of extruded feed products (Hernandez-Izquierdo and Krochta, 2008).

The interaction between protein source and starch level also affects density and expansion (Sorensen, 2007). For example, soy protein isolate and gluten protein affect extrudate expansion in identical ways due to their different protein hydration and starch gelatinization, as reported by Faubion et al. (1982). According to Sorensen (2007), starch and protein competition in absorbing water causes differences in starch gelatinization. The effect of graded levels of wheat starch combined with certain amounts of soybean meal (SBM) and FM proteins on starch gelatinization in extruded aquafeed was also investigated by Perez (2004). Increasing the amount of starch resulted in a significant decrease in starch gelatinization in FM-based diets, while it greatly increased the extent of starch gelatinization in SBM-based diets. Extrusion of proteins is likely more complex than extrusion of starch because of protein contributions in a wide range of chemical and biochemical reactions (Samuelsen et al., 2013).

Reducing particle size improves starch gelatinization and thus increases extrudate expansion; however, it may not be economically feasible on a large scale (Obaldo et al., 1998). High levels of fat and vegetable protein in the feed formula result in slow-sinking feeds with reduced expandability. Appropriate retention time during preconditioning of the raw ingredients, which can be adjusted by associated paddle configuration and shaft RPM (Riaz, 2003), positively affects extrudate expandability and water stability as it facilitates feed blend ripening and enhances starch gelatinization (Moscicki, 2011). According to Riaz (2009b), extrusion moisture and specific mechanical energy are the foremost operating parameters affecting aquafeed buoyancy (Table 14.3). Due to the high extruder temperature and the lubricant effect of water, molecular mobility increases, and the melted dough exhibits pseudoplastic shear thinning behavior; consequently, this reduces the degree of pressure release and negatively affects the expansion ratio and floatability of fish feed extrudates (Chevanan et al., 2007a; Mjoun and Rosentrater, 2011). The feed's floating and sinking properties also depend on water temperature and salinity. In general, bulk

Table 14.3 Relationship Relationship between bulk density and buoyancy characteristics of extruded aquafeed pellets in seawater at 20 °C (3% salinity).

Pellet buoyancy	Sea water at 20°C (3% salinity)		
Fast sinking	>640 g/l	Slow sinking Neutral buoyancy Floating	580–600 g/l 520–540 g/l <480 g/l

Density ranges are presented as representative engineering guidelines for floating, neutral, slow-sinking, and fast-sinking extruded aquafeeds, synthesized and reformulated from published literature on aquafeed pellet density and buoyancy behavior, including Riaz (2009b).

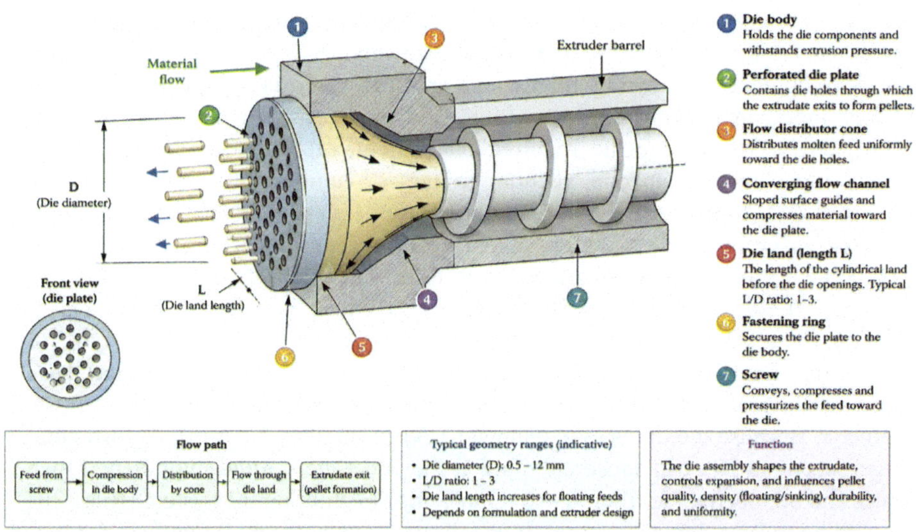

Fig. 14.3 Representative schematic of an extrusion die plate assembly used in aquafeed processing, illustrating material flow distribution, die geometry, and guidance of the extrudate toward perforated die openings. Illustration developed based on published aquafeed extrusion literature and industrial extrusion die design references, including Jauncey et al. (2008) and Bouvier and Brisset (2006).

density is <480 g/l for floatable feed and>640 g/l for sinking feed (Table 14.3) (Riaz, 2009b). Fish feed density can also be controlled by changing die plate design and die geometrical dimensions (Table 14.2, Fig. 14.3). To produce floatable feed, the die size should be 1.5 to 2.5 times smaller than the diameter of the final extrudate. The restriction at the extrusion discharge can be regulated on the spot using a back-pressure valve mounted on the end of the extruder before the die nozzle. The operator can then control extrudate density without changing the screw configuration or the die. Through this multiposition valve, shear stress inside the barrel can be controlled, and specific mechanical energy can be adjusted, which leads to changes in bulk density, control of starch gelatinization, adjustment of extrudate cell structure, and improvement in fat and water absorption (Riaz, 2009b).

14.7 Factors to be considered in extrusion processing of aquafeed

Backward flow and blocking screw rotation: Controlling the temperature in the cooking/compression zone with a cooling system is a key point to avoid the dough moving backwards and blocking the screw rotation. On the contrary, low oil/lipid content in extrusion processing, combined with relatively high starch content and low moisture content, can cause severe clogging due to increased friction heat and excessive starch degradation (Ilo et al., 2000). **High lipid content:** High fat content in some ingredients (>12%) may cause difficulty due to slippage inside the barrel, hampering the development of shearing forces and pressure, decreasing specific mechanical energy input, preventing agglomeration, and thus resulting in improper cooking (Guy, 2001). To achieve the required fat content, oil may be added in the final processing stage *via* special vacuum mixers or tumbling to coat the extrudate (Munz, 2008). **Noncohesive fish feed extrudate**: High protein and fat in the formula decrease space for binders (starch, gums, etc.) and reduce cohesion. Increased moisture content produces cohesive feeds with high water stability and durability (Rolfe et al., 2001). Raw ingredients with large particles reduce the effective superficial area, leading to undesirable blend conditioning, less starch gelatinization, and thus a nonintact final structure (Liu and Rosentrater, 2011). **Microsized aquafeeds:** In microsized aquafeeds, the dry ingredient particle size should be one-third of the die diameter. Because the die nozzle diameter is very small (< 1.2 mm), any irregular or larger-sized particle can cause clogging at the end zone. Liquid ingredients, such as water and oil, need to be filtered (<0.6 mm) prior to entering the system (Riaz, 2012). According to Riaz, low throughput is a common issue with small fish feed production due to the small die diameter. This can be resolved by increasing the number of orifices in the die area. Sinking feeds require more orifices per die per extrudate diameter than expanded floating feeds. The final extrudate texture is influenced by dough pseudoplastic behavior and die configuration. **Cutting device**: An appropriate cutting device tightly mounted on the die plate of the extruder is important for the formation of the final feed texture. The number of knives, rotation speed, and elasticity should be taken into consideration. **Extruder cleaning:** When changing from large-size to microscale feed production, it is important to clear the large fiber and particle residuals from the path flow to prevent clogging. The best strategy is to clean the entire system after each batch. Depending on the extruder type, steam, water, or air flushing may be used to clean the path flow and other parts of the extruder. There may not be a solid rule of thumb for extrusion processing of biological materials due to the different nature of these substances and their varying behaviors during the complex process of extrusion.

14.8 Protein alternatives in aqua feeds, challenges in extrusion processing, and extrudate quality

The sustainability of aquaculture depends on partially or fully replacing marine-derived protein in the feed (Fallahi et al., 2013b). Scientists favor increased utilization of alternative protein sources, including animal protein (i.e., fishery and poultry waste products), bacterial single-cell protein (i.e., waste product of brewery plants), or plant proteins (i.e., ethanol plant coproducts, soybean oil processing by-products, etc.) (Ayadi et al., 2012a; Hardy, 2010). In manufacturing aquafeeds with alternative protein sources, having knowledge about protein, starch, and fat contents of the protein source and understanding its role in extrusion processing are a MUST (Riaz, 2008). Utilization of plant-based proteins in aquafeeds has become attractive to aquaculture scientists and technologists. Some of the advantages of FM replacement with plant-based proteins in aquafeed extrusion that can be pointed out are: less-preprocessing; containing both starch and proteins; contributing to better-floatability; enhancing the binding properties and water stability of the extrudates; allowing for better lipid absorption; producing fewer mineral deposits in screw and die areas, and thus reducing die plugging (Riaz, 2008). However, using these protein sources is challenging due to antinutritional factors, lack of essential nutrients (Murray et al., 2010), and occasionally poor functionality during production (Fallahi et al., 2012). In particular, carnivorous species such as rainbow trout (Olli and Krogdahl, 1994) and Atlantic salmon (Krogdahl et al., 2003; Bæverfjord and Krogdahl, 1996) are extremely intolerant to plant-based proteins. Ayadi et al. (2012a) extensively reviewed the pros and cons associated with the utilization of various alternative protein sources in aquafeed for different species. The range of nutrient concentrations for a viable protein alternative to fish meal is summarized in Table 14.4 Gatlin et al. (2007) discussed the bioactive compounds in and/or nutritional limitations of plant-based protein products. Intensive interactions among water, heat, and macromolecules (i.e., starch, protein, and lipids) change the chemistry and structure of the materials inside the barrel, which are then restructured during evaporation at the die exit, leading to severe physical changes like expansion or texturization (Mercier et al., 1989). These physiochemical alterations affect the nutritional value and physical properties of the plant protein-based extrudate, which may enhance their quality.

The general effects of extrusion processing variables (temperature, screw speed, moisture content, ingredient, etc.) on extrudates' physical, chemical, and nutritional properties were discussed in the previous subheadings. Sørensen et al. (2002) observed that extrusion temperatures from 100°C to 150°C had no significant impact on protein, amino acids, and energy digestibility of single-screw extruded FM-based diets for rainbow trout. Similar results were reported for twin-screw extruded rainbow trout diets at extrusion temperatures from 100°C to 140°C; however, increasing temperature improved growth rate (Sørensen et al., 2005). In another study, Aarseth et al. (2006) found that feed produced at 140°C was significantly less durable and hard compared to feed produced at 100°C. Romarheim et al. (2006) examined the

Table 14.4 Proximate and nutrient content (as-fed basis) of fish meal, along with targeted ranges in alternative ingredients derived from grains and oilseeds.

Category/nutrient	Fish meal (menhaden)	Target range for alternative ingredients
Crude protein (%)	65–72	48–80
Crude lipid (%)	5–8	2–20
Fiber (%)	<2	<6
Ash (%)	7–15%	4–8
NFE (%)	<1	<20
Starch (%)	<1	<20
Nonsoluble CHO (%)	None	<8
Arginine (%)	3.75	>3.0
Lysine (%)	4.72	>3.5
Methionine (%)	1.75	>1.5
Threonine (%)	2.5	>2.2
Omega-3 fatty acids	~2%	1

Source: Data compiled from (Gatlin, D.M., Barrows, F.T., Brown, P., Dabrowski, K., Gaylord, T.G., Hardy, R.W., Herman, E., Hu, G., Krogdahl, A., Nelson, R., Overturf, K., Rust, M., Sealey, W., Skonberg, D., Souza, E.J., Stone, D., Wilson, R., Wurtele, E., 2007. Expanding the utilization of sustainable plant products in aquafeeds: a review.Aquac. Res. 1365–2109. 38 (6), 551–579. doi:10.1111/j.1365-2109.2007.01704.x.).

effect of extrusion processing on trypsin inhibitor activity in rainbow trout diets with the inclusion of white flakes and toasted SBM. Twin-screw extrusion processing of SBM-based trout diets drastically reduced trypsin inhibitors (<2000 TIU/g); extruder temperature did not influence feed intake, but increasing temperature improved feed conversion ratio. Decreased screw speed reduced feed intake, while a combination of high extruder temperature and high screw speed significantly enhanced feed intake (Barrows et al., 2007). Inclusion of 20%–27% extrusion pretreated pea protein concentrate in Atlantic salmon parr diets did not affect fish growth performance (Øverland et al., 2006). Severe extrusion conditions, such as low moisture content and high temperature, create new disulfide bonds (Liu and Hsieh, 2008) and negatively impact the bioavailability of essential amino acids (such as cysteine) in FM, SBM (Ljøkjel et al., 2000), as well as trout diets (Sørensen et al., 2002). Rokey and Huber (1994) suggested that wet extrusion processing with a moisture content range of 25% to 30% would preserve essential amino acid content. Some researchers suggest that supplementing plant-based fish feed ingredients with enzymes (such as phytase) (Cain and Garling, 1995; Rodehutscord and Pfeffer, 1995; Vielma et al., 1998; Sugiura et al., 2001; Cheng and Hardy, 2002; Storebakken et al., 1998; Papatryphon et al., 1999; Vielma et al., 2004) and essential amino acids (lysine, cysteine, etc.) (Cheng et al., 2003) prior to extrusion processing can improve the phosphorus bioavailability, protein digestibility, and amino acid accessibility of the feed. The functionality of proteins from animal, plant, and microbial sources and their contributions to the physical properties of extruded feeds, including density, water stability, durability, and floatability, are influenced by their amino acid composition, molecular structure, and interactions with other feed components (Verbeek and Van Den Berg, 2010; Samuelsen et al., 2013). Several studies have demonstrated that ingredient composition plays a critical role in determining extrusion behavior and the resulting feed characteristics. For example, the presence of non-starch polysaccharides in plant protein ingredients may reduce extrudate expansion (Ainsworth et al., 2007), whereas starch source has been shown to significantly influence feed expansion and pellet physical properties. In extruded fish feeds containing legumes, oilseeds, and cereal ingredients, starch type affected feed expansion and related physical characteristics, while total starch content had little effect on feed floatability, whereas starch source has been shown to significantly influence feed expansion and pellet physical properties. In extruded fish feeds containing legumes, oilseeds, and cereal ingredients, starch type affected feed expansion and related physical characteristics, while total starch content had little effect on feed floatability (Kraugerud et al., 2011). Similarly, differences among cultivars of lupin kernel meal influenced extrusion process parameters, melt viscosity, and the physical properties of salmonid feeds (Glencross et al., 2010). These extrusion-induced changes in feed structure can subsequently affect nutrient accessibility and digestibility. Furthermore, pretreatment of plant protein ingredients may alter protein conformation, modify interactions with starch and water, and influence ingredient functionality during extrusion by exposing hydrophobic groups and reducing water-binding capacity (Camire et al., 1990).

In the search for sustainable protein sources for aquafeeds, distillers dried grains with solubles (DDGS), a co-product of the dry mill corn-ethanol industry; SBM, the residual meal after oil extraction from the soybean; canola meal concentrate, protein isolate from canola meal residual of canola oil extraction; and corn gluten meal have potential to partially replace FM as a protein source (Gatlin et al., 2007; Cheng et al., 2003; Ayadi et al., 2011b). Numerous studies have investigated the dietary replacement of FM with these alternative proteins for different species (Pongmaneerat and Watanabe, 1992; Kaushik et al., 1995; Refstie et al., 2005; Zhou et al., 2011; Yuan et al., 2013; Floreto et al., 2000; Olli and Krogdahl, 1994; Naylor and Burke, 2005). Some have shown positive performance characteristics with the complete or partial inclusion of alternative proteins (Fallahi et al., 2014). The physical properties of the extruded feeds give insight into the processing functionality of the plant-based proteins, which are important for developing a viable fish feed. To our knowledge, until the last decade, research on the effect of extrusion processing and alternative proteins on the physical properties of extruded aquafeeds has been scarce. The effect of extrusion cooking on DDGS and soy-based diets for Nile tilapia and channel catfish was extensively studied by (Chevanan et al., 2009; Chevanan et al., 2007a; Chevanan et al., 2007; Chevanan et al., 2008, 2010; Chevanan et al., 2007a; Chevanan et al., 2007c; Kannadhason et al., 2011). Their findings were reviewed and summarized by Liu and Rosentrater (2011). Later on, Ayadi et al. (2012a), Ayadi et al. (2011a, 2011b), and Ayadi et al. (2012b) examined the effect of DDGS inclusion on the physical properties of the extruded yellow perch and Rainbow trout diets. In another study, Mjoun and Rosentrater (2011) studied the effect of increasing levels of DDGS (up to 50%) on physical properties of twin-screw extruded yellow perch diets and process parameters. Also, Fallahi et al. (2013a), investigated the effect of screw speed (350, 400, and 450 rpm), extruder water (0.236, 0.271, and 0.302 kg/min), and conditioner steam (0.1, 0.12, and 0.15 kg/min), using a twin-screw extruder with a single Nile tilapia feed blend containing 20% (db) DDGS. They suggested that the ranges of screw speed, extruder water, and conditioner steam used appeared to be close to optimum for this specific formulation. Ayadi et al. (2013) conducted a single screw extrusion study to model the effects of various levels of raw blend properties and processing conditions on physical properties of Nile tilapia extruded diets and extruder processing parameters (i.e. resulting temperatures, die pressure, extruder torque, mass flow rate, apparent viscosity, and specific mechanical energy). They observed that DDGS, moisture content, die dimension, and extrusion conditions had the biggest impact on most of the extrudate physical properties and processing conditions. In a feeding trial conducted by Schaeffer et al. (2011), the effect of SBM and DDGS combinations on the physical properties of single-screw extruded yellow perch diets was investigated. They determined that up to 49.5% combined inclusion of DDGS and SBM yielded suitable yellow perch feeds. The effects of DDGS, high-protein distillers dried grains (HPDDG), soy protein concentrate (SPC), and microbial fermented soybean meal (FSBM) and extrusion processing parameters on production of yellow perch and rainbow trout feeds, using a twin-screw extruder were studied by Fallahi et al. (2012), Fallahi et al. (2012b), Fallahi et al. (2013a), Fallahi et al. (2013b), 2014). Overall, they determined that up to 40% HPDDG inclusion, in contrast to DDGS, had no effect on bulk density of the rainbow trout diet; increasing levels of HPDDG did not change expansion ratio of the extruded feeds; inclusion of 20% FSBM-30%HPDDG and 20% SPC-30%HPDDG, substantially increased the expansion ratio of the yellow perch feeds; SPC inclusion substantially increased thermal properties of the feeds; total FSBM inclusion could produce more expanded rainbow trout feeds compared to 40% HPDDG inclusion; FM elimination with FSBM inclusion in trout diets resulted in a significant reduction in bulk density but combination of 20% FSBM and 40% DDGS, could raise bulk density of the yellow perch feed; all produced feeds in these studies appeared to have high durability index indicating that the products had long storage stability and high resistance against external mechanical forces. Also, in an unpublished study conducted under common commercial conditions, the effect of defatted SBM and untoasted white flakes incorporations on aquafeed properties and extrusion processing parameters, was evaluated (Sorensen, 2007). They observed that soy protein inclusion increased hardness and bulk density of the feeds compared to those of the FM-based diets; white flakes inclusion increased specific mechanical energy and bulk density compared to the other diets. The effect of DDG and soy products inclusion and extrusion processing on the physical properties of extruded aquafeeds for some common species is also reviewed and summarized in Figs. 14.4, 14.5 and Table 14.5.

Protein Sources (%)	Extruder & Species	Extrusion Conditions: Temperature profile (°C)	CS (kg/hr)	EW (Kg/hr)	SS (RPM)	MC (db%)	Die (mm)	SCR (-)	Physical Properties: UD (g/cm³)	BD (g/cm³)	ER (-)	PDI (%)	Ref.
DDGS													
0	Single-screw & Yellow perch	40-90-100	NA	NA	230	20	295	3:01	878.70	-	1.09	96.18	Ayadi et.al (2011a)
10									876.34	-	1.09	97.63	
20									973.10	-	1.04	96.43	
30									975.75	-	1.07	96.83	
40									1016.24	-	1.10	97.29	
50									1030.67	-	1.09	96.31	
0	Twin-Screw & Rainbow trout	25-15-15-15-15-15-75-75	7.19	6.45	250	5..7	1.9	NA	927.02	458.05	1.49	82.95	Ayadi et.al (2011b)
10		25-10-15-15-15-15-75-75	7.47	5.88					887.29	434.44	1.61	91.01	
20		25-10-15-15-15-15-75-75	7.64	5.10					890.87	444.27	1.45	88.54	
30		25-10-15-15-15-15-75-75	7.65	4.32					938.29	456.27	1.41	87.74	
40		25-10-15-15-15-15-75-75	7.53	4.32					971.09	466.67	1.40	91.54	
50		25-10-15-15-15-15-75-75	7.7	4.32					933.97	470.29	1.42	95.39	
0	Twin-Screw & Yellow perch	25-35-15-65-65-90-80-80	6.86	9.15	226-298	5..7	1.9	NA	855.10	427.59	1.43	93.35	Ayadi et.al (2012b)
10		25-35-15-65-65-90-80-80	6.88	11.1					917.05	447.19	1.35	96.26	
20		25-35-15-65-65-90-80-80	6.96	13.14					911.19	469.89	1.35	96.77	
30		25-10-15-15-15-15-81.5-81.5	9.7	6.72					867.89	434	1.44	89.89	
40		25-10-15-15-15-37.5-65-65	7.13	9.60					990.09	470.29	1.36	96.27	
20	Twin-Screw & Nile tilapia	25 to 90	7.2	16.26	350	20	3	NA	1015.1	529.38	1.10	97.70	Fallahi et.al (2013a)
			7.2	16.26	450				917.40	536.99	1.07	97.20	
			7.2	18.12	400				1020.6	465.02	1.13	94.36	
			7.2	14.16	400				1000	550.19	1.03	98.53	
			9.0	16.26	400				990	525.42	1.10	97.06	
			6.0	16.26	400				951	515.84	1.06	97.54	
			7.2	16.26	400				932.63	498.16	1.11	96.28	

Fig. 14.4 Summarizing extrusion conditions and physical properties of DDGS based aquafeeds, listing extruder type, species, processing settings, measured densities, expansion, and durability.

Protein Sources (%)	Extruder & Species	Extrusion Conditions: Temperature profile (°C)	CS (kg/hr)	EW (Kg/hr)	SS (RPM)	MC (db%)	Die (L/D)	SCR (-)	Physical Properties: UD (g/cm³)	BD (g/cm³)	ER (-)	PDI (%)	Ref.
DDGS-SBM													
20-50	Single-screw & Nile tilapia	80-100-100	NA	NA	100	20	5	2:1	See the reference for exact values				Ayadi et.al (2013)
30-40		100-80-120			150	30	9	2:1					
40-30		80-140-140			200	40	13	3:1					
DDGS													
20	Twin-Screw & Rainbow trout	25 to 90	6.60-9.60	6.60-11.14	230-300	20	3.0	NA	886.65	666.91	1.04	99.33	Fallahi et.al (2013b)
40									740.58	601.13	1.13	99.54	
HPDDG													
20									755.72	575.89	1.16	99.59	
40									750.92	583.98	1.16	99.64	
DDGS-FSBM													
20-20	Twin-screw & Yellow perch	25 to 90	6.60-9.60	6.60-11.14	230-300	20	3.0	NA	791.6	607.93	1.10	98.74	Fallahi et.al (2014)
40-20									895.55	654.5	1.08	98.91	
HPDDG-SPC-FSBM													
30-20-0	Twin-screw & Yellow perch	25-90	6.60-9.60	6.60-11.14	230-300	20	3.0	NA	601.43	478.78	1.23	99.23	Fallahi et.al (2012a)
30-0-20									504.02	417.5	1.34	99.13	
FSBM													
80	Twin-screw & Rainbow trout	25-90	6.60-9.60	6.60-11.14	230-300	20	3.0	NA	682.11	544.02	1.24	99.59	Fallahi et.al (2012b)
100									658.91	530.16	1.39	99.54	

Fig. 14.5 Summarizing extrusion conditions and physical properties of DDGS and soy-based aquafeeds, listing extruder type, species, processing settings, measured densities, expansion, and durability.

Table 14.5 Correlation of extrusion operating parameters and fish feed floatability.

Diet category	Low fat	Medium fat	High fat	Ultra-high fat
Total fat (%)	<15	15–25	25–35	>35
Density before	>594	594–520	520–480	<480
Extrusion moisture (%)	25–29	21–25	19–23	20–24
SME (kWhr/t)	<25	25–30	30–35	>35
Coating process	Atmospheric or Vacuum	Atmospheric or Vacuum	Vacuum	Vacuum
Aquatic species	Cod	Cod, Trout	Yellowtail Sea	Salmon
	Shrimp	Yellowtail Sea Bass	Trout Salmon	

14.9 Future

More research is needed to understand the correlation between the physical and nutritional quality of extruded aquafeed. An added concern about the inclusion of soy products in aquafeeds is the presence of lectin, a protein that is highly resistant to proteolytic enzymes. Studying the effect of extrusion processing on lectin activity would help enhance the inclusion of soy-protein-based ingredients in aquadiets. The effects of extrusion cooking on the physical and nutritional properties of feeds containing oilseed meals have not been explored yet.

References

Aarseth, K.A., Sørensen, M., Storebakken, T., 2006. Red yeast Xanthophyllomyces dendrorhus inclusion in diets for salmonids improves pellet strength. Anim. Feed Sci. Technol. 126, 75–101.

Ainsworth, P., Ibanoğlu, S., Plunkett, A., Ibanoğlu, E., Stojceska, V., 2007. Effect of brewers spent grain addition and screw speed on the selected physical and nutritional properties of an extruded snack. J. Food Eng. 0260-8774. 81 (4), 702–709. doi:10.1016/j.jfoodeng.2007.01.004.

Alonso, R., Rubio, L.A., Muzquiz, M., Marzo, F., 2001. The effect of extrusion cooking on mineral bioavailability in pea and kidney bean seed meals. Anim. Feed Sci. Technol. 0377-8401. 94 (1-2), 1–13. doi:10.1016/S0377-8401(01)00302-9.

Anonymous. 2017. The preparation of floating and sinking fish feeds by extrusion. Engormix Technical Articles – Aquafeed Extrusion (online technical publication).

Ayadi, F.Y., Fallahi, P., Rosentrater, K.A., Muthukumarappan, K., 2013. Modeling single-screw extrusion processing parameters and resulting extrudate properties of DDGS-based Nile Tilapia (*Oreochromis niloticus*) feeds. J. Food Res. 1927-0895. 2 (2), 11. doi:10.5539/jfr.v2n2p11.

Ayadi, F.Y., Muthukumarappan, K., Rosentrater, K.A., Brown, M.L., 2011a. Twin-screw extrusion processing of rainbow trout (*Oncorhynchus mykiss*) feeds using various levels of corn-based distillers dried grains with solubles (DDGS). Cereal Chem. 0009-0352. 88 (4), 363–374. doi:10.1094/CCHEM-08-10-0120.

Ayadi, F.Y., Muthukumarappan, K., Rosentrater, K.A., Brown, M.L., 2011b. Single-screw extrusion processing of distillers dried grains with solubles (DDGS)-based yellow perch (*Perca flavescens*) feeds. Cereal Chem. 0009-0352. 88 (2), 179–188. Available from: http://cerealchemistry.aaccnet.org/doi/pdfplus/10.1094/CCHEM-08-10-0118. doi:10.1094/CCHEM-08-10-0118.

Ayadi, F.Y., Rosentrate, K.A., Muthukumar, K., 2012a. Alternative protein sources for aquaculture feeds. J. Aquac. Feed Sci. Nutr. 2070-1667. 4 (1), 1–26. doi:10.3923/joafsnu.2012.1.26.

Ayadi, F.Y., Rosentrater, K.A., Muthukumarappan, K., Brown, M.L., 2012b. Twin-screw extrusion processing of distillers dried grains with solubles (DDGS)-based Yellow Perch (*Perca flavescens*) feeds. Food Bioprocess Technol. 1935-5149. 5 (5), 1963–1978. doi:10.1007/s11947-011-0535-5.

Bæverfjord, G., Å, K., 1996. Development and regression of soybean meal induced enteritis in Atlantic salmon distal intestine. A comparison with the intestines of fasted fish. J. Fish Dis. 19, 375–387.

Bandyopadhyay, S., Rout, R.K., 2001. Aquafeed extrudate flow rate and pellet characteristics from low-cost single-screw extruder. J. Aquat. Food Prod. Technol. 1049-8850. 10 (2), 3–15. doi:10.1300/J030v10n02_02.

Barrett, A.H., 2003. Characterization of macrostructures in extruded products. In: Kaletunç, G., Breslauer, K.J. (Eds.), Characterization of Cereals and Flours: Properties, Analysis and Applications. Marcel Dekker, New York, NY, pp. 369–386.

Barrows, F.T., Stone, D.A.J., Hardy, R.W., 2007. The effects of extrusion conditions on the nutritional value of soybean meal for rainbow trout (Oncorhynchus mykiss). Aquaculture. 0044-8486. 265 (1-4), 244–252. doi:10.1016/j.aquaculture.2007.01.017.

Bhandari, B., D'Arcy, B., Young, G., 2001. Flavour retention during high temperature short time extrusion cooking process: a review. Int. J. Food Sci. Technol. 0950-5423. 36 (5), 453–461. doi:10.1046/j.1365-2621.2001.00495.x.

Bhattacharya, M., Hanna, M.A., 1985. Extrusion processing of wet corn gluten meal. J. Food Sci. 0022-1147. 50 (5), 1508–1509. doi:10.1111/j.1365-2621.1985.tb10517.x.

Bhattacharya, S., Das, H., Bose, A.N., 1988. Effect of extrusion process variables on in-vitro protein digestibility of fish-wheat flour blends. Food Chem. 0308-8146. 28 (3), 225–231. doi:10.1016/0308-8146(88)90054-4.

Björck, I., Asp, N.-G., 1983. The effects of extrusion cooking on nutritional value—a literature review. J. Food Eng. 0260-8774. 2 (4), 281–308. doi:10.1016/0260-8774(83)90016-x.

Bouvier, J.M., Brisset, J., 2006. Aquafeed twin screw extrusion processing. Aquafeed Twin Screw Extrusion Processing Manual. Clextral, Firminy, France.

Cain, K.D., Garling, D.L., 1995. Pretreatment of soybean meal with phytase for salmonid diets to reduce phosphorus concentrations in hatchery effluents. Prog. Fish-Cult. 0033-0779. 57 (2), 114–119. doi:10.1577/1548-8640(1995)057<0114:POSMWP>2.3.CO;2.

Camire, M.E., Camire, A., Krumhar, K., 1990. Chemical and nutritional changes in foods during extrusion. Crit. Rev. Food Sci. Nutr. 1040-8398. 29 (1), 35–57. doi:10.1080/10408399009527513.

Camire, M.E., 2000. Chemical and nutritional change in food during extrusion. In: Raiz, M.N., Ed., Extruders in Food Applications. CRC Press, Boca Raton, FL, pp. 124–147.

Cavalcanti, W.B., Behnke, K.C., 2005a. Effect of composition of feed model systems on pellet quality: a mixture experimental approach. I. Cereal Chem. 0009-0352. 82 (4), 455–461. Available from: http://onlinelibrary.wiley.com/journal/10.1002/(ISSN)1943-3638/issues. doi:10.1094/CC-82-0455.

Cavalcanti, W.B., Behnke, K.C., 2005b. Effect of composition of feed model systems on pellet quality: a mixture experimental approach. II. Cereal Chem. 0009-0352. 82 (4), 462–467. Available from: http://onlinelibrary.wiley.com/journal/10.1002/(ISSN)1943-3638/issues. doi:10.1094/CC-82-0462.

Chang, Y.K., Wang, S.S., 1999. Advances in extrusion technology: aquaculture/animal feeds and foods. In: Proceedings of the International Symposium on Animal and Aquaculture Feedstuffs by Extrusion Technology and the International Seminar on Advanced Extrusion Technology in Food Applications, March 9-14, 1998. Technomic Publishing Company, Technomic Publishing Company, Áquas de Lindóia, São Paulo, Brazil, Available from: https://books.google.co.in/books?id=fulJAAAAYAAJ.

Cheng, Z.J., Hardy, R.W., 2002. Effect of phytase on apparent digestibility coefficients of nutrients and energy availability of barley, canola meal, wheat and wheat middlings measured in vivo using rainbow trout (*Oncorhynchus mykiss*). Aquac. Nutr. 8 (4), 271–277.

Cheng, Z.J., Hardy, R.W., Usry, J.L., 2003. Effects of lysine supplementation in plant protein-based diets on the performance of rainbow trout (*Oncorhynchus mykiss*) and apparent digestibility coefficients of nutrients. Aquaculture 215 (1-4), 255–265. doi:10.1016/S0044-8486(02)00166-7.

Chevanan, N., Muthukumarappan, K., Rosentrater, K.A., 2009. Extrusion studies of aquaculture feed using Distillers dried grains with solubles and whey. Food Bioprocess Technol. 1935-5149. 2 (2), 177–185. doi:10.1007/s11947-007-0036-8.

Chevanan, N., Muthukumarappan, K., Rosentrater, K.A., Julson, J.L., 2007a. Effect of die dimensions on extrusion processing parameters and properties of DDGS-based aquaculture feeds. Cereal Chem. 0009-0352. 84 (4), 389–398. Available from: http://onlinelibrary.wiley.com/journal/10.1002/(ISSN)1943-3638/issues. doi:10.1094/CCHEM-84-4-0389.

Chevanan, N., Muthukumarappan, K., Rosentrater, K.A., Julson, J.L., 2007c. Effect of die dimensions on extrusion processing parameters and properties of DDGS-based aquaculture feeds. Cereal Chem. 0009-0352. 84 (4), 389–398. doi:10.1094/CCHEM-84-4-0389.

Chevanan, N., Rosentrater, K.A., Muthukumarappan, K., 2007b. Twin-screw extrusion processing of feed blends containing distillers dried grains with solubles (DDGS). Cereal Chem. 0009-0352. 84 (5), 428–436. Available from: http://onlinelibrary.wiley.com/journal/10.1002/(ISSN)1943-3638/issues. doi:10.1094/CCHEM-84-5-0428.

Chevanan, N., Rosentrater, K.A., Muthukumarappan, K., 2008. Effect of DDGS, moisture content, and screw speed on physical properties of extrudates in single-screw extrusion. Cereal Chem. 0009-0352. 85 (2), 132–139. Available from: http://onlinelibrary.wiley.com/journal/10.1002/(ISSN)1943-3638/issues. doi:10.1094/CCHEM-85-2-0132.

Chevanan, N., Rosentrater, K.A., Muthukumarappan, K., 2010. Effects of processing conditions on single screw extrusion of feed ingredients containing DDGS. Food Bioprocess Technol. 1935-5149. 3 (1), 111–120. doi:10.1007/s11947-008-0065-y.

Fallahi, P., Muthukumarappan, K., Rosentrater, K.A. 2012b. Twin-screw extrusion processing of vegetable-based rainbow trout (*Oncorhynchus mykiss*) feeds using graded levels of high protein fermented soybean meal (FSBM), Paper Presented at the ASABE Annual International Meeting, Dallas, TX. 29 July 29–1 August, 2012. Paper Number: 1213 37565.

Fallahi, P., Muthukumarappan, K., Rosentrater, K.A., Brown, M.L., 2012. Twin-screw extrusion processing of vegetable-based protein feeds for Yellow Perch (*Perca flavescens*) containing distillers dried grains, soy protein concentrate, and fermented high protein soybean meal. J. Food Res. 1927-0895. 1 (3), 230. doi:10.5539/jfr.v1n3p230.

Fallahi, P., Rosentrater, K.A., Muthukumarappan, K., Brown, L.M., 2013b. Twin-screw extrusion processing of rainbow trout (*Oncorhynchus mykiss*) feeds using graded levels of high protein corn-based distillers dried grains (HP-DDG) and conventional distillers dried grains with solubles (DDGS). J. Food Res. 2 (1), 118–130. doi:10.5539/jfr.v2n1p118.

Fallahi, P., Rosentrater, K.A., Muthukumarappan, K., Brown, M.L., 2014. Characteristics of vegetable-based twin-screw extruded yellow perch (*Perca flavescens*) diets containing fermented high-protein soybean meal and graded levels of distillers dried grains with solubles. Cereal Chem. 0009-0352. 91 (1), 79–87. Available from: http://cerealchemistry.aaccnet.org/doi/pdfplus/10.1094/CCHEM-08-12-0100-R. doi:10.1094/CCHEM-08-12-0100-R.

Fallahi, P., Rosentrater, K.A., Muthukumarappan, K., Tulbek, M., 2013a. Effects of steam, moisture, and screw speed on physical properties of DDGS-based extrudates. Cereal Chem. 0009-0352. 90 (3), 186–197. Available from: http://cerealchemistry.aaccnet.org/doi/pdf/10.1094/CCHEM-08-12-0102-R. doi:10.1094/CCHEM-08-12-0102-R.

Fallahi, P., Rosentrater, K.A., Muthukumarappan, K., Tulbek, M., 2013b. Effects of steam, moisture, and screw speed on physical properties of DDGS-based extrudates. Cereal Chem. 0009-0352. 90 (3), 186–197. doi:10.1094/CCHEM-08-12-0102-R.

Faubion, J.M., Hoseney, R.C., Seib, P.A., 1982. Functionality of grain components in extrusion. Cereal Foods World 27, 212–216.

Floreto, E.A.T., Bayer, R.C., Brown, P.B., 2000. The effects of soybean-based diets, with and without amino acid supplementation, on growth and biochemical composition of juvenile American lobster, Homarus americanus. Aquaculture 189 (3-4), 211–235. doi:10.1016/S0044-8486(00)00363-X.

Food and Agriculture Organization (FAO). 1985. Protein and Amino Acid Requirements of Fishes (comparative requirement tables for carp, trout, catfish and tilapia). FAO, Rome.

Furuya, W.M., Furuya, V.R.B., 2007. Nutritional requirements of tilapia. In: Tilapia: Biology, Production and Nutrition. Haworth Press.

Gatlin, D.M., Barrows, F.T., Brown, P., Dabrowski, K., Gaylord, T.G., Hardy, R.W., Herman, E., Hu, G., Krogdahl, A., Nelson, R., Overturf, K., Rust, M., Sealey, W., Skonberg, D., Souza, E.J., Stone, D., Wilson, R., Wurtele, E., 2007. Expanding the utilization of sustainable plant products in aquafeeds: a review. Aquac. Res. 1365-2109. 38 (6), 551–579. doi:10.1111/j.1365-2109.2007.01704.x.

Glencross, B., Hawkins, W., Maas, R., Karopoulos, M., Hauler, R., 2010. Evaluation of the influence of different species and cultivars of lupin kernel meal on the extrusion process, pellet properties and viscosity parameters of salmonid feeds. Aquac. Nutr. 1353-5773. 16 (1), 13–24. doi:10.1111/j.1365-2095.2008.00636.x.

Guy, R., 2001. Extrusion Cooking. Technologies and Applications, CRC Press Inc, Boca Ration, FL.

Hardy, R.W., 2010. Utilization of plant proteins in fish diets: effects of global demand and supplies of fishmeal. Aquac. Res. 1365-2109. 41 (5), 770–776. doi:10.1111/j.1365-2109.2009.02349.x.

Hernandez-Izquierdo, V.M., Krochta, J.M., 2008. Thermoplastic processing of proteins for film formation – A review. J. Food Sci. 0022-1147. 73 (2), R30. doi:10.1111/j.1750-3841.2007.00636.x.

Hurber, G.R., 2000. Twin screw extruders. In: Riaz, M.N. (Ed.), Extruders in Food Applications. CRC Press, 81–114.

Ilo, S., Schoenlechner, R., Berghofe, E., 2000. Role of lipids in the extrusion cooking processes. Instituto de la Grasa, Austria Grasas Y Aceites. 0017-3495. 51 (1-2), 97–110. Available from: http://grasasyaceites.revistas.csic.es/index.php/grasasyaceites/issue/archive. doi:10.3989/gya.2000.v51.i1-2.410.

Insta Pro International. 2016. Die land length. Insta Pro International, Des Moines, IA (technical bulletin).

Iwe, M.O., van Zuilichem, D.J., Ngoddy, P.O., Lammers, W., 2001. Amino acid and protein dispersibility index (PDI) of mixtures of extruded soy and sweet potato flours. LWT Food Sci. Technol. 0023-6438. 34 (2), 71–75. doi:10.1006/fstl.2000.0731.

Iwe, M.O., Van Zuilichem, D.J., Stolp, W., Ngoddy, P.O., 2004. Effect of extrusion cooking of soy-sweet potato mixtures on available lysine content and browning index of extrudates. J. Food Eng. 62 (2), 143–150. doi:10.1016/S0260-8774(03)00212-7.

Jauncey, K., Sorensen, P., Areola, F., 2008. A Short Handbook – Catfish Feed For Nigeria Materials Storage, Handling, Stock and Quality Control, Ingredients, Feed Formulation, Feed Use (Feeding Husbandry) and Technical Aspects of Feed Manufacture. Sponsored by the CDE, Brussels, September 2007. [online]. http://www.aquafeed.com/documents/1221089997_1.pdf (Accessed 18 December 2013).

Kannadhason, S., Muthukumarappan, K., Rosentrater, K.A., 2011. Effect of starch sources and protein content on extruded aquaculture feed containing DDGS. Food Bioprocess Technol. 1935-5149. 4 (2), 282–294. doi:10.1007/s11947-008-0177-4.

Kaushik, S.J., Cravedi, J.P., Lalles, J.P., Sumpter, J., Fauconneau, B., Laroche, M., 1995. Partial or total replacement of fish meal by soybean protein on growth, protein utilization, potential estrogenic or antigenic effects, cholesterolemia and flesh quality in rainbow trout, *Oncorhynchus mykiss*. Aquaculture. 0044-8486. 133 (3-4), 257–274. doi:10.1016/0044-8486(94)00403-b.

Kaushik, S.J., Médale, F., 1994. Energy requirements, utilization and dietary supply to salmonids. Aquaculture 124, 81–97.

Kazemzadeh, M., 1991. Engineering fish foods. Extra Commun. 1, 12–13.

Kearns, J., 2012. Extrusion cooking systems for feed production. In: Feed Pelleting Reference Guide. WATT Global Media, Rockford, IL.

Killeit, U., 1994. Vitamin retention in extrusion cooking. Food Chem. 0308-8146. 49 (2), 149–155. doi:10.1016/0308-8146(94)90151-1.

Kraugerud, O.F., Jørgensen, H.Y., Svihus, B., Storebakken, T., 2011. Physical properties of extruded fish feed with inclusion of different plant (legumes, oilseeds, or cereals) meals. Anim. Feed Sci. Technol. 163, 244–254. doi:10.1016/j.anifeedsci.2010.11.010.

Krogdahl, Å., Bakke-McKellep, A.M., Baeverfjord, G., 2003. Effects of graded levels of standard soybean meal on intestinal structure, mucosal enzyme activities, and pancreatic response in Atlantic salmon (*Salmo salar* L.). Aquac. Nutr. 1353-5773. 9 (6), 361–371. doi:10.1046/j.1365-2095.2003.00264.x.

Lan, W., Ma, L., 2012. On producing small expanded pellets by single-screw aquafeed extruder. My View 5 (1), 2–6.

Lim, S.J., Lee, K.-J., 2009. Partial replacement of fish meal by cottonseed meal and soybean meal with iron and phytase supplementation for parrot fish, Oplegnathus fasciatus. Asian-Australas. J. Anim. Sci. 22, 1036–1043.

Liu, K., Rosentrater, K.A., 2011. Distillers Grains: production, Properties, and Utilization. In: Liu, K., Rosentrater, K.A. (Eds.), Technology & Engineering. CRC Press, pp. 1–8.

Liu, K.S., Hsieh, F.H., 2008. Protein–protein interactions during high-moisture extrusion for fibrous meat analogues and comparison of protein solubility methods using different solvent systems. J. Agric. Food Chem. 56, 2681–2687. doi:10.1021/jf073343q.

Ljøkjel, K., Harstad, O.M., Skrede, A., 2000. Effect of heat treatment of soybean meal and fish meal on amino acid digestibility in mink and dairy cows. Anim. Feed Sci. Technol. 84 (1-2), 83–95. doi:10.1016/S0377-8401(00)00104-8.

Makine, S.S., 2022. Die design in feed extrusion: The variables and how they affect the end product. Sarmak Sanayi Makine, Konya, Turkey (technical article).

Mercier, C., Linko, P., Harper, J.M., 1989. Extrusion cooking of starch and starchy products. In: Mercier, C., Linko, P., Harper, J.M. (Eds.), Extrusion Cooking. American Association of Cereal Chemists, Inc., St. Paul, MN, pp. 247–319.

Mjoun, K., Rosentrater, K.A., 2011. Extruded aquafeeds containing distillers dried grains with solubles: effects on extrudate properties and processing behaviour. J. Sci. Food Agric. 1097-0010. 91 (15), 2865–2874. doi:10.1002/jsfa.4536.

Moscicki, L., 2011. Extrusion-Cooking Techniques: Applications, Theory and Sustainability. Wiley-VCH Verlag GmbH & Co. KGaA, Weinheim, Germany.

Mottaz, J., Bruyas, L., 2001. Optimised thermal performance in extrusion. In: Guy, R., Ed., Extrusion Cooking Technologies and Application. Woodhead publishing limited, Cambridge, England, pp. 51–82.

Munz, K., 2008. Pet food production'. printed materials of the Workshop Petfood/Aquafeed. 24 September. Uzwil, Switzerland.

Murray, H.M., Lall, S.P., Rajaselvam, R., Boutilier, L.A., Blanchard, B., Flight, R.M., Colombo, S., Mohindra, V., Douglas, S.E., 2010. A nutrigenomic analysis of intestinal response to partial soybean meal replacement in diets for juvenile Atlantic halibut, *Hippoglossus hippoglossus*, L. Aquaculture. 0044-8486. 298 (3-4), 282–293. doi:10.1016/j.aquaculture.2009.11.001.

National Research Council (NRC). 2011. Nutrient Requirements of Fish and Shrimp. National Academies Press, Washington, DC.

Naylor, R., Burke, M., 2005. Aquaculture and ocean resources: raising tigers of the sea. Annu. Rev. Environ. Resour. 1543-5938. 30, 185–218. doi:10.1146/annurev.energy.30.081804.121034.

Obaldo, L.G., Dominy, W.G., Terpstra, J., Cody, J., Behnke, K., 1998. Does size matter? Int. Aquafeed 1, 29–32.

Olli, J.J., Krogdahl, A., 1994. Nutritive value of four soybean products as protein sources in diets for rainbow trout (*Oncorhynchus mykiss*, Walbaum) reared in fresh water. Acta Agric. Scand. A Anim. Sci. 1651-1972. 44 (3), 185–192. doi:10.1080/09064709409410896.

Opstvedt, J., Nygård, E., Samuelsen, T.A., Venturini, G., Luzzana, U., Mundheim, H., 2003. Effect on protein digestibility of different processing conditions in the production of fish meal and fish feed. J. Sci. Food Agric. 83 (8), 775–782. doi:10.1002/jsfa.1396.

Øverland, M., Romarheim, O.H., Hovin, M., Storebakken, T., Skrede, A., 2006. Apparent total tract digestibility of unprocessed and extruded diets containing basic and autolyzed bacterial protein meal grown on natural gas in mink and rainbow trout. Anim. Feed Sci. Technol. 0377-8401. 129 (3-4), 237–251. doi:10.1016/j.anifeedsci.2005.12.017.

Papatryphon, E., Howell, R.A., Soares, J.H., 1999. Growth and mineral absorption by striped bass *Morone saxatilis* fed a plant feedstuff based diet supplemented with phytase. J. World Aquac. Soc. 0893-8849. 30 (2), 161–173. Available from: http://www.blackwellpublishing.com/aims.asp?ref=0893-8849. doi:10.1111/j.1749-7345.1999.tb00863.x.

Perez, V., 2004. Extrusion process and the effect on starch. (MSc Thesis). Norwegian University of Life Sciences, pp.29.

Pongmaneerat, J., Watanabe, T., 1992. Utilization of soybean meal as protein source in diets for rainbow trout. Nippon Suisan Gakkaishi. 1349-998X. 58 (10), 1983–1990. doi:10.2331/suisan.58.1983.

Pritchard, G.I., 1976. Structured aquaculture development with a Canadian perspective. Fisheries Research Board of Canada Report No. 6, pp. 1–16.

Raw Materials and Their Impact on the Extrusion of Aqua Feeds. 2007.

Refstie, S., Sahlström, S., Bråthen, E., Baeverfjord, G., Krogedal, P., 2005. Lactic acid fermentation eliminates indigestible carbohydrates and antinutritional factors in soybean meal for Atlantic salmon (Salmo salar). Aquaculture. 0044-8486. 246 (1-4), 331–345. Available from: https://www.sciencedirect.com/science/journal/00448486. doi:10.1016/j.aquaculture.2005.01.001.

Riaz, M.N., 1977. Using extrusion to make floating and sinking fish feed: controlling the water stability of feed. Feed Manag. 48 (1), 21–24.

Riaz, M.N., 2000. Extruders in Food Applications, CRC Press.

Riaz, M.N., 2003. Extrusion of aquafeed using alternative protein sources. Alternative Protein Sources for Aquafeeds Workshop.

Riaz, M.N., 2009b. Advances in Aquaculture Feed Extrusion'. In: Paper Presented at the 17th Annual ASAIM SEA Feed Technology and Nutrition Workshop. 15–19 Hue, Vietnam.

Riaz, M.N., 2012. Extrusion technology for the production of micro-aquatic feeds and shrimp feeds. Int. Aquafeed 15 (6), 13–19.

Riaz, M.N., 2013. Maintaining Ingredient Quality in Extruded Feeds [online] http://en.engormix.com/MA-feed-machinery/manufacturing/articles/maintaining-ingredient-quality-extruded-t2754/801-p0.htm (Accessed 12 December 2013).

Riaz, M.N., Asif, M., Ali, R., 2009. Stability of vitamins during extrusion. Crit. Rev. Food Sci. Nutr. 1549-7852. 49 (4), 361–368. doi:10.1080/10408390802067290.

Riaz, M.N., 2008. Extrusion of aquafeeds formulated with alternative protein sources. In: Lim, C., Webster, C.D., Lee, C.-S. (Eds.), Alternative Protein Sources in Aquaculture Diets. The Haworth Press, New York, NY, pp. 523–546.

Rodehutscord, M., Pfeffer, E., 1995. Effects of supplemental microbial phytase on phosphorus digestibility and utilization in rainbow trout (oncorhynchus mykiss). Water Sci. Technol. 0273-1223. 31 (10), 143–147. doi:10.2166/wst.1995.0371.

Rokey, G., Huber, G., 1994. 'Extrusion processing of aquaculture feeds'. In: McEllhiney, R.R., Ed., Feed Manufacturing Technology IV. AFIA, Inc, Arlington, VA, pp. 509–515.

Rolfe, L.A., Huff, H.E., Hsieh, F., 2001. Effects of particle size and processing variables on the properties of an extruded catfish feed. J. Aquat. Food Prod. Technol. 10 (3), 21–34. doi:10.1300/J030v10n03_03.

Rolland, M., Lall, S.P., Tibbetts, S.M., 2023. Essential amino acid requirements of fish and crustaceans: A meta-analysis. Reviews in Aquaculture.

Romarheim, O.H., Skrede, A., Gao, Y., Krogdahl, A., Denstadli, V., Lilleeng, E., Storebakken, T., 2006. Comparison of white flakes and toasted soybean meal partly replacing fish meal as protein source in extruded feed for rainbow trout (*Oncorhynchus mykiss*). Aquaculture. 0044-8486. 256 (1-4), 354–364. doi:10.1016/j.aquaculture.2006.02.006.

Rosentrater, K.A., 2006. Some physical properties of distillers dried grains with solubles (DDGS). Appl. Eng. Agric. 1943-7838. 22 (4), 589–595. doi:10.13031/2013.21212.

Rosentrater, K.A., Muthukumarappan, K., Kannadhason, S., 2009a. Effects of ingredients and extrusion parameters on properties of aquafeeds containing DDGS and corn starch. J. Aquac. Feed Sci. Nutr. 1 (2), 44–60.

Rosentrater, K.A., Muthukumarappan, K., Kannadhason, S., 2009b. Effects of ingredients and extrusion parameters on aquafeeds containing DDGS and potato starch. J. Aquac. Feed Sci. Nutr. 1 (1), 22–38.

Rumsey, G.L., Ketola, H.G., 1975. Amino acid requirements of rainbow trout. J. Nutr. series of papers providing lysine anchored essential amino acid profiles.

Samuelsen, T.A., Mjøs, S.A., Oterhals, Å., 2013. Impact of variability in fishmeal physico-chemical properties on the extrusion process, starch gelatinization and pellet durability and hardness. Anim. Feed Sci. Technol. 0377-8401. 179 (1-4), 77–84. doi:10.1016/j.anifeedsci.2012.10.009.

Santiago, C.B., Lovell, R.T., 1988. Amino acid requirements for growth of Nile tilapia. Aquaculture 70, 1–13.

Schaeffer, T.W., Brown, M.L., Rosentrater, K.A., 2011. Effects of dietary distillers dried grains with solubles and soybean meal on extruded pellet characteristics and growth responses of juvenile yellow perch. N. Am. J. Aquac. 1548-8454. 73 (3), 270–278. doi:10.1080/15222055.2011.593461.

Shukla, C.Y., Muthukumarappan, K., Julson, J.L., 2005. Effect of single-screw extruder die temperature, amount of distillers dried grains with solubles (DDGS), and initial moisture content on extrudates. Cereal Chem. 0009-0352. 82 (1), 34–37. doi:10.1094/cc-82-0034.

Singh, S., Gamlath, S., Wakeling, L., 2007. Nutritional aspects of food extrusion: a review. Int. J. Food Sci. Technol. 1365-2621. 42 (8), 916–929. doi:10.1111/j.1365-2621.2006.01309.x.

Sorensen, M., 2007. Ingredient formulation and extrusion processing parameters interferes with nutritional and physical quality of aqua feeds. Feed Technol. Update 2, 17–20.

Sørensen, M., Ljøkjel, K., Storebakken, T., Shearer, K.D., Skrede, A., 2002. Apparent digestibility of protein, amino acids and energy in rainbow trout (*Oncorhynchus mykiss*) fed a fish meal based diet extruded at different temperatures. Aquaculture. 0044-8486. 211 (1-4), 215–225. doi:10.1016/s0044-8486(01)00887-0.

Sørensen, M., Stjepanovic, N., Romarheim, O.H., Krekling, T., Storebakken, T., 2009. Soybean meal improves the physical quality of extruded fish feed. Anim. Feed Sci. Technol. 0377-8401. 149 (1-2), 149–161. doi:10.1016/j.anifeedsci.2008.05.010.

Sørensen, M., Storebakken, T., Shearer, K.D., 2005. Digestibility, growth and nutrient retention in rainbow trout (*Oncorhynchus mykiss*) fed diets extruded at two different temperatures. Aquac. Nutr. 11 (4), 251–256. doi:10.1111/j.1365-2095.2005.00347.x.

Storebakken, T., Shearer, K.D., Roem, A.J., 1998. Availability of protein, phosphorus and other elements in fish meal, soy-protein concentrate and phytase-treated soy-protein-concentrate-based diets to Atlantic salmon, *Salmo salar*. Aquaculture 161, 365–379. doi:10.1016/S0044-8486(97)00284-61-4.

Sugiura, S.H., Gabaudan, J., Dong, F.M., Hardy, R.W., 2001. Dietary microbial phytase supplementation and the utilization of phosphorus, trace minerals and protein by rainbow trout [*Oncorhynchus mykiss* (Walbaum)] fed soybean meal-based diets. Aquac. Res. 1355-557X. 32 (7), 583–592. doi:10.1046/j.1365-2109.2001.00581.x.

Tacon, A.G.J., 1992. Nutritional fish pathology. FAO Fisheries Technical Paper (includes comparative amino acid requirement tables for common carp, catfish and tilapia). FAO, Rome.

Tumuluru, J.S., Sokhansanj, S., 2008. Optimization of aqua feed pellet properties using genetic algorithm. In: Paper Presented at the ASABE Annual International Meeting. American

Society of Agricultural and Biological Engineer. Providence, Rhode Island. 29 June–2 July. Paper No. 084218

U.S. Grains Council. 2018. Amino acid requirements of fish. In: Aquafeed Production, Chapter 13. U.S. Grains Council, Washington, DC.

Umar, S., Kamarudin, M., Ramezani-Fard, E., 2013. Physical properties of extruded aquafeed with a combination of sago and tapioca starches at different moisture contents. Anim. Feed Sci. Technol. 183 (1-2), 51–55. doi:10.1016/j.anifeedsci.2013.03.009.

Verbeek, C.J.R., Berg, L.E.V.D., 2010. Extrusion processing and properties of protein-based thermoplastics. Macromol. Mater. Eng. 1439-2054. 295 (1), 10–21. Available from: http://www3.interscience.wiley.com/cgi-bin/fulltext/122674997/PDFSTART. doi: 10. 1002/mame.200900167

Vielma, J., Lall, S.P., Koskela, J., Schöner, F.J., Mattila, P., 1998. Effects of dietary phytase and cholecalciferol on phosphorus bioavailability in rainbow trout (*Oncorhynchus mykiss*). Aquaculture 163 (3-4), 309–323. doi:10.1016/S0044-8486(98)00240-3.

Vielma, J., Ruohonen, K., Gabaudan, J., Vogel, K., 2004. Top-spraying soybean meal-based diets with phytase improves protein and mineral digestibilities but not lysine utilization in rainbow trout, *Oncorhynchus mykiss* (Walbaum). Aquac. Res. 1355-557X. 35 (10), 955–964. doi:10.1111/j.1365-2109.2004.01106.x.

Wilson, R.P., 1989. Amino acids and proteins. In: Halver, J.E. (Ed.), Fish Nutrition, second ed. Academic Press, San Diego, CA, pp. 111–151.

Wood, J.T., 1994. Selecting equipment for producing farm-made aquafeeds. In: New, M.B., Tacon, A.J.G., Csavas, I. (Eds.), Farm-made Aquafeeds. FAO Fisheries Technical Paper No. 343, FAO, Rome, Italy, pp. 135–145.

Yuan, Y.C., Lin, Y.C., Yang, H.J., GONG, Y., Gong, S.Y., Yu, D.H., 2013. Evaluation of fermented soybean meal in the practical diets for juvenile Chinese sucker, *Myxocyprinus asiaticus*. Aquac. Nutr. 1353-5773. 19 (1), 74–83. doi:10.1111/j.1365-2095.2012.00939.x.

Zhou, F., Song, W., Shao, Q., Peng, X., Xiao, J., Hua, Y., Owari, B.N., Zhang, T., Ng, W.K., 2011. Partial replacement of fish meal by fermented soybean meal in diets for black sea bream, *Acanthopagrus schlegelii*, Juveniles. J. World Aquac. Soc. 1749-7345. 42 (2), 184–197. doi:10.1111/j.1749-7345.2011.00455.x.

Development of pet foods by extrusion processing

15

Ed de Souza
Wenger Mfg, Inc., Sabetha, KS, United States

15.1 Introduction

In the realm of pet food manufacturing, the cooking extruder stands as a pivotal piece of equipment. It is defined by Rokey (2000) as "a continuous pressure cooker apparatus." This device is integral to the process, utilizing moisture, pressure, temperature, and mechanical shear to cook moistened, starchy, and proteinaceous materials. The primary objective is to produce pet food that is not only nutritious but also consistent in quality, while optimizing energy use and minimizing waste.

15.2 Key operating parameters of extruders

To ensure the efficient operation of an extruder, it is essential to understand its operational range and capabilities. Typically, extruders are designed to operate most effectively within 70% to 90% of their maximum capacity. This range allows for optimal processing conditions, where the feed rate must be kept steady to maintain a stable conveying process and achieve the desired product characteristics. Adjustments in the open area of the die may be necessary when changing feed rates to ensure consistent product quality. Operating an extruder beyond its designed capacity or at minimal loads can lead to instability, resulting in products that do not meet specifications or are defective.

15.2.1 Components of the extrusion cooking system

A comprehensive understanding of the extrusion system is vital for the configuration and maintenance of the equipment to achieve maximum performance (Fig. 15.1). The main components of an extrusion cooking system include the following.

15.2.2 Feeding system

While there are other feeding system designs, there is a need to focus on the design that fits most of the pet food cooking extruders in the industry, which is the bin and feeder screw design. The feeding system's role is to deliver a steady and uniform flow

Extrusion Cooking for Food and Feed. https://doi.org/10.1016/B978-0-443-29987-2.00015-4

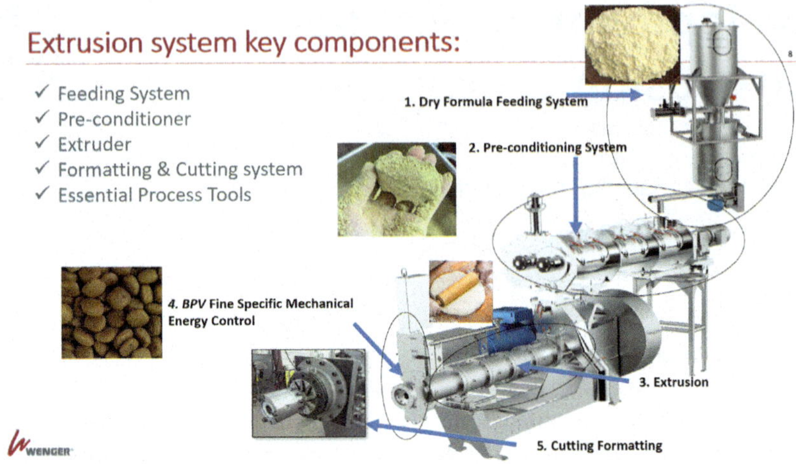

Fig. 15.1 Diagram of the extrusion process with key components. Courtesy of Wenger Manufacturing, LLC.

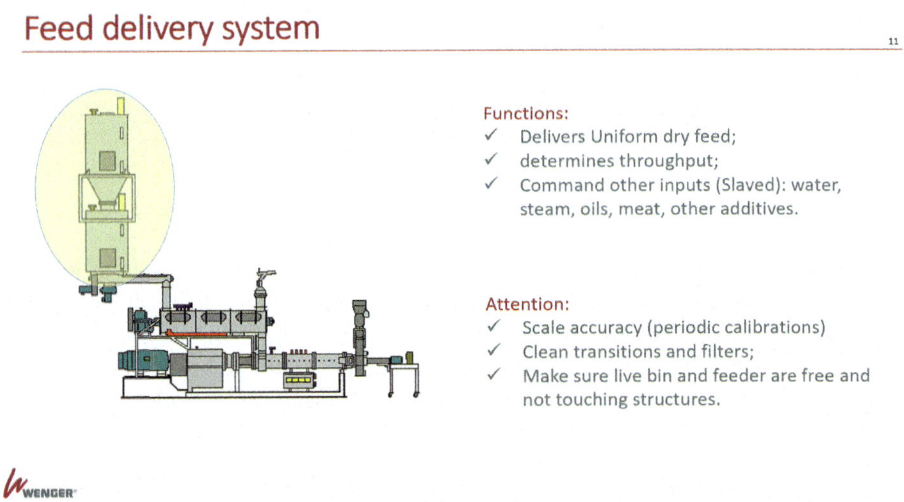

Fig. 15.2 Functions and attention points of feed delivery systems in the extrusion process. Courtesy of Wenger Manufacturing, LLC.

of finely ground mixed ingredients to the extruder inlet or preconditioner. It typically consists of a surge bin, a holding bin (also known as a feeding bin or live bottom bin), and a feeder screw (Fig. 15.2).

- **Surge bin:** Ensures a continuous supply of material to the feeding live bottom bin, maintaining a ready-to-extrude formula.
- **Live bin and feeder screw:** The live bottom bin, commonly referred to as the live bin, features a rotating cone at its base. This design helps prevent materials from bridging and ensures a consistent weight over the feeder screw. The live bottom bin feeds material to the

feeder screw, which accurately controls the feed rate into the preconditioner or extruder barrel. Feeding systems can be designed to operate either volumetrically or gravimetrically:

- **Volumetric feeding systems:** These systems control the feed rate based on the volume delivered, calculated from the feeder screw speed, the bulk density of the formula, and the screw design characteristics. The operator sets the desired speed of the feeder screw using a variable frequency drive; the speed is given in hertz, revolutions per minute (RPM), or both. The volumetric feeding system, at times called just a manual system, could present variations due to the bulk densities of raw materials, as well as changes in head weight pressure on the bin above the feeding screw as it increases or decreases. For example, when the bin feeding the feeder screw is full, there is more pressure on the feeder screw turning at a fixed speed, and the compressible nature of the materials in the formulation will pack more product per revolution of the feeder screw, thus delivering slightly more feed per revolution when compared to a half-full or nearly empty bin.
- **Gravimetric feeding systems:** These more sophisticated systems measure the weight of the materials in the bin and feeder through load cells, adjusting the feed rate based on the weight loss over time of the mash delivered by the feeder screw to further process. This method, known as the loss-in-weight (L-I-W) feeding system, allows for precise control of the feed rate, often achieving accuracy within $\pm 1\%$, and sometimes as precise as $\pm 0.25\%$ variation around the desired feed rate. Sophisticated L-I-W control software can measure the bulk density of the feeds over time and compensate for product weight during refills by slowing down the feeder speed to maintain a consistent delivery of feeds downstream.

15.3 Understanding feed delivery systems to measure and optimize the extrusion process in pet food production

To ensure the precise and efficient extrusion of pet foods, it is crucial to monitor and optimize several key parameters. A list of key parameters includes the stability of all inputs into the process for moisture and heat, ingredient selection and preparation, system configuration, and upkeep. Among these, we explore the dry feed rate, formula bulk density, and the feeder screw RPM, as these play pivotal roles at the beginning of the extrusion step.

15.3.1 Determining the screw factor

The screw factor is an essential parameter in the extrusion process, representing the relationship between the feed rate, bulk density, and screw feeder RPM. It is possible to calculate the screw factor using the following equation:

$$\text{Screw factor} = \frac{\text{Feed Rate}}{\text{Bulk Density} \times \text{Screw Feeder RPM}} \tag{15.1}$$

where:

- **Feed rate**: This is the mass flow rate of dry ingredients into the extruder, typically measured in kilograms per hour (kg/h). Accurate measurement of the dry feed rate is essential for maintaining the consistency of the product.

- **Bulk density**: This refers to the mass of the dry formula per unit volume, often typically expressed in grams per liter (g/L) or, at times, in kilograms per cubic meter (kg/m^3). It is a critical property that influences the volume, mass flow, handling, and processing characteristics of the feed material.
- **Screw feeder RPM**: The rotational speed of the screw feeder, measured in revolution per minute (RPM), dictates the rate at which materials are moved through the system. This can be measured using a tachometer, a device designed to measure rotational speed.

By calculating the screw factor, understanding the equation components, and monitoring the performance of the feeding system, operators can gauge the efficiency and accuracy of the feeding delivery system. A precise screw factor ensures that the correct amount of material is consistently delivered into the extruder, which is vital for producing high-quality pet food.

15.3.2 Importance of a steady dry feed rate

The foundation of a stable and efficient extrusion process is an accurate and steady dry feed rate. This consistency is crucial for achieving uniform cooking and texturing of pet food, as fluctuations in the feed rate can lead to variations in product quality. A well-calibrated feeding system, therefore, is the cornerstone of reliable extrusion operations. All subsequent additions to the process (water, steam, additives) will typically be adjusted as a percentage of the dry feed rate.

15.3.3 Monitoring and performance optimization for feeding system

Regular monitoring of the dry feed rate and associated parameters allows for continuous assessment and optimization of the extrusion process. This proactive approach helps in identifying potential issues early, such as wear in feeder components or inconsistencies in material feeding properties, and enables timely corrective actions. This calculation is essential for monitoring the accuracy and performance of the feeding system. The foundation of a consistent and efficient extrusion cooking process for pet food lies in a precise and steady dry feed rate delivery system. This ensures that the extrusion equipment operates under optimal conditions, producing high-quality pet food with consistent nutritional and physical properties.

15.3.3.1 Practical example

Consider a practical example with a dry feed rate of 1000 kg per hour and a bulk density of 550 g/L on the dry components of a standard dog food formula that might include ingredients like corn (for starch, protein, fiber, and fats), meat and bone meal (for proteins and fats), corn gluten meal (for protein and fibers), as well as other sources of starch, protein, and fats mainly. Accurate control of the feed rate is crucial in this scenario to ensure that the extruder operates efficiently and produces high-quality pet food consistently.

15.3.3.2 Adult dog food standard formulation

The standard formulation, as an example, is for adult dog food and includes the following ingredients:

- Corn at 40% of total: 400 kg
- Meat and bone meal at 30%: 300 kg
- Corn gluten meal at 15%: 150 kg
- Wheat by-products at 10%: 100 kg
- Tallow, salt, and vitamins at 5%: 50 kg

Batch total: 1000 kg

Disclaimer: The formula for the process learning example is not nutritionally balanced.

For a production rate of 1000 kg/h, based on a dry bulk density of 550 g/L, a screw feeder with a diameter of 4 inches (101.6 mm) is used. The screw feeder has a predefined screw factor (SF) of 0.031 m^3 per revolution.

To verify the feeder's accuracy in delivering the dry feed rate of 1000 kg/h, the equation is used:

$$\text{Dry Feed Rate} = \text{RPM} \times \text{Bulk Density} \times \text{Screw Factor} \tag{15.2}$$

Substituting the known values

$$1000 \ \text{kg/h} = \text{RPM} \times 550 \ \text{g/l} \times 0.031 \ \text{m}^3/\text{rev}$$

Solving for RPM

$$\text{RPM} = \frac{1000}{550 \times 0.031}$$

$$\text{RPM} \approx 60.6 \ \text{or} \ 61 \ \text{RPM}$$

Thus to achieve a dry feed rate of 1000 kg/h with a bulk density of 550 g/L, the feeder screw with a diameter of 101.6 mm should operate at approximately 61 rpm.

If there is a dry feed rate, the bulk density of the formula, and the screw feeder RPM, which can be measured with a tachometer, then the screw factor can be calculated using Eq. 15.1.

By using these calculations, one can monitor the accuracy and performance of a feeding delivery system. Reinforcing this, the first crucial step toward achieving a consistent and efficient pet food cooking process through extrusion is having an accurate and steady dry feed rate delivery system.

The equation is beneficial when a system operates volumetrically or during startup and refills in an L-I-W feeding system. During these situations, particularly, the feeding system is running in volumetric mode.

15.4 Preconditioner

The preconditioner serves as a crucial step in pet food extrusion, where the dry feed, delivered by the feeding system through the feeder screw, is mixed with water and steam, leading to hydration, heating, partial cooking of the starchy components of the formula, and homogenization of all the ingredients. This step prepares the material for further processing in the extruder barrel, ensuring uniformity and consistency in the final product. The preconditioner, also known as the conditioner cylinder, is the second component in an extrusion cooking system. It is important to note that a preconditioning system is not necessary for all extrusion cooking processes. If the materials being processed require 17% or less total in-barrel moisture, the preconditioning step is generally not needed. However, when cooking pet foods, moisture levels of 25%–30% are often used during the preconditioning phase for 28%–32% average total in-barrel moisture. We discuss the calculations for preconditioner and in-barrel process moisture in detail later in this chapter.

15.4.1 Evolution of preconditioners in pet food extrusion

15.4.1.1 Single-shaft preconditioners

In the early years of pet food extrusion technology, single-shaft preconditioners emerged as a pioneering design. As the name suggests, these preconditioners feature a single shaft equipped with beaters or paddles threaded along it. Developed in the early 1950s, this design introduced a new dimension to the cooking of pet foods by allowing for mixing, hydration, and heating of the extrudate before it entered the extruder barrel. However, single-shaft preconditioners typically had a low retention time (RT), generally around 30 s, when compared to newer designs that typically exceed 120 s. The relatively short RT is due to their simple structure and the limited material holding capacity of the vessel equipped with a single-pass shaft with beaters or paddles.

The single-shaft design posed some challenges in balancing mixing efficiency and RT. Increasing the shaft speed of the single-shaft preconditioner to enhance mixing often results in reduced RT, as materials travel more quickly through the vessel. In addition, the coefficient of variation (CV), which measures the uniformity of the mixture, was often high due to these limitations.

To address these issues, manufacturers experimented with increasing the volume of the preconditioner housing or stacking multiple single-shaft units. This approach aimed to extend RT and improve mixing and cooking. However, these modifications brought their own challenges, especially with high-starch diets, which could become sticky and difficult to handle, leading to process inefficiencies and potential losses.

Once excessive energy is built up, it takes time to reduce the process energy, causing process losses. These challenges prompted the pioneers in extrusion hardware to look for improvements in the design of these early preconditioners to overcome the identified processing issues.

15.4.1.2 Dual-shaft preconditioners

The dual-shaft preconditioner, also known as the double-shaft cylinder or dual-conditioning cylinder (DCC), was an early improvement over the single-shaft design. This innovation introduced two shafts into a single vessel, significantly enhancing RT and improving the mixing, hydration, heating, and cooking of starchy carbohydrates and functional proteins.

The dual-shaft preconditioner features two intermeshing shafts, each equipped with beaters or paddles threaded along their length. These shafts rotate inside an eight-shaped housing with dual, identical-diameter chambers. The design ensures that the beaters are positioned at specific angles (beater pitch) on both shafts, which vary along the length of the preconditioner to optimize different stages of the process and overall preconditioning RT.

At the inlet area, the beaters are angled forward, helping to move the product from the inlet toward the middle of the vessel. In the middle section, the beaters are positioned neutrally to maximize mixing, while in the last third, they are angled in reverse to hold the materials inside the vessel, increasing RT. This design typically results in an RT of 45–60 s, an important improvement over the single-shaft design.

Facing the discharge end of the dual-shaft design preconditioner, the left shaft turns counterclockwise and the right shaft clockwise, with the two shafts converging at the bottom of the preconditioner. This rotation pattern helps bring the product toward the center and bottom, facilitating the injection of steam from the bottom up, thereby enhancing the conditioning process (Fig. 15.3).

Fig. 15.3 A modern thermal-twin TT-3630 extruder for high-capacity pet food production with expanded capabilities, equipped with a HIP-1500 pre-conditioner.
Courtesy of Wenger Manufacturing, LLC.

The shaft speeds in these early dual-shaft designs were relatively low, generally below 200 rpm and often <100 rpm. This limitation was largely due to the mechanical and engineering challenges associated with gearing the shafts in a dual design. While increasing the shaft speeds could potentially enhance mixing, it also risked reducing dwell time and spreading out the first-in–first-out curve, leading to a higher CV and a potentially nonhomogeneous mixture, hydration, heating, and cooking.

In summary, while both single- and dual-shaft preconditioners represent significant advancements in the extrusion cooking of pet foods, they also come with inherent challenges. Balancing the need for mixing, RT, and uniformity remains a critical consideration in the design and operation of these systems. As technology advances, continuous improvements aim to address these challenges, optimizing the extrusion process for better quality and efficiency in pet food production.

15.4.1.3 Differential diameter preconditioners

Differential diameter preconditioners (DDCs) represent a significant advancement in preconditioning technology. The innovative design of the DDC features a large shaft that is divided into three distinct phases, each serving a specific purpose in the preconditioning process. It seamlessly integrates a bold design that allows the mixing shaft, positioned on the smaller diameter side of the preconditioner, to rotate at precisely twice the speed of the retention shaft, optimizing performance and efficiency through enhanced mixing and longer dwell time.

15.4.1.4 Three distinct areas of the larger diameter side

Inlet area

The first phase is located at the inlet, where materials enter the preconditioner. This area features a large-diameter shaft equipped with beaters angled forward. These beaters efficiently move the dry, incoming materials forward into the second phase, ensuring smooth passage through the vessel.

Mid-section area

The second phase, positioned halfway down the preconditioner body on the large-diameter shaft, is designed for thorough mixing and retention. In this zone, neutral beaters are employed to halt the forward movement of materials, allowing them to be chopped and mixed thoroughly. This chopping motion is crucial to ensure uniform blending and preparation of materials for the next phase.

Discharge area

The final phase, located at the discharge end of the preconditioner, features beaters configured in reverse. This reverse configuration increases the RT of materials within the preconditioner, enhancing the overall treatment process.

15.4.1.5 Smaller diameter side

On the smaller diameter side, the shaft is primarily configured for maximum mixing and additional RT, except for a few forward-positioned beaters at the inlet area designed to move materials into the preconditioner chamber. Most beaters on this shaft are set at a reverse angle. This configuration directs materials back toward the inlet, promoting additional mixing and extending RT.

15.4.1.6 Shafts speed and efficiency

One of the notable advancements in the DDC's design is the ability to operate the small-diameter mixing shaft at twice the speed of the large-diameter holding shaft. This speed differential, enabled by a special gearbox, significantly enhances mixing efficiency and RT (Fig. 15.4). In the early 1980s, this design innovation allowed for RT s of up to 240 s and a starch gelatinization rate of around 45%, a benchmark in the industry.

The DDC's fixed input speed, with differential speeds for the large and small shafts, provides flexibility in adjusting the preconditioner's operation. Speed adjustments can affect RT and the uniformity of the mixture, influencing factors like the CV and the homogeneity of the final product.

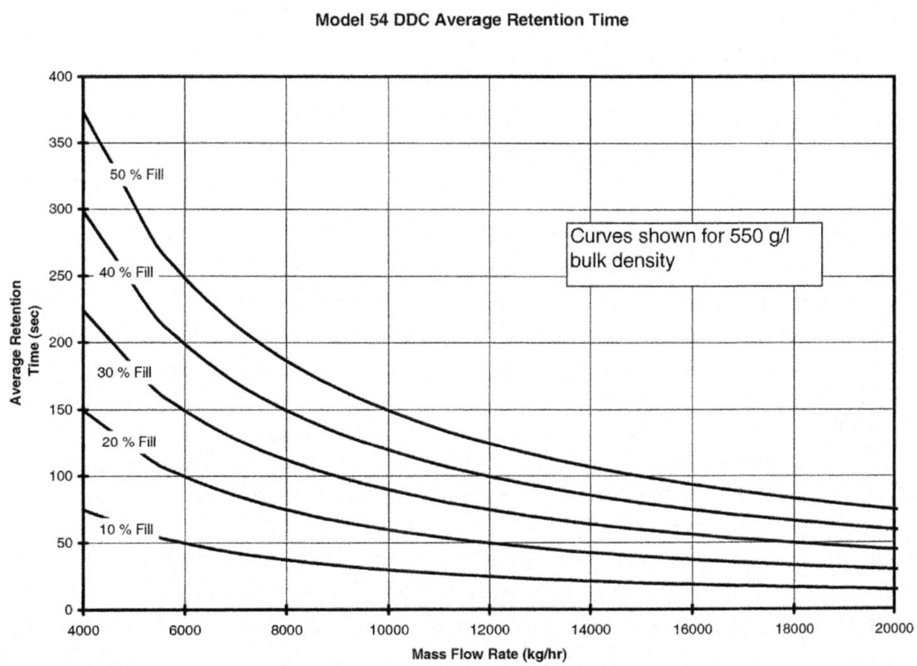

Fig. 15.4 Average retention times of a DCC 54 preconditioner (1500 cubic liters total volume) relative to different mass flow rates using a 550 g/L bulk density recipe.
Courtesy of Wenger Manufacturing, LLC.

The DDC design has become a standard in the preconditioner industry, often serving as a reference model for other brands' preconditioners. Its innovative features and efficiency improvements have set new benchmarks in material preconditioning, making it a crucial component in various industrial processes.

15.4.2 Recent developments in preconditioning technology

15.4.2.1 Conical single-shaft preconditioners

All equipment referenced in this chapter belongs to Wenger Manufacturing, LLC. The high-shear preconditioner (HSC) represents a significant advancement in preconditioning technology. Utilizing a conical, single-shaft design, this preconditioner excels in mixing, heating, and hydrating materials without relying heavily on RT. The design operates by increasing the speed at which materials are moved into a smaller volume area. This action, combined with beaters in the smaller diameter section of the housing, allows for fine control over the mixing process. This results in enhanced cooking precision through controlled adjustments to shaft speed. The high-speed single shaft creates shear and cooking effects, which are particularly effective for recipes high in nonsoluble fiber or that require minimal moisture, reducing the likelihood of stickiness during the preconditioning phase.

15.4.2.2 High-intensity preconditioners

The high-intensity preconditioner (HIP) was developed to meet the growing demand for handling higher moisture content, increased meat inclusion, and a broader range of ingredient viscosities. This advanced preconditioner features two independently driven shafts, offering unparalleled control over the mixing, hydration, and heating of the dry formula ingredients. Inspired by the DDC, the HIP is designed to handle three distinct phases: the first phase involves moving materials from the inlet with slightly forward beaters; the second phase uses neutral beaters for a chopping motion; and the third phase retains materials with reversed beaters, increasing RT. The smaller diameter shaft focuses on mixing, ensuring thorough blending of ingredients.

The HIP's beaters are pressed and welded to the shafts for precision, with a polished food-grade finish, maintaining correct pitch and configuration even at high mixing speeds. Independent drive controls allow for precise adjustments, making it possible to fine-tune the process for various formulations, from high raw meat slurry additions to grain-free diets. The design's capacity for significant steam addition (11%–13% of the dry feed rate) enables it to achieve temperatures up to 95°C, which is crucial for high-demand cooking scenarios, especially when handling sticky recipes or recipes that require high moisture, such as meat slurries. In the HIP Preconditioner, as the name implies, the mixing intensity can be dialed in for a variety of recipes by controlling the speeds of the mixing shaft and the retention shaft individually. Combining the best of both worlds, an increase in mixing intensity will also increase RT. The ability to control mixing intensity typically results in low mixing intensity for recipes that are easier to cook, such as those with tuberous starch sources, commonly found in grain-free diets, as well as high-starch formulas for grain-based diets. On the opposite spectrum, medium to high intensity is typically used for high-moisture recipes and recipes that require longer RT and higher mixing to be properly hydrated and heated, such as those with high meat inclusion or high insoluble fibers.

15.4.2.3 Steam mixing injection (SMI) system

Both the HSC and HIP are equipped with a steam mixing injection (SMI) system, which integrates water and steam before injection into the preconditioned materials. This system helps prevent issues like condensate formation and ensures consistent moisture levels by effectively eliminating steam pipe condensation at the beginning of a run, which can be detrimental to the quality of the product. The SMI system also reduces steam escape and eliminates the need for complex manifolds and multiple injectors, simplifying the process and enhancing efficiency.

In addition, preconditioners may include a condensation unit at the vent to improve overall efficiency and recyling of excessive steam. These modern preconditioners are typically open to the atmosphere to avoid pressurization, which can deteriorate amino acids and vitamins in the materials being processed. These advancements in preconditioning technology offer greater control and efficiency, meeting the diverse needs of modern food processing industries.

15.5 Extruder barrel

The extruder barrel is where the final cooking and shaping occur by means of shear generated by the rotating elements, and, in most cases, additional steam and water are added to the extruder barrel (Fig. 15.5). It conveys the preconditioned material through various stages, where it undergoes further mixing, cooking, and expansion.

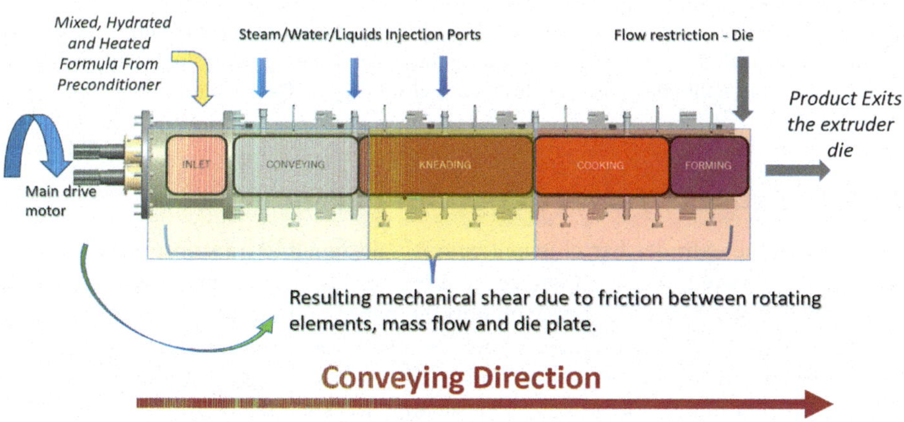

Fig. 15.5 Sections within the extruder barrel according to their functions on the extrudate. Courtesy of Wenger Manufacturing, LLC.

The barrel's design and configuration, including the die setup, play a significant role in defining the texture, size, and shape of the final product. The extruder barrel is generally divided into three distinct sections in the direction of flow: the inlet section, the kneading section, and the final cooking zone.

1. **Inlet section:** The first area of the extruder barrel, typically comprising the first third of the barrel length, is designed to quickly move the product away from the inlet. Ideally, the material at this stage is a free-flowing, heated, mixed, and hydrated mash prepared during the preconditioning stage. As these materials enter the extruder barrel, they are swiftly conveyed toward the midsection.
2. **Kneading zone on the mid-section of the extruder barrel:** After passing through the inlet section, the extrudate enters the mid-section of the extruder barrel. This zone typically features conveying screws designed to ensure that the extrudate continues to move uniformly and positively toward the discharge end, which is also known as the final cooking zone. In the mid-section, the kneading process begins, preparing the extrudate for the melting phase, located at the end of the extruder barrel. The barrel configuration often includes single-flight conveying screws followed by round locks, usually of small diameter. Under the load of the extrudate, a small plug forms before each round lock, which creates a minor pressure drop as the extrudate passes through. This pressure drop area, with its reduced extrudate volume, is an optimal point for steam injection, which enhances the absorption and mixing of steam into the extrudate. Typically, the diameter of the round locks increases as the extrudate moves toward the discharge end of the barrel. This design feature helps create controllable shear forces, which are crucial for consistent kneading and mixing. The round locks serve as barriers that not only keep the extrudate moving forward but also contribute to kneading and mixing as the material overcomes each lock, forming a reference small plug before the lock and experiencing a pressure drop immediately after the lock's position, where steam can be added.
3. **Final cooking zone:** As the extrudate exits the mid-section, it enters the final cooking zone. Ideally, the extrudate flows smoothly into this zone without experiencing surges or stickiness, which can be monitored by the stability of the main drive amperage or power consumption. The correct die setup and open area are crucial for forming a plug in the final zone, ensuring proper barrel fill, and maintaining process stability, where shear forces combine all ingredients through dispersive mixing. This process ensures that the extrudate melts sufficiently to flow out of the die openings uniformly, contributing to product homogeneity. It is crucial to avoid cooking the material too early in the barrel, as this can lead to inconsistent conveying and affect the quality of the final product.

15.5.1 Extruder barrel design for pet food

In summary, the extruder barrel is meticulously designed to transport preconditioned materials—prepared, mixed, hydrated, and heated—to the die area. Throughout this process, the energy input increases due to the barrel and die configuration, extruder screw speed, mass flow, and other variables. These factors collectively contribute to the specific mechanical energy (SME) input into the extrudate. In addition, water and steam are often added to the barrel to account for specific thermal energy (STE), further influencing the extrudate's properties.

15.5.1.1 Single-screw extruder barrel design

The single-screw extruder barrel features a heavy-duty shaft in a simple and robust design. This model generally offers lower capital and maintenance costs. It also provides, typically, a large screw volume, with capacities for standard pet foods reaching up to 30.000 kg/h. However, the single-screw design has some limitations, particularly in conveying extrudates with either highly slippery and/or sticky ingredients when cooked. The single shaft and stack of screws and locks, rotating inside a spirally grooved liner, can generate friction during conveyance. While this design offers good conveying capabilities, it operates best within a narrower range of pet food recipes and ingredients when compared to the twin-screw extruder barrel design, as further described.

Due to the larger screw volume, the single-screw model may require less power for similar capacities compared to a typical twin-screw extruder barrel design. However, in the single-screw design, the increased gap between the screw's outer diameter and the liner's inner diameter can lead to more slippage. As the gap widens with wear, conveying efficiency can decrease sooner than it would in a twin-screw design. Despite these challenges, the single-screw extruder remains a workhorse in the pet food industry, especially for standard formulations. Nevertheless, new ingredients, such as plant proteins, tuberous starch sources, and meat slurries, pose challenges for this design, particularly with high-fat, high-moisture formulations.

15.5.1.2 Twin-screw extruder barrel design

The twin-screw extruder barrel typically features two parallel shafts, though tapered or conical shaft designs also exist. These shafts are equipped with intermeshing screws and shear locks that are fully self-wiping and corotating inside a smooth, segmented barrel. The precision in the construction and metallurgy of the parts, along with tighter tolerances, ensures excellent conveying and high precision in most pet food extrusion applications.

The twin-screw design can handle a wide range of products, offering superior conveying capabilities with less dependence on formula-specific process requirements. It can efficiently process challenging formulations, including those with high moisture, viscosity, or fat content, as well as those prone to stickiness under heat and friction. This design is known for its reliability, excellent wear life, and increased uptime. However, it generally incurs higher capital and maintenance costs and requires more conveying power due to its typically reduced screw volumetric capacity compared to a similar-capacity single-screw extruder barrel design.

15.5.2 Advanced extrusion systems

Cutting-edge technologies are now possible due to significant advances in hardware and software design, machining, and fabrication technology, coupled with pet food processing. Knowledgeable researchers have developed advanced extrusion systems.

Let us take a deeper look at each point, breaking down the technical advantages and potential benefits for pet food applications.

15.5.2.1 Superior uptime and durability

The use of high-wear-resistant metallurgy on a deep-flight twin-screw design and a typically longer barrel, along with the sturdy rotating parts, minimizes maintenance needs, reduces downtime, and extends the service life of the machine. This means less frequent parts replacement, which lowers operating costs. In high-demand settings, such as large-scale pet food production, these durability improvements help meet production targets more reliably and quickly. This feature addresses an up-to-date industry demand for high-efficiency outfits producing high-quality pet foods.

15.5.2.2 User-friendly process control software

- An intuitive design simplifies operation, making it easier for technicians to learn and master the system. This reduces training time and operator error, contributing to smoother and more efficient production processes.
- Operators can adjust settings quickly and with precision, which is useful in situations where product specifications need to be changed frequently, such as a large product portfolio, short production runs, significant ingredient variability, or minor adjustments needed to optimize production.

15.5.2.3 Fast-acting advanced process control tools and responses

- The quick response of a high-end cooking extruder machine to advanced process control tools enables faster tuning, stabilization, and refinement of processes. This responsiveness allows it to achieve consistent product quality more quickly, with fewer trial-and-error adjustments. Faster responsiveness here means fewer resources are consumed during setup, which could give it an edge in operational efficiency.
- In complex operations or highly variable production environments, such as high meat or high starch diets, this feature could reduce waste and improve overall yield by ensuring that optimal settings are reached more quickly. As examples of these modern process tools, the Back Pressure Valve (BPV) is an adjustable valve located at the end of the extruder barrel and before the final die, designed to control the residence time of the extrudate inside the extruder barrel independent of the final die open area, by simply diminishing or enhancing the aperture of the extrudate to the die plate.

15.5.2.4 Flexibility in processing

- Due to a twin-screw barrel design, as noted, it enhances positive and uniform conveying in the barrel configuration, providing flexibility and allowing for uniformity in the extrudate regardless of the product type. This enables an advanced machine to handle different materials and formulations, adapting to various production demands. This feature makes it easier to switch between different types of recipes or product types without sacrificing quality. For companies producing a wide range of products, this flexibility means faster setup and changeovers, improving production efficiency and reducing downtime between product lines.

15.5.2.5 Efficient thermal energy utilization

- By coupling with the newest high mixing intensity preconditioner, these advanced extruders maximize thermal energy use, improving the cooking process through gentle, continued, and positive conveying of the extrudate. This not only saves energy but also enhances product flavor and helps form resistant starch, which is known for supporting gut health in dogs. This machine's ability to integrate thermal energy effectively, especially in pet food production, is a competitive advantage.
- For pet food manufacturers, the enhancement of flavor and gut health benefits through resistant starch formation (Peixoto et al., 2018; Ribeiro et al., 2019; Jackson et al., 2020; Corsato Alvarenga et al., 2021) aligns with current market trends toward health-focused pet foods. This feature supports differentiation in the pet food market by offering value-added nutritional benefits.

15.5.3 Configuring the function of a single-screw cooking extruder barrel

As depicted earlier, an extruder barrel typically consists of three zones, from the inlet to the discharge; however, it is all about conveying, and preferably, ensuring it is positive and constant. Here is a breakdown of how to configure each zone for optimal performance:

15.5.3.1 Barrel inlet opening

At the inlet of the barrel, it is crucial to have a generously sized opening. This prevents the creation of a choking zone, allowing preconditioned, free-flowing, heated, mixed, and hydrated materials to enter the barrel freely and gravimetrically, and to be moved quickly out of the inlet and further into the extruder barrel.

15.5.3.2 Extruder barrel first-zone configuration technique

Under the inlet opening, in the first zone of the extruder barrel, typically, there will be screws with wider pitch flights and deep flighting, high-conveying-featured screws to move the preconditioned product quickly away from the inlet and enhance extrusion capacity. Typically, just before the inlet screw or screws, some extruders may use a configuration technique that generates a slight product plug. This is achieved by installing a specific screw/lock stack configuration halfway between the inlet and the following barrel midsection kneading zone. This setup typically forms a small product plug, locking in-barrel steam injected further down the barrel and enhancing the thermal processing.

15.5.3.3 Kneading zone

The second zone, located around the middle of the barrel, is designed for positive and uniform conveying of the extrudate. This zone typically features single- or double-flight screws with a larger pitch, promoting thorough kneading. Here, the extrudate gains higher temperatures and pressures through moderate shear action, contributing to dispersive mixing and cooking. At the beginning of this zone, steam injectors are normally positioned to add steam to the extrudate, where the internal pressure is still low and the barrel fill is still partial. Generally, a well-designed and configured single-screw extruder barrel should allow the injection of 2%–3% steam into the extrudate as a percentage of the dry feed rate, which significantly enhances thermal energy due to higher steam pressure compared to preconditioner steam.

15.5.3.4 Final cooking zone

Toward the discharge end, before the die configuration, lies the final cooking zone, critical for products like direct-expanded pet foods. This zone may feature tighter-pitch screws and tapered final screws, frequently with a cut-flight design to increase shear and promote product expansion. The configuration can vary based on product requirements, such as for fully cooked and dense products.

15.6 Screw design and terminology

Outside diameter: This is the measurement taken from tip to tip of the screw flighting, indicating the total diameter of the screw. It typically corresponds to the extruder's size classification (e.g., a 144 mm outside diameter screw is used in a TX-144 twin-screw Wenger model).

Root diameter: This refers to the internal diameter where the flights begin. The height of the flight impacts the shear applied to the extrudate. The deeper the flighting or the smaller the root diameter, the more gentle and lower the shear the screw generates; the opposite applies.

Screw flighting: The number of flights or helical turns on a screw can vary, single, double, or triple. Increasing the number of flights enhances RT and the shear applied by the screw.

Screw pitch: This is the distance a flight on the screw covers from the beginning to the end of one complete revolution around the root diameter. If this distance equals the screw diameter, the screw is called a "full pitch" screw; if it's half the diameter, it's called a "half pitch" screw. Full pitch screws are used for high conveying capacity, typically in the inlet area, while tighter pitches are used in the final cooking zone to increase RT and shear.

15.6.1 Versatility of cooking in advanced extruder screw designs

A properly configured screw extruder, particularly single screw models, can produce high-quality pet food products. While single-screw extruders are versatile, twin-screw extruders offer more positive conveying and are generally more forgiving. Twin-screw extruders are capable of precise pressure building and are ideal for producing small, intricate products with precise shapes, as well as handling sticky or slippery recipes.

Traditional twin-screw extruders, such as those pioneered by Wenger Mfg., are reliable and versatile, and are used widely in the pet food industry. The latest generation of twin-screw extrusion systems incorporates advanced features like increased steam injection and volumetric capacity, enhancing their thermal cooking capabilities.

This new family of twin-screw extruders can accommodate a wide variety of product formulations, including high meat content, and features improved steam injection capabilities. These machines can produce up to 13.000 kg per hour, making them suitable for a wide range of food and feed applications. The improvements are related to barrel length, also known as "L/D" or length over diameter, and configuration techniques to allow more volumetric capacity, steam injection mixing, and positive conveying.

15.7 Understanding energy in extrusion processes

The extrusion cooking process relies on two primary energy sources: the electrical energy driving the rotational movement of the shaft inside the extruder barrel, and the thermal energy, mainly from steam injection. The electrical energy generates the rotation of the shaft or shafts with the screws and locks configured on the shaft or shafts, generating friction and mechanically cooking the starches in cereal grains, pulses, and tubers, while the steam ensures the thermal cooking of these ingredients. Balancing these energy inputs is crucial for achieving the desired product characteristics, making extrusion cooking a precise science rather than an art.

SME: It represents the electrical energy converted into mechanical energy at the extruder shaft (single-screw design) or shafts (twin-screw design), driven by the main motor. This energy is transferred to the extrudate through rotating screws, locks, and inside the extruder barrels, which create shear forces. SME is measured in kilowatt-hours per ton (kW-h/ton) and is a crucial factor in determining the amount of energy required per unit of product, indicating the energy required per hour to produce one ton of product (dry feed rate; normally matches dried and cooled product before coating).

1. **Measurement**: Energy consumption is measured in kW-h/ton.
2. **Conversion**: 1 horsepower is approximately 0.750 kW.
3. **Typical requirements**: Producing pet food typically requires 20–30 kW-h/ton of SME. With enhanced STE, the SME can be as low as a single digit.
4. **Extruder capacity**: The energy needs of a formulation determine the capacity of a specific extruder model.

Importance of SME: SME is vital for cooking functional proteins and starchy ingredients, contributing to the product's expansion and shaping. Excessive SME with low in-barrel moisture can degrade product quality, reducing expansion and structural integrity (Baller et al., 2018), leading to fines and breakage. Understanding and controlling SME for different ingredient combinations is crucial for a stable and efficient extrusion process.

STE: It is the energy introduced to the extrusion process, mainly from steam added during preconditioning or directly into the extruder barrel. This energy is essential for cooking the extrudate and is typically measured in kilocalories per kilogram (Kcal/kg), often converted to kW-h/ton for consistency with SME measurements. For SME calculations, refer to Baller et al. (2018).

Equipment knowledge: Knowing the extruder's hardware and operating it within its design limits (typically 70%–90% of its maximum capacity) is essential. Maintaining a steady feed rate with uniform process inputs and adjusting the die's open area as needed ensures a stable process. Overloading or underloading the extruder can lead to inconsistent product quality.

Calculating capacity: As an example, a standard adult dog food may require an average of 22 kW-h/ton of SME. For instance, a typical extruder with a 220-mm screw diameter and a 400-horsepower motor—hence 300 kW (400 × 0.750)—operated at 90% of its motor nominal power [at nominal extruder shaft speed (300–400 rpm for a larger-scale production extruder), optimum motor consumption is typically around 90% of its nominal power] would produce (300 kW × 90%)/22 kW-h/ton = 12 tons/h. Understandably, there might be other limitations to the extruder system, for instance, the volumetric capacity of the screw configuration in the extruder barrel, die open area, formula restrictions, dry extrudate moisture, improper preconditioner or no-preconditioner, wear on screw elements, wear on barrel elements, wear on beaters in the preconditioner, etc.

Variable frequency drives (VFDs): They allow frequency adjustments to optimize motor speed and energy use. For instance, the extruder example above might feature a nominal speed of 400 rpm on the extruder shaft. With the use of a VFD, the speed can be changed from 250 to 500 rpm. Increases in shaft speed will typically generate higher friction, higher cook, expansion, and lower bulk density of the final product. Lowering the shaft speed diminishes the friction and lowers the SME transferred to the extrudate.

Maintenance and quality: High SME is often a standard in the production of pet foods, either due to process requirements or a lack of knowledge. Worn-out extruder components can compromise product quality and capacity, leading to increased moisture requirements and reduced efficiency. Implementing a preventive maintenance program, which is a small investment compared to the overall cost, can significantly reduce losses and maintain product quality. Most process issues are generally caused by worn-out elements, improper extruder barrel configuration, and/or die open area, leading to high mechanical energy in the extrudate.

15.8 Extrusion dies: A critical component in pet food production

Extrusion is a widely used process in the pet food industry. It allows for the efficient production of a variety of kibble shapes and sizes. A key component in this process is the extrusion die, which shapes the extrudate as it exits the extruder barrel.

15.8.1 Understanding extrusion dies

Extrusion dies are specially designed tools that define the shape and size of the final product. In the context of pet food production, dies play a key role in determining the texture, size, and appearance of the kibble. The die consists of a metal plate with one or more holes through which the dough-like mixture is forced under pressure. The size and shape of these holes can be customized based on the desired product specifications.

15.8.2 Key considerations in die design

1. **Material selection**: Dies are typically made from high-strength materials, such as stainless steel or alloy steel, which offer durability and resistance to wear and corrosion. This is critical in maintaining consistent product quality and ensuring a long die lifespan.
2. **Die geometry**: The geometry of the die, including the shape and size of the openings, directly influences the product's characteristics. In pet food, common shapes include round, square, and various novelty shapes, catering to different market preferences.
3. **Flow dynamics**: The design must ensure uniform flow through the die, preventing issues like uneven cooking or structural weaknesses in the final product. This is achieved through careful consideration of the die land (the length of the die holes), which affects the residence time and pressure within the die.

15.8.3 The role of die open area

The open area of the die is an important parameter in extrusion processes. For standard pet food production, a common guideline is to use a die open area of 200 mm^2 per ton/hour of production capacity (dry feed rate). This recommendation is based on several factors:

- **Throughput capacity**: An ideal open area of the die allows for correct throughput, making it possible to produce uniform kibbles. The 200 mm^2 per ton/hour standard provides a balance between throughput and quality, ensuring that the extrudate is properly cooked and shaped.
- **Product quality**: A well-sized die ensures that the extrudate is uniformly cooked and formed, contributing to consistent quality in terms of texture, density, and appearance.
- **Energy efficiency**: Proper die sizing can optimize energy consumption by reducing the resistance encountered by the extrudate, thereby lowering the mechanical and thermal energy required for extrusion.

15.8.4 Practical implications

Implementing the 200 mm^2 per ton/hour guideline requires careful planning and consideration of the entire extrusion system, including the extruder's capabilities (condition) and the specific requirements of the pet food product. Operators must monitor die wear and replace the dies as needed to maintain the recommended surface area. Wear can lead to changes in die geometry, as well as scratches and grooves on the die's surface, which may affect product quality and process efficiency.

15.8.5 The role of back die plates and cone-shaped flow devices

When configuring the final die to produce extruded pet foods, achieving a consistent, high-quality product relies heavily on the control of various factors within the extrusion process. One decisive aspect, as discussed, is the die configuration, which directly influences the texture, shape, and density of the final product. This section explores the benefits of using a back die plate with a single hole, a spacer, and a cone-shaped flow device to optimize streamlined flow during extrusion.

15.8.6 The back die plate

The back die plate serves as a preliminary shaping component in the extrusion process. By using a back die plate with a single hole, typically twice the open area of the final die's open area, the extrudate experiences an initial reduction in flow resistance. This design choice helps in preforming the extrudate, allowing for a more controlled flow as it approaches the final die.

15.8.7 Spacer: creating the optimal distance

Positioning a spacer between the back die plate and the final die is another key consideration. A spacer, ranging from 25 to 50 mm, provides a critical buffer zone. This spacing allows the extrudate to stabilize before passing through the final die, which is particularly important for products requiring precise shapes and textures. The exact spacer length can be adjusted based on the specific flow properties of the extruded material and the desired product characteristics. This intermediate space is central to enhancing the homogeneity of the flow, preventing irregularities and inconsistencies in the final product.

15.8.8 Cone-shaped flow device

Integrating a cone-shaped flow device behind the final die further enhances the streamlined flow. This device, positioned just before the extrudate exits the die, helps to distribute the material evenly across the die opening. The conical shape guides the flow, reducing turbulence and ensuring that the material exits uniformly. This not only improves the surface quality of the extrudate but also minimizes variations in density, size, and texture, which are critical for maintaining product quality and consumer satisfaction.

15.9 Role of ingredients in the extrusion process

In the extrusion process of pet foods, ingredients play a critical role in determining the final product's texture, nutritional content, and overall quality. The Guy Classification System (Guy, 1994) is a useful framework for understanding how different ingredients interact during extrusion and impact the final product.

15.9.1 Starch

Starches are essential in pet food extrusion, primarily acting as binding agents that help shape and structure the kibble. They are polymers with film-forming abilities, capable of withstanding the expansion of water vapor as the product exits the extruder. Starches influence various aspects of the extrusion process, such as:

1. **Degree of expansion**: Starches affect how much the kibble expands during processing.
2. **Moisture requirements**: They play a role in determining the necessary moisture levels.
3. **Thermal energy needs**: The amount of heat required during extrusion is influenced by the starch content.
4. **Viscosity**: Starches contribute to the thickness and flow properties of the kibble mixture.

15.9.1.1 Contribution to kibble formation

As a main component in pet food formulas, starches contribute significantly to the formation, expansion, and adhesion of kibble. They can retain air when cooked and expanded, which helps reduce the bulk weight of the final product. Starches are commonly derived from various sources.

15.9.1.2 Cereal sources

- **Examples**: Corn, rice, sorghum, millet, wheat.
- **Characteristics**: Cereal starch granules are typically polyhedral and smaller in size when compared to tubers, have pores and channels on their surface, and usually contain more amylopectin than amylose, which contributes to their gelatinization properties (Corsato Alvarenga et al., 2021). Generally speaking, higher amylopectin promotes the most expansion but produces weaker extrudate cell walls (Vanier et al., 2016), whereas higher amylose reduces expansion, water absorption, and hardens the texture of extrudates (Hellemans et al., 2020). During extrusion, these starches gelatinize at relatively high temperatures, providing a firm and cohesive structure to the kibble.

15.9.1.3 Tuberous sources

- **Examples**: Potatoes, sweet potatoes, tapioca.
- **Characteristics**: Tuber starches generally have a higher amylopectin content and phosphate monoesters, and their starch structure is different when compared to cereals (tubers are type B, and cereals are type A) (Yoo et al., 2009), making them highly sticky during extrusion. This stickiness helps in creating a smooth texture in the kibble but also poses challenges, such as potential sticking to the extruder and difficulties in maintaining a consistent kibble shape. This is especially important in grain-free diets, where alternative binding agents are needed.

15.9.1.4 Legume sources

- **Examples**: peas, soybeans, lentils, chickpeas.
- **Characteristics**: Legumes' starch molecular structure is type C (a mixture of types A and B), and these ingredients have a thick cell wall and proteins tightly attached to starch granules, which might limit starch gelatinization (Bhattarai et al., 2017). Legumes, or pulses, are also good nutritional sources of vegetable protein that are functional/structure-forming in the extrusion process (Guy, 1994).

In summary, starches play a critical role in the extrusion process of pet food, influencing the texture, shape, expansion, and stability of the final product. The choice of starch sources, whether cereal, tubers, or legumes, can affect the extrusion process and the properties of the kibble.

15.9.2 Protein

Proteins are vital for the nutritional content of pet foods and play a significant role in the extrusion process. They can come from both animal and plant sources, with animal meals and fresh meats being commonly used. Typically, if the protein source has been through a heating and cooking step, such as rendered meals, it will lose partial or all functionality and essentially become dispersed-phase filling materials.

Animal meals: These are concentrated protein sources derived from rendering processes. They include meat and bone, chicken, fish, and lamb meals. Animal meals have a high protein content and a lower moisture level and are typically high in oils and fats. The extended shelf life contributes to making them suitable for extrusion. In an ideally balanced formula, they contribute to the kibble's firmness and higher density, essential for maintaining shape and texture; however, these sources of protein are not as functional to the extrusion process and will act like dispersed-phase fillers. Fillers can influence the expansion and strength of kibbles if the starch content of the formula is not sufficient to compensate for these proteins.

Meat slurries: Incorporating frozen or fresh meats, such as chicken, beef, or fish, provides high-quality protein that is usually functional to the process. Raw meats can enhance palatability but also add complexity to the extrusion process. The high moisture content requires careful balancing with other dry ingredients to ensure proper kibble formation. Raw proteins are usually functional and can help maintain the structural integrity of the kibble; however, they can also make the recipe sticky.

Plant proteins provide functional properties such as water holding capacity and emulsification, which are important in the extrusion process. These properties can affect the moisture content and cohesiveness of the extrudate in a similar way to starches or structure-forming materials.

15.9.3 Fats and fibers

Fats and fibers are crucial for the nutritional balance and functional properties of pet foods.

Fats: They are primarily included for energy density and flavor. Common sources include poultry fat, fish oil, and vegetable oils. Fats can affect the extrusion process by influencing the texture and expansion of the kibble. Excess fats can lead to lubrication issues, affecting the flow and shape consistency, creating bonds with starch and proteins, and weakening the bonds and kibble structure. Fats also interfere with expansion and increase the final product density. At low and controlled inclusions, fats can act as a process aid, controlling the stickiness of the extrudate, improving shape, and controlling the expansion and density of the extruded kibbles. Pet foods usually have intrinsic fats in the ingredients that cause little to no disruption of the process, and most fats will be added at the coating step. The internal fat content in pet food formulations (before coating) typically ranges from 5% to 12% of the total formula, depending on the specific product type, nutritional requirements, and the target species (e.g., dogs or cats). Producers include, after extrusion, additional fats (like animal fats or flavor enhancers) added during coating to enhance palatability, bringing total fat levels to 12%–22% in the final product, ensuring it does not interfere as a lubricant.

Fibers: These include both soluble and insoluble types, sourced from ingredients as examples: co-products from cereals, beet pulp, pea fiber, and cellulose. Fibers are essential for digestive health but also impact the extrusion process. They can influence the kibble's water absorption capacity, density, and mechanical strength. High fiber content may lead to a fragile texture, requiring adjustments in the extrusion parameters and typically much higher SME requirements. Unless super finely ground, below 400 μm, insoluble fibers will decrease expansion and reduce throughput. Insoluble fibers are normally dispersed-phase filler materials. This means they are not easily soluble in water and will interfere with the structure-forming materials for the kibble structure and strength. They reduce expansion and make the cell structure matrix weaker, yielding a rough surface product and generating fines downstream or at the extruder die and cutting system. Meanwhile, soluble fibers can be somewhat functional, providing more expansion and a lower extrudate density when compared to a formula with mostly insoluble fibers (Robin et al., 2012).

15.10 Extrusion software: Precision in process control

Advanced process controls: The integration of advanced software systems for process control is essential. These systems allow for precise adjustments of energy inputs and other critical parameters, ensuring consistent product quality and energy efficiency. Fine-tuning these controls enables the optimization of the extrusion process, reducing variability and improving overall product consistency.

This section outlines the importance of various roles in a process team, including formulators, maintenance crew, decision-makers, quality control, and operators. It emphasizes that the process conditions must be stable, smooth, and continuous, with reliable software generating data for process management. The goal is to produce and replicate high-quality products using a scientific approach, scaling up proportionally and monitoring key performance indicators (KPIs).

15.11 Extrusion process variables

15.11.1 Input variables: These are affected by operator interference

1. Dry recipe feed rate to the preconditioner and mass flow into the extruder barrel
2. Water is injected into the preconditioner and the extruder barrel
3. Steam is injected into the preconditioner and the extruder barrel
4. Preconditioner mixing intensity (shaft speeds)
5. Extruder preconditioner, screw, and barrel configuration
6. Extruder speed
7. Die configuration

15.11.2 Output variables: These are affected by input variables

1. RT in the preconditioner and extruder barrel
2. Temperature in the preconditioner and extrudate temperature at the discharge of the extruder
3. Moisture in the preconditioner and in-barrel moisture
4. Mixing quality of the product exiting the preconditioner, in the absence of clumps

15.11.3 Crucial parameters: Results of the input variables

Moisture, thermal energy input, mechanical energy input, and RT

15.11.4 Product responses: Reaction of the cooked ingredients in the extrusion cooking process

1. Bulk density
2. Cook
3. Size/shape
4. Expansion
5. Oil absorption
6. Hardness/durability
7. Digestibility
8. Palatability

15.12 Ingredients and recipes preparation for the extrusion processing top of form

To optimize nutrition for targeted species, it is crucial to carefully consider the processing of ingredients. This includes ensuring uniform grinding and mixing with a low coefficient of variation, ideally using a double-shaft mixer as the final mixer.

Proper preparation of raw materials, such as sifting, is also essential. Kibble homogeneity guarantees that pets are receiving their appropriate nutrition in every single bite. The techniques outlined in this chapter, when applied to formulations before extrusion, can enhance the extrusion processing stability and reduce energy consumption. This leads to improved efficiency, higher throughput, better product quality, and minimized losses.

Remember, knowing your ingredients is key; they are your best allies in achieving the desired results.

15.13 Formula and grinding

The grind should be fine enough to promote a proper product texture and avoid process instability and fines. All particles should be sized so there are little to no oversize or undersized particles. The use of a hammermill-style grinder is recommended to reduce the particle size of the ingredients in general. After grinding, a particle size distribution test should be performed. Typically, limiting the particles to 800 μm by using a sifter after the hammer mill is standard for pet foods, while 500–600 μm is preferred for a finer and smoother texture on the kibbles, especially for feline foods. Particularly with single-screw design extruders, powdery materials should also be avoided, as these fine materials would compete for time, moisture, and temperature, and usually hydrate, heat, and cook faster, whereas coarse particles could become difficult to convey, causing process instability.

In this example, the previous standard is examined with an easy-to-process dog food formula, featuring a production rate of 1000 kg/h.

- Corn at 40% of total: 400 kg
- Meat and bone meal at 30%: 300 kg
- Corn gluten meal at 15%: 150 kg
- Wheat by-products at 10%: 100 kg
- Tallow, salt, and vitamins at 5%: 50 kg

Batch total: 1000 kg

Disclaimer: The formula for the process learning example is not nutritionally balanced.

15.13.1 Premises

- The dry feed contains 8% moisture, equating to 80 kg of water and 920 kg of dry feed in a 1000 kg/h dry feed rate.
- To ensure adequate thermal energy is introduced, 8% steam (2% saturation) will be added during the preconditioning stage. The steam is to be delivered at a pressure of no higher than 2 kg/cm^2 through the preconditioner injectors.
- Water will also be added during the preconditioning stage at a rate of 15% of the dry feed weight. This practice aims to achieve an average moisture content of 25.2% and an estimated temperature of 85°C, with a maximum of 95°C, in the preconditioned materials.

15.13.2 Key data

- **Dry feed rate:** 1000 kg/h
- **Moisture in dry feed:** 80 kg (8%)
- **Bone dry feed:** 920 kg
- **Steam addition to preconditioner:** 80 kg/h (8%)
- **Water addition to preconditioner:** 150 kg/h (15%)
- **Steam addition to the extruder barrel:** 30 kg/h (3%)
- **Water addition to the extruder barrel:** 30 kg/h (3%)

Moisture calculation: To calculate the moisture content in the preconditioner and/or in-barrel moisture, add all sources of moisture to the numerator and all mass flows to the denominator:

15.13.3 Additional moisture added

3% water addition = 30 units (assuming 3% of 1000, which is the total moisture plus dry content).

3% steam addition = 30 units.

15.13.4 Total moisture content in the barrel

$$\text{Moisture content} = \frac{80+80+150+30+30}{80+80+150+30+30+920}$$

Let us calculate these percentages:

15.13.5 Preconditioner moisture calculation

$$\text{Pre-conditioner moisture} = \frac{80+80+150}{80+80+150+920} \times 100\%$$

15.13.6 In-barrel moisture calculation

$$\text{In-barrel moisture} = \frac{80+80+150+30+30}{80+80+150+30+30+920} \times 100\%$$

Now, let us calculate these values.

The moisture calculations yield the following results:

Preconditioner moisture 25.2%

In-barrel moisture 28.7%

These percentages represent the moisture content in the preconditioner and the total moisture in the barrel after adding steam and water to the barrel. With the above moisture conditions, including steam injection to the preconditioner of 8% or 80 kg/h, it is expected to yield 85°C–95°C in the preconditioned materials heading to the extruder barrel, ideal conditions for the standard adult dog food being processed.

Final moisture consideration: Due to the flash-off effect, where moisture is lost as the extrudate moves from a pressurized, heated area to atmospheric pressure, there can be a 3% to 7% moisture reduction. This results in a final moisture content of approximately 22% when the product is collected at the discharge of the extruders in the standard dog food recipe example.

This systematic approach ensures precise control over the moisture content and thermal energy during the production process, which is crucial for maintaining the quality and consistency of the dog food.

Remember, moisture content is essential for gelatinization and proper cooking of the starch content in the formulas, as well as for the uniform formatting of the kibbles. A rule of thumb is: If there is expansion of kibbles without stickiness (clumping of kibbles together at the exit of the extruder), with additional moisture added to the extrusion process, there will be an overall improvement to the process. For example, this includes an increase in starch cook, an increase in the firmness of kibbles, diminished fines, and increased wear life of extruder components.

15.13.7 Setting up the die assembly for producing standard pet food at the calculated moisture level above

Open area to be used: 200 mm^2-h/ton

Round final product target of 15 mm diameter × 5 mm cut length.

Consider an ideal expansion of 50% or 1.5 times the orifice diameter.

Orifice diameter of 10 mm.

Open area per hole: $A = \pi r^2$

78.54 mm^2/orifice = 5 × 2 × 3.1416

2.5 orifices per ton, rounding up to 3 orifices of 10 mm in diameter each, for a total of 236 mm^2-h/ton.

As discussed, you might consider using a back die with a center hole for uniform and stabilized flow to the final die.

491 mm^2/orifice = (12.5) × 2 × 3.1416

Back die orifice diameter of 25 mm.

Ratio of back die open area to final die: 2.07 times.

Use a 25–50-mm spacer between the back die and the final die, and a cone-flow distributor behind the final die.

In the production of dog food via extrusion, the design of the die orifice plays a crucial role in determining the quality and consistency of the final kibble. Two specific aspects of this design are particularly important: maintaining a maximum land length of 5 mm and incorporating a tapered inlet, especially if smaller in diameter, while considering a smaller land length.

The land length of a die orifice refers to the distance between the beginning of the orifice and its end, where the extrudate exits. Maintaining a shorter land length is essential for several reasons:

- **Uniformity and quality of kibble**: A shorter land length helps minimize shear and frictional forces on the extrudate as it passes through the die. Excessive shear could lead to overheating and degradation of the pet food's nutritional components, potentially impacting its quality.
- **Efficient production**: A longer land length can increase resistance and pressure within the extruder, requiring more energy and potentially slowing down the production process. By optimizing the land length, manufacturers can maintain efficient production speeds while ensuring the kibble's structural integrity. A longer land length is typically applied to produce compact and dense treats, such as dental chews. A typical high-conductivity material, such as synthetic high-wear and high-temperature-resistant materials, can be used as a die orifice liner for a smooth surface.

15.13.8 Tapered inlet of the orifice

The tapered inlet design, where the orifice gradually narrows toward the orifice, is another critical factor in the extrusion process. It helps streamline the flow and promotes axial expansion of the kibble, making it more of a ball shape with a smoother finish.

15.13.9 Maintaining smooth die surfaces

1. **Resurfacing**: Over time, dies can become worn or damaged, leading to rough surfaces that affect product consistency. Regular resurfacing is necessary to restore the die's smoothness. This process involves polishing the die's surface to remove any imperfections, ensuring that the extrudate flows evenly and maintains its intended shape.
2. **Proper open area**: The open area of a die, which refers to the total open area of the openings, must be maintained according to design specifications. This ensures uniform pressure and flow of the product through the die. If the open area changes due to wear or debris buildup, it can lead to inconsistent product dimensions and quality.
3. **Monitoring wear and tear**: Continuous monitoring of dies for wear and tear is essential. Visual inspections and measurements can help detect early signs of damage or wear, allowing for timely maintenance or replacement. Regular inspections help prevent unexpected downtime, rework, and maintain consistent production quality.

15.13.10 Technician and die assembly room

A typical pet food manufacturing facility will have one or two dedicated extrusion operators; however, it is recommended to have expert technicians both for the drying process and specifically for die and knife care, setup, and assembly. To ensure that these maintenance tasks are performed correctly and efficiently, it is recommended to have a dedicated technician and a well-prepared die assembly room:

1. **Technician expertise**: A skilled technician with experience in die maintenance and assembly can significantly improve the longevity and performance of dies. The technician should be familiar with maintaining the proper open area, techniques for monitoring wear and tear, cleaning, and caring for the die assemblies.
2. **Die assembly room**: A dedicated space for die preparation and maintenance is essential. This room should be equipped with the necessary tools and equipment for cleaning, inspecting, lubricating, setting up, and even resurfacing dies. A controlled environment

helps prevent contamination and ensures that the dies are handled properly, reducing the risk of damage.

15.13.11 Knife assembly and speeds

Rotating knife assemblies are used to efficiently cut directly expanded, extruded pet foods. This process ensures uniformity in product size, quality, and texture, which is essential for both esthetic appeal and proper nutrient distribution in the pet food.

15.13.12 Key considerations for maintaining knife speeds

1. **Knife sharpness and maintenance**
   - **Importance**: Sharp knives provide clean cuts, reducing the likelihood of product deformation, ensuring consistent piece sizes, and avoiding tails and fines when cutting extruded products.
   - **Maintenance**: Regularly check for dullness or damage, and sharpen or replace knives as needed. Dull blades can cause uneven cuts and increase the load on the assembly, potentially leading to speed fluctuations. Knives must be replaced regularly to prevent them from affecting product quality and creating reworks.
2. **Speed control systems**
   - **Role**: Modern knife assemblies often come equipped with speed control systems, including VFDs or electronic speed controllers.
   - **Function**: These systems allow for precise adjustments to the knife speed, ensuring it remains within the desired range of 800–1600 rpm. Consistent speed is critical for maintaining the uniformity and quality of the cuts.
3. **Material consistency**
   - **Importance**: The consistency of the extruded material affects cutting efficiency. Variations in density or moisture content can impact the ease with which the knife cuts through the product.
4. **Calibration and monitoring**
   - **Calibration**: Regularly calibrate speed control systems to ensure accurate speed settings. This includes checking sensors, actuators, and control interfaces.
   - **Monitoring**: Continuous monitoring of the knife's speed is crucial. Install sensors to provide real-time feedback and alerts if the speed deviates from the desired range. This enables prompt adjustments to maintain consistent cutting performance.
5. **Temperature and environmental factors**
   - **Impact**: External factors, such as ambient temperature and humidity, can affect both the machinery and the material being cut.
   - **Management**: Maintain a controlled environment around the knife assembly to mitigate these effects. This includes proper ventilation, temperature control, and ensuring that the machinery is not exposed to excessive dust or moisture.
6. **Safety protocols**
   - **Importance**: Operating at high speeds necessitates stringent safety measures to protect operators and equipment.
   - **Protocols**: As for the knife area, for all components of the extrusion system, the Original Equipment Manufacturer (OEM) should implement safety guards, emergency stop mechanisms, and regular safety training for operators to handle machinery correctly and respond to any issues swiftly. Users must understand and keep all safety devices functioning in proper condition.

7. **Regular maintenance and inspection**
   Routine checks: Conduct regular inspections and maintenance of the knife assembly, as well as all other components, including checking for wear and tear, lubricating moving parts, and ensuring the alignment of blades.

Maintaining knife speeds within the optimal range of 800–1600 rpm is essential for producing high-quality, uniform pet food products. Through regular maintenance, precise speed control, and adherence to safety and operational protocols, manufacturers can achieve consistent results and minimize product waste.

15.14 Application of new technologies and extrusion knowledge

If it is considered 22 kW-h/ton as the average SME required for standard adult dog food, what might occur if significantly more mechanical energy is applied during the process?

Excessive SME in cooked extrusion products can be a common issue in producing extruded pet foods. Factors contributing to this include worn-out rotating elements, sleeves, or inner barrels; worn-out die plates; inadequate preventive maintenance; incorrect configuration; extrusion-unfriendly ingredients; insufficient process knowledge; and setting the kill-step extrudate temperature too high (typically, it should be typically greater than 90°C).

If the applied SME significantly exceeds the norm, say, reaching 25–37 kW-h/ton, what symptoms or issues might arise?

High SME can lead to inconsistent and poor-quality products, potentially resulting in lost capacity and compromised uptime. In addition, extrudates from worn-out extruders usually require higher moisture levels to facilitate conveyance, which can be problematic. A very low investment, typically under 1% of the total cost, in an adequate preventive maintenance program and wear management for an extruder can prevent significant losses, especially since raw materials constitute 80%–90% of the total cost.

15.14.1 Low-shear extrusion cooking

The newer technologies in extrusion cooking have brought a substantial change to processing, which is the ability to precisely contain and use thermal energy to cook pet foods and still convey the sticky or slippery extrudate with ease and control (Fig. 15.6). These new-generation extruders have increased the typical ratios of 2:1 or 3:1 thermal to mechanical energy on traditional extrusion systems, both single- and twin-shaft designs, to up to 8:2 in pet foods with advanced thermal cooking extrusion systems. These systems take advantage of all the discussions in the chapter: Precisely designed and maintained hardware to promote gentle, positive, low-shear conveying.

The products below were processed using the Thermal Twin TT-760 Extruder, favoring higher steam (thermal cook) and low-shear processing. As noted, advancements in preconditioning and barrel design, as well as steam injection hardware, steam flow control, and process software on the advanced extrusion systems, have pushed the energy ratio further. Figs. 15.7 and 15.8 show the process conditions for high meat (70% of dry feed) or 700 kg of properly ground and emulsified raw meat slurry in

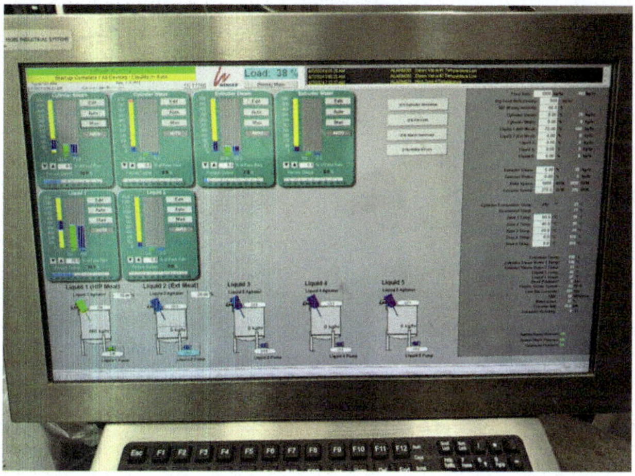

Fig. 15.6 Process conditions for low-shear processing.
Courtesy of Wenger Manufacturing, LLC.

Thermal Cooked 70% Meat Grain Free Recipe

Recipe	% DRY
Chicken Meal Low Ash	35.000
Whole potato flour (WPF100)	30.000
Tapioca Starch	15.000
Pea Protein Concentrate (50% min -75% max)	24.700
Oat Fiber	7.300
Flax meal	2.000
Total	100.000

- ✓ Low Low Shear Processing
- ✓ Grain free
- ✓ 700 kg of meat to 1000 kg of dry formula (70% meat)
- ✓ Fine grind
- ✓ Dried & Oil Coated

WENGER

Fig. 15.7 Adult pet food, grain-free kibbles with 70% meat slurry added to the preconditioner, processed at low shear (16 kW-h/ton).
Courtesy of Wenger Manufacturing, LLC.

Fig. 15.8 Adult pet food grain-free kibbles processed at low shear, 16 kW-h/ton. Courtesy of Wenger Manufacturing, LLC.

1000 kg/h of the dry grain-free formula. The SME was reduced to 16.0 kW-h/ton as the thermal energy was enhanced.

15.15 Final product specifications

As the product exits the extruder, measurements should be performed: Measuring bulk density must be carried out by carefully collecting the hot product without getting burnt or harmed. Some companies use a small perforated stainless steel tray to collect the product from the product stream. The operator should allow the surface vapor to release and then fill, preferably always in the same manner, a tared, known-volume cup and weigh the contents of the cup. Typically, for adult dog foods, the average bulk density is 400 g/L of wet product off the extruder. Density is a major control factor for most producers of pet foods; automatic devices are being improved to measure and control the extruder to achieve a given wet bulk density. Other quality controls to be measured include product moisture, flow rate, temperature, and pressure at the die, which can be automated and are currently offered by Original Equipment Manufacturers. The operator can also collect product samples for size analysis, moisture, expansion indices, and starch cooking.

Postextrusion, the product is typically transferred to an air-heated dryer, then enrobed and cooled before final packaging.

15.16 Postextrusion pet foods

Complete diets: Typically, in kibble format, these categories include many shapes, formats, expansion ratios, textures, moistures, and appearances. Fig. 15.9 shows the products produced via extrusion on low shear with no meat and 70% meat added to the

Guy Classification System

grouping based on functionality

	Group	Characteristics	General Effects in Extrusion
1	Structure-Forming Materials	Polymers with film-forming ability that withstands the expansion of water vapor on exiting the extruder.	May Increase or Decrease Expansion May Increase or Decrease Moisture Req. May Increase or Decrease Thermal Req. May Increase or Decrease Viscosity
2	Dispersed-Phase Filling Materials	Proteins and fibrous materials located within the continuous starch structure. physical presence disrupts the cell walls of structure-forming film.	Decrease Expansion Increase or Decrease Moisture Req
3	Plasticizers and Lubricants	Transforms polymers from solids to deformable plastic fluids. Plasticizers reduce the dissipation of mechanical energy and reduces the heat input. Lubricants reduce the applied shear forces within the dough mass and along the metal surfaces of the screw and barrel.	Increase Bulk Density Decrease Expansion Decrease Frictional Heat Decrease Specific Mechanical Energy
4	Soluble Solids	low molecular weight materials, such as sugars or salts, may be added to a recipe for flavoring or humectant properties.	Reduces Melt Viscosity & Retention Time Increase Water Holding Capacity
5	Nucleating Substances	Increase the number of bubbles present in the expanding extrudate	Increase Expansion
6	Coloring Substances	Contribute to color in the extrudates.	Little to no effect
7	Flavoring Substances	Contribute to flavor formation in thermal reactions.	Little to no effect

Fig. 15.9 Guy's classification system is based on the characteristics and functionality of different ingredient groups during extrusion processing.

grain-free-based recipe (Fig. 15.9). These products are usually dried to around 8% final moisture and below 0.60 water activity for shelf stability. After being coated with fat and flavors and cooled within 5°C of room temperature, the product can be stored or bagged. In the category of complete diets, extrusion can be performed to cook pet foods for soft semimoisture. Normally, the recipe will receive around 15% humectants, such as glycerin (which traps moisture), and approximately 1.5% of an acidification agent, such as phosphoric acid, along with a bit of mold inhibitor to keep water activity under 0.60 a_w. The soft semi-moisture products are usually air-cooled and handled with care for further storage and packing.

15.17 Specialty products

Treats: Extruded pet treats can be further divided into two categories, like kibbles on complete diets (cut at the extruder die), called SHORT GOODS, or cut further down after being produced on ropes or spaghetti, referred to as LONG GOODS:

15.17.1 Short goods

A large variety of short goods can be produced with all extrusion technologies available (Fig. 15.10). Attention to the fact that most pet treats are made from floury ingredients makes positive conveying crucial for perfect formatting. Typically, a smooth surface is desired, especially on dental treatments, which receive quite a bit of SME to become chewy. By default, a small twin-screw or advanced (thermal) twin-screw extruders might be considered.

Fig. 15.10 Short goods treats are cut at the die, much like complete diet kibbles. Courtesy of Wenger Manufacturing, LLC.

Fig. 15.11 Short goods coextruded treats, cut at the die much like complete diet kibbles. Courtesy of Wenger Manufacturing, LLC.

Short goods are mainly manufactured as single components (Fig. 15.11); however, two extruders feeding a single die generate the so-called coextrusion, where a piece of pet treat or kibble contains two different colors, either separated or intermeshed (marbled).

As mentioned, long goods are both manufactured as single components and coextruded (Fig. 15.12). The ropes, as they come off the extruder die, fed by one or two extruders, travel a distance on a belt conveyor that could be chilled to settle the structure of the product, to later be cut with a guillotine or rotary cutting device. These products are typically cooled and hand-handled into small packages for single or a few servings.

Fig. 15.12 Long goods, single components, and coextruded treats from the extruder die. Courtesy of Wenger Manufacturing, LLC.

15.17.2 Modern and reliable extrusion lines for pet food production

As exposed in this chapter, the present moment of extrusion processing relies on systems that offer great flexibility with the most intuitive, user-friendly, and reliable control inputs, produce quality and sellable products with the longest processing time, minimize the typical process losses, and promote the use of smart energy. In a short time, there will be artificial intelligence and machine learning to assist producers in maintaining the best set of parameters for a given recipe and tuning into the sweet spot.

15.18 How AI will revolutionize extrusion

Automated process optimization: AI algorithms will continuously analyze data, adjusting parameters like temperature, pressure, and moisture to ensure optimal extrusion conditions.

Predictive maintenance: Machine learning models will detect early signs of wear and tear, preventing costly downtime by scheduling maintenance proactively.

Real-time quality control: AI-powered vision systems and sensors will monitor kibble shape, texture, and density, making instant adjustments to maintain consistency.

Energy efficiency and sustainability: AI will fine-tune energy consumption, reducing waste and optimizing resource usage, contributing to a more sustainable production process.

Adaptive learning and customization: AI-driven systems will learn from historical data, adapting recipes and process settings to meet specific product requirements without manual recalibration. With these innovations, extrusion lines will become smarter, more autonomous, and highly efficient. The future is near—AI is set to revolutionize the industry, making extrusion smarter, faster, and more sustainable.

15.19 Conclusion

Extrusion technology is a cornerstone of modern pet food production. By focusing on equipment design, ingredient optimization, advanced process control, and consistent training, producers can achieve high-quality, nutritious, and sustainable pet foods. Implementing these practices ensures efficient resource use, reduced waste, and long-term industry success.

References

Alvarenga, I.C., Keller, L.C., Waldy, C., Aldrich, C.G., 2021. Extrusion processing modifications of a dog kibble at large scale alter levels of starch available to animal enzymatic digestion. Foods. 2304-8158. 10 (11). Available from: https://www.mdpi.com/2304-8158/10/11/2526/pdf. doi:10.3390/foods10112526.

Baller, M.A., Pacheco, P.D.G., Peres, F.M., Monti, M., Carciofi, A.C., 2018. The effects of in-barrel moisture on extrusion parameters, kibble macrostructure, starch gelatinization, and palatability of a cat food. Anim. Feed Sci. Technol. 0377-8401. 246, 82–90. Available from: http://www.journals.elsevier.com/animal-feed-science-and-technology. doi:10.1016/j.anifeedsci.2018.10.003.

Bhattarai, R.R., Dhital, S., Wu, P., Chen, X.D., Gidley, M.J., 2017. Digestion of isolated legume cells in a stomach-duodenum model: three mechanisms limit starch and protein hydrolysis. Food Funct. 2042-650X. 8 (7), 2573–2582. Available from: http://pubs.rsc.org/en/journals/journal/fo. doi:10.1039/c7fo00086c.

Corsato Alvarenga, I., Aldrich, C.G., Shi, Y.C., 2021. Factors affecting digestibility of starches and their implications on adult dog health. Anim. Feed Sci. Technol. 0377-8401. 282. Available from: http://www.journals.elsevier.com/animal-feed-science-and-technology. doi:10.1016/j.anifeedsci.2021.115134.

Guy, R.C.E., 1994. Raw Materials for Extrusion Cooking Processes. Springer Science and Business Media LLC, pp. 52–72. doi:10.1007/978-1-4615-2135-8_2.

Hellemans, T., Nekhudzhiga, H., Van Bockstaele, F., Wang, Y.J., Emmambux, M.N., Eeckhout, M., 2020. Variation in amylose concentration to enhance wheat flour extrudability. J. Cereal Sci. 0733-5210. 95, 102992. doi:10.1016/j.jcs.2020.102992.

Jackson, M.I., Waldy, C., Cochrane, C.Y., Jewell, D.E., 2020. Consumption of identically formulated foods extruded under low and high shear force reveals that microbiome redox ratios accompany canine immunoglobulin A production. J. Anim. Physiol. Anim. Nutr. 1439-0396. 104 (5), 1551–1567. Available from: http://onlinelibrary.wiley.com/journal/10.1111/(ISSN)1439-0396. doi:10.1111/jpn.13419.

Peixoto, M.C., Ribeiro, M., Maria, A.P.J., Loureiro, B.A., Di Santo, L.G., Putarov, T.C., Yoshitoshi, F.N., Pereira, G.T., Sá, L.R.M., Carciofi, A.C., 2018. Effect of resistant starch on the intestinal health of old dogs: fermentation products and histological features of the intestinal mucosa. J. Anim. Physiol. Anim. Nutr. 1439-0396. 102 (1), e111. Available from: http://www3.interscience.wiley.com/journal/118535294/toc. doi:10.1111/jpn.12711.

Ribeiro, É.D.M., Corrêa Peixoto, M., Putarov, T.C., Monti, M., Pacheco, P.D.G., Loureiro, B.A., Pereira, G.T., Carciofi, A.C., 2019. The effects of age and dietary resistant starch on digestibility, fermentation end products in faeces and postprandial glucose and insulin responses of dogs. Arch. Anim. Nutr. 1745-039X. 73 (6), 485–504. doi:10.1080/1745039x.2019.1652516.

Robin, F., Schuchmann, H.P., Palzer, S., 2012. Dietary fiber in extruded cereals: limitations and opportunities. Trends Food Sci. Technol. 0924-2244. 28 (1), 23–32. doi:10.1016/j.tifs.2012.06.008.

Rokey, G.J., 2000. Single-screw Extruders Extruders in Food Applications. CRC Press, pp. 25–50.

Vanier, N.L., Vamadevan, V., Bruni, G.P., Ferreira, C.D., Pinto, V.Z., Seetharaman, K., Zavareze, E.D.R., Elias, M.C., Berrios, J.D.J., 2016. Extrusion of rice, bean and corn starches: extrudate structure and molecular changes in amylose and amylopectin. J. Food Sci. 1750-3841. 81 (12), E2932. Available from: http://www3.interscience.wiley.com/journal/118509799/issueyear?year=2008. doi:10.1111/1750-3841.13545.

Yoo, S.H., Perera, C., Shen, J., Ye, L., Suh, D.S., Jane, J.L., 2009. Molecular structure of selected tuber and root starches and effect of amylopectin structure on their physical properties. J. Agric. Food Chem. 0021-8561. 57 (4), 1556–1564. Available from: http://pubs.acs.org/doi/pdfplus/10.1021/jf802960f. doi:10.1021/jf802960f.

Development of starch/polymeric foams by extrusion processing

16

Akash Kusum Maiti[a], *Poonam Singha*[a], *Sushil Kumar Singh*[a], *and Kasiviswanathan Muthukumarappan*[b]
[a]Department of Food Process Engineering, National Institute of Technology Rourkela, Rourkela, Odisha, India
[b]Agricultural and Biosystems Engineering, South Dakota State University, Brookings, SD, United States

16.1 Introduction

Plastics are fossil-fuel-derived synthetic polymers that are an essential part of the food packaging sector. These synthetic plastics in the food value chain are used for a single purpose; therefore discarding these packages after consumption generates a large amount of waste. The most commonly used polymers are mainly polyethylene, polypropylene, polystyrene, and polyvinyl chloride, with many favorable properties such as low cost, easy availability, lightweight, and easy processability (Kshitij et al., 2025). The quadrupled growth of plastic consumption is expected to generate roughly 12 billion tons of plastic waste by 2050. However, the mass production of these nonbiodegradable plastics substantially contributes to a serious environmental problem due to their disposal and end-of-life options (Peshkam, 2022). This threat has urged the search for renewable and environmentally friendly alternatives to conventional petroleum-based polymers. The research process to find an alternative led researchers to biobased packaging materials. According to European Bioplastics, bioplastic material should be biobased, biodegradable, or both. Bioplastic is produced from biomass, e.g., cellulose, poly(lactic acid) (PLA), starch, etc. Among these, the most adaptive and cost-effective biopolymer, starch, has shown to be a sustainable solution (Karan et al., 2019).

Starch is abundantly available in nature in different forms. However, the current focus is on the sourcing of starch because it determines the physicochemical properties, thereby leading to a sustainable solution. Thus emphasis has been placed on the sources, especially from renewable agricultural products or seafood waste (Razza et al., 2015). Starch is often combined with different polymers to enhance its mechanical properties and water resistance. These include cross-linking agents, fillers, plasticizers, and the blending of starch with different biopolymers such as polyvinyl alcohol (PVA), poly lactic acid, and chitosan (Debiagi et al., 2011).

Extrusion Cooking for Food and Feed. https://doi.org/10.1016/B978-0-443-29987-2.00016-6

Extrusion is a melting and solidification technique used to transform starch into foam with the addition of other materials. It has emerged as a highly effective and scalable method for producing starch foams. This technique involves the continuous formation of a material through the application of heat, pressure, and mechanical shear, making it ideal for the production of foams with controlled cellular structures (Su et al., 2009).

This chapter aims to discuss current advancements in the development of starch/polymeric foams produced via extrusion processing. The chapter provides a comprehensive review of recent advancements in this field, with a thorough discussion on material composition, extrusion parameters, foam characteristics, and an outline of future directions for research. It explores the selection of starch–polymer blends influencing foam structure, mechanical strength, moldability, water resistance, lightness, and thermal stability. The impact of extrusion conditions on foaming morphology and properties is also reviewed. Furthermore, recent innovative strategies, including variations in plasticizer concentration, cross-linking agents, mixing with natural polymers, integrating natural fibers, and adding nanofillers and biobased additives, are intensively reviewed. The chapter also highlights both the opportunities and challenges associated with the extrusion processing of starch/polymeric foams. It outlines future directions for research by reviewing a few advanced research attempts on scalable and high-performance foams. The insights of this chapter aim to review all the research innovations and industrial implementations by highlighting their potential to contribute to sustainability and environmentally friendly polymer materials in the food industry.

16.2 Materials and extrusion

16.2.1 Starch

Starch is a biodegradable polysaccharide, mainly available in granular form in various plants. A few examples of sources are tubers, cereals, seeds, and roots. The diameter of granules of starch ranges from 2 to 100 μm. These granules are synthesized during the photosynthetic process in the presence of water, carbon dioxide, and sunlight (Laftah, 2017).

The size and shape of starch granules can vary significantly depending on the plant source. The granules typically contain 10%–29% moisture by weight, with trace amounts of proteins, lipids, ash, and inorganic materials. Starch is a natural carbohydrate made up of carbon, hydrogen, and oxygen. The molecular structure of starch consists of amylose and amylopectin. Amylose is a linear polymer primarily composed of 6000 glucose units linked by α-(1→4) glycosidic bonds. Amylopectin is a highly branched polymer composed of 24–30 glucose units connected by both α-(1→4) glycosidic bonds and α-(1→6) glycosidic bonds (Hoover, 2001). The typical composition of starches varies around 20–25 wt.% amylose and 75–80 wt.% amylopectin. The ratio of these components influences its solubility and mechanical strength. Amylose primarily forms the amorphous regions of the granule. It plays an important role in determining the structural characteristics of starch, affecting the formation of short-range ordered structures, crystals, and the lamellar structure. In addition, amylose contributes to resistance against enzymatic degradation. Amylopectin, on the contrary, contributes to the crystalline portions (Figueiró et al., 2024; Cazón et al., 2017).

Since starch consists of anhydroglucose, it is hydrophilic in nature due to the presence of numerous hydroxyl groups. Amylopectin makes starch resistant to the diffusion of larger molecules by imparting a smooth surface to it. However, the granules begin to hydrate and swell upon heating in water, resulting in an increase in the viscosity of starch pastes. The behavior of starch depends on temperature, moisture content, and the thermomechanical treatment applied during extrusion (Kshitij et al., 2025).

Due to its renewable, cost-effective, and biodegradable nature, starch polymer is widely used in the production of a sustainable alternative to fossil-based packaging materials. It emerges as an attractive option for packaging applications due to the good oxygen barrier properties provided by the ordered and tightly packed hydrogen bond structures of amylose and amylopectin, which give starch its unique characteristics (Singh and Genovese, 2021). However, starch has certain limitations as a plastic alternative due to its hydrophilic nature and brittleness. Nevertheless, the modification of starch or the addition of different additives makes it a promising candidate for sustainable packaging.

16.2.2 Thermoplastic starch

The environmental issues at the global level regarding plastic waste and biopolymers have promoted alternatives to replace petrochemical-based polymers. The composition of bioplastic consists of starch as the primary biopolymer component. However, pure starch has several limitations due to its tendency to dissolve in water, brittleness, low melting point, and limited mechanical properties. To find a promising solution, various physical and chemical modifications, such as blending with additives and plasticization, have been explored. Among all possibilities, thermoplastic starch (TPS) has gained considerable attention. The physicochemical properties of TPS closely resemble petroleum-based polymers, while it is nontoxic and biodegradable in nature. TPS is produced by disrupting starch granules through heat and mechanical pressure, along with a plasticizer. TPS is commonly obtained from corn, potato, and wheat. The composition of TPS typically ranges from 50 to 90 wt.% starch, 10–50 wt.% plasticizer, and varying proportions of fillers. Other additives, such as plant fibers, emulsifiers, and cellulose, have also been studied to improve the mechanical properties of TPS. Extrusion is the most preferred choice to produce TPS due to its continuous production. The production of TPS involves several physical and chemical transformations, such as absorption of water, expansion of granules, starch gelatinization, melting, and crystallization. The final properties of TPS depend on factors such as the amylose-to-amylopectin ratio, processing conditions, type, and concentration of the plasticizer (Diyana et al., 2021).

16.2.3 Plasticizer

Plasticizers weaken the internal hydrogen bonds between polymer chains and increase the free space in starch matrices. This modification improves the behavior of starch, resulting in enhanced flexibility, processability, and mobility of molecular chains. The efficiency of plasticizers depends on their bonding with the starch matrices. Plasticizers are typically used in forms such as glycerol, sorbitol, urea, fructose, sucrose, and glycol. Among these, glycerol and sorbitol, both polyols, are the most commonly used plasticizers. Plasticizers with low molecular weight play a crucial role by occupying the intermolecular spaces between polymer chains to enhance the mobility of the macromolecular chains. The most commonly used plasticizers in research work are glycerol and sorbitol. Various studies have found that sorbitol forms a stronger bond than glycerol with starch (Montilla-Buitrago et al., 2021). Plasticizers are added to starch to disrupt its internal hydrogen bonds by replacing the interactions between starch molecules. This occurs as hydrogen bonds form between the polar groups of the plasticizer and the hydroxyl groups in the glucose units of the starch. The starch transforms into a thermoplastic material as the mobility of the starch macromolecules increases. Water present in starch also acts as a plasticizer. However, the formation of undesirable bubbles limits the upper temperature for processing starch. To mitigate this, water is typically combined with other plasticizers. A low concentration of plasticizer prevents plasticization, while excessive plasticizer can result in undesirable mechanical properties in the TPS (Singh and Genovese, 2021).

16.2.4 Other additives

Starch contains many hydrophilic hydroxyl groups, which absorb moisture when used in foam packaging. To improve its water barrier properties, various methods have been explored. Among them, one promising method involves cross-linking starch to enhance water resistance. Cross-linking of starch significantly reduces its swelling capacity, solubility, water retention ability, and water affinity, thereby making it more resistant to water absorption. Dialdehydes have the potential to cross-link starch by reacting with its hydroxyl sites, creating intermolecular bridges between polysaccharide chains. Glutaraldehyde ($C_5H_8O_2$) has been mostly used for this purpose. Glyoxal ($C_2H_2O_2$) is recognized as a safe chemical for use as a food-grade polymer in packaging by both the Federal Institute for Risk Assessment in Germany and the Food and Drug Administration in the United States (Singh and Genovese, 2021; Hassan and Fowler, 2022).

Starch can be cross-linked with varying concentrations of glyoxal to develop a food packaging tray foam. Uslu and Polat (2012) concluded that it reduced the time for baking and water uptake. Trays treated with glyoxal at 0.126 and 0.269 g/kg reduced foam density and water absorption but improved the flexibility of the foam. These findings suggested that cross-linked foam possesses a well-developed structure and strong mechanical properties. The authors suggested that the properties of these trays could be further improved by incorporating nanofillers and fibers.

Natural fibers have been utilized due to their versatility, durability, and easy integration into various aspects of daily life. The incorporation of natural fiber in starch-based biocomposite systems can show a significant enhancement of the mechanical properties of the foam. This is due to the similar chemistry of both polymers. Therefore natural fibers should be chosen depending on their compatibility and strong interaction with starch. Natural fibers have more advantages over synthetic alternatives, as shown in Table 16.1. Natural fiber offers comparable tensile strength, effective insulating properties, and lower density. These properties make them an attractive option in various applications (AL-Fakih et al., 2024).

Various studies have explored the incorporation of natural fibers into extruded starch foam. For example, sugarcane bagasse, cotton linter, hemp, cellulose, and wheat bran have been extruded with different types of starches. The various studies concluded that fiber content can decrease the expansion index of foam and increase the density. However, these effects can be reversed. These conflicting results suggest that natural fibers can both increase the viscosity of the starch and serve as nucleating agents to promote bubble cell growth on the surface (Soykeabkaew et al., 2015).

To address the hydrophilicity and low mechanical performance of starch (Yang et al., 2024), a novel approach was developed to create a blend with PLA, cellulose, and chitosan to produce a waterproof film, which was then incorporated into foam to enhance its water resistance properties. The resulting starch–fiber foam showed a 14.4% decrease in vapor permeability, from 766.83 $g/m^2 \cdot 24$ h to 664.89 $g/m^2 \cdot 24$ h. The water absorption rate also dropped from 12.3% to 7.9%. Furthermore, the hysteresis angle decreased from 15.57 degrees to 8.59 degrees within the first 5 min of moisture exposure, indicating improved moisture resistance. The inclusion of PLA films also increased tensile strength from 2.09 to 3.53 MPa and compressive strength from 7.5 to 11.4 MPa.

Table 16.1 Comparison of natural fiber and synthetic fiber (AL-Fakih et al., 2024).

	Natural fiber	**Synthetic fiber**	
Density	Light	Twice than natural fiber	Density
Cost	Low	High	Cost
Renewability	Yes	No	Renewability
Recyclable	Yes	No	Recyclable
Energy consumption	Low	High	Energy consumption
Distribution	Wide	High	Distribution
CO_2 neutral	Yes	No	CO_2 neutral
Health risk when inhaled	No	Yes	Health risk when inhaled
Disposal	Biodegradable	Not biodegradable	Disposal

16.2.5 Foaming process

Foam is a material that has spherical gaseous voids within a dense polymeric matrix. The foaming process involves four key stages: dissolution in the polymer, cell nucleation due to supersaturation, cell growth, and cell stabilization as the polymer solidifies. In starch foam, rheological characteristics, distribution of components, and material crystallinity play a critical role in determining foam morphology, expansion, and shrinkage. These factors influence both the mechanical and functional characteristics of foam. Starch foams need to exhibit high expansion, improved mechanical strength, lower density, and enhanced water resistance properties for food packaging applications.

Different methods of making foams result in distinct product characteristics. Among them, the extrusion process is mostly used in the packaging industry. This process is particularly suitable for producing starch-based foams that are lightweight, durable, and moisture-resistant materials (Tapia-Blácido et al., 2022).

16.2.6 Extrusion

Extrusion technology is widely used for preparing TPS blends and their composites due to its capability to process high-viscosity materials and its functionality. During the extrusion process, starch is combined with water above 100°C to generate voids through steam expansion as the material is forced through the die under high pressure. The extrusion process involves four key stages: (1) material feeding, (2) mass transport, (3) die flow, and (4) exit and downstream processing (Singh and Muthukumarappan, 2017; Singha et al., 2018). Single-screw extruders, commonly used for starch-based foams, are cost-effective but less efficient than twin-screw extruders. Extrusion temperatures typically vary between 120°C and 170°C, with screw speeds ranging between 70 and 400 rpm. A two-stage extrusion process typically produces TPS blends. The process begins with primary extrusion and then proceeds to secondary extrusion. Starch pellets are fed into a screw extruder along with plasticizers and additives in the first stage of the extrusion process, known as primary extrusion. They are then subjected to intense shear mixing at elevated temperatures across several heating sections. The mixing process disrupts the starch crystallites through the interaction of water and plasticizer, thereby facilitating the starch to melt and improve its flow. The resulting plasticized starch is blended with toughened polymers or reinforced with fillers/fibers in the next step of secondary extrusion. The extrudate is then molded into products or processed into films through casting or blowing. The properties of starch-based foam can be tailored by optimizing polymer content, water levels, temperature profile, screw speed, screw geometry, pressure, shear force, and cooling rate. The extrusion process involves three critical parameters: extrusion pressure, temperature, and extrusion time. High extrusion pressure reduces the density of the foam and creates small pores, thus creating a denser and more stable structure. Temperature and extrusion time also influence foam texture and mechanical properties (Figueiró et al., 2024; Singha et al., 2019a, 2019b; Surendren et al., 2022; Tapia-Blácido et al., 2022).

16.3 Characteristics of starch foam

Starch foam can be characterized based on different parameters to judge its performance as a packaging material. Some of the important parameters (Fig. 16.1) are discussed in the following.

16.3.1 *Morphology*

The morphology of the foam is analyzed using scanning electron microscopy (SEM) in a low-vacuum mode by applying a specific tension of kW. For sample preparation, foams are fractured into small pieces and coated with either platinum or gold (Zhang et al., 2022; Yun et al., 2023).

16.3.2 *Viscosity*

A rheometer is used to determine the shear viscosity of the foam at room temperature. A diameter of 25 mm parallel plate is used for the measurement of the rheological properties of the foam at a shear rate of 0.01–25 s^{-1} (Gu et al., 2024).

16.3.3 *Density*

The foam density is measured by considering the relationship between the weight and volume of the foam. The sample is cut into a standard size, and averages are taken for the final density determination (Trongnit et al., 2024).

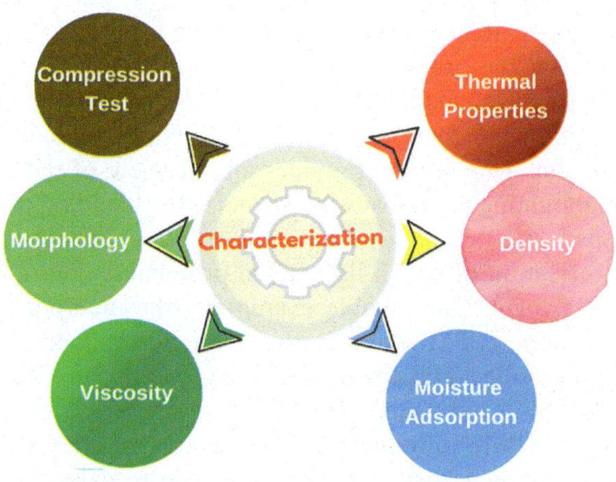

Fig. 16.1 Key characterization parameters of starch foam.

16.3.4 Moisture adsorption

It can be assessed by tracking the changes in the weight of the developed foam during a specified period of storage. The sample is first dried in a hot air oven to make it moisture-free. Then, it is placed in a sealed desiccator with 95% relative humidity (RH). The weight changes are tracked on a daily basis to calculate the moisture adsorption kinetics (Zhang et al., 2022).

16.3.5 Compression test

The sample should first be kept in 95% RH for at least the specified time. Then, a universal material testing machine is used to test the 50% deformation of the foam at a specified rate of pressure (Zhang et al., 2022).

16.3.6 Thermal characterization

The sample foam is preheated within a range of 30°C–850°C, as per the requirement, in a nitrogen environment. The heating rate is maintained at 10°C per minute with a continuous flow rate. A thermal analyzer is used to prepare a thermogravimetric curve and to perform thermal characterization of the foam (Han et al., 2024; Quilez-Molina et al., 2024).

16.4 Advancement

The biodegradable starch foams have gained significant attention in recent years. However, there are a few primary challenges that restrict the advancement of these foams. Starch foam has poor mechanical strength, a high water absorption rate, and limited thermal stability. Recent advancements in the blending of starch foams have focused on modifying the properties of these materials by incorporating various natural fibers, biopolymers, and additives (Fig. 16.2). Various approaches have been explored to enhance starch foam characteristics, including silane coupling agents, plant-based fibers, etc. This section provides an overview of the latest research on starch foam by highlighting the key advancements in the blending of starch foams and their modifications.

Chemical modification of starch has shown promising results in improving the properties of starch foam. One notable example is the incorporation of 3-chloropropyl trimethoxysilane (CPMS) and methyl trimethoxysilane (MTMS) as silane coupling agents to produce TPS foam. The foam was made with modified starch, gelatinized starch, PVA, and water. The silylation improved water resistance and mechanical strength. CPMS introduced bulkier groups into the starch by increasing foam density and reducing porosity. MTMS led to a more stable structure with better reaction efficiency. Foams with higher silane content had higher density and mechanical properties. The flexural moduli increased from 127.62 MPa

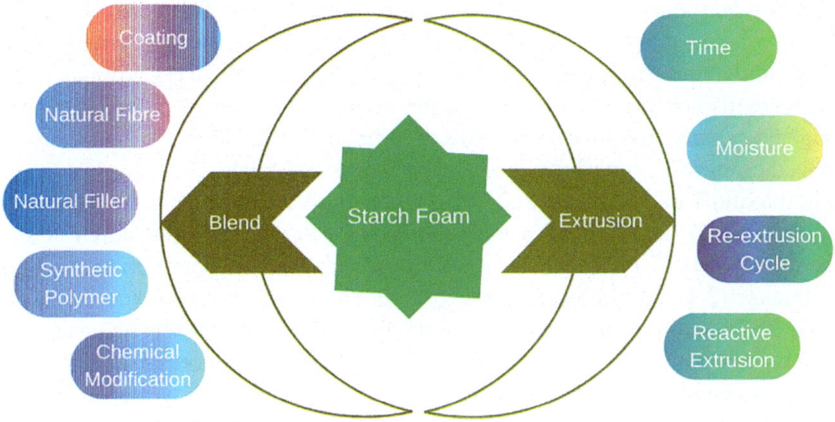

Fig. 16.2 Overview of modification stratagies and processing variables for starch foam modification.

(unmodified TPS) to 253.45 MPa for TPS-CPMS and 235.26 MPa for TPS-MTMS. Impact resistance was also improved, with 18.71 J/m and 17.27 J/m for TPS-CPMS and TPS-MTMS, respectively. Overall, silylation enhanced the performance of TPS foams and made them suitable for applications in biodegradable packaging (Bergel et al., 2020). The addition of chemicals has a significant effect on improving the starch molecular structure. SEM analysis showed that 50 μm $CaCO_3$ particles resulted in pores with an approximate diameter of 800 μm, while the addition of 50 nm $CaCO_3$ particles produced pores around 333 μm in size. The 58% reduction in smaller $CaCO_3$ particles contributed to more uniform and denser vesicle structures. The reduction of $CaCO_3$ as a nucleating agent also resulted in improved water resistance, thermal stability, and corrosion resistance. The 50 nm $CaCO_3$ composite demonstrated the lowest degradation rate (10.5% after 30 days) and moisture absorption (12.6%). The research revealed that using smaller $CaCO_3$ particles helped to improve tensile strength (1.72 MPa) and bending strength (0.58 MPa) (Wang et al., 2024b).

However, incorporating green-based polymers to improve the functionalities of starch is still a favorite choice for researchers pursuing a sustainable approach. Zhang et al. (2024) investigated the use of citrus fiber and citrus pectin (Zhang et al., 2025) to enhance the properties of starch foam. FTIR analysis confirmed that the incorporation of insoluble fiber promoted cell nucleation. This modification increased the cell density of the foam from 83.37 to 123.44 cells/$cm^3 \times 10^4$ and decreased the density of the developed foam from 0.457 to 0.142 g/cm^3. The cellular structure of the starch foam was also altered by the addition of 4–8 wt.% pectin. The soluble pectin enhanced the strength of the starch matrix by forming ester bonds. FTIR and DSC analyses showed that viscosity was increased during the foaming process. The water uptake and water migration of the foam were also reduced. This resulted in enhanced mechanical properties of the foam during wet storage. The foam with 4 wt.% pectin showed the best compression–recovery ratio (76.7%) and a lower moisture content

(19.22%) when exposed to 95% RH. The resulting starch foam incorporated with pectin displayed a high compression strength (0.9 MPa), good recovery capacity (>72%), and reduced moisture content (15.11% under 95% RH). Similar improvements were also observed with the addition of 20 wt.% citrus fiber, which resulted in the highest expansion ratio of 5.20. This modification of the foam structure resulted in enhanced compression performance and moisture resistance.

Aguilar and Tapia-Blácido (2023) had taken another approach to develop composite foam trays by using blended avocado seed (AS) with cassava starch at varying concentrations (0, 5, 10, and 15% w/w). Increasing AS concentration made the foam trays thicker (2.1–2.3 mm) and denser (0.8–0.9 g/cm^3), but they became less porous (25.6%–35.2%) compared to the control. Higher AS concentrations reduced puncture resistance and flexibility, while the tensile strength (2.1 MPa) remained similar to the control. The presence of AS also reduced the trays' hydrophilicity by enhancing their water resistance due to its lipid, fiber, and amylose-rich starch content. Thermal analysis revealed that AS decreased the thermal decomposition peak of cassava starch foam trays from 316°C to 303°C.

Few researchers have focused on integrating starch with synthetic polymers to support the industry during this transition phase. Willett and Shogren (2002) worked on blending starch with biodegradable thermoplastic resins such as PLA, poly (hydroxybutyrate-co-valerate), and poly(hydroxyester ether) (PHEE). These blends exhibited lower densities and greater radial expansion compared to pure starch foams. Polymers such as PLA contributed to smaller domain sizes (1–10 μm) in the foam. Surface concentrations of polymers increased foam expansion and resistance to fragmentation. The compressive strength of these foams is predominantly determined by their density, not the type of starch or polymer used. In addition, the resins reduced the water sensitivity of the foams by extending the time required for dissolution. When high-amylose starch, wheat starch, or potato starch was blended with PLA or PHEE, foam density decreased and expansion was enhanced. These foams displayed phase-separated morphologies, with the polymer occupying spherical or elongated inclusions within the starch matrix. Amorphous polymers or those with low melt viscosity and moderate glass transition temperatures led to highly expanded, durable foams, confirming the feasibility of creating biodegradable, low-density foams without modifying the starch itself. This work was extended by Mitrus and Moscicki (2014), who developed biodegradable loose-fill foams from TPS. Their research focused on utilizing locally sourced potato, corn, and wheat starches. They used a one-step extrusion cooking method. The glass transition temperature (Tg) of TPS decreases with increasing glycerol content, with potato TPS showing a Tg range from 187.7°C (7% glycerol) to 18.1°C (30% glycerol). The foams contained over 80% open-cell content, with density varying between 18.7 and 30.5 kg/m^3. Corn TPS-based foam with 3% plasticizer and 1% PVA was identified as the optimal formulation, having a compressibility of 70.54 kPa and a spring index of 0.83.

The widely used PVA polymerization can also affect the moisture content of extruded starch foams. The optimal foam formulation, with 12% moisture content and a PVA polymerization degree of 1700, achieved the highest foaming ratio (45.62), minimal water absorption (6.31%), and an improved recovery rate (88.95%). The composite foam showed the lowest apparent and bulk densities, i.e., 16.77 kg/m^3 and 11.36 kg/m^3, respectively. In addition, the composite foam exhibited excellent water resistance under various RH conditions, with water absorption rates of 6.31%, 12.81%, and 15.74% at 33%, 58%, and 85% RH, respectively (Liu et al., 2025).

Few studies have focused on the blending of starch with different materials, which shows a promising option to use starch foam in the packaging sector. However, TPS shows a strong affinity for water, which limits its effectiveness as a polymer. This issue can be solved by coating it with a hydrophobic material to prevent water from coming into contact with starch (Bergel et al., 2018). Yang et al. (2024) developed biodegradable composites using starch, plant fibers, and PLA to enhance the water resistance and mechanical properties of starch-based foams. The composites were prepared by foaming starch with plant fibers at 260°C. Then, the PLA-based hydrophobic film was used to coat the foam. The PLA significantly reduced water vapor permeability from 766.83 g/m$^2 \cdot$24 h to 664.89 g/m$^2 \cdot$24 h. It also decreased the hysteresis angle from 15.57 degrees to 8.59 degrees within the first 5 min of moisture exposure. The water absorption rate was also decreased from 12.3% to 7.9%. Furthermore, the PLA increased the tensile strength from 2.09 to 3.53 MPa and the compressive strength from 7.5 to 11.4 MPa. The inclusion of cellulose in the PLA film also contributed to the overall performance. Similar work on coating was also done to enhance the water resistance of starch foam. A beeswax coating was used in this case. The novel composite foam was developed by Chaireh et al. (2020) from native cassava starch and water hyacinth (WH) powder. The WH powder content was varied in different concentrations at 0, 3, 5, 7, and 10 wt.%. The formulation with 5 wt.% WH powder exhibited the best mechanical performance. It could withstand a flexural stress at a maximum load of 3.42 MPa, a flexural strain of 3.52%, and a modulus of 232.00 MPa. It also showed the lowest moisture content (6.77%) and the most uniform cell size distribution (0.44$\pm$0.09 mm). The beeswax coating helped the foam retain its shape after immersion in water and showed significantly reduced water solubility. Trongnit and Kaewtatip (2025) focused on alginate coatings on starch foam for food packaging. Starch foam was coated with alginate solutions at concentrations of 0, 0.5, 1.0, 1.5, and 2.0% w/v. The foam coated with 1.0% w/v alginate showed the highest flexural stress at maximum load (4.72 MPa), hardness (96), and the lowest mass loss in water (5.51%). This coating effect significantly outperformed the uncoated foam. Thermal stability was slightly improved by the alginate coating at 335°C (0.5, 1.0, and 1.5% w/v) compared to 333°C for the uncoated foam. The 1.0% w/v alginate-coated foam also had the highest hardness and water resistance, with a mass loss of 5.51% in water compared to 10.74% for the uncoated foam. Higher concentrations of alginate (1.5% and 2.0%) formed clumps on the foam surface, reducing their effectiveness in water resistance. The authors concluded that a 1.0% w/v alginate coating enhanced mechanical properties, water resistance, and thermal stability for starch foam as a sustainable food packaging material.

Starch is suitable for foam production because it can expand at high temperatures during extrusion (Georges et al., 2018). Extrusion is the most common method to develop starch foams. During extrusion, starch can effectively combine with other ingredients to achieve the desired expansion and structure (Soykeabkaew et al., 2015). Jebalia et al. (2019) studied the morphological structure of starch–legume protein composites developed through the extrusion process. Pea flour and pea starch–protein blends were extruded at various specific mechanical energies (SME) ranging from 100 to 2000 kJ/kg in a controlled temperature environment to avoid expansion. Higher SME increased starch swelling, resulting in larger areas of the continuous starch phase at the expense of the dispersed protein aggregates. It also demonstrated greater flexibility and higher rupture stress of the foam. Lower SME exhibited brittle behavior in the foam.

Recent studies have explored various methods to enhance the performance of starch-based foams by optimizing extrusion conditions. For example, extrusion temperature and moisture content control the density, expansion ratio, and overall structure of the foam. Cha et al. (2001) studied the impact of extrusion temperature on the moisture content of starch-based foam. The foams were made from a blend of 49% wheat and corn starch (CS), 33% synthetic polymers, 10.5% water, 7% blowing agents, and 0.5% nucleating agent. Extrusion temperatures ranged from 100°C to 160°C, with a screw speed of 100 rpm. The highest expansion ratio of the developed foam was noted at an extrusion temperature of 140°C. The moisture adsorption isotherms of the foams exhibited sigmoidal curves, typical of grains and grain products. As the extrusion temperature increased, the bulk density of the foams decreased. However, the starch-based foam was found to be 4–8 times denser than polystyrene foam. To improve the extrusion process, Wang et al. (2024a) investigated how multiple reextrusion cycles affect the material properties of starch-based biodegradable composites (SDC). They developed SDC using TPS, PLA, and poly(butylene adipate-co-terephthalate). The composites were subjected to three extrusion cycles, one by one. The reextrusion cycles reduced the TPS phase in the composite. The foams became more water-resistant after going through two extrusion cycles (SDC-2). However, the study also found that doing this repeatedly decreased the thermal stability and crystallinity of the foam.

Zhang et al. (2018) developed a novel method to enhance the thermal and mechanical properties of TPS. They incorporated calcium gluconate (CG) using reactive extrusion to achieve chelation between the calcium ions (Ca^{2+}) and the hydroxyl groups in the starch matrix. CS and glycerol were mixed with 5–20 wt.% CG, and the mixtures were extruded to create TPS-CG composites. The Fourier transform infrared (FTIR) and XRD analyses confirmed that CG formed a chemical bond with starch. The thermal stability of the TPS-CG composites was significantly improved, as shown by thermogravimetric analysis and FTIR. The impact strength, tensile strength, and melt velocity of the foam were increased by 52.4%, 44.2%, and 360%, respectively. These improvements were attributed to the chelation structure that enhanced molecular chain entanglement and stiffness. The structure acted as a bridge to transfer external energy between starch chains, which also increased the impact toughness of the material. The mild acidity of the gluconate group in CG helped to maintain the biodegradability of starch. This study provides a new approach to improve the properties of starch foam and make it more suitable for packaging.

16.5 Limitations and opportunities

Starch-based foam is a promising alternative to conventional petrochemical foams due to its biodegradability, renewability, and low cost. However, several challenges and limitations (Fig. 16.3) hinder its widespread adoption. The key challenge is its inherent moisture sensitivity, which leads to high water vapor permeability and reduced mechanical strength in humid conditions. This hydrophilic nature compromises its durability and stability. Advancements in surface treatments and coatings can effectively reduce moisture sensitivity (Jiang et al., 2020).

Polymer coating offers a simple solution to serve as a barrier by preventing water from coming into contact with the TPS matrix. This helps to improve the foam's water resistance where moisture sensitivity is a concern (Trongnit and Kaewtatip, 2025). Srisuwan and Baimark (2021) coated TPS foams with poly(L-lactide)-b-polyethylene glycol-b-poly(L-lactide)/poly(D-lactide) (PLLA-PEG-PLLA/PDLA) blends or stereo complex polylactide-b-polyethylene glycol-b-polylactide (scPLA-PEG-PLA). The coating improved the affinity between the coating phase and the surface of the foam compared to scPLA alone. They found that higher PDLA content enhanced water resistance in the foams. The coatings significantly reduced moisture absorption and improved dimensional stability in both cool and hot water. scPLA-PEG-PLLA coatings had no effect on mechanical properties but improved water resistance. The cell structure of the foams remained significantly unchanged after dip coating, which indicates that the coatings mainly affected the foam's surface properties and water resistance.

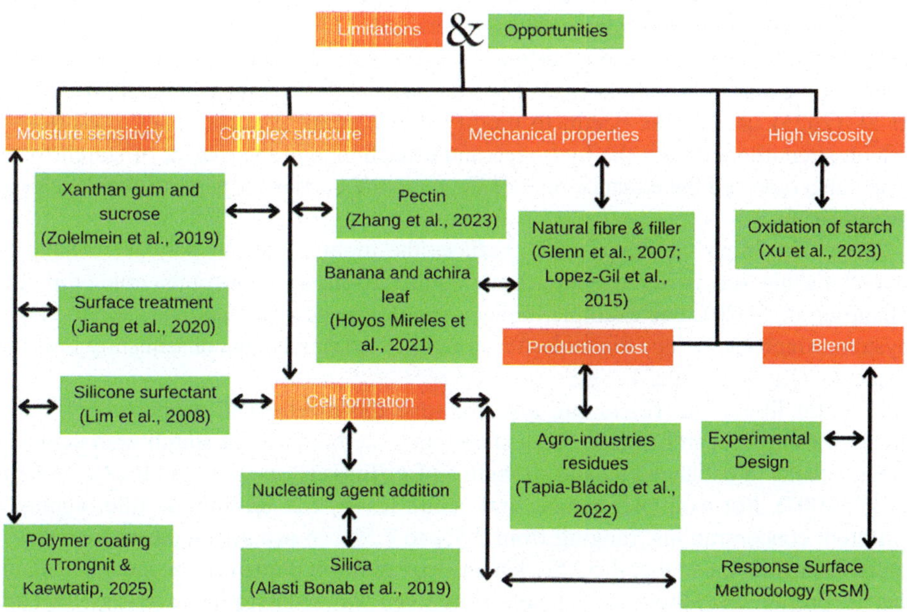

Fig. 16.3 Limitations and improvement stratagies of starch foam.

The surface properties of starch foam can be altered by the addition of surfactant. It basically forms intermolecular bonds and reduces surface tension (Yang et al., 2024). Polyurethane (PU) foams utilize silicone as a surfactant agent. The results demonstrated that silicone enhances the foam structure by facilitating cell formation (Lim et al., 2008).

Nucleating agents also promote the formation of cells during expansion by influencing the foam's morphological properties. Cell nucleation in polymeric foams takes place either through a homogeneous or a heterogeneous process. In homogeneous nucleation, aggregation and orientation happen spontaneously; therefore it requires super-cooling, making it thermodynamically unfavorable. In heterogeneous nucleation, nucleating agents act as impurities, reducing the free energy required for nucleation and accelerating crystallization. These impurities are intentionally added to promote nucleation and provide activation energy, resulting in a higher number of nuclei and faster crystallization. Silicate-based materials, such as silica (SiO_2), are commonly used as nucleating agents due to their good affinity with polymers and abundant natural availability. Bonab et al. (2019) incorporated 1 wt.% silica (aerogel) into PU foams and reported an enhanced foam cellular structure. The incorporation of silica increased cell density and reduced average cell size. This improvement in structure led to increased homogeneity and enhanced mechanical strength of the PU foam.

The complex microstructure of starch often restricts its applications. The ratio of composition of amylose and amylopectin plays a crucial role in choosing the starch. Therefore thermal processing requires precise control of temperature and moisture content to achieve proper gelatinization of starch. Incorporation of both xanthan gum and sucrose into starch-based composites leads to the creation of strong hydrogen bonds, which increase the transition temperature. This leads to an enthalpy change during starch pasting. These modifications improve the structural and thermal properties of the starch-based materials (Zolelmein et al., 2020). In another study, Zhang et al. (2025) used citrus pectin to modify the foaming process. They used pectin, which strengthened both the hydrogen and ester bonds in the matrix of starch foam. They also reported the enhancement of the moisture barrier property of the developed starch foam.

Starch has high viscosity, which reduces the mobility of the polymer chains. To overcome this, specialized techniques are required to ensure uniform material properties (Jiang et al., 2020). For example, tapioca starch can be oxidized using a Cu^{2+}/H_2O_2 system, which can reduce the molecular weight of the starch and consequently lower its viscosity (Xu et al., 2023).

Natural fibers and fillers can also improve the mechanical properties of starch foam. Various research studies have shown that the blending of starch with cellulose fibers results in enhanced tensile strength and stiffness (Lopez-Gil et al., 2015; Glenn et al., 2007). For example, banana and achira leaves can be used as filler agents at different concentrations, ranging from 2.5% to 7.5%, in oca starch-based foams. The study results revealed that the 2.5% banana leaf addition improved the mechanical and barrier properties of the foam. The resulting foam was found to be water-repellent and puncture-resistant, with good tensile strength (Hoyos Mireles et al., 2021).

Current research focuses on various factors such as foam formulation, extrusion process conditions, and foam properties. However, key factors such as starch concentration, additives, and process parameters must be optimized using advanced-level tools. Experimental design and response surface methodology can be used for the betterment of this area. Recent studies have shown that the production cost of starch-based foams is expensive. To reduce production costs, researchers are exploring low-cost raw materials such as residues from agroindustries. These residues may also impart additional properties, such as antimicrobial and antioxidant properties, which will make the foam more sustainable and functional (Tapia-Blácido et al., 2022).

16.6 Future perspectives

Starch is emerging as a sustainable polymer due to its abundance, biodegradability, and renewable nature. Various research works have been conducted to develop starch-based foam. However, the foaming properties of the foam have not improved yet. Besides that, the extrusion parameters should also be refined to make starch a suitable and effective option for sustainable packaging solutions (Wang et al., 2023). This section discusses the advancements in ongoing research and innovation in processing techniques and material modifications. The latter portion also provides a direction for further research in this area by identifying the key gaps that have not been addressed yet. Further research is required on these gaps to enable starch-based foam to fulfill its potential in the food packaging industry.

Starch foam can be incorporated with green foaming agents derived from sustainable sources like seaweed and agricultural by-products. Agar is a hydrocolloid used for its gelling and stabilizing properties. The seaweed-based agar was extracted from red-colored seaweed *Gracilaria fisheri*. The results concluded that all formulations yielded smooth and disposable foam trays without any cracks. The seaweed-based agar slightly altered the foam's off-white color but significantly reduced red and yellow color values. The agar content also decreased foam density. The optimal performance was observed in a 3:7 ratio of glycerol to agar. It resulted in a flexural stress of 3.23 MPa at maximum load, a modulus of 194.46 MPa, a hardness of 97.50, and a maximum weight loss temperature (T_{max}) of 337°C. This research highlights the underexploited application of agar in biodegradable packaging, which can enhance foam expansion, mechanical strength, and water resistance (Trongnit et al., 2024). The significant improvement can also be seen by incorporating fillers in cassava starch foams. Eggshells and shrimp shells were used as fillers. The addition of shrimp shells produced denser foams, while eggshells created a more expanded, less dense foam. The foam with eggshells also resulted in thinner cell walls and a narrower cell size distribution in the foam. This difference is attributed to the eggshell's nucleating effect during foaming. The incorporation of 10% eggshell powder resulted in low density (0.2056 g/cm^3) and a tighter cell size distribution in the foam. This work suggested that eggshell powder-incorporated starch foam is suitable for loose-fill packaging, where lightweight and efficient material use are important. This work has provided an idea for using natural fillers in foam production (Kaewtatip et al., 2018). The study shows that using fillers from agroindustrial waste can boost the development process to a higher scale. Therefore researchers are now looking for low-cost integration for further improvement. For example, kimchi cabbage powder (KCB-3) was added to CS to develop biodegradable foam containers. The powder is a byproduct of the kimchi manufacturing industry. The addition of KCB showed a denser porous microstructure in corn starch based foam containers. It also increased the density and tensile properties, as well as the water resistance property. However, a higher concentration of KCB-3 (above 20 wt.%) reduced mechanical strength and led to cracks in the containers. This study demonstrates KCB as an alternative waste to produce biodegradable packaging material (Yun et al., 2023).

Natural fibers are another effective method to improve the mechanical strength and water resistance of the starch matrix, resulting in more durable and functional packaging materials. Hassan and Fowler (2022) developed fully biodegradable starch foams reinforced with microfibrillated cellulose fibers (MFC-SFs). The MFC-SFs were cross-linked using two polysiloxane-based agents to improve their mechanical strength and water barrier properties. The two agents are PAEAPS-co-PDMS (cationic) and TMS-t-PDMS (nonionic). The mechanical properties of the MFC-SFs were significantly improved. The tensile strength increased from 1.78 MPa to 2.76 MPa with 5.32% PAEAPS-co-PDMS and to 2.53 MPa with TMS-t-PDMS. The TMS-t-PDMS cross-linked foams showed superior hydrophobicity compared to PAEAPS-co-PDMS. FTIR analysis confirmed the reduction in hydrophilicity of the foams after cross-linking, which enhanced their water resistance. The polysiloxane-cross-linked MFC-SFs hold great promise for further research.

Additive manufacturing of starch-based foams has revolutionized the way biodegradable materials can be used in packaging. Starch-based foams have also benefited from this technology. This advanced technique offers greater customization and efficiency in producing eco-friendly packaging. The 3D printing method can significantly affect the natural fibers in starch polymer matrices. Siddiqui et al. (2024) highlighted the advantages associated with the integration of starch with fibers, particularly hemp, jute, flax, and bamboo, for improvement in the mechanical properties of such polymer matrices. The review focused on extrusion-based 3D methods, particularly fused deposition modeling and fused filament fabrication (FFF). The findings suggested a promising future for natural polymers and fibers in additive manufacturing. These methods can help current research trends, which are aimed at developing stronger interfacial adhesion, improved cross-linking methods, and standardized testing protocols to contribute to a more sustainable and innovative manufacturing landscape. To increase the functionality of the fibers, they are often treated with FFF. Soybean hull-derived fibers treated with FFF were used to develop thermoplastic copolyester composites. The fiber treatment improved the quality of the printed parts, including surface finish, printing defects, and bonding between beads/layers. Composites reinforced with dilute acid-hydrolyzed fibers achieved a relative density of 99%. The pore size was also reduced from 81 to 39 μm. A fiber matrix without any defects was formed, which had an increased elastic modulus from 36 to 54 MPa. The toughness increased by up to 30%, and the stress of the composite also increased by 50% at 50% strain. The study highlights the potential of low-cost soybean hulls as an abundant agricultural by-product for reinforcements in natural fiber-reinforced composites (Balla et al., 2020). To support the current development, Dananjaya et al. (2024) reviewed the evolving connection of 3D printing with machine learning in relation to environmentally friendly polymers and their composites. The review states the importance of material preparation, design adaptation, and postprocessing techniques to enhance the structural integrity and performance of starch polymers. There is a research gap regarding the life cycle of biodegradable starch polymers in 3D printing. It also highlighted that the use of machine learning can potentially optimize manufacturing processes. The integration of 3D printing with machine learning can revolutionize sustainable manufacturing.

Nanotechnology is another potential approach that can enhance both the foam properties and its performance. Synthetic nanoparticles like silica nano particles (SNPs) can also improve the mechanical properties, thermal stability, and flame retardancy of jute/TPS composites. 1–4 wt.% SNPs were incorporated into the composite, which led to the following enhancements: tensile strength increased by 23%, flexural strength by 97%, and impact strength by 70%. FTIR spectroscopy confirmed a uniform dispersion of SNPs and the formation of strong Si–O–Si bonds between silica and the jute/TPS matrix. This work opens a new path to using a simple and effective strategy for developing high-strength, biodegradable, flame-retardant composites to expand their potential applications in the packaging sector (Azad et al., 2022). However, the search for green sources led to cellulose nanofibers (CNFs) that can affect the thermal and mechanical properties, density, and water absorption of TPS-based foam composites. The incorporation of CNFs enhances the thermal stability, storage modulus (E'), loss modulus (E''), and damping factor ($\tan \delta$) of the foams. The addition of 1.5 wt.% CNF significantly increased the glass transition temperature (T_g) and provided superior heat resistance. However, the nanofiber reduced the apparent density and water absorption of foam composites. The nanoparticles (1.0 or 1.5 wt.%) decreased cell size and increased cell density of the composites, as seen in SEM images. The study further found that 1.5 wt.% CNF loading significantly enhanced thermal stability, decomposition temperature, residual content, and viscoelastic properties of the TPS foam composites due to hydrogen bonding with the TPS matrix. This study seeks to address the existing gap in the literature regarding TPS-based foams reinforced with CNFs by utilizing extrusion and compression-molding techniques for preparation (Ghanbari et al., 2018). A similar study was also conducted by Adeyi et al. (2021), where they incorporated nanoscale natural fiber into cassava starch thermoplastic (CSTP) composites. The fiber was derived from *Momordica augustisepala*. The nanofiber-incorporated CSTP achieved a tensile strength of 6.35 MPa and a tensile modulus of 245.22 MPa at 9 wt.% fiber fractions. Fuzzy logic artificial intelligence method (FLAIM) was employed for modeling and predicting the mechanical properties of these composites and was found effective. The model showed strong accuracy with a high level of precision in the coefficient of determination (R^2). The average model accuracy and R^2 were found to be between 92.08% and 97.18% and 0.7981 and 0.9969, respectively. The potential of utilizing micro- and nano-momordica augustisepala stem natural fibre is cost-effective and environmentally friendly and can also enhance the mechanical properties of TPS.

The application of FLAIM for property prediction opens a new path for an innovative approach to material design, offering insights for both technical and economic development in composite production.

16.7 Conclusion

The global concern over plastic degradation has led us to find an alternative solution for the food packaging sector. The main requirements are finding an environmentally friendly and biodegradable polymer with good performance. Starch-based foams have gained huge attention as a promising solution in this case due to their biodegradability and renewability. They can be a sustainable alternative to petroleum-based plastics. Among various manufacturing techniques, extrusion emerges as an effective method for foam production. Extrusion produces larger-sized cell foam for the production of low-density foam. Thus it is mainly used for lightweight foams suitable for transport packaging. However, pure starch-based foams have limitations such as low water resistance, brittleness, and limited flexibility. Modified starches can be a solution to improve mechanical properties. In addition, blending with hydrophobic polymers, latex, and soy protein can also increase the mechanical properties of the foam, as reported by a few researchers. The continuous research on starch-based foams is promoting innovations such as nanocomposites and self-reinforced composites. Reinforcing starch with natural fibers shows enhanced strength and flexibility. Nanotechnology can also offer new opportunities to improve its properties, even with minimal filler content. The hydrophilic nature of starch makes it susceptible to water uptake, which affects the water vapor permeability and tensile strength of the foam. Coating with acrylate epoxidized soybean oil (AESO) has been found to be the most effective method to reduce moisture absorption and improve gas permeability. Polyethyleneimine has also been applied to facilitate the adhesion of a hydrophobic AESO coating to the starch matrix to provide moisture resistance.

So far, the significant potential of starch-based materials in the packaging industry has been reported. However, some hurdles need to be tackled before they can fully replace traditional plastic foams. More optimized mechanical properties of starch foams are needed to overcome these challenges while meeting the requirement of biodegradability. Therefore further studies are needed to optimize the performance of starch blends, especially in terms of moisture sensitivity and thermomechanical properties. The future of starch-based foams lies in improving their processing techniques, reducing costs, and exploring new applications. The use of agrowaste for starch production offers cost-effective and sustainable packaging solutions. Furthermore, multilayer films with hydrophobic coatings can reduce starch's hydrophilic nature, thereby improving packaging performance. Future research should focus on developing TPS-based multilayer foams with enhanced processing, permeability, and biodegradability to meet a wide range of packaging needs. Recent research has made significant improvements in starch foam. Research on innovative processing techniques and material modifications is still ongoing.

However, there are still some gaps that need to be addressed. The processing parameters and life cycle of biodegradable polymers are yet to be fully explored. A few research efforts are now focusing on machine learning and artificial intelligence, which can significantly enhance these processes in the packaging industry to create a sustainable future.

References

Adeyi, A.J., Durowoju, M.O., Adeyi, O., Oke, E.O., Olalere, O.A., Ogunsola, A.D., 2021. Momordica augustisepala L. stem fibre reinforced thermoplastic starch: mechanical property characterization and fuzzy logic artificial intelligent modelling. Results Eng. 2590-1230. 10. Available from: https://www.journals.elsevier.com/results-in-engineering. doi:10.1016/j.rineng.2021.100222.

Aguilar, G.J., Tapia-Blácido, D.R., 2023. Evaluating how avocado residue addition affects the properties of cassava starch-based foam trays. Int. J. Biol. Macromol. 1879-0003. 240. Available from: www.elsevier.com/locate/ijbiomac. doi:10.1016/j.ijbiomac.2023.124348.

AL-Fakih, G.O.A., Ilyas, R.A., Huzaifah, M.R.M., El-Shafay, A.S., 2024. Recent advances in sago (Metroxylon sagu) fibres, biopolymers, biocomposites, and their prospective applications in industry: a comprehensive review. Int. J. Biol. Macromol. 1879-0003. 269. Available from: https://www.sciencedirect.com/science/journal/01418130. doi:10.1016/j.ijbiomac.2024.132045.

Azad, M.M., Ejaz, M., Shah, A.U.R., Afaq, S.K., Song, J.I., 2022. A bio-based approach to simultaneously improve flame retardancy, thermal stability and mechanical properties of nano-silica filled jute/thermoplastic starch composite. Mater. Chem. Phys. 0254-0584. 289. Available from: http://www.journals.elsevier.com/materials-chemistry-and-physics. doi:10.1016/j.matchemphys.2022.126485.

Balla, V.K., Tadimeti, J.G.D., Kate, K.H., Satyavolu, J., 2020. 3D printing of modified soybean hull fiber/polymer composites. Mater. Chem. Phys. 0254-0584. 254. Available from: http://www.journals.elsevier.com/materials-chemistry-and-physics. doi:10.1016/j.matchemphys.2020.123452.

Bergel, B.F., Araujo, L.L., Dos Santos da Silva, A.L., Campomanes Santana, R.M., 2020. Effects of silylated starch structure on hydrophobization and mechanical properties of thermoplastic starch foams made from potato starch. Carbohydr. Polym. 0144-8617. 241. Available from: http://www.elsevier.com/wps/find/journaldescription.cws_home/405871/description#description. doi:10.1016/j.carbpol.2020.116274.

Bergel, B.F., da Luz, L.M., Santana, R.M.C., 2018. Effect of poly(lactic acid) coating on mechanical and physical properties of thermoplastic starch foams from potato starch. Prog. Org. Coat. 0300-9440. 118, 91–96. Available from: www.elsevier.com/locate/porgcoat. doi:10.1016/j.porgcoat.2018.01.029.

Bonab, S.A., Moghaddas, J., Rezaei, M., 2019. In-situ synthesis of silica aerogel/polyurethane inorganic-organic hybrid nanocomposite foams: characterization, cell microstructure and mechanical properties. Polymer. 0032-3861. 172, 27–40. doi:10.1016/j.polymer.2019.03.050.

Cazón, P., Velazquez, G., Ramírez, J.A., Vázquez, M., 2017. Polysaccharide-based films and coatings for food packaging: a review. Food Hydrocoll. 0268-005X. 68, 136–148. Available from: https://www.journals.elsevier.com/food-hydrocolloids. doi:10.1016/j.foodhyd.2016.09.009.

Cha, J.Y., Chung, D.S., Seib, P.A., Flores, R.A., Hanna, M.A., 2001. Physical properties of starch-based foams as affected by extrusion temperature and moisture content. Ind. Crops Prod. 0926-6690. 14 (1), 23–30. doi:10.1016/s0926-6690(00)00085-6.

Chaireh, S., Ngasatool, P., Kaewtatip, K., 2020. Novel composite foam made from starch and water hyacinth with beeswax coating for food packaging applications. Int. J. Biol. Macromol. 0141-8130. 165 (Pt A), 1382–1391. doi:10.1016/j.ijbiomac.2020.10.007.

Dananjaya, S.A.V., Chevali, V.S., Dear, J.P., Potluri, P., Abeykoon, C., 2024. 3D printing of biodegradable polymers and their composites—current state-of-the-art, properties, applications, and machine learning for potential future applications. Prog. Mater. Sci. 0079-6425. 146, 101336. doi:10.1016/j.pmatsci.2024.101336.

Debiagi, F., Mali, S., Grossmann, M.V.E., Yamashita, F., 2011. Biodegradable foams based on starch, polyvinyl alcohol, chitosan and sugarcane fibers obtained by extrusion. Braz. Arch. Biol. Technol. 1678-4324. 54 (5), 1043–1052. Available from: http://www.scielo.br/pdf/babt/v54n5/a23v54n5.pdf. doi:10.1590/S1516-89132011000500023.

Diyana, Z.N., Jumaidin, R., Selamat, M.Z., Ghazali, I., Julmohammad, N., Huda, N., Ilyas, R.A., 2021. Physical properties of thermoplastic starch derived from natural resources and its blends: a review. Polymers. 2073-4360. 13 (9), 5–20. Available from: https://www.mdpi.com/2073-4360/13/9/1396/pdf. doi:10.3390/polym13091396.

Figueiró, C.D.S., Calcagno, C.I.W., Santana, R.M.C., 2024. Starch foams and their additives: a brief review. Starch/Staerke. 1521-379X. 76 (3-4). Available from: http://onlinelibrary.wiley.com/journal/10.1002/(ISSN)1521-379X. doi:10.1002/star.202300012.

Georges, A., Lacoste, C., Damien, E., 2018. Effect of formulation and process on the extrudability of starch-based foam cushions. Ind. Crops Prod. 0926-6690. 115, 306–314. doi:10.1016/j.indcrop.2018.02.001.

Ghanbari, A., Tabarsa, T., Ashori, A., Shakeri, A., Mashkour, M., 2018. Thermoplastic starch foamed composites reinforced with cellulose nanofibers: thermal and mechanical properties. Carbohydr. Polym. 0144-8617. 197, 305–311. doi:10.1016/j.carbpol.2018.06.017.

Glenn, G.M., Klamczynski, A., Holtman, K.M., Chiou, B.-S., Orts, W.J., Wood, D., 2007. Cellulose fiber reinforced starch-based foam composites. J. Biobased Mater. Bioenergy. 1556-6560. 1 (3), 360–366. doi:10.1166/jbmb.2007.010.

Gu, Y., Shen, Y., Wu, T., Hu, F., Wang, T., 2024. Comprehensive enhancement of flame retardant starch/cellulose/diatomite composite foams via metal-organic coordination. Int. J. Biol. Macromol. 0141-8130. 266 (Pt 2), 131313. doi:10.1016/j.ijbiomac.2024.131313.

Han, C., Yuan, Z., Sun, B., Qiu, H., Huang, Y., Zhang, M., 2024. Effects of carboxymethyl cellulose concentration on mechanical, viscoelastic properties, and thermal properties of starch/plant fiber foaming tableware materials with foam structure. Ind. Crops Prod. 0926-6690. 215, 118586. doi:10.1016/j.indcrop.2024.118586.

Hassan, M.M., Fowler, I.J., 2022. Thermal, mechanical, and rheological properties of microfibrillated cellulose-reinforced starch foams crosslinked with polysiloxane-based crosslinking agents. Int. J. Biol. Macromol. 1879-0003. 205, 55–65. Available from: www.elsevier.com/locate/ijbiomac. doi:10.1016/j.ijbiomac.2022.02.017.

Hoover, R., 2001. Composition, molecular structure, and physicochemical properties of tuber and root starches: a review. Carbohydr. Polym. 0144-8617. 45 (3), 253–267. doi:10.1016/s0144-8617(00)00260-5.

Hoyos Mireles, B.J., Del Rocío Cieza Díaz, E., Castro Medina, R.S., 2021. Ecological trays based on banana (*Musa paradisiaca*) and achira (*Canna indica*) leaf blades: physical, mechanical and chemical characteristics. Agroind. Sci. 2226-2989. 11 (1), 87–96. doi:10.17268/agroind.sci.2021.01.11.

Jebalia, I., Maigret, J.-E., Réguerre, A.-L., Novales, B., Guessasma, S., Lourdin, D., Valle, G.D., Kristiawan, M., 2019. Morphology and mechanical behaviour of pea-based starch-protein composites obtained by extrusion. Carbohydr. Polym. 0144-8617. 223, 115086. doi:10.1016/j.carbpol.2019.115086.

Jiang, T., Duan, Q., Zhu, J., Liu, H., Yu, L., 2020. Starch-based biodegradable materials: challenges and opportunities. Adv. Ind. Eng. Polym. Res. 2542-5048. 3 (1), 8–18. doi:10.1016/j.aiepr.2019.11.003.

Kaewtatip, K., Chiarathanakrit, C., Riyajan, S.A., 2018. The effects of egg shell and shrimp shell on the properties of baked starch foam. Powder Technol. 1873-328X. 335, 354–359. Available from: www.elsevier.com/locate/powtec. doi:10.1016/j.powtec.2018.05.030.

Karan, H., Funk, C., Grabert, M., Oey, M., Hankamer, B., 2019. Green bioplastics as part of a circular bioeconomy. Trends Plant Sci. 1360-1385. 24 (3), 237–249. doi:10.1016/j.tplants.2018.11.010.

Kshitij, M., Dhal, M.K., Banerjee, A., Prasannavenkadesan, V., Katiyar, V., Kumar, A., 2025. Thermoplastic starch based melt processed films for packaging applications: experimental and computational studies. Mater. Today Commun. 2352-4928. 42. Available from: https://www.sciencedirect.com/science/journal/23524928. doi:10.1016/j.mtcomm.2024.111475.

Laftah, W.A., 2017. Starch based biodegradable blends: a review. Int. J. Eng. Res. Technol. 6, 19.

Lim, H., Kim, S.H., Kim, B.K., 2008. Effects of silicon surfactant in rigid polyurethane foams. Express Polym. Lett. 1788-618X. 2 (3), 194–200. doi:10.3144/expresspolymlett.2008.24.

Liu, F., Cao, Y., Ren, J., Xie, Y., Xiao, X., Zou, Y., Bai, H., Zhang, X., Chen, Y., 2025. Optimization of starch foam extrusion through PVA polymerization, moisture content control, and CMCS incorporation for enhanced antibacterial cushioning packaging. Carbohydr. Polym. 0144-8617. 347, 122763. doi:10.1016/j.carbpol.2024.122763.

Lopez-Gil, A., Silva-Bellucci, F., Velasco, D., Ardanuy, M., Rodriguez-Perez, M.A., 2015. Cellular structure and mechanical properties of starch-based foamed blocks reinforced with natural fibers and produced by microwave heating. Ind. Crops Prod. 0926-6690. 66, 194–205. doi:10.1016/j.indcrop.2014.12.025.

Mitrus, M., Moscicki, L., 2014. Extrusion-cooking of starch protective loose-fill foams. Chem. Eng. Res. Des. 0263-8762. 92 (4), 778–783. doi:10.1016/j.cherd.2013.10.027.

Montilla-Buitrago, C.E., Gómez-López, R.A., Solanilla-Duque, J.F., Serna-Cock, L., Villada-Castillo, H.S., 2021. Effect of plasticizers on properties, retrogradation, and processing of extrusion-obtained thermoplastic starch: a review. Starch/Staerke. 1521-379X. 73 (9-10). Available from: http://onlinelibrary.wiley.com/journal/10.1002/(ISSN)1521-379X. doi:10.1002/star.202100060.

Quilez-Molina, A.I., Meins, J.F.L., Charrier, B., Dumon, M., 2024. Starch-fibers composites, a study of all-polysaccharide foams from microwave foaming to biodegradation. Carbohydr. Polym. 0144-8617. 328. Available from: https://www.sciencedirect.com/science/journal/01448617. doi:10.1016/j.carbpol.2023.121743.

Razza, F., Innocenti, F.D., Dobon, A., Aliaga, C., Sanchez, C., Hortal, M., 2015. Environmental profile of a bio-based and biodegradable foamed packaging prototype in comparison with the current benchmark. J. Clean. Prod. 0959-6526. 102, 493–500. doi:10.1016/j.jclepro.2015.04.033.

Siddiqui, M.A.S., Rabbi, M.S., Ahmed, R.U., Billah, M.M., 2024. Biodegradable natural polymers and fibers for 3D printing: a holistic perspective on processing, characterization, and advanced applications. Clean. Mater. 2772-3976. 14. Available from: https://www.sciencedirect.com/science/journal/27723976. doi:10.1016/j.clema.2024.100275.

Singh, A.A., Genovese, M.E., 2021. Green and sustainable packaging materials using thermoplastic starch. In: Sustainable Food Packaging Technology. Wiley, Italy, 133–160. Available from: https://onlinelibrary.wiley.com/doi/book/10.1002/9783527820078. doi:10.1002/9783527820078.ch5.

Singh, S.K. and Muthukumarappan, K., 2017. Rheological characterization and CFD simulation of soy white flakes based dough in a single screw extruder. J. Food Process. Eng. 40 (2). doi:10.1111/jfpe.12368.

Singh, S.K., Singha, P. and Muthukumarappan, K., 2019b. Modeling and optimizing the effect of extrusion processing parameters on nutritional properties of soy white flakes-based extrudates using response surface methodology. Anim. Feed Sci. Technol. 254, 114197. doi:10.1016/j.anifeedsci.2019.06.001.

Singha, P., Singh, S.K. and Muthukumarappan, K., 2019a. Textural and structural characterization of extrudates from apple pomace, defatted soy flour and corn grits. J. Food Process. Eng. 42 (4). doi:10.1111/jfpe.13046.

Singha, P., Singh, S.K., Muthukumarappan, K., Krishnan, P., 2018. Physicochemical and nutritional properties of extrudates from food grade distiller's dried grains, garbanzo flour, and corn grits. Food Sci. Nutr. 6 (7), 1914–1926. doi:10.1002/fsn3.769.

Soykeabkaew, N., Thanomsilp, C., Suwantong, O., 2015. A review: starch-based composite foams. Compos. Part A Appl. Sci. Manuf. 1359-835X. 78, 246–263. doi:10.1016/j.compositesa.2015.08.014.

Srisuwan, Y., Baimark, Y., 2021. Improvement of water resistance of thermoplastic starch foams by dip-coating with biodegradable polylactide-b-polyethylene glycol-b-polylactide copolymer and its blend with poly(D-lactide). Prog. Org. Coat. 0300-9440. 151, 106074. doi:10.1016/j.porgcoat.2020.106074.

Su, B., Xie, F., Li, M., Corrigan, P.A., Yu, L., Li, X., Chen, L., 2009. Extrusion processing of starch film. Int. J. Food Eng. 1556-3758. 5 (1). doi:10.2202/1556-3758.1617.

Surendren, A., Mohanty, A.K., Liu, Q., Misra, M., 2022. A review of biodegradable thermoplastic starches, their blends and composites: recent developments and opportunities for single-use plastic packaging alternatives. Green Chem. 1463-9262. 24 (22), 8606–8636. doi:10.1039/d2gc02169b.

Tapia-Blácido, D.R., Aguilar, G.J., de Andrade, M.T., Rodrigues-Júnior, M.F., Guareschi-Martins, F.C., 2022. Trends and challenges of starch-based foams for use as food packaging and food container. Trends Food Sci. Technol. 0924-2244. 119, 257–271. Available from: http://www.elsevier.com/wps/find/journaldescription.cws_home/601278/description#description. doi:10.1016/j.tifs.2021.12.005.

Trongnit, J., Kaewtatip, K., 2025. Performances of starch foam improved by an alginate coating. Ind. Crops Prod. 0926-6690. 223, 120155. doi:10.1016/j.indcrop.2024.120155.

Trongnit, J., Mayakun, J., Kaewtatip, K., 2024. The effect of agar from the seaweed *Gracilaria fisheri* on properties of biodegradable starch foam. Int. J. Biol. Macromol. 0141-8130. 273 (Pt 1), 132952. doi:10.1016/j.ijbiomac.2024.132952.

Uslu, M.K., Polat, S., 2012. Effects of glyoxal cross-linking on baked starch foam. Carbohydr. Polym. 0144-8617. 87 (3), 1994–1999. doi:10.1016/j.carbpol.2011.10.008.

V Peshkam, M., 2022. Eradicating plastic pollution globally by 2030. Academicus Int. Sci. J. 2079-3715. 26, 60–77. doi:10.7336/academicus.2022.26.04.

Wang, L., Li, D., Ye, L., Zhi, C., Zhang, T., Miao, M., 2024a. Starch-based biodegradable composites: effects of in-situ re-extrusion on structure and performance. Int. J. Biol. Macromol. 0141-8130. 266 (Pt 1), 130869. doi:10.1016/j.ijbiomac.2024.130869.

Wang, S., Li, J., Li, F., Li, J., Zhang, C., Ji, M., Man, J., Peng, S., 2024b. Study on the influential effect of different $CaCO_3$ particle sizes on the internal pore structure of starch/fiber foaming composite materials. Mater. Today Commun. 2352-4928. 38, 107867. doi:10.1016/j.mtcomm.2023.107867.

Wang, S., Zhang, P., Li, Y., Li, J., Li, X., Yang, J., Ji, M., Li, F., Zhang, C., 2023. Recent advances and future challenges of the starch-based bio-composites for engineering applications. Carbohydr. Polym. 0144-8617. 307, 120627. doi:10.1016/j.carbpol.2023.120627.

Willett, J.L., Shogren, R.L., 2002. Processing and properties of extruded starch/polymer foams. Polymer. 0032-3861. 43 (22), 5935–5947. doi:10.1016/s0032-3861(02)00497-4.

Xu, R., Yu, H., Dong, H., Ye, Y., Xie, S., 2023. Preparation and properties of modified starch-based low viscosity and high consolidation foam dust suppressant. J. Hazard. Mater. 0304-3894. 452, 131238. doi:10.1016/j.jhazmat.2023.131238.

Yang, J., Li, Y., Li, X., Ji, M., Peng, S., Man, J., Zhou, L., Li, F., Zhang, C., 2024. Starch-fiber foaming biodegradable composites with polylactic acid hydrophobic surface. Int. J. Biol. Macromol. 0141-8130. 267 (Pt 1), 131406. doi:10.1016/j.ijbiomac.2024.131406.

Yun, S.M., Kang, M., Park, S.Y., Eun, J.B., Chun, H.H., 2023. Characterization of biodegradable corn starch-based foam container incorporating kimchi cabbage (*Brassica rapa L. pekinensis*) by-product. LWT. 0023-6438. 188. Available from: https://www.journals.elsevier.com/lwt. doi:10.1016/j.lwt.2023.115432.

Zhang, S., Lin, Z., Jiang, G., Wang, J., Wang, D.Y., 2018. Construction of chelation structure between Ca^{2+} and starch via reactive extrusion for improving the performances of thermoplastic starch. Compos. Sci. Technol. 0266-3538. 159, 59–69. Available from: http://www.journals.elsevier.com/composites-science-and-technology/. doi:10.1016/j.compscitech.2018.02.027.

Zhang, Y., Bao, X., Yang, W., Liu, H., Yu, L., 2025. Impact of citrus pectin on the foaming behavior, structures, and barrier properties of the starch foam blocks. Food Chem. 0308-8146. 463 (Pt 4), 141500. doi:10.1016/j.foodchem.2024.141500.

Zhang, Y., Li, C., Fu, X., Ma, N., Bao, X., Liu, H., 2022. Characterization of a novel starch-based foam with a tunable release of oxygen. Food Chem. 0308-8146. 389, 133062. doi:10.1016/j.foodchem.2022.133062.

Zhang, Y., Li, Z., Bao, X., Zeng, J., Gao, H., Liu, H., 2024. Evaluating how citrus fiber affects the cell structure and functions of starch-based foam. Ind. Crops Prod. 0926-6690. 218, 118984. doi:10.1016/j.indcrop.2024.118984.

Zolelmein, A., Movahhed, S., Azizi, M.H., Chenarbon, H.A., 2020. Assessment of simultaneous addition of sucrose and xanthan effects on the thermal, pasting, and rheological behavior of corn starch. J. Texture Stud. 1745-4603. 51 (3), 453–463. Available from: http://onlinelibrary.wiley.com/journal/10.1111/(ISSN)1745-4603. doi:10.1111/jtxs.12497.

Extrusion processing of foods: Product and process standardization

17

Suvendu Bhattacharya
Food Engineering Department, CSIR-Central Food Technological Research Institute, Mysore, India

17.1 Introduction

In the process of extrusion cooking, an extruder (single- or twin-screw) is employed to process the moist, powdery feed that gets compressed, mixed, sheared, heated/cooked, and finally exits through a restriction called a die. The process thus offers shaped continuous strands that are ultimately cut to an appropriate size and then subjected to finishing steps like drying, flattening, and the addition of flavoring/nutritional ingredients. The extrusion shaping process is a simplified version of food extrusion technology, where thermal (or cooking) treatment is omitted, while the objective is to shape the product; pasta products are the typical examples, wherein the finished product requires adequate cooking, baking, or frying at the consumers' end to make them edible.

The popularity of the extrusion/extrusion cooking process is due to several advantages that have made it an expanding food processing technology over the last 40 years. The worth-mentioning advantages are the continuous processing facility, an energy-efficient process, high production capacity, the ability to perform several unit operations (like mixing, cooking, and shaping) in one step, and the reduced cost of production.

Many designs are available for a food extruder. The cost-effective single-screw cooking extruders are difficult to control, though this limitation can be minimized in the costlier version of the twin-screw extruders. The coextrusion facility is required for developing a sophisticated product like filled snacks, employing a twin-screw extruder, as the nature of raw materials is different and needs to be fed through two input devices. Fully intermeshing corotating twin-screw extruders are preferred by food industries for their improved operational advantages and efficient control of the machine. A two-stage extrusion process in series, used for the production of half-products or products with unique shapes, is also practiced. In this case, the first extruder is used for cooking/gelatinization of starch-rich feed, while the second extruder provides the desired shape. Another improved extruder design is called turbo extrusion, which operates at a very high rotational speed to enhance production capacity.

Extrusion Cooking for Food and Feed. https://doi.org/10.1016/B978-0-443-29987-2.00017-8

Industrial applications of food extrusion technology include the manufacture of several ready-to-cook (RTC) and ready-to-eat (RTE) snacks, flat breads, breakfast cereals, baby foods, texturized vegetable proteins (TVP) or texturized plant proteins (TPP), and pasta products. In recent years, the high-moisture extrusion process for producing wet meat analogs/fake meat has been in use. As these extruded products have different characteristics, the raw materials vary from high starch-rich to protein-rich feed ingredients, and extrusion processing conditions also differ. At the same time, postextrusion treatments are equally important for improved consumer acceptance, and thus postextrusion processes also need to be standardized.

17.2 Diversity of extruders and the extrusion process

Extrusion technology, applied to food systems, has now diversified into many areas to develop RTE and RTC products, as well as specialized raw materials that are ultimately used to create other convenience foods. The extruders used also differ widely, such as short-length single-screw extruders for developing expanded snacks like corn balls and cheese balls, longer-length high-shear single-screw extruders for producing TVP or TPP, forming extruders (pasta) where grain flour-based products are shaped, twin-screw extruders to produce paste-filled snacks, and long-die low-temperature extrusion systems where a plant-based raw material is given the size and shape of a high-moisture-containing imitation/fake/mock meat.

Collet extruders are typically short-length single-screw extruders [length-to-diameter (L/D) ratio, is generally <10] that are usually operated at a relatively high rotational speed (screw speed) of 400 rpm or above. Collet extruders are high-shear machines that are used to make directly expanded snacks by gelatinizing starch at a high temperature (melt temperature is typically between 130°C and 140°C) from a relatively dry feed (12%–14% moisture content) of ingredients, mostly from corn grits or a mixture of corn, rice, and legume grits. These extruders are low-cost machines and are often used to prepare highly expanded [bulk density (BD) is around 50 kgm^{-3}] RTE snacks like corn balls, corn curls, cheese balls, and small rod-shaped products. After extrusion cooking, the products may be dried and flavored with salt-spice powder, cheese flavor, oleoresin, etc. The details of flavoring snack foods are available. A limitation of collet extruders is the high wear and tear of the extruder components, such as the barrel lining, extruder screw, and die nozzle. Often, hardened steel is used for the construction of the components of collet extruders. Other types of single- and twin-screw extruders are also available (Frame, 1994; Harper, 1981a).

The products of food extrusion technology have a wide range of textures, extents of cooking, expansion characteristics, density, structure, and moisture content. The corn ball, a typical expanded product with a microstructure containing several air cells or cavities, is a very soft-textured product with low moisture content (3%–6%) that can melt easily in the mouth. The dried TVP is hard in texture with aligned fibers and exhibits a compact microstructure after extrusion cooking and subsequent drying. It is an RTC product requiring adequate cooking at the consumer's end. The expansion of the product is minimal, and the moisture content typically varies between 5% and 9%. On the contrary, a simple pasta product is a typical example of the forming extrusion process. It is an uncooked and shaped product with a moisture content of about 30%. Postextrusion drying is mandatory to make the product shelf-stable. The product needs to be cooked thoroughly before consumption. The high-moisture, plant-protein-based feed delivers products like imitation meat with a moisture content of 60%–70% and is a soft product. The product requires low-temperature storage.

Many types of raw materials are used for extrusion processing. The ingredients may be derived from plant sources like cereals, legumes, oilseeds, starches, protein isolates and concentrates, tuber and root crops; animal sources like fish and meat, dairy and eggs; and aids for extrusion like plasticizers, emulsifiers, and nucleating agents. Furthermore, salt, sugar, permitted color and flavor, oil, antioxidants, nutritional ingredients, taste improvers, binders, and fortification agents are also added during the postextrusion step, which consists of flavoring and coating. Thus the feed that enters the extruder is a heterogeneous mixture of several ingredients that undergo conveying, compression, shearing, and heating inside the extruder, and finally comes out of the die end.

Thus standardization of the extrusion cooking process is a necessity for developing the product and manufacturing it on a large scale. The ultimate objectives are, therefore, to produce a consistent quality product, ensure the safety of the machine and product, and maximize the output while minimizing energy expenditure. It is worth mentioning here that extrusion cooking practice is an energy-intensive process characterized and quantified by specific mechanical energy (SME), which is defined as the net energy expenditure for producing one kilogram of the extruded product (Guha et al., 1997). The SME value (Eq. 17.1) is an important index with regard to scaling up an extrusion process from laboratory to pilot plant scale and, finally, to the production stage.

$$\mathrm{SME} = \frac{\text{rpm of screw (run)}}{\text{rpm of screw (rated)}} \times \frac{\text{\% torque (run)}}{100} \times \frac{\text{motor power (rated)}}{\text{production capacity}} \tag{17.1}$$

The residence time distribution, or the average residence time of feed inside the extruder cooker, is another index that can help with scaling up. This index is kept the same or close to that of the small-scale experiments and large-scale industrial trials, and is thus considered to be a useful approach.

17.3 Complexity of the extrusion cooking process

A large number of variables are associated with the extrusion cooking process. These variables are linked to

(a) Extruder hardware variables include the type and configuration of the extruder screw and screw elements, the geometry of the die and die nozzle, the gap between the inner surface of the barrel and the tip of the extruder screw, the pitch of the screw, the L/D ratio, the presence of grooves in the barrel, etc. Furthermore, the extent of compression the feed undergoes during extrusion depends on the die opening and screw profile (presence of a mixing disk, kneading block, reverse pitch screw element, and cut flight), and even the presence of a tapered screw and barrel near the discharge end.

(b) Processing system variables like temperature at different zones of the extruder barrel and at the die, pressure profile in different zones and at the die end, feed rate, torque generated on the extruder screw, rotational speed of the screw, extent of the shear force acting on the feed, etc.

(c) Feed variables like moisture, fat, and protein contents of the feed, particle size distribution, use of permitted emulsifying, stabilizing, and nucleating agents, and the extent of pretreatment or preheating of the feed just before entering the extruder barrel.

(d) The extent of heat input can occur through the addition of steam, hot water, and heat transmitted through the barrel. The dissipation of mechanical energy due to the rotation of screw(s) may also serve as a source of heat input during extrusion cooking. However, all these variables are not strictly independent and, on many occasions, they are interdependent. For instance, if the screw speed is increased, the shearing action will also increase.

The extruder cooker has numerous product variables that depend on the product being processed or developed. These variables include the extent of expansion, density, extent of cooking required, gelatinization of starch, texture, etc. The variables concerning postextrusion treatments (like drying, baking, toasting, or frying) are also to be considered while developing a flavored or coated extruded product, such as a salt-spicy RTE expanded snack or sugar/chocolate-coated breakfast cereals. The product variables may include the required moisture, sugar, oil, and salt contents of the finished product, as well as the method of flavoring or coating, such as sprinkling or spray coating and external surface characteristics.

17.4 Methods to handle large number of variables

Process standardization is dependent on the products to be manufactured by the extruders. The extruded products, based on the type of processing conditions, may be broadly classified into three categories. The first group is the extrusion-cooked popular products like breakfast cereals, snacks, TVP, modified/gelatinized flour/starch, etc. Thermal input occurs from the dissipation of mechanical energy (due to the rotation of the extruder screw) into thermal energy, external heat input (using steam or electrical heating through barrels), or a combination of both systems. Temperature control is a critical issue for these products to achieve the required level of cooking. The second product category is extrusion-shaped produce, like pasta-like products that are formed without intentionally added thermal inputs, as cooking is not the target of these products. The temperature rise is minimal during extrusion and the extent of cooking; these products thus need a reasonable extent of drying after extrusion and a final cooking in water at the consumer's end. The third group is a low-temperature, high-moisture extrusion process where fake meat/mock meat is produced from feed containing plant proteins. The characteristics of this process are the use of a low extrusion temperature (like 10°C), employing a cooled long die, and using a feed with relatively high moisture content, like 30%–60%. However, there exists a variation in the extrusion processing of fake/mock meat, where a still higher temperature is applied as per various patented literature. A detailed mention of these extrusion processes is shown in Table 17.1.

Some of the extrusion processes mentioned in Table 17.1 involve standardization. Various raw materials were used in these extrusion processes to develop different products, where the effects of various process variables were investigated. The following section discusses the important results from the trials.

17.4.1 Dry and wet extrusion processes

The meat analog is a type of textured plant protein that can mimic animal meat; the product has a fibrous structure. After developing the product using extrusion or other technologies, the desired color and flavor are added. If the final product is dried, as in the case of TVP or TPP, the analog must be hydrated and boiled in water before consumption. Often, the hydrated-cooked product is chopped and incorporated into regular meat formulations along with beef, ham, and poultry meat to reduce the content of animal meat. Extrusion cooking is also employed to obtain texturized products from protein-rich flours like peanut, cottonseed, and sunflower, but these products are less commercialized than soy products.

Noguchi, 1990 reported a wet extrusion process involving the application of low temperatures, such as a 10°C barrel temperature, along with the use of a cooled long die to develop textured protein or meat analogs having a fibrous structure. Soy flour/isolate was used as the major raw material, and feed with high moisture content was extruded. The realigned fibrous products were cut/shaped, colored, and flavored to look and taste like shrimp or a piece of beef. A combined process of extrusion and wet-spinning was reported to produce a fibrous structure (Huang et al., 1995). Soy protein was subjected to this high-moisture extrusion process. A patented process for the production of structured plant protein products was claimed by Rolan et al., 2008) employing twin-screw extrusion cooking technology. The feed for extrusion cooking comprised soy isolate, wheat gluten, starch, etc. Water was added to the dry mix and mixed in a preconditioner, followed by extrusion cooking, wherein the temperatures of different barrel zones were kept at 25°C (feed end), 50°C, 95°C, 130°C, and 150°C (die end). The pressure at the feed zone of the extruder was maintained at 2800 kPa, while it was 10,000 kPa in the fifth zone (die end). Water was also injected into the barrel during the extrusion process. The extrudates coming out of the die had a protein-aligned fibrous structure that was cut to the desired size by flexible knives and finally dried to a moisture content of about 10%. However, the product required rehydration/cooking prior to consumption.

Ohtsu, 1990 claimed a patented process using processed soybean curd to develop a texture similar to meat. Soybean curd was prepared using isolated soybean protein and corn starch (3:1); the mix was subjected to single-screw extrusion cooking under pressure. The die temperature was maintained at 140°C, and the extrudates were cut

Table 17.1 Different extrusion processes with details.

Type of extrusion system	Major raw material used	Type of extruder	Operating conditions standardized	Other features	Product	Reference
Extrusion cooking	Cereal flour/grit	Single-screw (high-shear or collet) extruder	13%–16% moisture content and 135°C–145°C melt temperature	Feed contains 90%–95% corn grit	RTE snacks like flavored/spiced snacks	(Bhattacharya, 1994, 2007, 2011, 2022)
Coextrusion technology	Cereal dough as the outer layer, while viscoelastic filling as the inner material of the extruded filled product	Twin-screw (dual- and coextrusion technology) extruder	Rheological approach for standardization	RTE snacks	Filled RTE snacks	(Cindio et al., 2002)
Extrusion cooking	Germinated lentil flour and corn flour mix	Single-screw extruder	The optimized condition was 20% lentil flour, 20% feed moisture content, and a screw speed of 147 rpm	RTC snacks	Germination method of raw material is advantageous	(Nongmaithem et al., 2024)
Extrusion cooking	Defatted soybean flour	Single-screw (high-shear) extruder	About 14% moisture content and 135°C–150°C die temperature	To be rehydrated and cooked before consumption	RTC texturized vegetable protein (TVP) and granules	(Harper, 1981b)
Extrusion cooking	Soybean protein and corn starch	Single-screw extruder	Die temperature was 140°C	Water was added during extrusion cooking. Postextrusion processing was needed	Soybean curd-based patties	(Ohtsu, 1990)

Extrusion cooking	Fish muscle, corn flour, and other additives	Twin-screw extruder	15% silver carp fish muscle and 85% corn grit were suitable	Mixing corn flour and fish muscle, extrusion cooking, drying, and seasoning were performed	RTE snack	(Shahmohammadi et al., 2016)
Low-temperature extrusion	Pea/soy protein isolate and other ingredients	Single- and twin-screw extruder	Cooled long die maintained at 10°C	Product needed low-temperature storage. Used in burgers and sandwiches	Mock meat/fake meat requiring no further cooking	(Noguchi, 1990)
High-temperature extrusion	Soy protein isolate, wheat gluten, starch, and other additives	Twin-screw extruder	Last zone (die end) barrel temperature was 150°C, and the product was dried to a moisture content of 10%	Extrudates had aligned fibers to be used as a meat substitute	RTC product requiring rehydration/cooking at the consumer's end	(Rolan et al., 2008)

TVP, texturized vegetable protein or soy nugget; *RTE*, ready-to-eat; *RTC*, ready-to-cook.

into portions, each having a diameter of 5 mm and a length of 20 mm. The feed was heated and stirred at 85°C for 30 min, followed by dewatering to obtain fibrous soybean protein with a moisture content of 75%. The blend of soybean curd, fibrous soybean protein, diced onion, white pepper, table salt, and egg white powder was mixed and shaped into an oval form to make patties. These patties were sterilized at 115°C for 40 min.

Shahmohammadi et al. (2016) developed an expanded extruded corn-fish snack using a twin-screw extruder, where the feed contained 15% silver carp fish muscle and 85% corn grit. The shelf life of the snack was indicated to be 7–8 months when stored at room temperature and packaged in metallized biaxially oriented polypropylene film pouches with a thickness of 30 μm. Total volatile basic nitrogen and thiobarbituric acid were reported to be suitable indices for monitoring the product during storage.

17.4.2 Coextrusion technology

The coextrusion process is sometimes used when two raw materials possess different properties. One of the major feeds is powdery, while the minor ingredient may be a paste or a viscoelastic item, such as a paste of sugar, fat, cream, dyes, color, flavor, meat, jelly, jam, protein pastes, peanut butter, pastes of nuts, cheese, onion, and caramel. Typical examples include pillow-shaped cream-filled breakfast cereals and meat-flavored RTE snacks. The products of coextrusion technology are characterized by a hard shell and an inner core containing soft food ingredients.

Food industries employ two commonly used methods of coextrusion. In the first method, the filler or the minor ingredient is fed perpendicularly to the axis of the extruder. In the second method, the filler is fed through the axis of the die, meaning that both materials flow in the same horizontal direction. Here, the cooked cereal flour forms the outer tube, which is filled with salty-spicy soft cream. Finally, the continuous strands/pipes containing the filling material are cut into desired sizes; the product may be dried, if needed, to increase shelf life. The filler may contain light- and oxygen-sensitive material, which is also protected by the outer layer.

A batch laboratory-scale coextrusion method was reported to produce nutrient-containing alginate beads to be used as carriers for hydrophilic compounds (Piazza and Roversi, 2011). A dual jet of liquid core containing an aqueous solution of vitamin B_{12} or olive oil, and a liquid shell material containing an aqueous solution of alginates, was pumped through concentric extrusion nozzles to obtain droplets by employing a rotating fluid jet system. Furthermore, the coextrusion encapsulation technique is also possible for developing core–shell liquid-filled microcapsules for carrying nutrients.

17.4.3 Standardization and modeling of extrusion processes

Standardization of any process or product is a prerequisite for the manufacture of food products, with the overall intention of commercial success. This is also true for extrusion processes and extruded products. The initial trials at the laboratory scale are standardized, and later the process is scaled up to match the capacity of the pilot scale, and finally to industrial production.

The dissimilarity of extrusion processing is due to various types of (1) raw material or feed to be handled, (2) type of extruder, (3) product to be developed, (4) extent of cooking to be achieved in the developed product, (5) operating conditions of the extruder during processing, and (6) postextrusion treatments, including ingredients to be used. All these factors make the extrusion systems complex and difficult to generalize, as indicated in Section 17.3. The process of standardization thus needs to be linked to the products to be developed.

Extrusion processing deals with various raw materials, including grain flour/grit, dairy ingredients, and oilseed meals like cottonseed, soybean, mustard, rapeseed, etc. (Bhattacharya, 2011). These raw materials have low moisture content, making them suitable for use directly as extruder feed after adding water or steam. The feed for extrusion cooking typically has a moisture content below 20%, while forming extrusion is carried out at a moisture content of about 30%. Raw materials from muscle foods like fish and meat contain high moisture content and need to be partially dried for use in an extruder individually or along with a starch-rich raw material to develop snack foods. However, high-moisture extrusion (>30% moisture content) is also practiced to develop mock meat products, where pea or soy protein concentrates/isolates are usually used. The moisture content of the feed is the most critical factor among the feed variables, while protein content and particle size play less important roles. The presence of fat in the feed is detrimental to extrusion cooking as it affects expansion and results in noncontinuous strands and slippages in the extruder. In general, feed containing particles that are too fine or too coarse is undesirable for extrusion cooking to produce RTE snacks and breakfast cereals.

The required moisture content of an extrusion feed mainly depends on the product to be developed and also on the single- and twin-screw extruders. The minimum moisture content of feed for a twin-screw extruder required for extrusion-cooked products like RTE snacks and breakfast cereals is 11%–12%, while it is 13%–14% for a single-screw machine. Too low moisture content in the feed can result in incomplete cooking or gelatinization of starch, and high torque, leading to extruder jamming. The problem of jamming is markedly higher for a single-screw extruder. On the contrary, high moisture content, like 28%–30%, is used in the feed where only shaping is performed, as in the case of pasta products. It is thus generally accepted that the moisture content of the feed is a critical parameter for dictating the quality of the developed product and the smooth operation of the extruder.

The types of extruders used for food processing mainly fall under two categories: single-screw and twin-screw extruders. Single-screw extruders can be divided into low-shear forming extruders and high-shear cooking extruders. The former single-screw extruders are typically used for pasta products, where heating or cooking is not the objective. On the contrary, high-shear cooking extruders operate with the objective of heating/cooking and providing an environment of shearing, particularly for starch-rich extruder feed. Twin-screw extruders, though costlier than their counterpart single-screw machines, are employed to make several products, such as filled snacks and breakfast cereals. This is due to their suitability and technical superiority, particularly in providing better control of the extruder during operation. Corotating fully intermeshing twin-screw extruders are preferred by food industries due to their operational advantages.

The hardware components of the extruder vary with respect to screw design. Specific items, such as the length and pitch of the screw elements, the gap between the tip of the screw and the barrel, the presence of kneading blocks, cut flights, and mixing disks, as well as features of the die like the diameters of die nozzles and the number of die openings, play a crucial role. The intention of using a kneading block, cut flight, and mixing disk is to modify the residence time of the feed and the extent of shearing, heating, and mixing during extrusion processing. Thus the hardware components of an extruder are complex features that dictate the type of product to be produced and the quality attributes of the products, such as the extent of cooking and expansion, density, and texture of the finished product. The general rule for selecting a single- or twin-screw extruder for a particular product is mainly based on cost considerations. It is usually opined that a single-screw extruder should be selected for conventional extruded products, and twin-screw machines should be used only when the single-screw counterpart is unsuitable.

The extruded product to be developed may be a convenience raw material (like modified grain flour, pregelatinized starch, etc.), a ready-to-use/prepare product (like pasta and TVP, also called *texturized* soy *protein* or TSP, meat extender, soy meat, or soy chunks, requiring further processing like cooking before consumption), or RTE products like expanded snacks, breakfast cereals, flat breads, etc. The specialty products, like the high-moisture-containing meat analogs (mock/fake meat and shrimp/beef analogs), are prepared using protein-rich ingredients to achieve the desired level of texturization and appropriate size and shape.

The operating conditions of the extruder during extrusion processing are important factors for the standardization of processes, products, and scale-up. The key operating conditions are temperature (at different zones of the barrel and die or melt temperature), feed rate, and screw speed, while a less important issue is the pressure developed. However, all these variables affect the torque, average residence time, production capacity, and SME. A safety limit is often set for torque during extrusion, while the objectives are to ensure the safety of the machine, maximize production capacity, and minimize SME. An increase in feed rate increases torque and production capacity but decreases average residence time, provided other variables are not changed. It is a general practice to increase the feed rate to maximize production capacity, but simultaneously, the screw speed is increased to maintain the torque well below its upper or safe limit. The temperatures of different sections of the barrel are progressively increased from the feed end to the die end. The barrel temperatures at different sections influence the die temperature and the melt temperature, while the latter index is more important for the quality of the product, including the extent of cooking, density, and expansion. These factors are critical for expanded products like RTE snacks and breakfast cereals. If a twin-screw extruder has five barrel sections with different temperature zones to be set, the temperatures may be maintained at 60°C (feed zone), 80°C, 100°C, 120°C, and 140°C (die end) for developing RTE snacks obtained from a corn-grit-rich formulation.

The extrusion-cooked products possess a bland taste due to the flash-off of steam at the die exit. An appropriate postextrusion treatment is thus necessary to make the product attractive to consumers. Further, the extruded collets/balls that emerge at the die end of the extruder are dried, baked, or fried to attain the desired texture and safe moisture content, usually around 3%–6%. The postextrusion treatments, particularly the ingredients to be used for flavoring and coating, play a major role in the acceptance of the finished product, especially the RTE snacks. The common RTE snacks contain salt, spice powder, or spice fractions like oleoresin, antioxidant, microencapsulated flavor, organic acid, maltodextrin, modified starch, permitted color, etc. On the contrary, sugar, inverted sugar, corn syrup, and honey in liquid form, powder, or in solution are added to make sweetened breakfast cereals. However, several extruded products like pasta, TVP, baby food, uncoated corn flakes (as breakfast cereal), breadsticks, and flat bread may not require the postextrusion flavoring and coating processes, though drying is mandatory to bring the product to a safe level of moisture content suitable for packaging and storage. Thus standardization of the formulation for flavoring/coating is a challenge to the product developers that affects consumer acceptance.

The flavor of the high-moisture extrusion process for developing meat analogs arises from the flavor compounds present in the raw materials and flavor precursors (proteins, carbohydrates, and lipids), which undergo degradation, changes, and reactions. Furthermore, additives like spices, color, hydrolyzed proteins, cysteine, and yeast extracts are also incorporated during the high-moisture extrusion process. As the main raw material for high-moisture extrusion is proteins, several volatile flavor substances like aldehydes, ketones, furans, alcohols, and pyrazines are formed (Jiang et al., 2024).

As an empirical approach, extruded products are developed using the following major variables (keeping the extruder machine unchanged) as per their priority or importance: the moisture content of feed, barrel or melt temperature, die opening, feed rate, protein and fat contents of feed, etc. Thus standardization of these variables, particularly the moisture content of feed, is to be attempted first.

The process of standardization of an extrusion process may be approached in several ways, such as (1) previous experience and knowledge, and ad hoc attempts, (2) black-box modeling using various statistical methods like the design of experiments, fitting to suitable models, response surface methodology (RSM), and finding optimized conditions, (3) system analytical model and multiple-input multiple-output (MIMO) approach, and (4) using the concept of artificial intelligence (AI), such as artificial neural network (ANN) and genetic algorithm (GA).

17.4.4 Previous experience and knowledge, and ad hoc approaches

It is usual for a person working in the field of food extrusion with different products for some time to gather adequate knowledge. This knowledge concerns the effect of the processing variables on the operation of the extruder and the attributes of extruded products. Some of these are mentioned in the subsequent section.

1. For a starch-rich expanded snack, an extruder, such as a low-cost single-screw collet extruder or any short-length high-shear extruder cooker, is suitable. The moisture content of the feed is 13%–15%, and product examples include RTE snacks like corn balls, corn curls, cheese balls, etc. The raw feed materials consist of corn flour/grit, starch, rice, wheat flour, and similar ingredients. A small amount of a pulse, such as green gram, lentil, or Bengal gram, may be added to improve consumer acceptability. The melt temperature is approximately 140°C, and thus postextrusion drying is optional, while salt-spice-oil flavoring is also applied.
2. For producing TVP, extrusion cooking of defatted soy flour is carried out at a moisture content of 15%–16% using a high-shear single-screw extruder. The melt temperature is maintained at 130°C–135°C to ensure adequate cooking and expansion characteristics in the product. The extruder barrel may have grooves to increase the shearing action on the feed. The exudates coming out of the die are cut to the required small sizes, followed by drying to bring the moisture content to about 5%, while no other postextrusion treatment is generally needed.
3. Moisture content of feed plays a dominant role in the process of extrusion cooking and in determining the product characteristics. Extremely low moisture content (possibly below 11%) is not suitable for a single-screw extruder. Too low a moisture level is also inadequate for cooking starch-rich feed and may lead to extruder jamming. An increase in moisture content from 11% to around 17% enhances the expansion of a starch-rich feed in the case of a single-screw extruder. However, an increase in moisture content beyond 17% can result in nonuniform expansion in the product, exhibiting highly expanded portions alongside collapsed regions. Extruder surging is also possible with high-moisture feed.
4. The feed temperature, measured as the melt temperature or the temperature at the die exit, is kept around 140°C–145°C for developing expanded corn/wheat/rice-based snacks. However, even higher temperatures may be employed with a simultaneous reduction in the average residence time due to the shorter length of the screw or a higher screw speed. If the feed contains only a pulse, such as green gram or Bengal gram, a 5°C–10°C reduction in the melt temperature seems appropriate to minimize the overcooked flavor.
5. The feed rate, or the input to the extruder, should be as high as possible so that the productivity of the machine is increased. However, a high feed rate may decrease the residence time of the feed in the extruder, resulting in insufficient thermal treatment. The torque also increases, and the chance of the extruder jamming increases as well. Thus screw speed is also increased in synchronization with the feed rate. However, any changes in feed rate and/or screw speed must be made in small increments.
6. In the case of developing filled snacks, a twin-screw extruder with a separate feeding system for the filling material is required. In such a case, the screw profile may consist of a mixing unit, a kneading block, and a reverse-pitch screw element to decrease the pressure at the feeding point of the filling material.
7. The twin-screw extruders operate at a much lower pressure, such as 5–10 atmospheres, while the pressure at the die end of a single-screw extruder may exhibit a higher pressure, like 25–40 atmospheres, and thus the control of a twin-screw extruder is easier.
8. The torque of an extruder during operation should be within 80%–90% of the highest permissible torque of the machine to avoid extruder jamming and accommodate changes due to surging of the machine, if any. If the torque increases, an increase in screw speed and/or a decrease in feed rate is appropriate.
9. Most of the experienced extruder operators operate by adjusting variables like the moisture content of the feed, screw speed, feed rate, and marginal changes in the temperature. If it is not possible to standardize the process and the product in this manner, the other possible

options are changing the die geometry, such as the diameter of the openings and the number of openings, and varying the feed composition to include a permitted nucleating agent, emulsifier, stabilizer, and a protein-rich raw material.

10. In the case of breakfast cereals extrusion, marginally higher moisture content may be used compared to snack products to decrease the extent of expansion and obtain a uniform product that may be subjected to further flattening, drying/toasting, and coating operations.

11. The extrusion cooking process is an example of the high temperature short time (HTST) process, and marginal changes occur in nutrients. Thus the temperature of the feed during the extrusion cooking process and its residence time play major roles in the retention of the nutritional components of the finished products. Further, the postextrusion processes, like drying and toasting, are also responsible for nutrient degradation. It is, therefore, desirable to add heat-sensitive nutritional supplements, like minerals and vitamins, during the postextrusion steps, such as coating and flavoring.

17.4.4.1 Black-box modeling

The black-box model is an empirical approach for analyzing the effect of extrusion variables, where the expected changes can be quantified, but the mechanism of these changes is unknown to the researchers or extruded product developers (Fig. 17.1). The intention in this situation is to standardize the extrusion process or optimize the process and product. However, the scientific basis of the black box (the unknown item) appears poor, though it may be suitable for a complex system like extrusion cooking and the products developed. The black-box system is specific to the individual system and cannot be generalized.

The advantages of black-box models are their simplicity while offering handy relationships between system inputs and system or targeted outputs. The model parameters may not have a clear physical significance. However, if the aim is to merely represent some trends in process behavior, then the black-box modeling approach is an effective tool.

The approach for black-box modeling has been applied to understand the specific effects of selected important extrusion variables, such as the moisture content of the

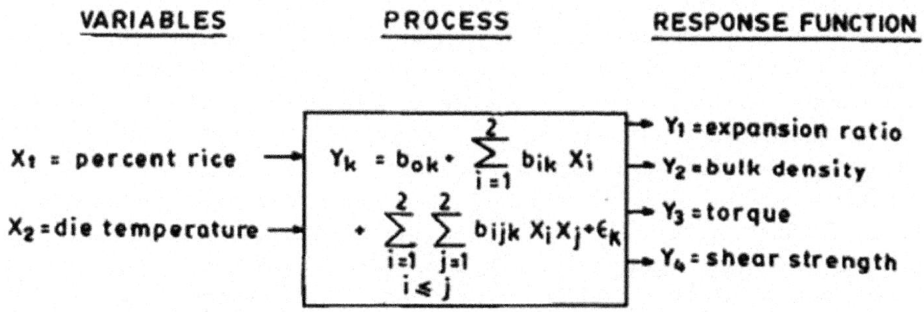

Fig. 17.1 Response surface "black-box" modeling of the extrusion process.
From Bhattacharya, S., Prakash, M., 1994. Extrusion of blends of rice and chick pea flours: a response surface analysis. J. Food Eng. 21(3), 315–330. https://doi.org/10.1016/0260-8774(94)90076-0

feed, screw speed, temperature at the die and/or barrel, feed composition, and feed rate. These variables of interest are commonly known as independent variables. The targeted or desired response functions may include the extent of product expansion, the density of the obtained product, the extent of cooking and starch gelatinization, torque during extrusion, energy expenditure, and nutrient loss.

Myat and Ryu (2014) indicated that the enzymatic hydrolysis of starch is an important industrial process. The process consists of three steps: gelatinization, liquefaction, and saccharification. The industrial process uses a jet cooker for gelatinizing starch by heating the starch slurry with steam. An increase in the substrate concentration during enzymatic hydrolysis was reported to achieve better yield and higher enzyme stability. The present investigation dealt with the effect of adding thermostable α-amylase on the properties of saccharification of extruded corn starch.

Screening experiments are performed when dealing with multiple independent variables to understand their relative importance to the selected response function. Thus screening experiments can significantly reduce the number of experiments. Screening experiments were performed using the Plackett–Burman experimental design, where 12 experiments were conducted (Table 17.2) as per the experimental design containing the high (+1) and low (−1) levels (Jain et al., 2010; Guha et al., 2003). Data obtained after conducting the experiments were fitted to first-order polynomials in the coded levels of variables (−1 and +1) for each response function. The coefficients of the polynomials could provide an indication for identifying the relative importance of each input independent variable (x_i) on the targeted response functions (y_i).

In the equation mentioned in Fig. 17.1, i and k were the integers, b_{0k} was a constant term, b_0, b_1, … ., b_k were coefficients of the polynomial indicating linear effects of the variables under study, and ϵ is the random error. The terms b_{11}, b_{22}, . … . were the quadratic effects of variables, while b_{12}, b_{13}, … … were the interaction terms associated with the variables. After conducting experiments, authors usually try to analyze the obtained results employing the concept of RSM. The effects of individual variables were then quantified through analysis of variables in terms of F-values or t-values. The resultant F-values or t-values were statistically analyzed to know the significance of the individual terms of the equation as linear, quadratic, and interaction effects when the probability (p) of their significance was judged at 0.05 and 0.01 levels. The multiple correlation coefficients (r) were calculated to determine the extent of suitability of the equation for the selected response functions. Optimization of the response functions was also frequently attempted, employing methods like graphical overlapping and other statistical procedures. The nature of optimum conditions, like maximization or minimization, and saddle points, was also determined to have an improved understanding of the RSM method and the extrusion process.

If the experimenter encounters a large number of independent variables to be investigated for their effects, a preliminary screening experiment might be conducted before proceeding with the actual experiments on extrusion. In such a situation, a specific screening experimental design, like the Plackett–Burman design, is employed to identify the most important variables; a first-order model is fitted. A few examples of the application of the Plackett–Burman experimental design are

Table 17.2 Research investigations concerning black-box modeling for extrusion cooking.

Feed material	Type of extruder	Variables investigated	Response function	Important findings	Reference
Rice flour	Twin-screw extruder without die	Screening experiments conducted with ten variables, including hardware, operating, and feed variables	Extrusion characteristics and product attributes were noted	Hardware variables, like the presence of a reverse pitch screw element and a mixing disk, offered the most prominent effects on the response functions	(Guha et al., 1997, 2003)
Preselected raw materials as the carrier of the drug	Single-screw extruder	Screening experiments conducted with nine variables	Release of drugs from the extrudates	• Poly (ethylene oxide) amount, ethylcellulose amount, and drug solubility had significant effects on the time required for 90% of the drug to be released	(Jain et al., 2010)
Finger, foxtail, and pearl millet flour	Twin-screw extruder	Moisture content of feed, screw speed, and barrel temperature	Physicochemical properties of extrudates, such as expansion ratio (ER), water solubility index, water absorption index, and structural features, were evaluated	• Foxtail samples showed the highest ER, followed by pearl millet and finger millet extrudates. • Foxtail millet gave the best results, having the highest ER values, achieved at 130°C with the least SME input	(Kharat et al., 2019)

Continued

Table 17.2 Research investigations concerning black-box modeling for extrusion cooking—cont'd

Feed material	Type of extruder	Variables investigated	Response function	Important findings	Reference
				• The results highlighted the potential use of whole millet flour for the development of directly expanded extruded foods	
Rice flour	Twin-screw extruder without die	Barrel temperature and screw speed	Torque, SME, bulk density, water absorption index, water solubility index, sediment volume, and in vitro starch digestibility	• A high extrusion temperature, in combination with a high screw speed, was associated with lower torque and SME values • Extrusion cooking without using a die was a preferred way to conduct extrusion cooking when shaping or sizing was not important, such as in the development of modified or pregelatinized starch and grain flour	(Guha et al., 1997)
Seed coat of Australian sweet lupin	Twin-screw extruder	Extrusion temperature, screw speed, and moisture content of feed	SME, soluble dietary fiber (SDF), insoluble dietary fiber (IDF), total dietary fiber (TDF), total polyphenol content (TPC), and individual free and bound polyphenols	• Optimal extrusion conditions, which achieved maximum levels of SDF, TPC, and total free individual phenolic contents, were obtained • The extrusion cooking largely had nil or marginal	(Zhong et al., 2021)

				effects on the bioaccessibility and bioavailability of the selected minerals and individual polyphenols • Extrusion-cooked lupin seed coat could be a natural antioxidant dietary fiber source for human consumption	
Rice and cowpea flour blend	Twin-screw extruder	Feed moisture content, barrel temperature, and blend composition	Mineral content of rice-cowpea extrudates	• Minerals present in the extrudates might be maximized when process conditions are carefully adjusted within the reported values • Instant porridge was the targeted product, employing a graphical optimization technique	(Danbaba et al., 2015)
Composite blend of cereal, millet, and pulses such as rice, corn, pearl millet, green gram, and cowpea bean	Twin-screw extruder	Feed moisture content, screw speed, barrel temperature, and particle size of feed	Product attributes and microstructure	Optimized parameters were 120°C (barrel temperature), 350 rpm (screw speed), 10% (moisture content), and a particle size of 0.5 mm	(Karun et al., 2023)

Continued

Table 17.2 Research investigations concerning black-box modeling for extrusion cooking—cont'd

Feed material	Type of extruder	Variables investigated	Response function	Important findings	Reference
Corn starch and thermostable α-amylase	Twin-screw extruder	Melt temperature with or without α-amylase	SME, density, color, WSI, WAI, and amylose, amylopectin, and reducing sugar contents	Possibility existed for extrusion with thermostable α-amylase injection at 115°C and 135°C as an effective pretreatment for modifying starch granules due to a significant increase in reducing sugar and saccharification yield	(Myat and Ryu, 2014)
Maize–mung (green gram) bean mix	Twin-screw extruder	Feed moisture content, screw speed, and barrel temperature	SME, bulk density, water absorption index, water solubility index, and degree of gelatinization	• Optimized results were feed moisture content of 14.3%, screw speed of 524 rpm, and barrel temperature of 174°C to produce extrudates having preferable functional properties with high protein and starch digestibilities • The optimized weaning mix contained maize–mung bean extrudates: 40%, skim milk powder: 35%, and sugar: 25%	(Ali et al., 2016)

shown in Table 17.2. The effect of several variables is studied by conducting 12 experiments, while only the lowest and highest levels of the variables (−1 and +1 in the coded level of variables) are selected. Later, those important variables might be studied in detail using conventional experiments or an RSM design of experiments. The conventional experiments use a one-variable-at-a-time approach, while the RSM designs use only the preselected levels of variables. In this case, central composite rotatable design, orthogonal design, and canonical design are employed.

A method to reduce the number of variables or select the most important variables was to perform a screening experiment (Guha et al., 2003) . A screening experiment, such as the Plackett–Burman experimental design, was performed on the extrusion cooking of rice flour; the number of independent variables was ten. The variables investigated included the hardware variables (such as mixing disk and reverse pitch screw elements), feed variables (like moisture, sugar, salt, amylose contents, and particle size of feed), and extrusion operating variables (consisting of feed rate, screw speed, and barrel temperature). The targeted response functions were the extrusion characteristics (SME and residence time) and product attributes [water absorption index, water solubility index (WSI), and viscoamylographic indices like peak viscosity, hot paste viscosity, and cold paste viscosity]. The hardware variables had the strongest effect on the response functions. It was concluded that the Plackett–Burman experimental design can serve as a useful tool for screening large numbers of extruder and extrusion variables and can markedly reduce the number of experiments.

Reviews are available concerning the process of extrusion cooking and the quality attributes of extruded products as affected by extrusion process parameters (Alam et al., 2016). The areas addressed included the potential of extrusion technology for the development of different types of products, as well as the role of operating conditions in the extrusion process and their effects on product development, resulting in quality changes (physical, chemical, and nutritional) and process standardization.

Chen et al. (2010) investigated the effect of the moisture content of feed, which is the most important factor in the process of extrusion cooking. Soybean protein isolate was extrusion-cooked using a pilot-scale twin-screw extruder at different moisture contents (28%–60%) and temperatures (140°C–160°C). The extrusion system parameters studied included in-line viscosity at the die, average residence time, SME, and textural attributes of the product such as tensile strength, hardness, chewiness, and degree of texturization, as well as the molecular weight distribution characterized by SDS–PAGE. The interrelationships between system parameters and product properties were also analyzed. The moisture content of the feed had a greater influence on the system parameters and product attributes than the cooking temperature. Higher moisture content resulted in decreased viscosity of the dough in the extruder, a shorter residence time, and a lower conversion ratio of mechanical energy into thermal energy. Consequently, it reduced product attributes such as strength, hardness, chewiness, and the degree of aggregation. The extrusion system parameters and product attributes were well correlated, which helped to explain and control the quality characteristics of the extrudates.

The torque, a safety index during extrusion cooking of an extruder, increased in the screw profile that had higher shear configurations due to the presence of kneading blocks, mixing disks, and/or reverse-pitch screw elements. This occurred due to an increase in the residence time of the feed and the level of filling of the extruder barrel. Furthermore, this type of configuration might require a marginally higher level (1%–2%) of moisture content in the extruder feed to avoid overheating of the raw material and product.

17.4.4.2 System analytical model and MIMO approach

A system analytical model is concerned with studying the procedure of an operation to identify the goals, such as process standardization and prediction of target parameters. It is a problem-solving approach and is part of operations research. In the area of complex food processing operations, system analytical models are infrequently proposed in areas like extrusion cooking and coating of foods (Bhattacharya and Patel, 2007; Ganjyal et al., 2003; Mueser and van Langerich, 1984).

Extrusion processing involves (1) process parameters (namely, moisture content of mixed feed, screw speed, feed rate, die diameter, and die or barrel temperature), which are input parameters; (2) system parameters (residence time, torque, rheological status, SME); and (3) product attributes (such as radial expansion, density, and chemical, nutritional, and mechanical properties, as well as sensory characteristics), which are the output parameters. Thus extrusion cooking is an example of a multiple-input and multiple-output (MIMO) process. General predictive modeling is a difficult task in the case of food processing applications because the ingredients are diverse, and their compositions can vary considerably. Physical and chemical changes also occur during the process, which can affect the flow inside the extruder and the attributes of the finished products. Thus food extrusion modeling tends to be product- and machine-specific, and new product development tends to rely mostly on a trial-and-error approach and good fortune.

The extrusion process has been described as a complex MIMO system (Ganjyal et al., 2003). Various crosslinked waxy maize starches were extruded at moisture contents of 20%, 24%, and 28% (dry basis) and screw speeds of 110, 150, and 190 rpm. Multiple-input single-output neural network models were developed for predicting selected product properties, and MIMO models were reported for predicting individual product properties at a time from selected input process parameters, employing the concept of ANNs. Models were developed by varying the number of hidden neurons from 8 to 14; the coefficient of determination (r^2) values were good ($\geq$0.89). The models thus could explain the process and the effect of process parameters on product properties.

Information on the modeling of plastics extrusion has significantly helped the food extrusion process, though the latter system is much more complex than the former. The heat-sensitive food ingredients undergo several changes in physical, chemical, and nutritional properties. Thus simple mathematical models to understand the extrusion cooking process of foods are difficult to develop. Dynamic and steady-state modeling approaches have been followed to model food extrusion processes. Tehrani et al. (2017) proposed a system analytical model for the extrusion cooking of starch. The developed model distinguished between process (input) and system parameters, which, in turn, influenced target (output) parameters. Process parameters were the operating parameters or conditions that could be controlled and manipulated directly. System parameters referred to the properties influenced by the process parameters, which ultimately affected the characteristics of the product (target/output parameters). An appropriate function was expected to describe the relationship between process parameters and system parameters or between system parameters and target parameters. It was claimed that the developed system was useful for the standardization of extrusion processes and scale-up.

The method of ANNs is reported to improve the predictability of a system more effectively than regression models. The noteworthy advantages are that an ANN system has the capability of learning from the input and output data, developing relationships between them, and subsequently predicting the outputs or target parameters if inputs are provided. A complex process like an extrusion cooking system can thus be handled by employing an ANN. In addition, if the process under examination changes, new examples can be added to the existing data, and retraining of the ANN is possible. The ANN system learns from examples through iteration without requiring any prior knowledge of the relationships between the variables. Thus it is possible to control the extruder/extrusion system by managing the system variables. Ganjyal et al. (2003) extruded waxy maize cross-linked starches, and selected properties of extrudates were investigated using ANN. Individual ANN models were developed to predict expansion ratio (ER), unit density and BD, water absorption index, and WSI. A MIMO model was proposed to predict all these properties of the product with good accuracy. These models were reported to serve as tools for determining the input parameter levels needed to develop the desired product from a specific system. The developed ANN configuration for the target parameter (radial expansion) was 3-13-1, where the input layer had 3 variables and 13 neurons in the hidden layer; the target parameter was 1 in number. The other attributes of the product, like unit density, BD, water absorption index, and WSI, were represented by the ANN structures of 3-14-1, 3-14-1, 3-8-1, and 3-10-1, respectively.

17.4.4.3 Optimization of extrusion process variables using AI and GAs

The concept of AI has been applied in food extrusion technology. The specific areas of application include the standardization of the extrusion cooking process, control and prediction of important process variables, and optimization of developed products. Two hybrid ANN chains were developed and tested for the configuration and control of the food extrusion process (Cubeddu et al., 2014). The process variables investigated were the moisture content of feed and the screw speed, while the target indices were SME, torque, pressure, shear stress, and temperature. The developed chains were feed-forward multilayer ANNs and trained using the back propagation algorithm. A total of 42 patterns (datasets) were used, of which 24 were for training, while the remaining 18 sets were for verification purposes. Two ANN chains were developed by these researchers for the prediction of optimal process variables (feed moisture content and screw speed) linking the desired product characteristics.

Process standardization and optimization are possible using ANN along with other analytical methods like RSM and GA. Jaya Shankar and Bandyopadhyay (2004) discussed the method of optimization of extrusion process variables using GA. Second-degree regression equations were developed for the attributes of the extrudates, such as ER, BD, hardness (H), and WSI. The independent variables were moisture and fish contents in the feed, barrel temperature, and screw speed. These equations were used as the objective functions to determine the optimum process conditions separately for each target parameter of extrudate properties. The regression equations of ER and WSI were maximized, while BD and H were minimized to obtain the optimum processing conditions. A maximum population of 100 and 100 iterations was found to be suitable for successful convergence. A high barrel temperature of about 200°C and screw speed of 110 rpm were identified as both individual and common optimum process conditions for all the properties of the extrudates except for WSI. Fish and feed moisture contents were identified as two interacting process variables. Under individual optimum process conditions, maximum ER was associated with a low fish content of 5%. The maximum WSI was obtained when fish content was high (41%–45%). On the contrary, the common optimum process conditions for all four properties of the extrudate were optimized at high fish content (41%–45%). The experimental extrudate properties were reported to be closer to the values predicted for common optimum conditions than those for individual optimum process conditions.

A few more studies on extrusion cooking dealing with AI have also been reported. Kowalski et al. (2018) optimized the twin-screw extrusion cooking of six pea flour varieties consisting of green and yellow peas, employing regression and GA. The extrusion conditions varied were screw speed and temperature, while response functions were product attributes (density, radial expansion, color, WAI, and WSI) and extrusion variables (pressure developed, torque, and SME). The GA process included selection, reproduction, crossover, and mutation. A strong fitness of >0.90 was reported. It was claimed that GA is an effective tool to optimize food extrusion cooking processes.

Tehrani et al. (2017) applied GA to optimize extrusion cooking conditions for the preparation of soy-based meat analogs. The defatted soybean flour was replaced with whole soybean meal at a 10% level, while variations were made in temperature (150°C–200°C), screw speed (270–300 rpm), and feed moisture content (20%–25%). The effect of barrel temperature was greater than that of feed moisture content and screw speed on the attributes of the product. Higher temperatures resulted in a higher rehydration ability and oil absorption capacity but decreased product brightness. The optimized processing conditions were reported.

17.5 Case studies

Case studies concerning extrusion and extrusion cooking are available in some published documents (Maskan and Altan, 2011; Frame, 1994). In the present chapter, two case studies are mentioned that deal with producing breakfast cereal pellets for flaking or shredding using corn flour, and instant modified rice flour using extrusion cooking.

17.5.1 Breakfast cereals

Breakfast cereals are becoming an important part of the diet in many countries. The extrusion of cereal-based products can play an important role in providing adequate nutrition to people of all ages. Breakfast cereal manufacturing, though it has existed for many years following traditional processing methods, has been partially replaced by extrusion technology. Thus the development of different varieties of breakfast cereals has occurred, and improvements have been made regarding the equipment used, especially with the increasing application of twin-screw extrusion technology.

Breakfast cereals, in the form of an extrusion-cooked product, are often manufactured using various ingredients like corn grit/flour, wheat flour, and starch. Interesting shapes like animals, hollow rings, stars, and pillow shapes are produced using special dies, mainly to attract children. The application of various colors and flavors can improve consumer attraction. Later, the extrudates coming out of the extrusion cookers are often flaked or shredded, toasted, dried, and coated with sweet ingredients like sugar powder, honey, chocolate, maltodextrin, and flavoring ingredients like fruit flavor and malt powder. A more sophisticated product, like filled breakfast cereal, is also manufactured, generally using two extruders. However, simple corn flakes are the most popular ones that are produced by extrusion cooking, followed by flaking, toasting, and the optional sweet coating.

Breakfast cereals, as the name implies, are prepared mainly from cereal grains like corn and wheat, though other cereals like rice, barley, and oats are also used. Generally, the protein and fat contents of the finished products are <5% though special products may contain higher levels. Though not mandatory, sweet ingredients are added during postextrusion treatments. The ingredients may include malt syrup, corn syrup, monosaccharides, disaccharides, honey, inverted sugars, fruit paste, and molasses. The optional ingredients added are dried pieces of nuts and fruits, protein sources, permitted colors and flavors, vitamins, minerals, etc.

The main types of extruded breakfast cereals are direct-expanded extrusion, puffed, and extrusion-cooked flaked products. The direct-expanded breakfast cereals are processed in the same manner as direct-expanded RTE snacks, though the raw materials and processing conditions may differ. However, these breakfast cereals have a marginally higher density, while the texture is slightly harder. The moisture content of the extrusion feed is also marginally higher, while the temperature at the die is kept lower to ensure the integrity of the product is maintained. The postextrusion toasting process is critical to prevent the product from becoming soggy immediately after being placed in milk.

Extrusion-processed breakfast cereals, like corn flakes, use a feed containing moist corn flour/grit to obtain marginally expanded, small spherical shapes (Fig. 17.2). Often, the feed contains other ingredients, like sugar and salt, which retard gelatinization. These are then flattened in a roller flaker to obtain a flake with a thickness well below 1 mm. The flakes are then toasted to decrease the moisture content to about 3%. At the same time, the toasting process develops a crisp texture that does not become soggy immediately when put in milk prior to consumption.

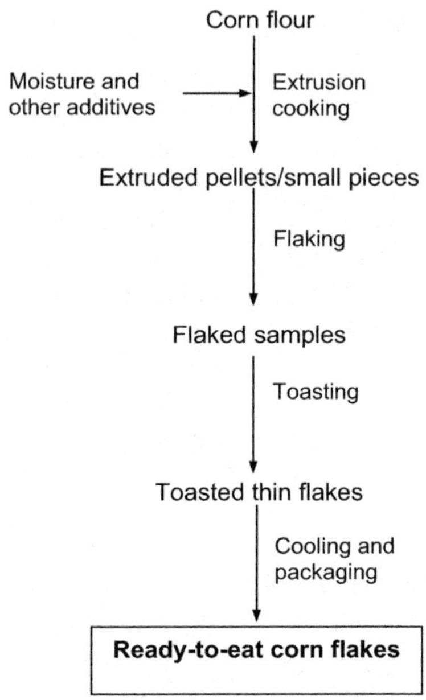

Fig. 17.2 Flow chart for the preparation of extruded corn flakes as breakfast cereals.

The traditional process of making corn flakes produces irregular-sized flakes and is associated with significant waste due to broken pieces. On the contrary, the important advantages of extruded corn flakes are the uniform product size, lower production cost, and reduced fines/wastage. However, the traditionally prepared flakes have a typical flavor, liked by many consumers; this flavor is absent in extrusion-prepared samples. Still, corn flakes prepared using both processes are available on the shelves. A feed moisture content of about 18%, an average residence time of 60 s, and a melt temperature at the die of 140°C–150°C for a corn grit-rich feed appear to be standardized conditions for extruded corn flakes.

Extrudates coming out of the die end need to possess enough plasticity so that the next flaking process is smooth. This happens when the extrudates are still warm and have a marginally higher moisture content. The flaking rolls consist of a pair of parallel, counterrotating horizontal cylinders, so that a small gap exists between them. Product pieces, fed from the top, are drawn downward into the gap, where they are compressed and flattened by friction with the roller surfaces. Roller surfaces should not be too smooth; they are usually somewhat roughened to engage the product pieces properly. The flattened product pieces are removed from the flaking rolls using doctor blades. The blades should be regarded as backup devices, as most of the flakes fall from the rollers. The overall target of flaking is to obtain a product with uniform thickness, where breakage is minimal.

The toasting of corn flakes is an important processing step that dictates the attributes of the final RTE breakfast cereals. Properly toasted flakes possess the appropriate moisture content, texture, flavor, and color. The toasted flakes typically have a moisture content of about 3%, which depends on the grain used, the toasting time, and the temperature. The toasting process is conducted at a low moisture content (usually below 10%) using high temperatures (250°C–300°C) for a short duration, typically below 5 s. Air toasting units are frequently used, though IR and microwave/radiofrequency ovens are possible alternatives. This sudden release of steam leaves behind an expanded and noncollapsing structure with air cells.

Flakes are conventionally toasted at a high temperature, employing several types of equipment, such as rotary ovens and high-velocity fluid bed dryers. The product thus develops a characteristic flavor and crisp texture, though a marginal change in color may occur. The texture of a toasted product depends on the thickness, porosity, and microstructure of flakes, in addition to the toasting process conditions (Sumithra and Bhattacharya, 2008). Some flaked cereals develop voids or air cells to reduce the density and create a crisp texture. Chaunier et al. (2007) discussed the relationships between texture, other mechanical properties, and the structure of corn flakes. The processing steps for the manufacture of traditional and extruded corn flakes were also mentioned.

Though not mandatory, breakfast cereals like cornflakes are often coated with sweet ingredients to improve consumer acceptability and create a different product variety. The formulation for coating, based on the mass of uncoated corn flakes, may include about 20% each for water and sucrose, 5% maltodextrin, 4% fat, and 0.2% salt. Lecithin may be added if permitted by law, and strawberry, vanilla, or other food flavors may be part of the formulation. Other optional ingredients include liquid honey or its powder, chocolate syrup, corn syrup, inverted sugars, permitted artificial sweeteners, hydrocolloids, encapsulated flavors, minerals and vitamins as nutritional supplements, pieces of dried fruits and nuts, etc. However, postcoating drying is necessary to bring the final moisture content to about 6%, after which fat/oil may be added as the finishing step to prevent moisture uptake by the coated flakes. A detailed description of the processes and machinery used for flavoring and coating is available (Cerqueira et al., 2015).

17.5.2 Instant modified rice flour using extrusion cooking

Extrusion cooking of moist grain flour can also be used to develop breakfast cereals and snacks and to prepare partially or completely cooked (gelatinized) modified flour. The developed extruded product can be used as an intermediate raw material for the preparation of different products like baby foods, convenient mixes, instant soups, etc.

There is a prominent role of the die in the process of extrusion and extrusion cooking. The presence of the die affects the extrusion process and the obtained products; it alters the residence time of the feed inside the extruder. The shaping of the product at the die is the main function of extrusion processes. On the contrary, the extent of heat treatment increases as the residence time of the feed also increases in the case of producing breakfast cereals and snacks; the extent of product expansion and density depends on the die geometry. Die heating and maintenance of an appropriate temperature at the die, as well as the melt temperature of the feed, are important variables for extrusion-cooked products. The current demand for plant protein-based meat analogs is being met through extrusion processing. Traditionally, low-moisture extrusion has been used to texturize protein into meat analogs like TVP or soy nuggets. However, high-moisture extrusion using a cooling die has become popular for producing products with a fibrous texture (Ryu, 2020).

Without using any die, Guha et al. (1997) extruded rice flour with 14% moisture content at different barrel temperatures (80°C–120°C) and screw speeds (200–400 rpm) employing a twin-screw extruder. A die restriction was reported to be optional if shaping of the product was not necessary. The targeted product was extrusion-cooked modified rice flour that could be used to prepare baby foods and various convenience mixes. An increase in production capacity and a marked decrease in SME could be possible if a restriction, like a die, was not placed. The system parameters (torque and SME), as well as the extrudate attributes (BD, water absorption capacity, WSI, sediment volume, and in vitro starch digestibility), were reported to be mainly dependent on temperature, whereas screw speed imparted a lesser effect (Fig. 17.3). The screw profile

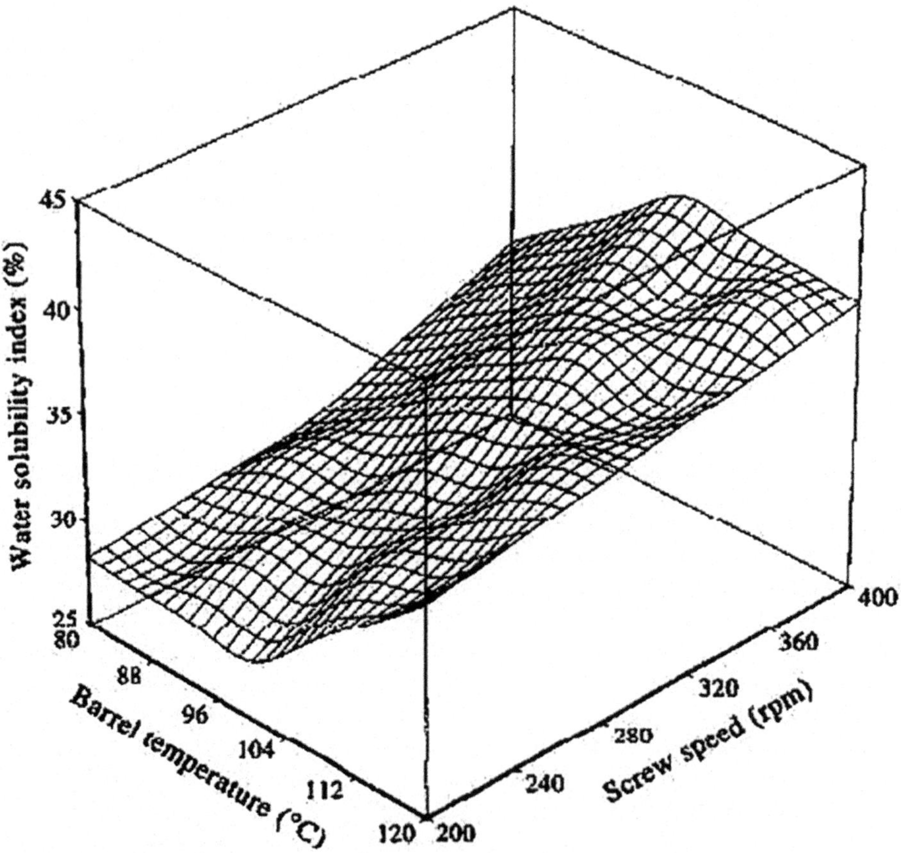

Fig. 17.3 Water solubility index of the extruded products.
From Guha, M., Ali, S.Z., Bhattacharya, S., 1997. Twin-screw extrusion of rice flour without a die: effect of barrel temperature and screw speed on extrusion and extrudate. J. Food Eng. 32, 251–267. With permission.

consisted of forward pitch screw elements of varying pitches, kneading blocks, and reverse pitch screw elements. The screw profile was selected to provide adequate shearing and increase the residence time of the feed in the extruder to achieve improved thermal treatment of the flour. The interrelations between system parameters and product attributes were also reported.

The desirable conditions for system parameters were that the torque needed to be low so that the feed rate could be increased, leading to an increase in production capacity. At the same time, the SME values should also be low to make the process energy-efficient and reduce the cost of production. Among the product attributes, water absorption capacity, WSI, and in vitro starch digestibility are also expected to be high if the extrusion-modified rice flour is intended to be used in baby foods and other convenience mixes, including instant soup powder. Optimum extrusion conditions for obtaining minimum torque, SME, and BD were reported. A temperature of 120°C required less torque (about 40%), which was associated with low SME (322–497 kJ kg^{-1}); the screw speed was between 300 and 400 rpm, but the product had low in vitro starch digestibility. On the contrary, extrusion cooking at 100°C and 300 rpm was associated with the least SME of 317 kJ kg^{-1} and torque (39%), and yielded a product with a desirable product profile. A positive linear (correlation coefficient $r=0.78$, significant at a probability level of $p \leq 0.01$) relationship existed between SME and BD, indicating that a low-density extrudate is possible to obtain with a low SME. Extrusion of rice flour without a die appears to be an alternative approach to produce processed/modified rice flour with a high water absorption index and in vitro starch digestibility.

17.6 Conclusions

The process of food extrusion has been successfully applied to manufacture several RTE and ready-to-prepare foods, as well as modified raw materials or specialty ingredients. Several advantages of extrusion technology, coupled with marked improvements in the design of extruders, have also allowed the development of several specialty food products. However, the presence of many variables makes the process of extrusion cooking a complex phenomenon. The variables include hardware, feed, and processing variables, while the postextrusion treatments also affect the operation of the extruder and the quality attributes of the developed products. The complex situation can be standardized or controlled using several approaches, including black-box modeling, AI, and ad hoc approaches based on previous expertise and knowledge.

References

Alam, M.S., Kaur, J., Khaira, H., Gupta, K., 2016. Extrusion and extruded products: changes in quality attributes as affected by extrusion process parameters: a review. Crit. Rev. Food Sci. Nutr. 15497852. 56 (3), 445–473. Available from: https://www.tandf.co.uk/journals/titles/10408398.asp. doi:10.1080/10408398.2013.779568.

Ali, S., Singh, B., Sharma, S., 2016. Response surface analysis and extrusion process optimisation of maize–mungbean-based instant weaning food. Int. J. Food Sci. Technol. 0950-5423. 51 (10), 2301–2312. doi:10.1111/ijfs.13186.

Bhattacharya, S., and Prakash, M., 1994. Extrusion of blends of rice and chick pea flours: A response surface analysis. J. Food Eng. 21 (3), 315–330. doi:10.1016/0260-8774(94)90076-0.

Bhattacharya, S., 2011. Raw Materials for Extrusion Cooking. Advances in Food Extrusion Technology. CRC Press.

Bhattacharya, S., 2022. Snack Foods Processing and Technology. London: Academic Press.

Bhattacharya, S., Patel, B.K., 2007. Simulation of coating process: rheological approach in combination with artificial neural network. J. Texture Stud. 17454603. 38 (5), 555–576. doi:10.1111/j.1745-4603.2007.00113.x.

Cerqueira, M.A., Costa, M.J., Rivera, M.C., Ramos, O.L., Vicente, A.A., 2015. Flavouring and coating technologies for preservation and processing of foods. In: Bhattacharya, S. (Ed.), Conventional and Advanced Food Processing Technologies. John Wiley and Sons, pp. 267–311.

Chaunier, L., Valle, G.D., Lourdin, D., 2007. Relationships between texture, mechanical properties and structure of cornflakes. Food Res. Int. 09639969. 40 (4), 493–503. doi:10.1016/j.foodres.2006.07.014.

Chen, F.L., Wei, Y.M., Zhang, B., Ojokoh, A.O., 2010. System parameters and product properties response of soybean protein extruded at wide moisture range. J. Food Eng. 02608774. 96 (2), 208–213. doi:10.1016/j.jfoodeng.2009.07.014.

Cindio, B.D., Gabriele, D., Pollini, C.M., Peressini, D., Sensidoni, A., 2002. Filled snack production by coextrusion-cooking: 1. Rheological modelling of the process. J. Food Eng. 02608774. 52 (1), 67–74. doi:10.1016/S0260-8774(01)00087-5.

Cubeddu, A., Rauh, C., Delgado, A., 2014. Hybrid artificial neural network for prediction and control of process variables in food extrusion. Innov. Food Sci. Emerg. Technol. 14668564. 21, 142–150. doi:10.1016/j.ifset.2013.10.010.

Danbaba, N., Nkama, I., Badau, M.H., 2015. Application of response surface methodology (RSM) and central composite design (CCD) to optimize minerals composition of rice-cowpea composite blends during extrusion cooking. Int. J. Food Sci. Nutr. Eng. 5 (1), 40–52.

Frame, N.D., 1994. The Technology of Extrusion Cooking. ASPEN Publisher.

Ganjyal, G.M., Hanna, M.A., Jones, D.D., 2003. Modeling selected properties of extruded waxy maize cross-linked starches with neural networks. J. Food Sci. 0022-1147. 68 (4), 1384–1388. doi:10.1111/j.1365-2621.2003.tb09654.x.

Guha, M., Ali, S.Z., Bhattacharya, S., 1997. Twin-screw extrusion of rice flour without a die: effect of barrel temperature and screw speed on extrusion and extrudate characteristics. J. Food Eng. 02608774. 32 (3), 251–267. Available from: http://www.sciencedirect.com/science/journal/02608774. doi:10.1016/S0260-8774(97)00028-9.

Guha, M., Zakiuddin Ali, S., and Bhattacharya, S., 2003. Screening of variables for extrusion of rice flour employing a Plackett–Burman design. J. Food Eng. 57 (2), 135–144. doi:10.1016/S0260-8774(02)00282-0.

Harper, J.M., 1981a. Extrusion of Foods, Vol. 1. CRC Press.

Harper, J.M., 1981b. Extrusion of Foods, Vol. 2. CRC Press.

Huang, H.C., Hammond, E.G., Reitmeier, C.A., Myers, D.J., 1995. Properties of fibers produced from soy protein isolate by extrusion and wet-spinning. J. Am. Oil Chem. Soc. 0003-021X. 72 (12), 1453–1460. doi:10.1007/bf02577837.

Jain, S.P., Singh, P.P., Javeer, S., Amin, P.D., 2010. Use of Placket-Burman statistical design to study effect of formulation variables on the release of drug from hot melt sustained release extrudates. AAPS PharmSciTech. 15309932. 11 (2), 936–944. doi:10.1208/s12249-010-9444-6.

Jaya Shankar, T., Bandyopadhyay, S., 2004. Optimization of extrusion process variables using a genetic algorithm. Food Bioprod. Process. 09603085. 82 (2), 143–150. doi:10.1205/096030804161 4855.

Jiang, W., Yang, X., Li, L., 2024. Flavor of extruded meat analogs: a review on composition, influencing factors, and analytical techniques. Curr. Res. Food Sci. 26659271. 8, 100747. doi:10.1016/j.crfs.2024.100747.

Karun, G., Aryasree Sukumar, G.N., Preetha, R., 2023. Development of multigrain ready-to-eat extruded snack and process parameter optimization using response surface methodology. J. Food Sci. Technol. 0022-1155. 60 (3), 947–957. doi:10.1007/s13197-022-05390-8.

Kharat, S., Medina-Meza, I.G., Kowalski, R.J., Hosamani, A., Ramachandra, C.T., Hiregoudar, S., Ganjyal, G.M., 2019. Extrusion processing characteristics of whole grain flours of select major millets (foxtail, finger, and pearl). Food Bioprod. Process. 09603085. 114, 60–71. Available from: http://www.elsevier.com/wps/find/journaldescription.cws_home/713883/description#description. doi:10.1016/j.fbp.2018.07.002.

Kowalski, R.J., Li, C., Ganjyal, G.M., 2018. Optimizing twin-screw food extrusion processing through regression modeling and genetic algorithms. J. Food Eng. 02608774. 234, 50–56. Available from: http://www.sciencedirect.com/science/journal/02608774. doi:10.1016/j.jfoodeng.2018.04.004.

Maskan, M., Altan, A., (Eds.), 2011. Advances in Food Extrusion. Boca Raton, FL: CRC Press (Taylor and Francis Group).

Mueser, F., van Langerich, B., 1984. System analytical model for the extrusion of starches. In: Zeuthen, P., Cheftel, J.C., Eriksson, C., Jul, M., Leniger, H., Linko. P., Varela, G., Vos, G. (Eds.), Thermal Processing and Quality of Foods. New York: Elsevier Applied Science, pp. 175–179.

Myat, L., Ryu, G.H., 2014. Effect of thermostable α-amylase injection on mechanical and physiochemical properties for saccharification of extruded corn starch. J. Sci. Food Agric. 10970010. 94 (2), 288–295. doi:10.1002/jsfa.6252.

Noguchi, A., 1990. High temperature/short time (HTST) processing, guarantee for high quality food with long shelf-life. In: Zeuthen, P., Cheftel, J.C., Eriksson, C., Gormley, T.R, Linko, P., Paulus, K. (Eds.), Processing and Quality of Foods, Vol. 1. London: Elsevier Applied Science, pp. 1.203-1.214.

Nongmaithem, R., Meda, V., Niranjan, T., Shukla, R.M., Hashmi, S., Dwivedi, M., 2024. Effect of extrusion process on germinated lentil (*Lens culinaris*) based snack foods: impact on starch digestibility, physio-functional, and antioxidant properties. Meas. Food. 27722759. 13. Available from: https://www.sciencedirect.com/science/journal/27722759. doi:10.1016/j.meafoo.2023.100132.

Ohtsu, H., 1990. Processed soybean curd food having a texture similar to meat. US Patent 4,897,280.

Piazza, L., Roversi, T., 2011. Preliminary study on microbeads production by co-extrusion technology. Procedia Food Sci. 2211601X. 1, 1374–1380. doi:10.1016/j.profoo.2011.09.204.

Rolan, T., Mueller, I., Carbon, G., Mertle, T.J., Swenson, K.J., Conley, C., Orcutt, M.W., Mease, L.E., 2008. Ground meat and meat analog compositions having improved nutritonal properties. US Patent 2008/02681 12 A1.

Ryu, G-H., 2020. Extrusion cooking of high-moisture meat analogues. In: Ganjyal, G.M. (Ed.), Extrusion Cooking. 2nd edn. Duxford, UK: Woodhead Publishing, pp. 205–224.

Shahmohammadi, H.R., Bakar, J., Russly, A.R., Noranizan, M.A., Mirhosseini, H., 2016. Studying the effects of fish muscle incorporation on storage stability of a novel corn-fish snack. J. Food Qual. 0146-9428. 39 (1), 45–53. doi:10.1111/jfq.12173.

Sumithra, B., Bhattacharya, S., 2008. Toasting of corn flakes: product characteristics as a function of processing conditions. J. Food Eng. 02608774. 88 (3), 419–428. doi:10.1016/j.jfoodeng.2008.03.001.

Tehrani, M.M., Ehtiati, A., Azghandi, S.S., 2017. Application of genetic algorithm to optimize extrusion condition for soy-based meat analogue texturization. J. Food Sci. Technol. 09758402. 54 (5), 1119–1125. Available from: http://www.springerlink.com/content/121580/. doi:10.1007/s13197-017-2524-9.

Zhong, L., Fang, Z., Wahlqvist, M.L., Hodgson, J.M., Johnson, S.K., 2021. Multi-response surface optimisation of extrusion cooking to increase soluble dietary fibre and polyphenols in lupin seed coat. LWT. 00236438. 140. Available from: https://www.journals.elsevier.com/lwt. doi:10.1016/j.lwt.2020.110767.

Novel applications of extrusion in food product development

18

Kasiviswanathan Muthukumarappan
Agricultural and Biosystems Engineering, South Dakota State University, Brookings, SD, United States

18.1 Introduction

Extrusion is a highly adaptable processing technology that integrates multiple unit operations into a single system. It is a continuous production method with the capability to transport, blend, apply shear, and thermally process (modify) food ingredients as they move through the system. Extrusion is a fascinating method often described as a "comprehensive processing technology," as it incorporates several operational steps into one streamlined process (Ek and Ganjyal, 2020).

During extrusion, the interaction of mass, energy, and momentum transfer is intricately linked with the complex physicochemical transformations occurring in the material that influence its properties. This means the food structure is not only a product of the process but also an integral part that impacts the evolution of material properties and the processing conditions within the extruder. For example, the mechanical stresses generated during extrusion are directly influenced by the rheological properties of the material, which are affected by the evolving structure that is continuously altered by these mechanical stresses along the extrusion process (Emin, 2021).

Extruders mainly consist of one or two screws covered by barrels, which have either a grooved or smooth surface. The screw is powered by a motor, often equipped with a gearbox to enable adjustable screw speeds during operation. At the barrel's exit, there is usually a die through which the processed material is extruded. Each component of the extruder plays a vital role in ensuring the system operates efficiently. Modern extruders are designed for high-temperature, short-time (HTST) cooking, providing a variety of product qualities, including improved texture, enhanced nutrient quality, better starch and protein digestibility, reduction of antinutritional factors, and increased functionality. Extrusion technology is extensively employed in the processing of cereal grains, which are transformed into a variety of products through extrusion. Typically, cereal grain flour serves as the primary ingredient for producing extruded goods. Many popular snacks and breakfast cereals are manufactured using these flours (Ganjyal, 2020).

Extrusion Cooking for Food and Feed. https://doi.org/10.1016/B978-0-443-29987-2.00018-X

The main objectives of this chapter are to explore the principles and mechanisms of traditional and novel extrusion technologies in food processing and to examine how extrusion enhances the nutritional, functional, and sensory properties of food products. It also aims to highlight the role of incorporating food industry by-products in extrusion for sustainable production and covers emerging techniques such as supercritical carbon dioxide (SC-CO_2) and hot melt extrusion (HME) for improving product quality and identifying current challenges while proposing future directions for innovation and optimization in extrusion-based food manufacturing.

18.2 Background on extrusion technology in food systems

The term "extrusion" originates from the Latin word "extrude," which means "to force out" or "to thrust out" (Prabha et al., 2021). The value of the food extrusion market was $73.1 billion globally in 2021and it is projected to reach $99.7 billion by 2026. The ongoing growth of the food extrusion market can be linked to the rise of the food industries (Lazou, 2024). Extrusion involves the forcing of raw material through a barrel with screws and a shaped die. It exposes the material to shear force along with high temperature and pressure. During this process, a physicochemical change occurs, transforming solids into a melt that stabilizes upon cooling (Dey et al., 2021). Although extrusion technology is widely used, improvements in equipment design, engineering, and process control are still needed. This idea could lead to new and innovative food product development. Extruders can be classified into two categories based on operation: cold extrusion and hot extrusion.

18.2.1 Cold extrusion

The high cooking temperatures normally used in conventional extrusion processes can cause issues such as whey protein discoloration due to the Maillard reaction, protein racemization during cross-linking, and degradation of sulfur-containing amino acids like cysteine and methionine, among other challenges (Onwulata et al., 2011). In contrast, cold extrusion involves processing food at temperatures up to 100°C, maintaining a stable temperature throughout the process. This technique is primarily used for shaping and mixing food products such as meat items and pasta. Temperatures below 100°C are also used in low-pressure extrusion, as seen in the production of pet foods. Preservation methods for cold-extruded products often include chilling, baking, or drying. Cold extruders are suitable for small-scale industries and domestic use due to their affordability, whereas large-capacity extruders are primarily used in large-scale industries due to their higher cost. A key application of cold extruders is in pasta production; different dies are used to change the shape of the dough into various forms (Shelar and Gaikwad, 2019).

18.2.2 Hot extrusion

Hot extrusion, also called extrusion cooking, is a process where food materials are heated to temperatures above 100°C. The continuous increase in temperature is achieved through frictional heating and other heating methods. After that, the heated food is transferred to the barrel equipped with small flights, which enhance the shear and pressure during the process. Later, the final shape of the product is formed after the material is extruded through a die under pressure and temperature, followed by rapid cooling. The final shape and size of the product depend on the type and aperture of the die (Shelar and Gaikwad, 2019). Examples of food products obtained from hot extrusion include snacks, breakfast cereals, pet foods, protein products (textured vegetable proteins and meat analog), noodles, and animal feed (aquaculture and livestock feed) (Kristiawan et al., 2020).

18.3 Overview of extrusion technology in the food industry

Extrusion is characterized as a process whereby a combination of blended components is propelled through a small aperture, known as a die, to shape and form the materials (Launay and Lisch, 1983; Shelar and Gaikwad, 2019). Extrusion technology is primarily used due to its superior product quality, multifunctionality, versatility, cost-effectiveness, high productivity, energy efficiency, and eco-friendliness. In the process of extrusion cooking, high-moisture, starchy, and protein-rich food substances are subjected to mechanical shear, thermal energy, and pressure. This operation is executed through a barrel and a screw accordingly (Shelar and Gaikwad, 2019). Extrusion food processing is extensively used to modify starch and protein-based food substances to produce a diverse variety of ready-to-eat (RTE) breakfast cereals, including pasta, bread alternatives, and animal feed (Adekola, 2016).

Extrusion technology has the capacity to produce new and innovative food products by improving and designing new processing methods, as well as enhancing existing ones. In relation to the design, development, and formulation of innovative and new food products characterized by superior quality and specific, unique shapes and colors, various types of novel raw materials may be used. Moreover, it is essential for food product manufacturers to adapt, modify, and improve current processes to adjust formulations for the creation of new products that satisfy consumer preferences and market requirements (Lazou, 2024; Prabha et al., 2021).

The growing demand for many varieties of food products, particularly extruded snacks, makes producers look for the necessary and needed raw materials in different parts of the world and obtain the extruded products from those locations. Some factors are affecting the production of raw materials, such as climate change, which causes a shortage of raw materials. This creates challenges in satisfying customers and meeting the growing demand. This could be the reason for the increasing price of products nowadays, and producers are focusing on developing new products (Lazou, 2024).

Innovation of new products is necessary to ensure the sustainability of the industry and to meet consumer preferences, as people nowadays are looking for healthier products rather than prioritizing taste. Concurrently, customers' demand for high-quality and safe food products has grown, which helps predict the development of nutrient-dense, fiber-rich food with distinct texture, color, and flavor. The food processing industry produces a huge quantity of waste yearly. These by-products create environmental issues due to their negative impact. Therefore the utilization of food by-products could enhance the nutritional profile of products for human consumption while reducing serious environmental effects (Dey et al., 2021).

18.4 Current applications and limitations in traditional extrusion methods

Food extrusion is a HTST processing method that converts the texture, flavor, fluid dynamics, and particle size of products. Extrusion cooking also offers key advantages, such as the ability to produce a wide variety of products from low-cost raw materials in a short time, allowing for consistent output through an efficient and continuous process (Dendegh et al., 2022). This method brings modifications through processes such as caramelization, mixing, cooking, heat and mass transfer, shearing, shaping, and forming (Liang, 2013). In modern food processing, the applications of extrusion technology have increased significantly due to its wide range of properties, such as shearing, separation, encapsulation, shaping, mixing and flavor development, heating or cooling, pasteurization or sterilization, volatile component or moisture removal, and coextrusion (Lazou, 2024). Extrusion technology is widely used for producing cereal-based products, including pasta, breakfast cereals, snacks, and modified starches, as well as protein-rich foods, texturized vegetable proteins (TVP), meat substitutes, fortified foods, and snacks. It is also used for convenience foods, such as RTE products, extruded chips, instant noodles, and by-products (Sule et al., 2024). Similarly, it is applied for the value addition of fruit and vegetable processing residues into dietary fiber supplements and protein-rich foods (Offiah et al., 2019; Prabha et al., 2021).

Traditional extrusion methods come with some challenges, despite their many benefits. One big issue is that high temperatures can break down essential nutrients, such as vitamins, antioxidants, and some proteins, making the final product less nutritious (Alam et al., 2016). On top of that, heat-sensitive ingredients, such as bioactive compounds, lose their effectiveness during the extrusion process. This limits the use of traditional extrusion for certain materials (Prabha et al., 2021). When it comes to creating complex food textures, conventional single-screw extruders often fall behind because they are not great at delivering precise and consistent results (Offiah et al., 2019). Finally, there is the environmental concern. Traditional petroleum-based extrusion methods are not exactly eco-friendly, which clashes with the growing push for sustainable food production (Xu et al., 2020).

18.5 Categorization of extruders

The food industry uses three types of extruders: piston extruders, roller-type extruders, and screw extruders. Screw extruders are the most common type of extruder used today, and they can be classified as single or twin-screw extruders.

18.5.1 Single-screw extruder

The barrel of a single-screw extruder, which frequently has a fluted or grooved design, adopts just one screw. To produce compression, the screw in a single-screw extruder is typically made with a decreasing pitch. The compression ratio is the measure of the amount of pitch decrease (Gu et al., 2017). A single-screw extruder is shown in Fig. 18.1. The performance of the extruder is dependent on the screw speed, barrel temperature, degree of fullness of the screw, melt characteristics of the product, and, most importantly, the pressure on the die. Melt characteristics depend on the temperature of the extrusion and the moisture content of the materials used (Adekola, 2016).

In the single-screw extrusion process, the extruders are divided into three regions based on the conversion of starch molecules, namely, swelling, conversion, and degradation or swelling. In the cooling region, the conveying and swelling regions are placed. Single-screw extrusion cooking is mainly used to produce a pasta-type product due to its low cost and feasibility (Alam et al., 2016).

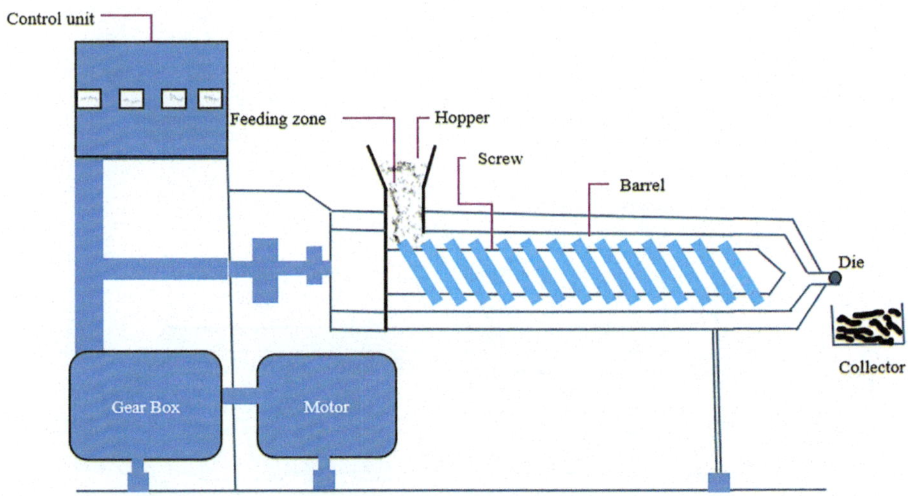

Fig. 18.1 Basic components of a single-screw extruder (Abeykoon, 2016).

18.5.2 Twin-screw extruder

It has two screws, either intermeshing or nonintermeshing, that make it a twin-screw extruder. Corotating or counterrotating screws are two possible configurations of a twin-screw extruder (shown in Fig. 18.2). Corotating is mostly used because it generates mechanical energy for the material. The screws of the corotating extruder turn in the same direction, whereas the screws run in opposite directions in the case of counterrotating extruders. The fully intermeshing and self-wiping design is the most commonly used configuration in corotating twin-screw extruders. In this system, the interlocking flights of the screws prevent the product from rotating within the system. Compared to single-screw extruders, corotating twin-screw extruders provide more efficient conveying, a tighter residence time distribution, and superior handling of sticky food materials (Adekola, 2016). Although twin-screw extruders have higher maintenance costs compared to single-screw extruders, they are widely used in the food industry due to their broad range of operating conditions and their capacity to produce a variety of food products (Gu et al., 2017).

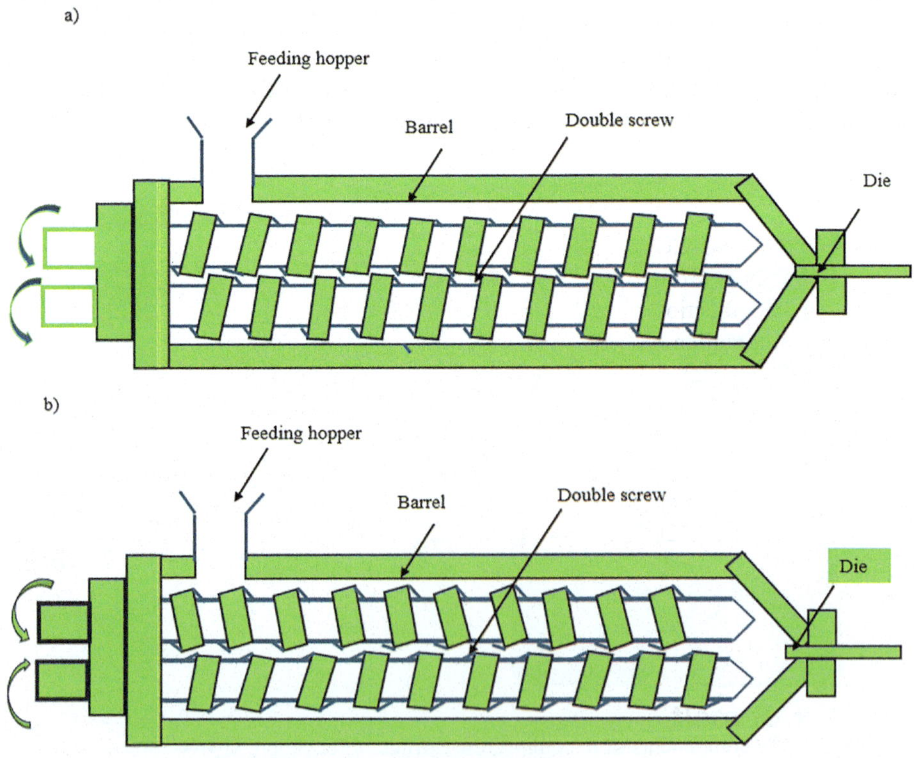

Fig. 18.2 A twin-screw extruder with a preconditioner.
Source: Ganjyal, G.M., 2020. Extrusion Cooking : Cereal Grains Processing. Woodhead Publishing & Wenger Manufacturing, Inc.

18.6 Recent developments in food extrusion

18.6.1 Supercritical carbon dioxide in extrusion

The concept of supercritical carbon dioxide (SC-CO_2) in extrusion was introduced by Rizvi and Mulvaney (1992) and was used to develop a uniform and smooth surface for the end product. SC-CO_2 has emerged as a transformative innovation in extrusion technology, providing a significant advantage over conventional steam-based methods. When operating at relatively low temperatures (60°C–100°C), SC-CO_2 extrusion minimizes the degradation of heat-sensitive nutrients, making it particularly suitable for fortifying foods with proteins, bioactive compounds, and essential vitamins (Mitra et al., 2022). The mechanism of the process involves the injection of SC-CO_2 into the extruder, where it acts as both a plasticizer and an expansion medium. The CO_2 dissolves in the food matrix under high pressure, and upon release at atmospheric pressure, it expands and forms a porous structure that enhances and improves the textural properties of the product. This process results in extrudates with increased crispness, reduced bulk density, and an enhanced expansion ratio compared to conventional extrusion techniques (Arora and Rizvi, 2021).

The significant application of SC-CO_2 extrusion lies in the production of high-protein snacks and cereals using milk protein concentrates. These products exhibit improved mechanical properties, such as reduced hardness and enhanced elasticity, satisfying the demand for functional, high-protein food extrudates (Mitra et al., 2022). In addition, SC-CO_2 extrusion facilitates the incorporation of food by-products like fruit pomaces, broken rice, and whey into extruded snacks, which helps increase dietary fiber and antioxidant retention while reducing environmental waste (Chauvet et al., 2017). Furthermore, studies have illustrated that the development of nutrient-rich snacks using SC-CO_2 extrusion retained up to 84% of total phenolics and 74% of antioxidant activity from apple pomace, demonstrating the ability of SC-CO_2 extrusion to enhance nutritional profiles while maintaining sensory acceptability (Lohani and Muthukumarappan, 2017).

Despite its potential, the industrial-scale adoption of SC-CO_2 extrusion faces challenges, which include the need for precise optimization of parameters such as CO_2 pressure, feed composition, and moisture levels to achieve consistent product quality. However, advancements in equipment design and process control are paving the way for wider implementation in the food manufacturing sector (Arora and Rizvi, 2021; Chauvet et al., 2017). Therefore SC-CO_2 extrusion offers a sustainable and adaptable solution for producing innovative, nutrient-rich, and heat-sensitive food products that represent advanced techniques with significant potential to address and fulfill consumer needs.

18.6.2 Hot melt extrusion

Hot-melt extrusion (HME) has emerged as a transformative technology with various applications in both the food and pharmaceutical industries. Originally developed for the packaging plastic industry, its utilization has since expanded to include food processing, especially for producing cereal-based food products. HME is widely known for its ability to form solid dispersions by changing the physicochemical properties, as well as the thermal and mechanical characteristics of materials, which leads to enhanced stability and bioavailability of food and pharmaceutical components. This process has become a well-established industrial method in food production because of its ability to improve solubility, gelatinization of starches, and denaturation of proteins. In addition, it plays a crucial role in improving digestibility, inactivating enzymes and microorganisms, and eliminating antinutritional factors (Alam et al., 2019).

HME is an eco-friendly and solvent-free process that allows for the encapsulation of bioactive compounds in nutraceutical food products. By using materials like maltodextrin, corn starch, and β-cyclodextrins, bioactive compounds such as flavors, vitamins, and essential oils can be protected during processing. The process involves the plasticization of wall materials, which reduces the glass transition temperature and protects heat-sensitive compounds (Prabha et al., 2021). This ensures high throughput with reduced oxygen exposure in the extrusion channel, which is essential for preserving sensitive nutrients.

A typical hot melt extruder is a technique that includes a drive system, a barrel for extrusion, a rotating screw, and an extrusion die to determine the final shape of the product. Extruders can have single or twin screws that rotate either in the same or opposite directions within the barrel. The barrel reduces residence time, with the sections either bolted or clamped together. A twin-screw extruder is widely used in HME and includes three functional zones: the feed zone for mixing, the compression zone for melting and homogenization, and the metering zone for achieving uniform temperature and composition in the final product. This systematic design allows for the production of materials in various shapes, such as rods, pellets, and tablets, depending on the die used (Prabha et al., 2021).

In the food industry, HME has been successfully applied to improve the functional properties of food products. Alam et al. (2019) utilized HME to produce insect-based protein pellets by blending insect flour and corn flour. The extrusion process improved the solubility, stability, and dispersion properties of the resulting pellets. Similarly, Ottoboni et al. (2018) illustrate the application of HME for producing snacks using insect flour blends. This sustainable method has the potential to address the shortages of protein. HME has also been used to develop nutraceutical foods that contain encapsulated bioactive compounds to enhance antioxidant capacity and reduce particle size. Despite its broad applications, further research is needed to evaluate the toxicity and bioaccessibility of HME products for food industry applications (Prabha et al., 2021).

An additional application of HME in functional food development is the combination of anthocyanin-rich potato flour with rutin-enriched buckwheat flour. This process increases antioxidant activity, phenolic content, and bioavailability (Azad et al., 2022). Moreover, HME has demonstrated its capability to improve the solubility and bio accessibility of less soluble compounds like phytosterol and naringenin, which makes it a powerful tool for increasing the efficacy of functional food ingredients (Ishimoto et al., 2022; Li et al., 2021). It has the ability to preserve the structural and functional integrity of sensitive compounds under controlled conditions, making it a valuable technique in 3D food printing. The process has been used to create complex food structures while maintaining the nutritional and sensory attributes of the ingredients (Kavimughil et al., 2022).

18.7 Incorporation of new ingredients and materials in extrusion processes

As previously mentioned, extrusion processing has been widely used to manufacture various food products, offering great potential for creating novel and innovative items. These products can contribute to the growth of both consumer and food processing industries. The incorporation of new ingredients (as shown in Table 18.1) into extrusion processes has enhanced the diversity and nutritional value of extruded products. Modern extrusion techniques enable the addition of functional ingredients such as proteins, fibers, antioxidants, and bioactive compounds to increase the health benefits of extruded products. Fruit and vegetable by-products, such as apple pomace, cranberry pomace, orange pulp, and sugar beet pulp, have been utilized as sources of dietary fiber and antioxidants in extruded snacks, addressing sustainability concerns while adding value to industrial waste (Offiah et al., 2019; Alam et al., 2016). Furthermore, advancements in extrusion technology have facilitated the use of a variety of materials, including cereals, pulses, and new protein sources like milk protein, to produce high-protein snacks,

Table 18.1 Ingredients used in extrusion and their function.

Component	Function in extrusion	References
Starch	Primary structural component; undergoes gelatinization, contributing to expansion, viscosity, and texture; acts as a matrix former during puffing.	Lazou (2024), Prabha et al. (2021), and Offiah et al. (2019)
Protein	Provides viscoelasticity, texturization, and structure (especially in high-moisture extrusion); denatures and cross-links under heat and shear.	Chakravorty and Das (2024), Lazou (2024), and See et al. (2025)
Lipids/fats	Lubricate the mass, reduce barrel friction and energy input, decrease expansion at high levels, and improve mouthfeel and flavor.	Lazou (2024)

Continued

Table 18.1 Ingredients used in extrusion and their function—cont'd

Component	Function in extrusion	References
Sugars	Effect on viscosity and browning reactions (Maillard) can inhibit starch gelatinization at high levels and contribute to sweetness and flavor.	Prabha et al. (2021) and Lazou (2024)
Salt	Enhances flavor, regulates moisture distribution, influences water activity, and may decrease product expansion.	Offiah et al. (2019) and Lazou (2024)
Dietary fiber	Reduces expansion due to interference with bubble formation, enhances nutritional value, and affects water absorption and solubility.	Robin et al. (2012) and Lazou (2024)
Water (moisture)	Critical plasticizer determines rheology, expansion, gelatinization, and residence time. It facilitates biochemical transformations.	Liang (2013) and Lazou (2024)
Vitamins and minerals	Often added postextrusion due to the risk of thermal degradation, which contributes to nutritional value. May require encapsulation.	Lazou (2024)
Bioactive compounds	Include antioxidants and phytochemicals, which are sensitive to thermal degradation. They can be protected through encapsulation or low-temperature extrusion.	Galanakis et al. (2021) and Lazou (2024)
Additives (e.g., emulsifiers, colors, flavors)	Improve product appeal, processing characteristics, and stability. It may affect viscosity and product structure.	Lazou (2024) and Prabha et al. (2021)

meat analogs, and fortified food products. These developments not only improve the nutritional profiles but also enhance the functional properties of extruded products, such as crispness, expansion, color, texture, and solubility, for large-scale food production scale (Ek and Ganjyal, 2020).

18.8 Value addition to food by-products and wastes

Use of by-products not only helps to increase the economic status of industries but also decreases the adverse environmental effects that may arise from the disposal of food by-products or wastes. Table 18.2 shows the use of these by-products and their effects on product quality.

Table 18.2 Food by-products/wastes and their effects on product quality.

By-products/wastes	Effect on extrudates	References
Fruits and vegetables (carrot pomace, tomato pomace and seeds, pineapple, cranberry pomace, avocado seed, orange peel, and grape seed)	Tomato pomace increases the antioxidant activity of the product, and its peel helps to increase the crude fiber and protein content of the extrudates. Similarly, avocado seed increases the water absorption capacity and water solubility index, cranberry pomace increases the flavanol and anthocyanin retention, and pineapple and carrot waste improve the overall acceptability of the final product.	Yagci et al. (2022) and Offiah et al. (2019)
Meat and seafood (fatty materials, skin, bones, liver, heart, kidneys, brain, intestines, spleen, and feet)	These by-products are nutritionally rich in protein, lipids, carbohydrates, and other bioactive compounds. For example, the liver and kidney contain 5–10 times more riboflavin compared to lean meat. Also, the intestine could be a useful source for sausage casings. In addition, extracted protein from seafood can be incorporated into a number of processing lines.	Bharat Helkar and Sahoo (2016), Jayathilakan et al. (2012), and Offiah et al. (2019)
Dairy (mainly whey from cheese production)	Whey consists of 20% of milk protein, including immunoglobulin, α- and β-lactalbumin, lactoferrin, bovine serum albumin, lactoperoxidase, and some other constituents present in whey. In addition, milk by-products have been used mostly in bakeries, confectioneries, and other health supplements, especially in dry powder form, where they serve as nutritional and functional ingredients in extrudate products.	Hoffman and Falvo (2004) and Offiah et al. (2019)

Continued

Table 18.2 Food by-products/wastes and their effects on product quality—cont'd

By-products/wastes	Effect on extrudates	References
Cereal by-products (germ, bran, hull, broken grains from rice, wheat, and sorghum, brewer's spent grain from brewing industries)	Rice germs and bran are richer in proteins and lipids and consist of vitamins E, B, and dietary fiber. Distillers' grains and brewers' spent grains are considered protein sources due to their high protein content. The use of BSG increases protein, phenolic content, antioxidant capacity, and dietary fiber in the product.	ElMekawy et al. (2013), Nascimento et al. (2017), and Offiah et al., 2019)
Root crops (sweet potato, cassava, and potato peel)	These by-products are rich in starch and sweeteners like glucose and fructose, and they help in the expansion of the product. The expansion of starch is especially dependent on its degree of gelatinization, which can be affected by feed moisture, temperature, and shear rate.	Téllez-Morales et al. (2022) and Offiah et al. (2019)
Oilseeds (defatted sunflower, soybean, canola, cotton seed, peanut, and ground nut)	Sunflower meals help to improve the functional properties of the extrudates; soybean meal improves the hardness, crispness, and expansion ratio; and defatted groundnut cakes enhance the color (dark color) due to the Maillard reaction, which occurs because of their high protein content. Flaxseed and cottonseed improve the fat, protein, and fiber content of the product.	Purohit and Rajyalakshmi (2011), Chaiyakul et al. (2008), Offiah et al. (2019), and Moure et al. (2006)

18.9 Nutritional/structural changes during extrusion

Extrusion techniques have varied effects on different food systems due to differences in components such as starch, proteins, moisture, and other ingredients that influence the structure and texture of the final products. The changes in the nutritional composition of the product are due to the high temperature, pressure, and shear, which leads to various reactions between the constituents, resulting in changes to the functional properties of the final product (Offiah et al., 2019).

Protein is the source of growth for animals and humans, composed of amino acids. The nutritional value of protein is influenced by its quantity, digestibility, and the bioavailability of essential amino acids. Based on these factors, digestibility is regarded as the most essential determinant of protein quality in adults. Extruded products exhibit higher protein digestibility compared to nonextruded products due to protein denaturation, deactivation of enzyme inhibitors (in vegetable protein) present in raw plant materials, and the inactivation of antinutritional compounds that hinder digestion. The inactivation of enzymes like lipoxygenase, a decrease in protein solubility, and improved digestibility occur due to the denaturation of protein after applying high temperature and shear force. Extrusion processing has significant impacts on protein quality and amino acid composition due to the combined effects of temperature, material type, and the physical and chemical properties of food components. Several reactions during extrusion, such as caramelization, dextrinization, and pigment degradation, also cause changes in product color, as well as Maillard reactions involving amino acids. In cereal-based products, lysine is the most limiting essential amino acid among all essential amino acids. Therefore ensuring lysine retention during the extrusion process is important. It is necessary to focus on how different processing parameters influence lysine retention (Singh et al., 2007).

Starch is the major carbohydrate found in plants, composed of amylose, amylopectin, and glucans. Among various cereal types, extruded products exhibit higher starch digestibility. High starch digestibility is particularly observed in waxy starches. Dextrinized starch has higher molecular mobility, which makes it more prone to retrogradation (Zhang et al., 2014). However, most research has focused on preventing retrogradation due to its negative effects on the texture and quality of starchy food products. Similarly, some studies have found that retrogradation after extrusion cooking can produce enzyme-resistant starch (RS) in cereal starches. Typically, extruded products are dried to a stable moisture level immediately after extrusion. Nevertheless, some researchers have allowed undried extrudates to rest for a period and observed an increase in retrograded starch and a decrease in starch digestibility (Gulati et al., 2020). Extrusion has also impacted the solubility and molecular weight distribution of polysaccharides in cereal grains, which subsequently influences starch digestion properties (Robin et al., 2016).

Lipids are an essential component of the extruder, as they add lubricity inside the barrel to decrease expansion. During extrusion, lipids bind to protein and starch molecules. Naturally present lipids can form a starchy–fatty complex when higher moisture content and high barrel temperatures are used. However, this makes the product less desirable due to the higher hardness and density of the extrudates. Most studies have shown that the interaction between starch and saturated fatty acids is highly desirable. In addition, the starch–lipid complex is essential from a nutritional point of view because it helps improve the oxidative stability of lipids, retaining more essential fatty acids during product storage compared to unbound lipids. This is particularly important for cereal grains and legumes used in extrusion, as they contain highly unsaturated oils. Furthermore, this lipid-starch binding slows down starch digestion in the small intestine, which can lower the glycemic index, that is, beneficial effect for individuals with type 2 diabetes (Thachil et al., 2014; Gulati et al., 2020).

Lipids can also bind to both proteins and starch during extrusion. A study by De Pilli et al. (2015) investigated the effects of adding oleic acid to mixtures of wheat starch and gluten, comparing them to mixtures of wheat starch and oleic acid without gluten. They found that adding gluten reduced the formation of starch–oleic acid complexes. In addition, the lubricating effect of oleic acid, which reduces the specific mechanical energy (SME), decreased when gluten was included. The researchers attributed these changes to the formation of gluten–oleic acid complexes. These complexes strengthen gluten, slightly increasing the viscosity of the mixture inside the extruder and raising the SME (Gulati et al., 2020).

For the functioning and normal growth of the human body, vitamins are essential and must be obtained through diet. Vitamins are divided into two categories based on their properties: water-soluble (vitamins B and C) and fat-soluble (vitamins A, D, E, and K). Most cereals contain B vitamins in their bran or outer pericarp. The stability of vitamins depends on the processing conditions during extrusion. Among all the vitamins, A and E are more sensitive to higher extrusion temperatures, but this sensitivity can be mitigated by increasing the screw speed, thereby reducing the residence time inside the barrel (Riaz et al., 2009). The increase or decrease in vitamin levels due to various parameters during extrusion is shown in Table 18.3.

Vitamin losses during the storage of extrudates are mainly influenced by oxidation, storage, and extrusion parameters, with vitamins A, C, and E being the most affected. In one study, β-carotene in a casein-stabilized oil-in-water emulsion was used during the extrusion of a wheat flour mix. After 27 days at 20°C, 50% of β-carotene remained with the casein-only emulsion, compared to 30% with the casein and gum Arabic emulsion. Despite retaining more β-carotene immediately after extrusion, the casein and gum Arabic emulsion demonstrated lower retention during storage (Caliskan et al., 2015). Extrusion has a significant impact on the retention of vitamin content in cereals, with high processing temperatures and prolonged thermal exposure being the primary causes of vitamin loss, as shown in Fig. 18.3.

Dietary fiber is found in the edible parts of plants, which are resistant to absorption and digestion in the human body (mainly in the small intestine). Extrusion cooking increases the total dietary fiber in barley flour. In waxy barley, only soluble dietary

Table 18.3 Change in vitamin levels due to extrusion parameters.

Vitamin level	Extrusion parameter
Increase	Barrel temperature
Increase	Screw speed
Increase	Specific energy input
Decrease	Feed moisture
Decrease	Die diameter
Increase	Throughput

From Kiileit, U., 1994. Effects of extrusion cooking on vitamins. In: Asp, N.-G. (Ed.), Nutritional Effects of Extrusion Cooking. Elsevier Applied Science, London, pp. 149–155; Riaz, M.N., Asif, M., Ali, R., 2009. Stability of vitamins during extrusion. Crit. Rev. Food Sci. Nutr. 15497852. 49 (4), 361–368. doi:10.1080/10408390802067290.

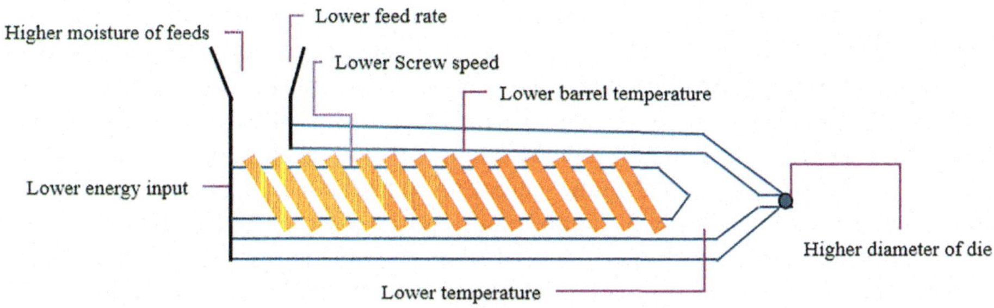

Fig. 18.3 Factors that generally result in increased vitamin retention in extruded products (Gulati et al., 2020; Athar et al., 2006).

fiber (SDF) increases as the total dietary fiber increases, while in regular barley flour, both soluble and insoluble dietary fibers increase (Vasanthan et al., 2002). The changes in dietary fiber during extrusion are mainly due to the conversion of insoluble fiber to soluble fiber and the formation of RS and enzyme-resistant glucans.

In mild or moderate extrusion conditions, the total dietary fiber content remains mostly unchanged, but some fiber components become soluble. Under more intense conditions, total dietary fiber increases, primarily due to higher levels of soluble fiber and enzyme-resistant starch (Singh et al., 2007). According to Zhang et al. (2011), SDF in extruded oat bran exhibited higher yields (14.2%), larger mean particle size, increased peak temperature, improved solubility, swelling capacity, solvent retention capacity, foaming ability, apparent viscosity, and consistency coefficient, along with a lower flow behavior index compared to SDF in untreated oat bran. These findings demonstrate that extrusion enhances certain properties of SDF from oat bran.

The addition of dietary fiber in extruded products results in a reduction of expansion ratio and texture, which are less favorable for consumers. However, soluble fiber tends to produce extrudates with a greater expansion ratio and has a smaller impact on the bulk density of extruded products compared to insoluble fiber. These differences in expansion and textural properties between soluble and insoluble fibers can be associated with their interactions with starch, variations in water sorption and plasticization behavior, as well as the physicochemical changes during extrusion. Pretreatment of insoluble fiber before extrusion can improve its expansion and textural characteristics (Robin et al., 2012).

Minerals obtained in minor quantities in food composition can be difficult to decompose. Iron (Fe) and calcium (Ca) are mostly added to food to improve its nutritional value (Singh et al., 2007). Minerals are divided into two categories: macro- and microminerals. Minerals are heat-stable and are not lost through steam at the die. Extrusion cooking mainly affects large molecules, but smaller molecules can also be impacted either directly by extrusion or indirectly through changes in larger molecules. It can enhance mineral absorption by reducing inhibitors like phytate, which forms insoluble complexes with minerals and inhibits their absorption. During extrusion, phytate is broken down, releasing phosphate molecules. For example, the extrusion of peas and kidney beans led to phytate breakdown, increasing mineral availability after processing at high temperatures (Alonso et al., 2001).

The increase in mineral absorption is due to the destruction of antinutritional factors such as phytates and tannins. During extrusion, phytate (an insoluble complex) breaks down and releases phosphate molecules. For example, the extrusion of peas and kidney beans has been shown to degrade phytate, which increases mineral availability after processing at high temperatures (Singh et al., 2007).

The nutritional quality of legume–cereal mixtures may be hindered by antinutritional factors like phytic acid, tannic acid, and trypsin inhibitors, which cause a reduction in protein digestibility. The extrusion process causes a decrease in antinutrient levels, with phytic acid content reduced by 40.64%–46.07% and tannic acid content by 40.46%–44.88% in blends of germinated, dehulled chickpea and corn. Antioxidants prevent the oxidation of molecules, particularly fats and fat-containing meals. The antioxidant activity of most meals is primarily due to phenolic compounds. The extrusion method significantly increased the total phenolic content and antioxidant content of extrudates by 1.92%–7.94% and 1.07%–5.55%, respectively. The extrusion method increases the release of bioactive compounds from the cell wall matrix, making them more accessible for extraction (Zieliński et al., 2006; Choton et al., 2020).

18.10 Applications of novel extrusion techniques in food systems

Extrusion technology has become one of the key areas in the food industry, offering a versatile and efficient method to develop novel food products. Modern advancements have focused on improving innovative extrusion techniques to develop specific food products with desired textures, flavors, and nutritional profiles. Many varieties of food items can be produced using extrusion technology, and they are as follows:

- Snacks foods
- Breakfast cereals
- RTE foods
- Meat analog
- 3D-printed food
- Oil expelling
- Use of industrial food waste

18.10.1 Snack foods

Extrusion technology is widely used in the production of snacks, providing flexibility in mixing diverse ingredients like whole grains, legumes, and some bioactive compounds. Mostly, twin-screw extrusion is used to make expanded, crispy snacks with added nutritional benefits nowadays, such as gluten-free or high-protein snacks (Singh et al., 2007) including extruded corn puffs like Cheetos. In addition, chickpea and lentil flour are also increasingly used in extruded snack formulations to cater to health-conscious consumers. The health benefits of consuming these snacks are due to fortification with vitamins and minerals to enhance their nutritional value (Wu et al., 2024).

18.10.2 Breakfast cereals

Breakfast cereals have been a popular product of extrusion technology for a long time. These food items include fibers, whole grains, and bioactive compounds with enhanced product quality. Advanced extrusion methods produce cereals like cornflakes with a variety of shapes, sizes, and sensory attributes. For example, oats and quinoa are often extruded to make gluten-free breakfast cereals. Breakfast cereals are convenient, RTE formats and provide desired flavors for consumer appeal (Woomer and Adedeji, 2021).

18.10.3 RTE foods

Convenience and trends have driven the growth of RTE foods like instant noodles and TVP chunks, where extrusion plays a critical role in ensuring shelf stability and enhanced sensory properties. In addition, extrusion cooking is used to produce instant noodles and fortified pasta, and it also allows the combination of functional ingredients like dietary fibers or omega-3 fatty acids. It has a long shelf life and retains nutrients during processing (Shah et al., 2019).

18.10.4 Meat analogs

High-moisture extrusion (HME) processing has been used in developing plant-based meat analogs, such as pea protein-based meat substitutes. This method imitates the fibrous texture of animal meat using plant-based proteins like soy, wheat, and peas. Plant-based products like burgers and sausages are now becoming popular in the market. Beyond Meat and Impossible Foods use advanced extrusion techniques to replicate meat-like textures and flavors. This approach has certain advantages over other varieties, such as reducing the environmental footprint and meeting the growing demand for sustainable protein foods (Zhang et al., 2023).

18.10.5 3D-printed foods

3D printing technology has transformed various industries due to its capabilities for customization, digitization, and personalization. In the area of food production, significant advancements have been made in optimizing printable materials and enhancing the precision of printing. However, its role in modifying material properties and improving the quality of printed food products is crucial. During the 3D printing process, the ink comes into contact with components such as the squeeze piston and pipes of the printer to make products like dough-based pizza or pizza shapes. In addition, the printed sample is often exposed to air, which increases the risk of microbial contamination. Therefore ensuring microbial safety throughout the 3D printing process is of utmost importance (Shen et al., 2023). This technology also helps in the minimization of food waste through precision manufacturing.

18.10.6 Oil expelling

Extrusion technology is also useful in oil expelling, where it facilitates the extraction of oil from different seeds, such as soybeans, flaxseeds, and sunflower seeds. Twin-screw extruders are more effective in improving oil yield and quality while reducing processing time.

The benefits of using extrusion for oil expelling are high efficiency in oil recovery and the retention of by-products for further use in animal feed or food products (Rajpurohit and Li, 2023).

18.10.7 Use of industrial food waste

Extrusion technology provides sustainable solutions for repurposing industrial food waste into value-added food products. For example, wheat bran, fruit peels, vegetable skins, and spent grain are converted into functional food ingredients using extrusion. These by-products are extruded into high-fiber snacks or used as fillers in other food formulations. Brewer's spent grain is extruded into high-protein snacks, reducing waste in the brewing industry, which helps to promote circular economic practices. The use of such waste also reduces food waste while creating nutrient-dense products (Grasso, 2020; Ainsworth et al., 2007).

18.10.8 Pet foods

Extrusion is also utilized to produce semi-moist and dry expanded pet foods, such as snacks or pellets, aquatic feeds, and laboratory animal diets. Cat and dog foods are typically extruded and then dried directly. Moreover, the extrusion process is used to manufacture feed for ornamental fish, high-quality complete feeds to support health, and specialized diets for exotic aquarium species. This technique allows for more efficient use of available cereal grains, as well as vegetable and animal proteins (Choton et al., 2020).

18.11 Consumer preference and safety of extruded products

For improvement in research and development, consumers' preferences play a key driving force, which leads to the growth of food industries. There is a growing demand for convenience foods, alongside an increasing preference for safe and healthy snacks. There has been an increasing consumer demand for safe, convenient, and high-quality products, which is anticipated to drive the growth of extruded foods that are nutritious, high in fiber, and tailored for specific textures, colors, and flavors, incorporating bioactive and functional ingredients. Protein and meat alternatives, as well as food processing by-products, are expected to remain key raw materials in the development of new products. The use of these materials has the potential to foster new innovations and technological disruptions in the food industry, while also reshaping the food sector's role within the bioeconomy and climate-neutral economy in the near future (Lazou, 2024; Šárka et al., 2020).

Another important aspect is to ensure the safety of extruded products. Generally, the products prepared from hot extrusion are known to be microbiologically safe due to the killing of microorganisms by applying high temperature and high pressure, but there might be chances of cross-contamination during postextrusion steps, which is why care needs to be taken. Materials used in the manufacture of extruded products should meet acceptable standards for microbial contamination (Lazou, 2024). It is well known that many plant-derived materials may contain food allergens. Soy, peanuts, wheat, milk, fish, and eggs contain allergens. This is a concern globally, but some researchers have found that extrusion processing effectively reduces the allergenic properties of proteins in soybean protein isolate and soybean flour-corn extrudates, with temperature being the main influencing factor (Lazou, 2024; Zheng et al., 2020). In addition, extrusion has been shown to denature undesirable enzymes and deactivate antinutritional factors, which helps to enhance protein and starch digestibility, sterilize the final product, and preserve the product's original color and flavor (Zheng et al., 2020; Lazou, 2024; Singha et al., 2018).

Baiano (2022) found that microbial contamination and toxic substance migration from printer components raised significant concerns in 3D-printed food production. Safety requires effective cleaning protocols and the use of food-safe materials that comply with regulations. Since high temperatures are not applied to control microorganisms, the heating and cooling processes during extrusion could promote microbial growth. Moreover, extrusion-based 3D-printed foods may contain allergens and antinutritional factors that pose a risk to consumer health. For proper handling, processing, and storage, it is necessary to follow national and international food safety regulations (Lazou, 2024; Tomašević et al., 2021).

18.12 Advantages of novel extrusion applications in food systems

Extrusion technology has several advantages over traditional food processing methods that make it highly adaptable and efficient. One of its main strengths is the versatility of products it can produce, including those that cannot be manufactured using other techniques. This adaptability is enhanced by the ability to easily change parts like dies and adjust operating conditions. The process is energy-efficient, operating at low moisture levels during cooking, which reduces the need for redrying of the product. It is also cost-effective, requiring less raw material, less labor, and a lower capital investment compared to other methods. Extrusion occupies less space, has a high production rate, and facilitates automated control systems for continuous and streamlined operations.

This technology is particularly suited for creating new snack foods, as it can modify proteins, starches, and other ingredients while retaining heat-sensitive components despite the HTST process. Extrusion also effectively destroys antinutritional compounds like trypsin inhibitors and undesirable enzymes while killing microorganisms due to its high extrusion temperature and pressure. In addition, it minimizes the degradation of food nutrients, improves protein digestibility through denaturation, and enhances starch gelatinization. With minimal water usage and lower process effluents, extrusion is environmentally friendly and a sustainable choice for food processing (Akhtar et al., 2015).

18.13 Challenges and future perspectives

Food extrusion continues to be a pivotal technology in modern food processing, offering versatility and efficiency in producing RTE snacks, cereals, meat analogs, and functional ingredients. However, the process still faces some considerable challenges, including a limited understanding of rheological behaviors and material transformation within the extruder, particularly when using underutilized or novel plant proteins and food by-products. Process modeling remains complex due to non-isothermal, multiphase conditions and short residence times, which affect consistency and scalability (Lazou, 2024). The adoption of HME for plant-based meats requires more predictive tools and optimized control strategies to manage protein structuring and fiber alignment (See et al., 2025). Furthermore, as extrusion applications expand into reactive extrusion, HME, and 3D/4D printing, new engineering and formulation challenges arise, demanding greater interdisciplinary innovation (Prabha et al., 2021; Chakravorty and Das, 2024).

Many studies have been conducted on the potential use of new raw materials for the development of innovative food products, focusing on the nutritional and sensory quality of these products. In addition, it is essential to evaluate their health benefits. A food product obtained from industrial waste or other less usable by-products needs to pass the risk assessment conducted by the food safety authority. Therefore any food items manufactured from food by-products are limited, so they are produced and tested on a pilot scale, making their commercialization challenging (Lazou, 2024).

The quality-by-design (QbD) approach, mostly used in the pharmaceutical and biotech industries, provides a systematic framework for enhancing product and process understanding in food extrusion. Unlike end-product testing, QbD focuses on building quality into the process through risk assessment, root cause analysis, and systematic experimentation (Rathore and Kapoor, 2017). In extrusion, the processing parameters include variables such as barrel temperature, screw speed, screw design, feed rate, and die characteristics. To achieve the desired product quality, it is necessary to have comprehensive knowledge of raw material properties, including physical, chemical, and microbiological attributes. This method is especially critical for upcycled materials, where rigorous risk assessment ensures product safety and facilitates consumer acceptance. By adopting QbD, food extrusion processes can ensure safety, compliance with regulations, and consistent product quality.

Extrusion cooking is a promising technique for developing innovative food products, ingredients, encapsulated and fortified products, and those with desired structures, textures, and functional properties. This method effectively uses plant-based by-products to create value-added, high-nutritional products using less costly raw materials. Extruded foods with low moisture content ensure microbial safety and an extended shelf life. The versatility of extrusion techniques allows the incorporation of a variety of materials, including agricultural by-products, reducing environmental effects by reutilizing waste. Proper selection of raw materials and optimization of process parameters are essential for achieving desired product characteristics, making extrusion a foundation of novel food production (Prabha et al., 2021).

18.14 Conclusion

Extrusion technology has become an advanced tool in food processing, combining versatility, efficiency, and adaptability to meet consumers' needs and the demands of the food industry. This book chapter focuses on the technological advancements and innovative applications of extrusion techniques, particularly in producing a variety of food products such as plant-based alternatives, nutrient-enriched snacks, and milk protein-based foods. These innovations not only enhance nutritional quality and sensory appeal but also address issues of sustainability through the efficient use of by-products and reduced resource consumption. Novel extrusion techniques, including the incorporation of functional ingredients and precise process controls, demonstrate the potential to create healthier, environmentally friendly food products. This focus on sustainability aligns with global efforts to reduce food waste and promote eco-friendly practices in the food sector.

Future advancements in extrusion technology will focus on optimizing process parameters to improve consistency, as well as exploring the utilization of novel raw materials. Addressing safety concerns and allergen risks will be critical to ensuring product quality and consumer preference. Interdisciplinary research and collaborative efforts will play a key role in the potential use of extrusion, establishing it as a pivotal innovation in food processing and contributing to the development of sustainable, health-oriented food systems.

References

Abeykoon, C., 2016. Single screw extrusion control: a comprehensive review and directions for improvements. Control Eng. Pract. 09670661. 51, 69–80. doi:10.1016/j.conengprac.2016.03.008.

Adekola, K.A., 2016. Engineering review food extrusion technology and its applications. J. Food Sci. Eng. 21595828. 6 (3). doi:10.17265/2159-5828/2016.03.005.

Ainsworth, P., Ibanoğlu, S., Plunkett, A., Ibanoğlu, E., Stojceska, V., 2007. Effect of brewers spent grain addition and screw speed on the selected physical and nutritional properties of an extruded snack. J. Food Eng. 02608774. 81 (4), 702–709. doi:10.1016/j.jfoodeng.2007.01.004.

Akhtar, J., Malik, S., Alam, M.A., 2015. Extrusion technology used for novel foods production 1. IJEDR 3, 379–383.

Alam, M.R., Scampicchio, M., Angeli, S., Ferrentino, G., 2019. Effect of hot melt extrusion on physical and functional properties of insect based extruded products. J. Food Eng. 02608774. 259, 44–51. Available from: http://www.sciencedirect.com/science/journal/02608774. doi:10.1016/j.jfoodeng.2019.04.021.

Alam, M.S., Kaur, J., Khaira, H., Gupta, K., 2016. Extrusion and extruded products: changes in quality attributes as affected by extrusion process parameters: a review. Crit. Rev. Food Sci. Nutr. 15497852. 56 (3), 445–473. Available from: https://www.tandf.co.uk/journals/titles/10408398.asp. doi:10.1080/10408398.2013.779568.

Alonso, R., Rubio, L.A., Muzquiz, M., Marzo, F., 2001. The effect of extrusion cooking on mineral bioavailability in pea and kidney bean seed meals. Anim. Feed Sci. Technol. 03778401. 94 (1-2), 1–13. doi:10.1016/s0377-8401(01)00302-9.

Arora, B., Rizvi, S.S.H., 2021. Process optimisation and product characterisation of milk protein concentrate extrudates expanded by supercritical carbon dioxide. Int. J. Dairy Technol. 14710307. 74 (4), 641–654. Available from: http://onlinelibrary.wiley.com/journal/10.1111/(ISSN)1471-0307. doi:10.1111/1471-0307.12801.

Athar, N., Hardacre, A., Taylor, G., Clark, S., Harding, R., McLaughlin, J., 2006. Vitamin retention in extruded food products. J. Food Compos. Anal. 08891575. 19 (4), 379–383. doi:10.1016/j.jfca.2005.03.004.

Azad, M.O.K., Adnan, M., Sung, I.J., Lim, J.D., Baek, J.S., Lim, Y.S., Park, C.H., 2022. Development of value-added functional food by fusion of colored potato and buckwheat flour through hot-melt extrusion. J. Food Process. Preserv. 17454549. 46 (5). Available from: http://onlinelibrary.wiley.com/journal/10.1111/(ISSN)1745-4549. doi:10.1111/jfpp.15312.

Baiano, A., 2022. 3D Printed Foods: a Comprehensive review on technologies, nutritional value, safety, consumer attitude, regulatory framework, and economic and sustainability issues. Food Rev. Int. 15256103. 38 (5), 986–1016. Available from: https://www.tandf.co.uk/journals/titles/87559129.asp. doi:10.1080/87559129.2020.1762091.

Caliskan, G., Lim, A.S.L., Roos, Y.H., 2015. Beta-carotene stability in extruded snacks produced using interface engineered emulsions. Int. J. Food Prop. 15322386. 18 (10), 2256–2267. Available from: https://www.tandfonline.com/loi/ljfp20. doi:10.1080/10942912.2014.973963.

Chaiyakul, S., Jangchud, K., Jangchud, A., Wuttijumnong, P., Winger, R., 2008. Effect of protein content and extrusion process on sensory and physical properties of extruded high-protein, glutinous rice-based snack. Nat. Sci. 42, 182–190.

Chakravorty, P., Das, A.B., 2024. Advances in high moisture extrusion cooking for meat analogue with 3D and 4D printing approaches: a comprehensive review. Discov. Food. 27314286. 4 (1). Available from: https://www.springer.com/journal/44187. doi:10.1007/s44187-024-00222-4.

Chauvet, M., Sauceau, M., Fages, J., 2017. Extrusion assisted by supercritical CO_2: a review on its application to biopolymers. J. Supercrit. Fluids. 08968446. 120, 408–420. doi:10.1016/j.supflu.2016.05.043.

Choton, S., Gupta, N., Bandral, J.D., Anjum, N., Choudary, A., 2020. Extrusion technology and its application in food processing: a review. Pharma Innov. 23498242. 9 (2), 162–168. doi:10.22271/tpi.2020.v9.i2d.4367.

De Pilli, T., Legrand, J., Derossi, A., Severini, C., 2015. Effect of proteins on the formation of starch–lipid complexes during extrusion cooking of wheat flour with the addition of oleic acid. Int. J. Food Sci. Technol. 0950-5423. 50 (2), 515–521. doi:10.1111/ijfs.12698.

Dendegh, T.A., Akpapunam, S.O., Yelmi, A., 2022. Extrusion technology and its application in food processing-an overview. Trop. J. Eng. 1, 25–53.

Dey, D., Richter, J.K., Ek, P., Gu, B.J., Ganjyal, G.M., 2021. Utilization of food processing by-products in extrusion processing: a review. Front. Sustain. Food Syst. 2571581X. 4. Available from: https://www.frontiersin.org/journals/sustainable-food-systems#. doi:10.3389/fsufs.2020.603751.

Ek, P., Ganjyal, G.M., 2020. Basics of Extrusion Processing. Elsevier BV, pp. 1–28. doi:10.1016/b978-0-12-815360-4.00001-8.

ElMekawy, A., Diels, L., De Wever, H., Pant, D., 2013. Valorization of cereal based biorefinery byproducts: reality and expectations. Environ. Sci. Technol. 0013-936X. 47 (16), 9014–9027. doi:10.1021/es402395g.

Emin, M.A., 2021. Key technological advances of extrusion processing. In: Food Engineering Innovations across the Food Supply Chain. Elsevier, Germany, pp. 131–148. Available from https://www.sciencedirect.com/book/9780128212929. doi:10.1016/B978-0-12-821292-9.00005-4.

Galanakis, C.M., Rizou, M., Aldawoud, T.M.S., Ucak, I., Rowan, N.J., 2021. Innovations and technology disruptions in the food sector within the COVID-19 pandemic and post-lockdown era. Trends Food Sci. Technol. 09242244. 110, 193–200. Available from: https://www.sciencedirect.com/science/journal/09242244. doi:10.1016/j.tifs.2021.02.002.

Ganjyal, G.M., 2020. Extrusion Cooking : Cereal Grains Processing. Woodhead Publishing.

Grasso, S., 2020. Extruded snacks from industrial by-products: a review. Trends Food Sci. Technol. 09242244. 99, 284–294. Available from: http://www.elsevier.com/wps/find/journaldescription.cws_home/601278/description#description. doi:10.1016/j.tifs.2020.03.012.

Gu, B.J., Kowalski, R.J. Ganjyal, G.M., 2017. Food extrusion processing: an overview. Agric. Food Sci., 1–7.

Gulati, P., Brahma, S., Rose, D.J., 2020. Impacts of extrusion processing on nutritional components in cereals and legumes: carbohydrates, proteins, lipids, vitamins, and minerals. In: Extrusion Cooking. Elsevier BV, pp. 415–443. doi:10.1016/b978-0-12-815360-4.00013-4.

Hoffman, J.R., Falvo, M.J., 2004. Protein-which is best? J. Sports Sci. Med. 3, 118.

Ishimoto, K., Shimada, Y., Ohno, A., Otani, S., Ago, Y., Maeda, S., Lin, B., Nunomura, K., Hino, N., Suzuki, M., Nakagawa, S., 2022. Physicochemical and biochemical evaluation of amorphous solid dispersion of naringenin prepared using hot-melt extrusion. Front. Nutr. 2296-861X. 9. doi:10.3389/fnut.2022.850103.

Jayathilakan, K., Khudsia Sultana, K.R., Bawa, A.S., 2012. Utilization of byproducts and waste materials from meat, poultry and fish processing industries: a review. J. Food Sci. Technol. 0022-1155. 49 (3), 278–293. doi:10.1007/s13197-011-0290-7.

Kavimughil, M., Leena, M.M., Moses, J.A., Anandharamakrishnan, C., 2022. Effect of material composition and 3D printing temperature on hot-melt extrusion of ethyl cellulose based medium chain triglyceride oleogel. J. Food Eng. 02608774. 329, 111055. doi:10.1016/j.jfoodeng.2022.111055.

Kristiawan, M., Chaunier, L., Sandoval, A.J., Valle, G.D., 2020. Extrusion-Cooking and expansion. In: Breakfast Cereals and How They are Made: Raw Materials, Processing, and Production. Elsevier Inc, France, pp. 141–167. Available from http://www.sciencedirect.com/science/book/9780128120439. doi:10.1016/B978-0-12-812043-9.00007-2.

Launay, B., Lisch, J.M., 1983. Twin-screw extrusion cooking of starches: flow behaviour of starch pastes, expansion and mechanical properties of extrudates. J. Food Eng. 02608774. 2 (4), 259–280. doi:10.1016/0260-8774(83)90015-8.

Lazou, A.E., 2024. Food extrusion: an advanced process for innovation and novel product development. Crit. Rev. Food Sci. Nutr. 15497852. 64 (14), 4532–4560. Available from: https://www.tandfonline.com/loi/bfsn20. doi:10.1080/10408398.2022.2143474.

Li, Y.L., Ruan, Q.J., Wang, J.M., Yang, X.Q., 2021. Fabrication and structural properties of water-dispersible phytosterol using hot melt extrusion. J. Food Sci. Technol. 09758402. 58 (6), 2447–2451. Available from: http://www.springerlink.com/content/121580/. doi:10.1007/s13197-020-04956-8.

Liang, Y., 2013. Extrusion Processing of Protein Rich Food Formulations. Library and Archives Canada, Canada.

Lohani, U.C., Muthukumarappan, K., 2017. Process optimization for antioxidant enriched sorghum flour and apple pomace based extrudates using liquid CO_2 assisted extrusion. LWT. 00236438. 86, 544–554. Available from: https://www.journals.elsevier.com/lwt. doi:10.1016/j.lwt.2017.08.034.

Mitra, P., Zhou, R., Rizvi, S.S.H., 2022. Supercritical carbon dioxide extrusion of milk protein concentrate and rice flour blend: a new dairy nutrition delivery platform. J. Food Process Eng. 17454530. 45 (8). Available from: http://onlinelibrary.wiley.com/journal/10.1111/(ISSN)1745-4530. doi:10.1111/jfpe.14056.

Moure, A., Sineiro, J., Domínguez, H., Parajó, J.C., 2006. Functionality of oilseed protein products: a review. Food Res. Int. 09639969. 39 (9), 945–963. doi:10.1016/j.foodres.2006.07.002.

Nascimento, T.A., Calado, V., Carvalho, C.W.P., 2017. Effect of Brewer's spent grain and temperature on physical properties of expanded extrudates from rice. LWT. 00236438. 79, 145–151. Available from: https://www.journals.elsevier.com/lwt. doi:10.1016/j.lwt.2017.01.035.

Offiah, V., Kontogiorgos, V., Falade, K.O., 2019. Extrusion processing of raw food materials and by-products: a review. Crit. Rev. Food Sci. Nutr. 15497852. 59 (18), 2979–2998. Available from: https://www.tandfonline.com/loi/bfsn20. doi:10.1080/10408398.2018.1480007.

Onwulata, C.I., Tunick, M.H., Qi, P.X., 2011. Extrusion texturized dairy proteins. processing and application. Adv. Food Nutr. Res. 62, 173–200. doi:10.1016/B978-0-12-385989-1.00005-3.

Ottoboni, M., Spranghers, T., Pinotti, L., Baldi, A., De Jaeghere, W., Eeckhout, M., 2018. Inclusion of Hermetia Illucens larvae or prepupae in an experimental extruded feed: process optimisation and impact on in vitro digestibility. Ital. J. Anim. Sci. 1828-051X. 17 (2), 418–427. doi:10.1080/1828051x.2017.1372698.

Prabha, K., Ghosh, P., S, A., Joseph, R.M., Krishnan, R., Rana, S.S., Pradhan, R.C., 2021. Recent development, challenges, and prospects of extrusion technology. Future Foods. 26668335. 3. Available from: https://www.journals.elsevier.com/future-foods. doi:10.1016/j.fufo.2021.100019.

Prathamesh Bharat Helkar, A.K. Sahoo, 2016. Review: food industry by-products used as a functional food ingredients. Int. J. Waste Resour. 22525211. 6 (3). doi:10.4172/2252-5211.1000248.

Purohit, C., Rajyalakshmi, P., 2011. Quality of products containing defatted groundnut cake flour. J. Food Sci. Technol. 0022-1155. 48 (1), 26–35. doi:10.1007/s13197-010-0125-y.

Rajpurohit, B., Li, Y., 2023. Overview on pulse proteins for future foods: ingredient development and novel applications. J. Future Foods. 27725669. 3 (4), 340–356. Available from: https://www.sciencedirect.com/journal/journal-of-future-foods. doi:10.1016/j.jfutfo.2023.03.005.

Rathore, A.S., Kapoor, G., 2017. Implementation of quality by design toward processing of food products. Prep. Biochem. Biotechnol. 15322297. 47 (5), 435–440. Available from: https://www.tandf.co.uk/journals/titles/10826068.asp. doi:10.1080/10826068.2017.1315601.
Riaz, M.N., Asif, M., Ali, R., 2009. Stability of vitamins during extrusion. Crit. Rev. Food Sci. Nutr. 15497852. 49 (4), 361–368. doi:10.1080/10408390802067290.
Rizvi, S.S., Mulvaney, S., 1992. Extrusion processing with supercritical fluids. U.S. Patent US5120559A.
Robin, F., Heindel, C., Pineau, N., Srichuwong, S., Lehmann, U., 2016. Effect of maize type and extrusion-cooking conditions on starch digestibility profiles. Int. J. Food Sci. Technol. 0950-5423. 51 (6), 1319–1326. doi:10.1111/ijfs.13098.
Robin, F., Schuchmann, H.P., Palzer, S., 2012. Dietary fiber in extruded cereals: limitations and opportunities. Trends Food Sci. Technol. 09242244. 28 (1), 23–32. doi:10.1016/j.tifs.2012.06.008.
Šárka, E., Sluková, M., Smrčková, P., 2020. New food compositions to increase the content of phenolic compounds in extrudates. Czech J. Food Sci. 12121800. 38 (6), 347–358. doi:10.17221/223/2020-cjfs.
See, X.Y., Chiang, J.H., Law, L.M., Osen, R., 2025. High moisture extrusion of plant proteins: advances, challenges, and opportunities. Crit. Rev. Food Sci. Nutr. 15497852. 65 (1), 143–164. Available from: https://www.tandfonline.com/loi/bfsn20. doi:10.1080/10408398.2023.2268736.
Shah, F.U.H., Sharif, M.K., Bashir, S., Ahsan, F., 2019. Role of healthy extruded snacks to mitigate malnutrition. Food Rev. Int. 15256103. 35 (4), 299–323. Available from: https://www.tandf.co.uk/journals/titles/87559129.asp. doi:10.1080/87559129.2018.1542534.
Shelar, G.A., Gaikwad, S.T., 2019. Extrusion in food processing: an overview. Pharma Innov. J. 562 (2), 562–568.
Shen, D., Zhang, M., Mujumdar, A.S., Li, J., 2023. Advances and application of efficient physical fields in extrusion based 3D food printing technology. Trends Food Sci. Technol. 09242244. 131, 104–117. Available from: http://www.elsevier.com/wps/find/journaldescription.cws_home/601278/description#description. doi:10.1016/j.tifs.2022.11.017.
Singh, S., Gamlath, S., Wakeling, L., 2007. Nutritional aspects of food extrusion: a review. Int. J. Food Sci. Technol. 0950-5423. 42 (8), 916–929. doi:10.1111/j.1365-2621.2006.01309.x.
Singha, P., Muthukumarappan, K., Krishnan, P., 2018. Influence of processing conditions on apparent viscosity and system parameters during extrusion of distiller's dried grains-based snacks. Food Sci. Nutr. 20487177. 6 (1), 101–110. Available from: https://onlinelibrary.wiley.com/journal/10.1002/%28ISSN%292048-7177. doi:10.1002/fsn3.534.
Sule, S., Okafor, G.I., Momoh, O.C., Gbaa, S.T., Amonyeze, A.O., 2024. Applications of food extrusion technology. MOJ Food Process. Technol. 2381182X. 12 (1), 74–84. doi:10.15406/mojfpt.2024.12.00301.
Téllez-Morales, J.A., Hernández-Santos, B., Juárez-Barrientos, J.M., Lerdo-Reyes, A.A., Rodríguez-Miranda, J., 2022. The use of tubers in the development of extruded snacks: a review. J. Food Process. Preserv. 17454549. 46 (7). Available from: http://onlinelibrary.wiley.com/journal/10.1111/(ISSN)1745-4549. doi:10.1111/jfpp.16693.
Thachil, M.T., Chouksey, M.K., Gudipati, V., 2014. Amylose-lipid complex formation during extrusion cooking: effect of added lipid type and amylose level on corn-based puffed snacks. Int. J. Food Sci. Technol. 13652621. 49 (2), 309–316. doi:10.1111/ijfs.12333.
Tomašević, I., Putnik, P., Valjak, F., Pavlić, B., Šojić, B., Markovinović, A.B., Bursać Kovačević, D., 2021. 3D printing as novel tool for fruit-based functional food production.

Curr. Opin. Food Sci. 22147993. 41, 138–145. Available from: http://www.journals.elsevier.com/current-opinion-in-food-science/. doi:10.1016/j.cofs.2021.03.015.

Vasanthan, T., Gaosong, J., Yeung, J., Jihong, L., 2002. Dietary fiber profile of barley flour as affected by extrusion cooking. Food Chem. 03088146. 77 (1), 35–40. doi:10.1016/s0308-8146(01)00318-1.

Woomer, J.S., Adedeji, A.A., 2021. Current applications of gluten-free grains–a review. Crit. Rev. Food Sci. Nutr. 15497852. 61 (1), 14–24. Available from: https://www.tandfonline.com/loi/bfsn20. doi:10.1080/10408398.2020.1713724.

Wu, J., Zhou, Q., Zhou, C., Cheng, K.W., Wang, M., 2024. Strategies to promote the dietary use of pigeon pea (*Cajanus cajan* L.) for human nutrition and health. Food Front. 26438429. 5 (3), 1014–1030. Available from: https://www.onlinelibrary.wiley.com/journal/26438429. doi:10.1002/fft2.381.

Xu, E., Campanella, O.H., Ye, X., Jin, Z., Liu, D., BeMiller, J.N., 2020. Advances in conversion of natural biopolymers: a reactive extrusion (REX)–enzyme-combined strategy for starch/protein-based food processing. Trends Food Sci. Technol. 09242244. 99, 167–180. Available from: http://www.elsevier.com/wps/find/journaldescription.cws_home/601278/description#description. doi:10.1016/j.tifs.2020.02.018.

Yagci, S., Calıskan, R., Gunes, Z.S., Capanoglu, E., Tomas, M., 2022. Impact of tomato pomace powder added to extruded snacks on the in vitro gastrointestinal behaviour and stability of bioactive compounds. Food Chem. 18737072. 368. Available from: https://www.elsevier.com/locate/foodchem. doi:10.1016/j.foodchem.2021.130847.

Zhang, M., Bai, X., Zhang, Z., 2011. Extrusion process improves the functionality of soluble dietary fiber in oat bran. J. Cereal Sci. 07335210. 54 (1), 98–103. doi:10.1016/j.jcs.2011.04.001.

Zhang, Y., Liu, W., Liu, C., Luo, S., Li, T., Liu, Y., Wu, D., Zuo, Y., 2014. Retrogradation behaviour of high-amylose rice starch prepared by improved extrusion cooking technology. Food Chem. 03088146. 158, 255–261. doi:10.1016/j.foodchem.2014.02.072.

Zhang, Z., Zhang, L., He, S., Li, X., Jin, R., Liu, Q., Chen, S., Sun, H., 2023. High-moisture extrusion technology application in the processing of textured plant protein meat analogues: a review. Food Rev. Int. 8755-9129. 39 (8), 4873–4908. doi:10.1080/87559129.2021.2024223.

Zheng, H., Yan, G.S., Lee, Y., Alcaraz, C., Marquez, S., de Mejia, E.G., 2020. Effect of the extrusion process on allergen reduction and the texture change of soybean protein isolate-corn and soybean flour-corn mixtures. Innov. Food Sci. Emerg. Technol. 14668564. 64. Available from: http://www.elsevier.com/locate/ifset. doi:10.1016/j.ifset.2020.102421.

Zieliński, H., Michalska, A., Piskuła, M.K., Kozłowska, H., 2006. Antioxidants in thermally treated buckwheat groats. Mol. Nutr. Food Res. 16134133. 50 (9), 824–832. doi:10.1002/mnfr.200500258.

Index

Note: Page numbers followed by "*f*" indicate figures, "*t*" indicate tables.

C

D

E

R

Printed and bound by CPI Group (UK) Ltd, Croydon, CR0 4YY
14/09/2026
02216408-0010